T0261224

The Biology of Sharks and Rays

By A. PETER KLIMLEY

With Illustrations by STEVEN OERDING

The Biology of
Sharks and Rays

The University of Chicago Press

Chicago & London

A. Peter Klimley is adjunct professor in the Department of Wildlife, Fish, and Conservation and director of the Biotelemetry Laboratory at the University of California, Davis.

The University of Chicago Press, Chicago 60637
The University of Chicago Press, Ltd., London
© 2013 by The University of Chicago
All rights reserved. Published 2013.
Printed in China

22 21 20 19 18 17 16 15 14 13   1 2 3 4 5

ISBN-13: 978-0-226-44249-5 (cloth)
ISBN-13: 978-0-226-92308-6 (e-book)
ISBN-10: 0-226-44249-7 (cloth)
ISBN-10: 0-226-92308-8 (e-book)

Library of Congress Cataloging-in-Publication Data

Klimley, A. Peter.
   The biology of sharks and rays / A. Peter Klimley ; with illustrations by Steven Oerding.
      pages cm.
   Includes bibliographical references and index.
   ISBN 978-0-226-44249-5 (cloth : alk. paper) — ISBN 0-226-44249-7 (cloth : alk. paper) — ISBN 978-0-226-92308-6 (e-book) — ISBN 0-226-92308-8 (e-book)
1. Sharks—Anatomy. 2. Sharks—Behavior. 3. Rays (Fishes)—Anatomy. 4. Rays (Fishes)—Behavior. I. Title.
   QL638.9.K568 2013
   567′.3—dc23
                                                          2012038999
⊗ This paper meets the requirements of ANSI/NISO Z39.48-1992 (Permanence of Paper).

This book is really not my book but is the fruit of the arduous work of many dedicated scientists who have worked with these wonderful creatures, the cartilaginous fishes. Some of these scientists, Frank Carey, Sonny Gruber, Art Myrberg, and Donald Nelson, were or have been good friends and close colleagues of myself. Others, such as Bashford Dean and Perry Gilbert, lived long before my time or were retiring when I was young student infatuated with sharks. And there are now those, such as Steve Kajiura, Salvador Jorgenson, Kevin Weng, and my current graduate student James Ketchum, who are much younger than I and will carry on the tradition of chondrichthyan research in the future. It is to these people, to the many other shark researchers, and to the creatures that have fascinated all of us that I dedicate this book.

Contents

CHAPTER 1

An Introduction
to the Cartilaginous Fishes

Top: White shark leaping out of water to seize a fur seal within its jaws in False Bay in South Africa. *Bottom:* Basking shark slowing cruising near the surface with its mouth wide open to swallow shrimp-like krill or copepods in Monterey Bay off Central California. The basking shark has mesh-like extensions from the gills that serve as strainers, retaining the small planktonic organisms as water enters the mouth and passes outward through its gill openings.

The cartilaginous fishes, in particular the sharks, inspire both awe and fascination. Why is this? The answer to this question may lie in the role of some, such as the white and tiger sharks, as apex predators in an alien world. In comparison to all of these species, humans are ill adapted for life in the seas. We swim awkwardly with clumsy arm strokes and kicking leg movements. We can dive underwater only for at most a few minutes before we must rush to the surface to breathe, often suffering panic due to our inability to see far in this dimly lit, blue-green world. Humans do not smell underwater or hear well in this eerily silent world. The cartilaginous fishes are masters of this alien environment. They swim effortlessly in this viscous medium and possess senses that enable them to find their prey in total darkness at great distances. Humans feel insecure in the presence of these fishes, of which a few of the largest feed on marine turtles, seals, sea lions, and dolphins. We hold these predatory species in awe and are keen to learn more about why they are masters of the sea. However, you will learn in this book that other cartilaginous fishes such as the whale shark, basking shark, and many rays are lower on the food chain, the former being filter feeders of plankton and the latter swallowing buried clams and worms.

This chapter is a brief introduction to this very diverse

group. The emphasis will be on describing the benchmark studies on each particular subject. It is impossible here to cover the enormous scientific literature on the biology of the cartilaginous fishes, which is reviewed in over a dozen compendiums on specific topics. The results of these key studies will be presented in diagrams and graphs in order to introduce you to authentic scientific data so that you can read them critically. "Spotlights" are included in each chapter to focus your attention on the sophisticated methods used by pioneer scientists to make important scientific discoveries. These features will help you understand the scientific method and learn to appreciate the adventure of being a scientist. Finally, each chapter ends with intellectually stimulating questions; you are encouraged to delve into the scientific literature to answer them. Also included are lists of keynote articles for your further reading. The purpose here is to provide you with a "road map" with which to pique your interest and motivate you to learn more about the biology of the cartilaginous fishes.

There are currently 503 species of sharks, 699 species of rays, and 49 species of chimaeras in the class Chondrichthyes according to FishBase, a web-based portal of information on fishes maintained by an international consortium of scientists. The name Chondrichthyes is formed from the Greek prefix *khondros* meaning "cartilage" and suffix *ikhthus* for "fish." These species are referred to as the cartilaginous fishes. They have a skeleton composed of cartilage, a substance less calcium-impregnated than bone. This class is in the subphylum Craniata, which includes all animals having a brain enclosed within a solid case. The cartilaginous fishes are separated into two subclasses, the Holocephali and Elasmobranchii. The species in the first subclass have a single gill opening on either side of the head whereas members of the second have multiple openings on either side. The name Holocephali is derived from *holos* meaning "whole" and *kephalē* meaning "head." It refers to a large, rounded head. These species are called chimaeras based on their appearance. Their large head with large eyes and protruding teeth, flexible body, and long tail make them vaguely resemble the fire-breathing Greek monster of that name. The monster had a lion's head, goat's body, and serpent's tail. The name of the second subclass, Elasmobranchii, is formed from *elasmo* meaning "plate" and suffix *branch* for "gill." This large taxonomic group is composed of the sharks, which have cylindrical bodies, and the rays, which have flattened bodies. Unlike the chimaeras, these species have multiple gill openings—the latter numbering from five to seven.

The long evolutionary history of the cartilaginous fishes is described in chapter 2. You will learn that the holocephalans were very diverse and abundant in the Paleozoic Era that ended roughly 250 million years ago (mya) and became far less so by the end of the Mesozoic Era 60 mya. Although sharks existed in the ancient seas with the early chimaeras, the species of sharks and rays radiated in the Mesozoic Era to occupy many new niches and become dominant fishes in the modern oceans. You will be introduced to species in the modern orders of sharks at the end of that chapter. Chapter 3 provides a description of the swimming styles of the sharks, rays, and chimaeras, their skeletons and fins, muscles, and external dermal denticles, and finally of how their caudal and pectoral fins propel them through the water with great agility. Ridged, tooth-like denticles are distributed over the entire body of sharks. However, they cover only part of the body of the less rapidly swimming rays and are absent on the bodies of the even slower-swimming chimaeras. The denticles serve to reduce drag, or the resistance to forward movement, as the viscous medium of water passes close to the body during forward movement. Competitive swimmers now wear suits covered with ridges similar to the rows of denticles on the skin of a shark or ray to increase hydrodynamic efficiency. Chapter 4 explains how the cartilaginous fishes balance the water and electrolyte content of their body fluids with that of the ocean. They are able to control the influx of the ions of sodium and chloride and the efflux of water using their rectal gland, kidney, and gills. They remain nearly isosmotic with saltwater because in addition to sodium and chloride ions, their tissues contain two other solutes in large concentrations, urea and trimethylamine oxide. Some species, such as the bull shark and freshwater stingray, enter fresh water environments. These species need instead to counteract the osmotic gain of water; they do this by producing copious urine. As water is drawn through the mouth, past the gills, and out the gill slits, heat from the body diffuses outward across the gills to warm the passing water, and oxygen diffuses inward from the water during the process of respiration. Furthermore, heat is lost from the muscles through the skin to a colder environment. Thus, it is not surprising that most chondrichthyans are ectothermic, or cold-blooded—their body temperature conforms to the temperature of the external environment. Despite these challenging obstacles to heat retention, the mackerel and thresher sharks and manta and mobula rays have evolved anatomical adaptations that enable them to warm their body in cold water. Specialized retes consisting of bundles of blood vessels in muscles, stomach, brain, and eyes enable these endothermic, or warm-blooded, species to maintain warm bodies. This important capability of mackerel sharks from the family Lamnidae and manta ray from the family Mobulidae will be discussed in chapter 5.

The cartilaginous fishes have a diversity of senses with varying ranges of detection that will be described in chapters 6 through 9. These fishes rely on their different senses to provide them with information that enables them to seek out their prey or mate or to flee from a predator. Chemicals are transported by currents in an odor corridor. These species detect chemicals as they

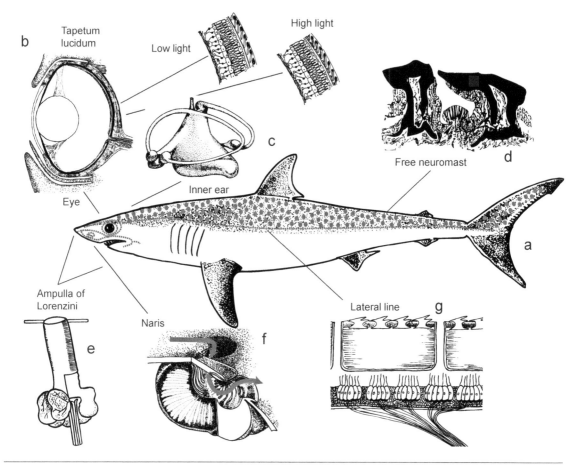

Fig. 1.1 The senses of a cartilaginous fish. (a) The shortfin mako shark (*Isurus oxyrinchus*) is shown with its free neuromasts, lateral line, and cranial canals, which are sensitive to water particle displacements, outlined in red. Surrounding the shark are expanded views of (b) the eye, with the dark pigments in the tapetum lucidum shown under low and high light levels, (c) the inner ear, which is sensitive to pressure oscillations, (d) a free neuromast, (e) an ampulla of Lorenzini, which detects electromagnetic impulses, (f) a naris, which is sensitive to waterborne chemicals, with the water flow direction indicated by (f) a red arrow, and (g) a segment of the lateral line.

pass through their nares (fig. 1.1f) and can follow a chemical gradient to its source. For example, a white shark can find a decaying whale carcass over a distance of several kilometers. Rays use this sense to locate living species buried within the substrate such as clams. They excrete chemicals into the water exiting their burrows. The anatomy of the olfactory receptor, how it differs among different species, the receptor's sensitivity to a myriad of chemicals, and the ability to localize the source of an odor source will be explained in chapter 6. Cartilaginous fishes can perceive with their inner ears (fig. 1.1c) the low-frequency pressure waves emitted by a struggling fish as far as half a kilometer away. At a closer distance, the tiny particle displacements propagated by the struggling fish can be detected with tiny cilia in lateral line and cranial canals and small neuromasts scattered over the body (figs. 1.1g & 1.1d). The structure of these sound pressure and water displacement organs, their spectral sensitivities, and the anatomical diversity among species will be described in chapter 7. The shark or ray sees its prey or mate upon approaching even closer—under the best conditions at a distance of 20 meters or so from it. The eyes of cartilaginous fishes have a tapetum lucidum (fig. 1.1b). This thin layer of reflective crystals located at the back of the retina reflects light back through it so that the light passes twice by the sensitive receptors providing two opportunities to absorb the photons. This increases the predator's ability to distinguish shapes in the low light levels present at twilight and nighttime. During daytime, dark tissue migrates over the exposed surface of the crystals and absorbs the impinging light so that it is not reflected backwards through the retina. In chapter 8, you will learn about the underwater photic environment, the anatomy of the eye, the spectral sensitivities of visual pigments, and the visual capabilities of the different cartilaginous fishes. All of the sharks, rays, and chimaeras have small pores on the underside of their heads connected to gel-filled tubules that lead to the nervous system, called the ampullae of Lorenzini (fig. 1.1e). These sense minute electrical fields, which are produced by fish, clams, and crabs while out of sight buried in the sand. This electromagnetic sense, described in chapter 9, enables sharks, rays, and chimaeras to find their way in the apparently featureless ocean by following the subtle patterns of magnetization on the sea floor or the earth's dipolar magnetic field.

The cartilaginous fishes are often described as "mindless feeding machines" in film documentaries. This is not really true. The size of their brains and various lobes will be described in chapter 10 along with their learning capabilities. The ratios of brain to body mass of more advanced sharks such as the scalloped hammerhead and bat eagle ray are comparable to those of birds and even mammals. Consistent with this is the ability of sharks to learn to discriminate between targets on the basis of their degree of illumination at a rate comparable to that of a mouse. The more advanced species of cartilaginous fishes have large and well-developed foliated forebrains, consistent with their possessing a diverse behavioral repertoire and complex social system comparable to those exhibited by birds and mammals.

The reproductive mode of the cartilaginous fishes promotes early feeding success. The male inserts one of its two claspers (fig. 1.2f), scrolled outside portions of their pelvic fins, into the female's cloaca to inseminate her in a manner analogous to human reproduction. Unlike most bony fishes, the embryos of cartilaginous fishes incubate for longer periods of up to two years. Some sharks, many rays, and all chimaeras produce embryos that develop within large leathery cases, yet the majority of sharks give birth to fully developed young. The newborn is almost a miniature replica of an adult and able to chase and capture small fishes immediately upon birth. Courtship behavior

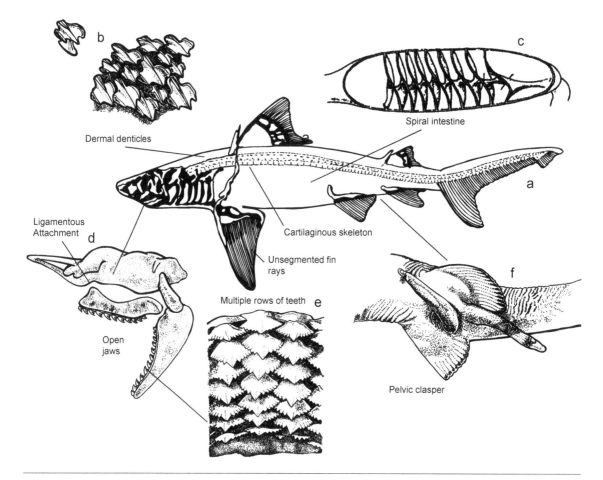

Fig. 1.2 The skeleton and organs of a cartilaginous fish. (a) The piked dogfish (*Squalus acanthias*) is shown with the chondrocranium encasing and protecting the brain, the vertebral column, and the cartilaginous fin supports that provide great flexibility when swimming. Surrounding the shark are expanded views of (b) the dermal denticles, which reduce drag when swimming, (c) the spiral intestine, which increases the surface area available for digestion, (d) the loosely attached upper and lower jaws, which can be opened wide enough to saw off large chunks of prey using its (e) multiple rows of teeth, and (f) the pelvic clasper, which is inserted into the female's cloaca to inseminate her during copulation.

and reproductive biology are described in chapter 11. Live bearing has both advantages and disadvantages relative to egg laying. The fully developed juvenile has a greater chance of surviving than a tiny larval fish, which will perish if it does not find food quickly after absorbing the nutrients within its yolk sac. However, fewer well-developed embryos can be produced by a live-bearing shark than the thousands of larvae produced by an egg-bearing fish, and the juvenile sharks are easily caught by hook and line or in the nets of fishermen.

In chapter 12, the anatomy of the jaws and teeth of the cartilaginous fishes will be described, as well as the role of the muscles and teeth in jaw extension, seizure, and handling of prey. Shark and ray jaws are particularly well-suited for handling their prey. Their upper jaw is attached to the cranium by a pair of ligaments; the lower jaw is attached to cartilaginous elements that support the gills (fig. 1.2d). The jaw's elastic connections enable a shark to open its jaws wider and swallow large prey items as well as to remove large chunks of meat from prey too large to swallow. There are multiple rows of serrated teeth in the upper and lower jaws (fig. 1.2e). Embedded in one jaw are pointed teeth for holding struggling prey in place whereas in the other serrated teeth are moved back and forth laterally to saw off a chunk of meat. With sharpness and thinness of the teeth comes a tradeoff—brittleness. Either fragments or whole teeth become dislodged from the shark's jaw when feeding. Embryonic teeth continuously develop in the back of the tooth fold along the forward edge of the jaw and move forward to take their place in a new row. The diminutive members of this row grow larger while moving forward to form the next outer row of teeth. The chimaeras possess jaws better suited to feeding upon crustaceans and mollusks by crushing their hard external skeletons with their multiple rows of hard, flattened teeth arranged within the jaws.

Chapter 13 will discuss what cartilaginous fishes feed upon, their frequency of feeding, their rates of digestion and growth, and to what age they live. The composition of the diet of the cartilaginous fishes often changes as they grow larger and migrate from one geographical region to another. A species must digest its prey efficiently. Most cartilaginous fishes have spiraling folds in their intestines that increase the digestive surface within a small space to maximize the rate of digestion (fig. 1.2c). Some scientists have suggested that cartilaginous fishes grow slower than bony fishes because the coiled stomach of the former is smaller than the straight stomach of the latter. The diminutive size of the intestine makes room for a large liver, needed to maintain buoyancy, and a large uterus, needed for the prolonged development of their young.

You will learn about types of movement patterns exhibited by sharks and rays in chapter 14. Little is known about the movements of the chimaeras. Some elasmobranchs are ambushers, and their movements are restricted to a confined area. These sharks and rays lie in wait on the bottom to ambush crabs and shrimps or fishes that walk or swim over them close to the bottom. Members of other species of sharks actively search for their prey over intermediate distances yet return to a single location either to rest or interact socially with other members of the species. Such is the case of the reef and hammerhead sharks. The planktivorous sharks are more nomadic, moving over large distances in the oceans and searching for areas where plankton is abundant such as at the mouths of rivers and seamounts and at the boundaries between currents. Generally, swimming is sustained and directional between resting and foraging locations and slow and nondirectional when at feeding grounds. During their migrations elasmobranchs display two characteristic swimming behaviors, oscillatory diving and surface swimming, whose functions are not yet known with certainty.

Chapter 15 is devoted to human interactions with sharks. In the beginning, you will learn about the two motivations, feeding and defense, behind attacks on humans. When animals feel threatened, they experience conflicting instincts—one is to escape and another is to fight. If they are unable to flee because their opportunity to escape is blocked, they do not always fight but often perform an aggressive display. A shark will likely exhibit a display when a diver first approaches closely but in the absence of a prompt retreat may quickly seize or slash with its jaws a body part of the diver without removing much flesh. This display can be recognized by the upward pointing of the snout, bend between the body and head, holding of the pectoral fins close together, arching of the back, and exaggerated tail beats. The risk of being attacked by a shark is truly negligible compared to other risks encountered by humans on a daily basis. The National Safety Council tallies and makes public statistics on the average number of deaths caused by various other injuries over three-year periods. The yearly average of 50 attacks worldwide on humans, of which eight were attributed to white sharks, can be compared to the average yearly rates of fatalities caused by other everyday risks over the same period. The number of fatalities in motor vehicle accidents within the United States averaged 42,593 per year over a three-year period from 1993–1995. The risk of dying from an automobile in the United States is six thousand times greater than that of being attacked by a shark in any place in the world. A wound inflicted by a stingray barb is much more common than an attack by a shark. Bathers step on stingrays while at the beach because they are hard to see on the bottom through the surface of the water, and upon stepping on the pectoral disk of the stingray cause it to reflexively lift its tail and stab the barb into the foot or lower leg of the bather. The venom along with fragments of the barb are often left at the penetration site, and they cause the victim great pain. The venom can be denatured by placing the ankle or hand in hot water. This chapter ends with a brief introduction to shark and ray ecotourism. This form of ecotourism has become immensely popular during the 1990s and 2000s and has become worldwide in its scale. There are opportunities to view both sharks and rays in North, Central, and South America, the Caribbean Islands, Europe, Africa, and Asia. The diversity of species that may be viewed underwater is great, and these species can be viewed in safety at localities in the waters off most of the continents on the earth.

Chapter 16 is devoted to the recent expansion of fisheries for sharks and rays and the precipitous decline of the populations of sharks and rays in the world's oceans. You will be introduced to a few historical examples of the collapse of shark fisheries, the methods of managing elasmobranch fisheries including recent regulations, as well as the recent establishment of reserves to protect sharks and rays. That which has made the sharks, rays, and chimaeras so successful over time also has made them very vulnerable today in a world where humans are the real masters. The acute senses, anatomy, and reproductive mode of the cartilaginous fishes have made them succeed over 400 million years as the dominant predators in the sea. Yet at the same time, the role of many as predators, high on the trophic scale, has limited their abundance to a small fraction of that of their prey lower on the food chain. This life history strategy with its tradeoffs worked well for sharks prior to the advent of human fisheries. However, humans began to fish intensively for sharks with long lines and gill nets in the twentieth century. The fisheries for basking sharks off the western coast of the United States for the high-grade oil used in military equipment during the Second World War resulted in a severe decline in their population. The fisheries for meat for human consumption during the 1990s targeted inshore species of sharks off the eastern coast of the United States and resulted in their decline. This caused the National Marine Fisheries Service to mandate that catch be recorded semiannually for the different species and that fishing effort must be reduced if the catch per time period declines consistently. Presently, there is keen international interest in fisheries for sharks. This is because the unsegmented rays in their fins are used as a thickener in soups throughout China and the cartilaginous vertebral column is processed into "chondroitin" to restore cartilage in arm and knee joints of older humans worldwide. The serving of soup made with shark fins has recently been made illegal by the legislature of the State of California.

Sharks and rays are vulnerable to overfishing due to their life history strategy, and close regulation is needed to sustain fisheries. This leads one to wonder whether any motivation might exist to discourage the continued exploitation of shark stocks. Perhaps such a countervailing motivation exists! The public has in the last dozen years developed a penchant for viewing sharks both in captivity and in the wild. This may be just the reprieve needed for the beleaguered shark populations worldwide. Ecotourists from all over the world are now visiting remote locations to view sharks in their own environment. Very successful ecotourism industries exist for watching white sharks at Guadalupe Island off the western coast of the Baja Peninsula or observing the schools of hammerhead sharks present at Wolf and Darwin Islands, the two northernmost islands of the Galapagos Archipelago. This book is intended to enhance your appreciation of the unique and complex biology of the sharks, rays, and chimaeras. Ideally, it will stimulate you to learn more about these wonderful creatures. You can do this not only by observing them in an aquarium and viewing them in the wild but by reading many of the wonderful articles in the scientific literature about them. If you do this, you too are likely to become an ardent advocate for the protection and conservation of these species that over geological time have been the masters of the sea.

CHAPTER 2

Evolutionary History

Cladoselache, whose appearance was similar to those of the modern sharks, lived in the late Devonian. It was only a meter long, and fed on small prey by seizing them with its pointed teeth, yet it lacked many of the other features of the modern sharks.

While a graduate student at the University of Miami's Rosenstiel School of Marine Sciences, I occasionally drove on the weekends to the phosphate mines north of Miami to search for fossil sharks' teeth. The excavating cranes at these mines dug deep into the earth to remove the phosphates, creating large pits and depositing orange-tinted phosphate-laden soil around the edges of the pits. These tailings were composed of soil and sedimentary rock from geological periods many millions of years ago. Scattered throughout the soil were rock-hard fossilized teeth of ancient sharks. The most common tooth had a large protruding central cusp with many smaller ones to either side on top of a broad base. These teeth, two centimeters high and two and a half centimeters across, belonged to "*Cladoselache*," the first common shark in the Paleozoic Era. These fossilized teeth didn't appear to me that different than pointed teeth of two living sharks, the sandtiger, *Carcharias taurus*, and the shortfin mako shark, *Isurus oxyrinchus*. After a rainstorm, many teeth were uncovered as the loose soil was washed away. If you visited the mines at this time, you might be lucky enough to find the huge tooth, almost the size of your hand, of the giant-tooth shark, an ancient relative of the great white shark. This shark was an imposing predator, which could reach a length of fifteen meters—five Volkswagen Beetles long. The huge teeth, some four to five centimeters high and across, are triangular with razor-sharp serrations on either side leading to a pointed tip. Finding so many of these teeth made me wonder about the life of these ancient predators. How did they swim? How did they capture their prey? How similar were they to the sharks I was studying now in the waters off Florida?

Our evidence of the first cartilaginous fishes is from fossil deposits over 455 million years ago (mya). At that time, there were two groups of these fishes, the Elasmobranchii, ancestors of the modern sharks and rays, and the Holocephali, ancestors of the modern chimaeras.[1] The majority of these Paleozoic elasmobranchs, who lived more than 251 mya, were predators.[2] The holocephalans, on the other hand, were more diverse in their feeding ecology. Some were grazers, feeding on benthic algae; others were suction feeders that extracted invertebrates out of the sediments. The two taxonomic groups shared the oceans with the members of two other taxa of jaw-bearing fishes.[3] The first comprised the placoderms, which were covered with a thick and often ornamented bony shield, or placo, over the front half of their bodies. The second group was composed of the acanthodians, which were unarmored but possessed large bony spines, or acanthi, in front of their many paired fins.

Some anatomical features better adapted chondrichthyans to the marine environment and enabled them to diversify into many niches by the Cretaceous Period.[4] First, they had a cartilaginous skeleton for body support. While the skeleton of placoderms and acanthodians was composed of stiff and hard bone, the skeleton of chondrichthyans was composed of more pliable and soft cartilage. Bone is a mineralized matrix composed of hydroxyapatite, a hard tissue impregnated with calcium phosphate interspersed with small spaces containing soft cells, or osteocytes. The skeleton of the primitive sharks, on the other hand, was composed of cartilage, which has a hard superficial layer of prismatic structures impregnated with calcium phosphate, each no more than one millimeter in diameter and shaped like a prism when viewed from above and an hour-glass when examined from the side.[5,6] This tissue is more flexible than bone. The flexibility of the cartilaginous skeleton may have enabled chondrichthyans to accelerate and change direction rapidly, resulting in success in pursuing and capturing prey.

A second anatomical innovation of the early sharks was their multiple rows of teeth. When an old, worn tooth was shed from the outer row in the jaw, a new tooth from the next row behind moved into the space left by the discarded tooth, and so forth. The outer teeth of the modern sharks remain in place for only a few weeks before they no longer receive nutrients and are then shed from the mouth. Due to the short period of tooth growth, the teeth in each successive outer row are only slightly larger than those in the inner row. The cusps on these teeth are usually sharp because there is little time to wear down. There are indications that tooth replacement was slow in the early sharks.[3] These teeth are rarer in fossil deposits than those of more modern sharks. Furthermore, the fossilized teeth are of very different sizes. These observations are consistent with the hypothesis that the early sharks possessed only a few tooth rows, each with teeth of a different size. These teeth likely developed over a longer period of time before migrating outward to the replace teeth lost from the outer row. Teeth recovered in fossil deposits also exhibit considerable wear and abrasion, and this is also consistent with their being present in the jaw longer than those of the modern sharks. Yet the ability to replace outer teeth as they became worn, albeit slowly, may have given an advantage to early sharks over the predatory placoderms and acanthodians of the Paleozoic age, which had a single row of teeth in their jaw.

The early sharks also reproduced through internal fertilization. The male possessed a pair of claspers. The axial elements at the base of the male's pelvic fin formed a scrolled organ that extended outward from the base of the fin. These claspers presumably functioned like those of modern sharks and were inserted within the female's cloaca to fertilize ova within her uterus. The fertilized egg developed to an advanced state within the female before being released into the environment either as an egg, protected by a tough outer case, or as a miniature replica of the adult. The young cartilaginous fish began its life as an active predator. This form of reproduction may have conferred an advantage over egg-laying species that produced a larva that had to feed on plankton soon after hatching and avoid predation by planktivores. However, at least one group of pacoderms, the arthrodires, also had claspers and performed chondrichthyan-like internal fertilization.[7]

These features were not the only anatomical innovations that enabled sharks to radiate throughout the seas and become the dominant predators during the second half of the Devonian. They also possessed a torpedo-shaped body propelled forward by a two-lobed fin at the posterior end with a mouth at the anterior end to seize its prey. Paired dorsal and anal fins on the vertical plane prevented yaw, the tendency to sway from side to side, and horizontal pectoral and pelvic fins prevented roll, the tendency to rotate when swimming quickly through the water. These adaptations have made them the most historically enduring of all fishes, permitting them to survive the environmental cataclysm that caused the mass extinction of species at the end of the Paleozoic era and to radiate in the Mesozoic and Cenozoic eras.

The jawless fishes and primitive jawed fishes flourished 443–417 mya during the Silurian Period. The placoderms became less abundant by the early Devonian and were virtually extinct by the end of this period. The agnathans, jawless hagfishes, and acanthodians became much less diverse during a 15 million year period from 377–362 mya during the Devonian.[4] Fifty-one of the 70 families of fishes present died out at this time, an extinction rate of 73%—one of the highest rates of extinction ever recorded during geologic time.[8] The mass extinction was likely a response to the widespread loss of marine plants and the consumers of them lower on the food chain. This was not the only period of extinctions at this time. There were three extinction events during the late Devonian (fig. 2.1). These resulted in the creation of new niches for mobile predators that could move to refugia characterized by benign environmental conditions. It was at this time that the chondrichthyans began to diversify.

Species Diversification

There were movements of the earth's crust at the end of the Devonian period that likely promoted the diversification of the chondrichthyans.[5] There were three separate continental masses during the early Devonian (fig. 2.2a).[9] They coalesced to form a single continent, Pangea, by the end of the Devonian. Pangea elongated along a north-south axis by the early Carboniferous 360 mya and extended from pole to pole (fig. 2.2b). This single continent was expansive with many rivers, swamps, lakes, and mountains. Surrounding Pangea was a broad shallow continental shelf at the margin of the deep oceans. Near the equator was a large inland sea with only a narrow opening leading out to the sea. The continent and its surrounding waters experienced four distinct climates characterized as tropical, arid, warm temperate, and cool temperate. The cartilaginous fishes likely radiated in response to the expansion of the continental shelf and growth in size of the inland sea as well as in response to the increase in diversity of climates.[2] The sharks certainly occupied new ecological niches in the freshwater habitats of the continent but also in its estuaries and bays, along the continental shelf, and in the open ocean.

The earliest evidence that chondrichthyans inhabited the seas is in the form of dermal denticles, the small toothlike plates that covered their skin. These are found in sedimentary rock deposits 455 mya from the late Ordovician[1] (see fig. 2.1). The earliest denticles consisted of an elongated cavity containing soft tissue surrounded by a cone of medium-hard, calcium-impregnated dentine that was covered on the outside with a harder, shiny enamel.[10] As additional

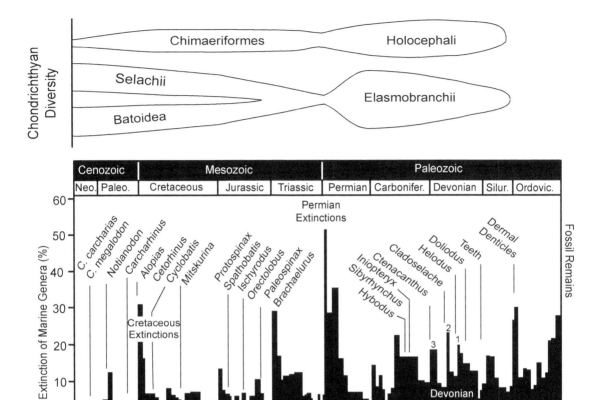

Fig. 2.1 The relative diversity and abundance of holocephalans and elasmobranchs indicated over geological time. The relative rates of extinction of marine genera with a scale of geological time below and the geological eras and periods indicated above. Denoted on the time record is when dermal denticles, teeth, or skeletal remains of the different genera of chondrichthyan fishes have been found in fossil deposits worldwide.

finger-like projections grew, they coalesced to form a flat plate with multiple ridges with separate pulp cavities underneath.

Sharks' teeth first appear in the fossil record 418 mya at the beginning of the Devonian.[1] They were clearly derived from dermal denticles, which had become concentrated by this time along the margins of the jaw.[2] The first teeth had a continuous and open pulp cavity surrounded by dentine covered by a hard shiny enamel layer on the surface. They were attached to the jaw by connective tissue. Most of the teeth have a cladodont pattern, "clado" meaning branched and "dont" denoting tooth. Each tooth consisted of a single primary cusp flanked on either side by many smaller secondary cusps above a broad base (fig. 2.3b). These multicusped teeth became common by the middle Devonian and are present in fossil deposits worldwide.

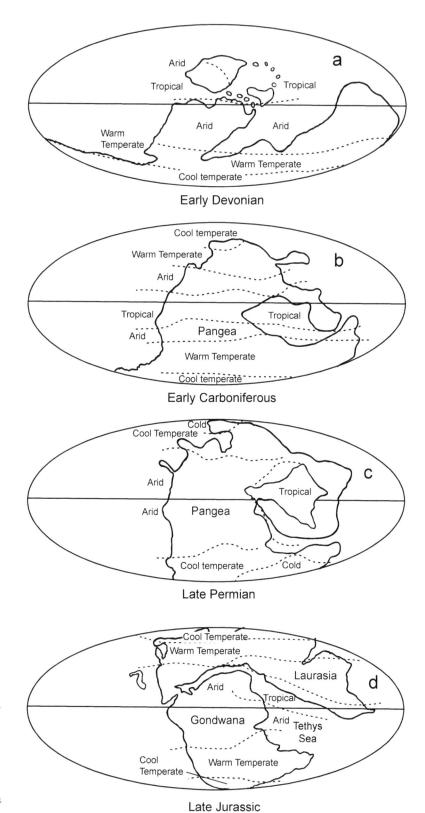

Fig. 2.2 The land masses and their climates during the early Devonian (a), early Carboniferous (b), late Permian (c), and late Jurassic (d). The cartilaginous fishes became extinct and subsequently diversified at times when there were large tectonic changes in the size, shape, and extent of the earth's continents.

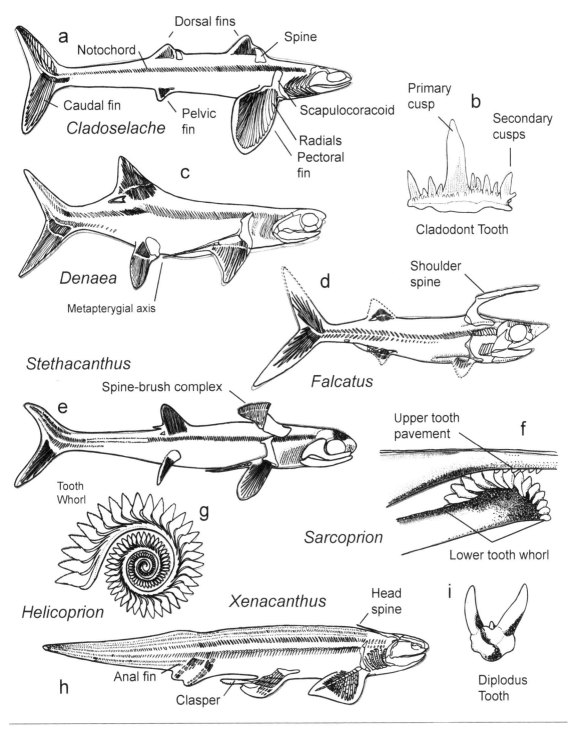

Fig. 2.3 Skeletal anatomy of the early Paleozoic sharks: (a) *Cladoselache*, (b) its multicusped tooth, (c) *Denaea*, (d) *Falcatus*, (e) *Stethacanthus*, (f) the tooth whorl of *Helicoprion*, (g) its alignment in the jaw of *Helicoprion*, (h) *Xenacanthus*, and (i) its two-cusped tooth.

Early Paleozoic Sharks

The oldest known intact skeleton of a chondrichthyan is of a small species of shark known as *Doliodus problematicus* that had large spines on its pectoral fins. It was found in sedimentary rocks formed 408 mya, during the early Devonian.[10] However, sharks did not become common until 380 mya, during the middle Devonian. Many specimens of *Cladoselache* have been found from that time in shale deposits in the vicinity of Cleveland, Ohio (fig. 2.4a).[3] Not only is the skeleton apparent in these fossils, but also the outline of the body as well as anatomical features such as muscle fibers and kidney tubules.[2] This ancestral shark was not that different from the sharks of today. Its body was streamlined and fusiform in shape with a terminal mouth.[11] The number and placement of fins was the same as in modern sharks. These early sharks had two triangular dorsal fins with spines, paired pectoral and pelvic fins, and a caudal fin with equal-sized upper and lower lobes. The shark also possessed five gill openings on either side of the head, as do most modern sharks.

Cladoselache differed from its modern counterparts in that it had a stiff notochord, made up of a core of closely packed cells distended with fluid-filled vacuoles with vertebrae spaced closely together to form a relatively rigid rod (fig. 2.3a). Radial elements supported the dorsal and anal fins along their entire bases, unlike the radials of modern sharks, which are slightly tapered inward at their base to enable the fins to move back and forth with a sculling motion. There was a subcutaneous dorsal spine in front of the first and second dorsal fins. These spines lacked the shiny enamel surface layer on the spines of more modern sharks. The absence of this layer implies that these primitive spines did not penetrate the skin and were not exposed to water.

The caudal fin of *Cladoselache* had similarly sized upper epichordal and lower hypochordal lobes, as do the fins of the modern lamnid sharks. Yet the tail's internal anatomy was asymmetrical. The cartilaginous elements in the upper lobe were segmented with hollow arches to protect the blood vessels while the elements in the lower lobe were unsegmented. As in the modern lamnid sharks, there were lateral keels on either side of the narrow base of the tail fin. The scapulocoracoid, or shoulder element supporting the pectoral fin, formed a broad arc extending from beside the notochord downward and curving inward but was never fused at the base of the body as in modern sharks. The fin was supported along its entire base by basals, small triangular plates, and from these extended the radials, long slender elements that reached the margins of the fins. The pectoral fin appeared to have had little capacity to alter its angle of contact with the water. Hence, this shark was probably not able to change its direction as quickly as do modern sharks. Furthermore, the radials of modern sharks extend only partway out on the fin and are replaced

Fig. 2.4 Fossil remains of three shark species: (a) *Cladoselache fyleri* from Devonian mineral deposits in Ohio, (b) *Symmorium reniforme* from the late Carboniferous deposits in North America, and (c) *Akmoniston* from middle Carboniferous strata in Glasgow.

by more flexible ceratotrichia extending to the fin's margin. *Cladoselache* was likely a fast swimmer, using sideway sweeps of its broad tail to propel itself forward with its pectoral fins preventing it from rotating to the side while swimming. However, it probably moved forward with a stiff motion unlike the sinuous swimming movements of modern sharks. This was due to its unsegmented notochord, wide connections at the base of its fins, and unsegmented radials within the fins.

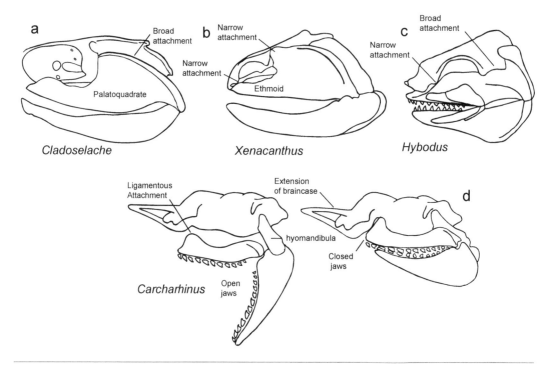

Fig. 2.5 Jaws of three Paleozoic genera of sharks: (a) *Cladoselache*, (b) *Xenacanthus*, (c) *Hybodus*, (d), and a generalized extant shark, *Carcharhinus*.

To date, no fossil of *Cladoselache* has yet been found with claspers at the base of its pelvic fins. This is contrary to other fossil species from the Paleozoic Period. One possible explanation is that all of the fossils are females.[5] The fossils might come from a particular oceanic habitat where females aggregated away from males. Female scalloped hammerhead sharks form schools at seamounts and islands in subtropical and tropical oceans.[12] Female *Cladoselache* may have formed aggregations at similar sites during the late Devonian Period.

The mouth of *Cladoselache* was located at the terminus of the head, unlike today's sharks, which have subterminal mouths.[11] The palatoquadrate, or upper jaw, was firmly attached by ligaments to the braincase (fig. 2.5a). The convex upper edge of the upper jaw fit snugly in the concave lower surface of the chondrocranium. This jaw support was characteristic of the other primitive sharks of the late Devonian and early Carboniferous Periods. The tight connection between the two elements limited the extent to which primitive sharks could open their jaws and limited them to swallowing small fishes.

Two other families of the Cladoselachiformes, the Symmoridae and Eugeneodontidae, were present by the end of the Devonian, as were members of another order, the Xenacanthiformes.[1,11] These sharks were common throughout the Carboniferous until the late Permian. However, no fossils of the *Clado-*

selache have been found later than the Devonian.[11] Members of these taxonomic groups resembled *Cladoselache* in that their upper jaw was bound closely to the braincase with ligaments. These had two narrow points of attachment to the chondrocranium, a posterior palatoquadrate and an anterior ethmoid one, as is evident in the upper jaw of *Xenacanthus* (fig. 2.5b).

The members of the family Symmoridae were unique in that they had a single dorsal fin above the pelvic region of the body. *Denaea* had a metapterygial axis, a whiplike extension of the pectoral fin (fig. 2.3c).[11] It is still unknown whether this skeletal element conferred a hydrodynamic advantage, aided in defense, or was used in courtship.[4] *Falcatus* had a long shelflike spine rising upward from roots deep in the dorsal musculature and extending over the head (fig. 2.3d). This was present only on males. They may have directed their spines towards other males in an attempt to drive them from a mating site to gain access to females. Alternatively, the spine may have been used to prod the female into position for clasper insertion during copulation. *Stethacanthus* and its relatives had a skeletal element extending upward from the shoulder, resembling a shaving brush with tooth-like denticles covering its upper anterior body surface (figs. 2.3e & 2.4c).[4] There was a matching denticular patch on its forehead. Males may have pitched their body downwards, orienting the two areas of denticles towards other males in an aggressive display that simulated the appearance of the teeth in the lower and upper jaws of the enormous mouth of a massive shark. However, it is equally possible that a male initiated courtship with a female by swimming upward and striking the underside of her anterior torso with these two rough surfaces as modern male sharks bite females to initiate courtship.

The Eugeneodontidae are known exclusively from their teeth, some of which grew in spiral shapes.[11] Traditionally these species have been classified as elasmobranchs,[4,11] but they have alternatively been considered to be holocephalans.[1] This indecision is understandable given the absence of any fossil evidence of body parts other than their jaws and teeth. The tooth whorl of members of the genus *Helicoprion* consists of a spiral of successively larger teeth with the smallest and newest teeth in the interior and the largest and oldest in the exterior of the whorl (fig. 2.3g).[11] Thus, new teeth always rotated into place when the older teeth were shed from the mouth. These teeth fit in the symphysis, or joint between the two sides of the lower jaw. The teeth within the whorl came in contact with smaller teeth held in a flat upper jaw when the upper and lower jaws were closed in the genus *Sarcoprion* (fig. 2.3f).[11]

The Xenacanthiformes inhabited freshwater, unlike the rest of the Paleozoic sharks.[12] They had an elongated skull with a large spine attached either to the shoulder girdle or dorsal surface of the braincase. They also had large paired pectoral and pelvic fins and an elongate dorsal fin that extended along most of the back and tapered into a narrow pointed tail (fig. 2.3h). The noto-chordal support for the caudal fin was only slightly upwardly curved with the supporting radials arranged symmetrically above and below the notochord. Unlike other primitive sharks, the members of this taxon had two stalked anal fins, one behind the other. Their teeth consisted of two large and laterally di-rected cusps with a small central cusp and a small button-like articulation that attached to the next tooth in the jaw (fig. 2.3i).[11] These diplodus teeth are com-mon in sedimentary deposits from the early Devonian to the end of the Tri-assic Period. It is likely that the elongated dorsal fin and specialized anal fins enabled these sharks to swim in a sinuous manner, ideal for moving around in the shallow, vegetation-choked swamps of the Carboniferous.

Late Paleozoic Sharks

Sharks of two new orders, the Ctenacanthiformes and Hybodontiformes, be-came common by 359 mya at the end of the Devonian. They possessed ana-tomical improvements over the earliest Paleozoic sharks. The members of the former order were the likely ancestors of the modern sharks.[4,11] Like its prede-cessors, the more advanced, or "derived," *Ctenacanthus* had an unsegmented notochord, a terminal mouth, and multicusped teeth (fig. 2.6a).[2] However, its supporting skeleton, the shapes of its fins, and the composition of its spine dif-fered from those of the primitive sharks. Attached to the shoulder girdle were three small triangular elements at the base of each pectoral fin. These ele-ments provided support for the radial elements that extended outward into the fin. In order from anterior to posterior, the first element is the propterygium, the second the mesopterygium, and third the metapterygium.[4] This narrow basal support enabled the pectoral fin to rotate on its axis with greater flexibil-ity than did the broad base of the pectoral fin of the more primitive sharks. An anal fin was present on the belly immediately anterior to lower lobe of the cau-dal fin. The vertebral column now was longer, extending all of the way to the tip of the upper lobe of the caudal fin. The spines of these less primitive sharks had an external layer of enamel-like dermal denticles covering their external surface. These spines likely penetrated the skin and provided protection to the shark. The dorsal spines were deeply set in the dorsal musculature and fit snugly against a large triangular element of cartilage. The surface of the body of ctenacanthids was entirely covered with dermal denticles unlike that of the more primitive sharks. The Ctenacanthiformes were most common 290 mya, at the end of the Carboniferous Period. They survived the mass extinctions at the end of the Permian, but only a few species were still present 251 mya, at the beginning of the Triassic.

Living together with the ctenacanthids were the hybodonts. They first ap-

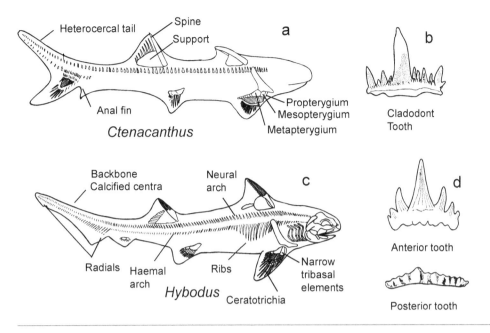

Fig. 2.6 Skeletal anatomy of the late Paleozoic sharks: (a) *Ctenacanthus* and (b) its multicusp tooth, as well as (c) *Hybodus* and (d) its anterior and posterior teeth.

peared in the late Devonian but became common in the Carboniferous and Permian. They barely survived the Permian extinctions but later diversified during the Mesozoic epoch from 248 to 65 mya.[11] Their upper jaw was attached to the chondrocranium on either side by ligaments at two points as in the more primitive sharks. However, unlike these sharks the area of contact on the posterior palatoquadrate was broad and the point of contact on the anterior palatoquadrate was narrow (fig. 2.5c).[11] Other anatomical innovations enabled the hybontids to supplant the ctenacanthids as the apex predators of the Mesozoic. The pectoral fins of the hybontids were better supported because the scapulocoracoids, or shoulder bones, on either flank of the shark were connected in the pelvic ventral region by a continuous cartilaginous plate (fig. 2.6c). The three elements that attached to the shoulder bone at the base of the pectoral fin were narrower and more elongated than those of the ctenacanthids, enabling the fin to rotate to different angles as the shark swam up and down in the water column. The radial elements did not extend to the margin of the fin but were replaced toward the margin by ceratotrichia, proteinaceous fin rays that were more flexible than the more primitive external radials. Muscles within the fin could curve the pectoral fin from front to back or from base to tip. The muscular control enabled these sharks to steer their bodies better when swimming.[2]

Hybodus also possessed a more functional caudal fin. Radials extended outward from the base to the tip of the caudal fin, and they were longer at the base and shorter at the tip (fig. 2.6c).[11] The radials never reached the margin of the fin but were supplanted distally by flexible ceratotrichia. The lower lobe of the fin was reduced in size. This style of tail fin is referred to as heterocercal, "hetero" meaning "different" and "kerkos" meaning "tail." The more numerous radial elements gave the fin more flexibility. In addition, the supporting musculature could control the shape of the fin better during movement. When the fin moved from side to side, the fin rotated so that its flexible lower lobe trailed behind the stiff upper lobe. This difference in the distribution of force resulted in more efficient forward thrust. Below the notochord were haemal arches that offered protection for the vein and artery below the notochord. Also, the hybontids had well-developed ribs in the midsection of the body.

The hybontids became common during the Carboniferous and diversified greatly in the Triassic, when they were the dominant predator in the marine waters off Europe and North America. They continued to be abundant until the late Cretaceous, 60 mya, and were present in the same waters as the euselachians, or modern sharks.[4] The majority of the hybontids were likely sluggish swimmers capable of short bursts of rapid swimming. Their paired pectoral fins would stabilize them so they did not pitch while swimming, and the dorsal and anal fins prevented yaw. The presence of a larger upper lobe on their caudal fin and a subterminal mouth indicates that they captured prey from above. They likely swam along the bottom, seizing crustaceans and molluscs with their pointed anterior teeth and passing them to the back of their mouth to be crushed by their pavement-like posterior teeth (fig. 2.6d).

Paleozoic Chimaeras

The ancestors of the modern chimaeras were common during the Paleozoic. *Helodus,* a member of the Helodiformes, first appeared in the freshwater habitats 380 mya in the late Devonian and was present until the end of the Permian (fig. 2.7a).[4] The shape of this small holocephalan's braincase and the firm attachment of its upper jaw to its skull resembles that of the modern chimaeras as do its unornamented dorsal fin spine attached to the skeleton near the chondrocranium. Furthermore, the two basal elements at the base of its pectoral fin are almost identical to those of modern chimaeras. However, *Helodus* did not have the tooth plates characteristic of modern chimaeras. These consist of two pavements of teeth, one on the upper jaw and another on the lower jaw. *Helodus* had ten small teeth in each side of its upper and lower jaw.

The Iniopterygiformes also resembled the modern chimaeras in many respects.[14] They appeared later than the Helodiformes—340 mya in the early Carboniferous. The skull of *Iniopteryx* (fig. 2.7b) was similar that of the mod-

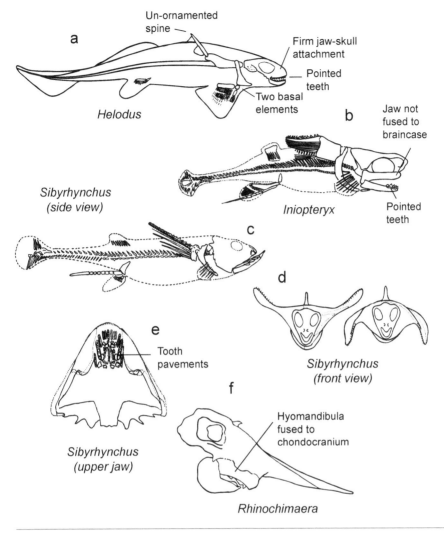

a

Un-ornamented spine —

Firm jaw-skull attachment

Pointed teeth

Two basal elements

Helodus

b Jaw not fused to braincase

Sibyrhynchus (side view)

Iniopteryx

Pointed teeth

c

d

e Tooth pavements

Sibyrhynchus (front view)

f

Hyomandibula fused to chondocranium

Sibyrhynchus (upper jaw)

Rhinochimaera

Fig. 2.7 Skeletal anatomy of the Paleozoic chimaeras: (a) *Helodus*, (b) *Iniopteryx*, (c) *Sibyrhynchus* and (d & e) its upper jaw, and (f) the jaw of an extant chimaera, *Rhinochimaera*.

ern chimaeras. However, the upper jaw had many delicate teeth unlike the modern chimaeras, which have pavement-like teeth, and its jaw was not fused to the braincase as in the modern ratfish *Rhinochimaera* (fig. 2.7f).[15] *Sibyrhynchus* possessed whorls of teeth around the margin of its jaw and pavements of star-shaped, fused dermal denticles in the center (fig. 2.7e). These crushing teeth were aligned in rows along the anterior-posterior axis of the jaw. Some have suggested that *Sibyrhynchus* propelled itself out of the water, using its pectoral fins like the wings of an airplane, thus resembling the modern flying fishes (figs. 2.7c & 2.7d). Putative evidence for this was the attachment of its pectoral fins to its shoulder bone toward the top of the body. However, the

You can make inferences about the behavior of the ancient sharks, rays, and chimaeras based on the shape of their body, fins, and teeth—a common practice of functional morphologists. The habitats in which these species lived can be identified by the nature of the rock deposits where they are found.[1] You can imagine what an ecosystem during the Paleozoic Era looked like dominated by members of these two groups. There were numerous cartilaginous fishes in the seas at that time (fig. 2.8). Some species such as *Stethacanthus* and *Falcatus* resemble the modern sharks in the family Lamnidae, possessing a caudal fin consisting of large upper and lower lobes attached to the trunk of the body by a narrow caudal peduncle. The two equal-sized lobes generated considerable forward thrust with the narrow peduncle minimizing drag so that these Paleozoic sharks could swim quickly and continuously for long periods of time. The jaws of these sharks were full of pointed teeth within a single large cusp flanked by two smaller cusps resembling the teeth of the extant salmon shark (*Lamna ditropis*). The Paleozoic sharks would have chased down small placoderms, acanthodians, elasmobranchs, and holocephalans. They likely seized them with their gripping teeth and swallowed them whole. *Stethacanthus*, the largest of these predators, was 3.5 m long. It is found in fossil marine deposits worldwide such as are present in Montana, Scotland, and Moscow. This large predator likely inhabited the upper waters of the ocean basins and could have migrated great distances in response to changes in water temperature during the changing seasons.[5] *Falcatus* were smaller, l.5 m long, but were also active swimmers like *Stethacanthus*. However, they likely lived on the continental shelf close to shore and may at times have entered the bays. A male is shown here initiating courtship with a female by striking her with his brush-like dorsal appendage. Two male *Falcatus* are shown here about to swim toward each other and engage their flat dorsal spines in a ritualized display with the purpose of gaining access to females for mating.

Ctenacanthus also had teeth with pointed cusps for seizing and swallowing its prey, but its caudal fin with an enlarged upper and smaller lower lobe was attached to the torso by a relatively wide caudal peduncle.[2] The movement of the peduncle from side to side would result in considerable drag, so its forward movements were likely confined to brief bursts of speed. *Ctenacanthus* likely swam not far off the bottom and accelerated downward to capture its prey—much as is done by the modern reef sharks of the family Carcharhinidae. This species may have inhabited the large tropical inland sea in Pangea (fig. 2.2b). *Hybodus* was a generalist. It could feed on both placoderms and acanthodians up in the water column and invertebrates buried within the muddy and sandy substrates covering the bottom of the continental shelf and bays of Pangea. *Hybodus* could seize a placoderm with its pointed teeth and crush its shield with hard and flattened teeth. It could also feed on actively swimming prey. Fig. 2.8 shows three *Hybodus* feeding upon members of a school of squid near a seamount. *Xenacanthus* predominated in the estuarine and freshwater environments. It had sharp two-cusped teeth for seizing prey and swallowing them. Its elongated dorsal fin ending in a spade-like caudal fin enabled this predator to move freely with eel-like swimming motions in the shallow swamps that were choked with vegetation.[13] Three holocephalans, *Helodus*, *Iniopteryx*, and *Sibyrhynchus*, were present with the elasmobranchs in the oceans during the Paleozoic. These were smaller in size and slower swimmers than the elasmobranchs.[2] The first three had pointed gripping teeth for seizing their prey. *Helodus* and *Iniopteryx* likely swam along the bottom, seizing slow-moving prey swimming either directly in front or above them. On the other hand, *Sibyrhynchus* may have patrolled near the surface, grasping slowly swimming prey swimming underneath it.

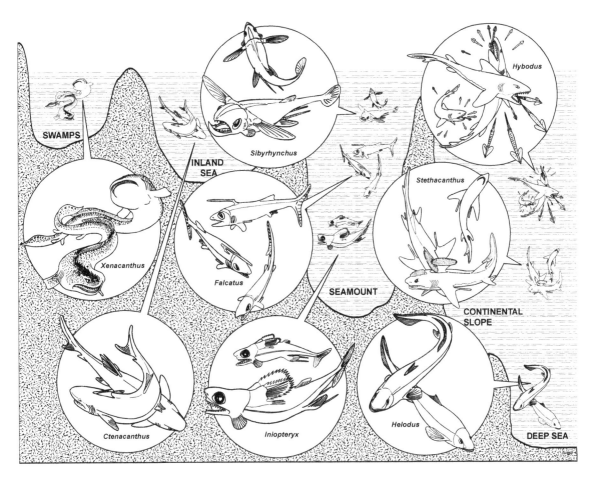

Fig. 2.8 The ancient sharks and chimaeras living in the ecosystems of the Paleozoic with estuarine and freshwater species shown on the left, coastal species in the middle, and oceanic species on the right.

fossil skeleton is not as delicate and streamlined as the skeletons of the modern flying fishes. Holocephalans of many diverse body shapes were common during the Carboniferous, but only a few survived the Permian to live in the Mesozoic, whereas the euselachians greatly diversified during this period.

Permian Extinctions

The elasmobranchs and holocephalans began to decline during the middle of the Permian. Most of the primitive sharks and holocephalans had disappeared by the end of this period. The reduction in species abundance and diversity may have been partly due to the formation of the supercontinent, Pangea (fig. 2.2c). The favored habitat of many chondrichthyans, the shallow continental shelf, was reduced in expanse as the subcontinents coalesced into a single continent, leading to the loss of species due to their inability to survive when faced with greater competition for fewer prey. However, many other causes have been proposed for this massive reduction in the number of marine genera (see fig. 2.1) such as the impact of large meteorites or increased volcanic activity producing dark clouds in the atmosphere that prevented light from reaching the seas, the sudden discharge of noxious methane compounds from the sea-floor, the marine environment becoming anoxic or deprived of oxygen, and a shift in ocean circulation driven by climate change. There was another major extinction event 251 mya, at the end of the Permian; 50–75% of the families in the sea were lost. This equates to 80–96% of the marine species becoming extinct over a period of 500,000 years. Considerable time lapsed before the marine ecosystems recovered from these extinctions. Coral reefs reappeared in the seas only after 10 million years. This long period to recovery implies that the disturbance was followed by long-term harsh environmental condi-tions that hindered the rate of faunal recovery. The extinction of the primary producers, the marine phytoplankton, that utilize sunlight and carbon dioxide to produce organic carbon likely caused widespread starvation of the primary consumers, the planktivores. The extinction of species on the lower trophic levels likely had a knockout effect on the species at the higher levels of the marine ecosystem, such as the Paleozoic elasmobranchs and holocephalans. Those that survived this period were likely migratory because this ability would have enabled them to seek out waters of more habitable temperatures and salinities. Furthermore, they were probably feeding generalists like *Hybo-dus*, feeding on whatever species survived the environmental disturbance. An ecological vacuum was likely created by so many species becoming extinct. This may have created opportunities for the surviving lineages to diversify and fill many of the emptied niches in the new ecosystem.

SHARKS AND RAYS OF THE MESOZOIC ERA

The movements of the earth's crust during the Permian and Triassic Periods coincided with a second major diversification of the cartilaginous fishes. The single continent of Pangea that was present during the Permian (fig. 2.2c) split into Gondwana, a southern continent, and Laurasia, a northern continent, separated by a narrow body of water at the mid latitudes in the northern hemi-sphere (fig. 2.2d). During the Permian, a large body of water straddled the equator within the single continent. The continental plate within the north-ern hemisphere moved counterclockwise away from the equator. This rotation created the triangular-shaped Tethys Sea in the southern hemisphere, which was tropical in climate. Not only was there an increase in the extent of conti-nental shelf, a favored habitat of cartilaginous fishes but also the range of tem-peratures in the waters surrounding the continent. This opened new niches

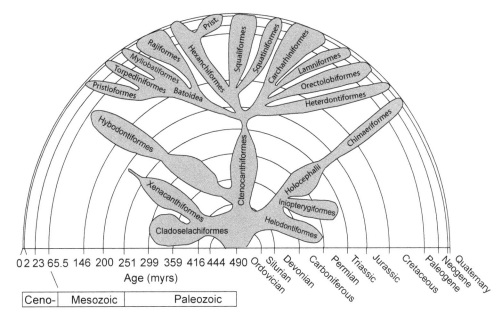

Fig. 2.9 The diversity of cartilaginous fishes over geological time is shown in an evolutionary tree with each branch denoting the orders of ancestral and modern chondrichthyans. The width of each branch is proportional to the diversity and abundance of a particular taxon at that time in geological history. The time scales for the geological periods used are those of the International Commission on Stratigraphy, and the beginnings and ends of the ages are based on the appearance and disappearance of planktonic species in cores of the earth's crust that are dated using radiometric methods.

for exploitation by the cartilaginous fishes. The oceans were cool at the poles, warm at the middle latitudes, and hot near the equator. The sharks radiated a second time likely in response to this increase in the variety of habitats, occupying the estuaries and bays, the continental shelf, and the open ocean. However, the chimaeras did not radiate at this time.

None of the earliest sharks, which flourished during the Carboniferous, survived the harsh conditions at the end of the Permian.[4] Only a few ctenacanths survived into the Triassic. This is apparent in the constriction in the main trunk of an evolutionary tree of the chondrichthyans (fig. 2.9). The hybodonts, which were common in the Carboniferous, did survive the Permian extinctions and continued to be common in the seas during the Triassic and Jurassic until they too became extinct at the end of the Cretaceous (see left branch of evolutionary tree). It is most likely that the ctenacanths gave rise to the modern sharks, or Euselachi. The earliest known euselachian was *Palaeospinax*, which was present at the beginning of the Jurassic (fig. 2.10a).[11] The first fossils of rays, which were intermediate in shape between torpedo-shaped sharks and pancake-shaped rays, are found in deposits from the early Jurassic in Europe.[3] These individuals resembled the angelsharks of the order Squatiniformes and the guitarfishes of the order Rajiformes. By the late Juras-

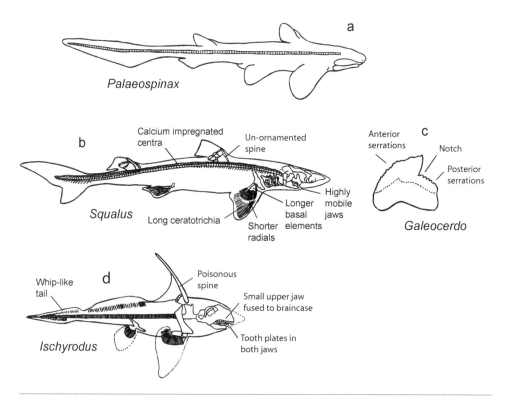

Fig. 2.10 Skeletal anatomy of modern sharks and chimaeras of the Mesozoic and Cenozoic Eras: (a) *Palaeospinax*, (b) *Squalus*, (c), a modern shark's tooth, and (d) *Ischyrodus*.

sic, ancestral guitarfish such as *Protospinax* and *Spathobatis* were common on the bottom in bays and on the continental shelf (figs. 2.11a & 2.11b). Only a few chimaeras such as *Ischyrodus* lived alongside the sharks and rays during the Jurassic (figs. 2.10d & 2.12c).[11] Its upper jaw was closely fused to the braincase, its gills lay beneath the braincase, and two tooth plates were present in the upper jaw and one pair in the lower jaw as in the modern chimaeras. Its tail was long and whip-like, the pectoral fins were wide, and the tall spine in front of the dorsal fin may have born a poison gland as is present in the extant chimaeras.

The euselachians showed considerable improvements to their anatomy over the ctenacanths, in particular to the parts of the skeleton associated with swimming and feeding.[11] This is apparent in the skeleton of *Squalus* (fig. 2.10b). Its vertebral centra were impregnated with calcium to resist the compression forces exerted on them during swimming. The two half-girdles supporting the pectoral fins and pelvic fins became fused to provide both pairs of fins with more support. The elements at the base of the pectoral fins became more elongate and closely articulated with each other so that a narrow isth-

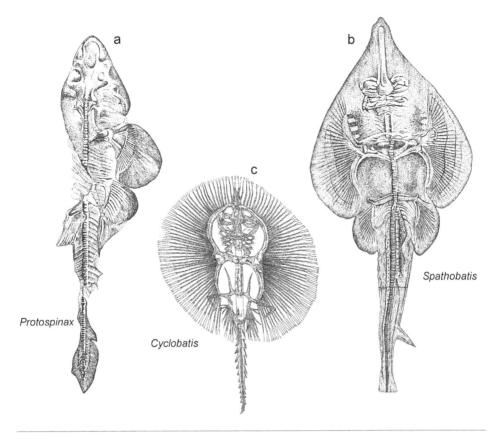

a

b

c

Spathobatis

Protospinax

Cyclobatis

Fig. 2.11 Skeletal anatomy of Mesozoic rays: (a) *Protospinax*, (b) *Spathobatis*, and (c) *Cyclobatis*.

mus attached each fin to the girdle. The radials were shorter than those of the Paleozoic sharks, but the ceratotrichia were longer and more flexible. They extended farther out to support the distal end of the fins. Finally, the euselachians had smaller, unornamented spines unlike the primitive sharks and rays.

The type of connection between the jaws and braincase of the euselachians enabled them to open their jaws wider than the selachians. This is apparent in the jaws of the modern reef sharks of the genus *Carcharhinus* (fig. 2.5d).[4] The upper jaw is narrow with a single anterior enlargement that attaches by a ligament to the anterior chondrocranium. The posterior lower jaw is attached by another ligament to the hyomandibula, a skeletal element derived from the first gill arch and attached by a ligament to the posterior chondrocranium. The upper jaw moves up and down relative to the posterior braincase, and the lower jaw slides forward and backward. This detachment of the posterior upper jaw from its prior close articulation with the braincase enables both jaws to open wider in order to swallow larger prey. The jaws are also protruded outward and withdrawn to gouge out large pieces from prey too large to swallow whole.

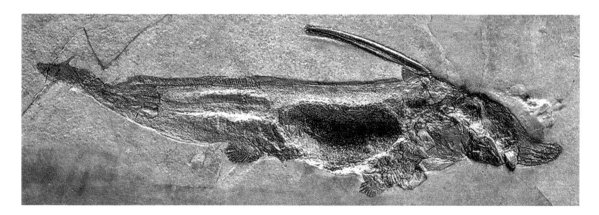

Fig. 2.12 Fossil remains of a Mesozoic shark, ray, and chimaera: (a) *Scyliorhinus elongatus* from the middle Cretaceous in mineral deposits in Lebanon, (b) *Heliobatis sp.* from the Eocene in deposits from Wyoming, and (c) *Ischyodus avitus* from the late Jurassic in deposits from Solnhofen, Germany.

The teeth of these modern sharks have multiple serrations along the edges of a flattened cusp such as the tooth of the tiger shark. These enable them to saw through flesh (fig. 2.10c). These teeth are quite unlike the teeth with one to five conical cusps of the primitive sharks that were better for simply seizing and swallowing prey (see figs. 2.3b & 2.3i).[4] The modern shark can also open its jaws rapidly. In some species, this rapid opening of the jaws creates enough of a vacuum to draw in crustaceans and small fishes into the mouth. Finally, the braincase extends forward in the modern sharks to form a rostrum. The mouth is situated below the snout, affording more space to accommodate larger nasal capsules. This is evidence that the sense of smell is more important in the modern sharks.

Evolutionary Relationships

It is difficult to solely rely on cladistic analysis, in which unique shared anatomical characters from fossils are used, to describe the evolutionary history of the cartilaginous fishes. To start with, their skeletons preserve poorly. There are therefore relatively few fossils available in recent geological time to examine in order to find uniquely shared primitive features indicating common ancestry. Furthermore, the similar ecological roles of many nonrelated species have led to convergent evolution with the independent development of similar anatomical features. For these two reasons, molecular studies must also be used to elucidate the evolutionary relationships among modern sharks. The similarity of sequences of mitochondrial and nuclear genes can be used to indicate common ancestry. This technique, of course, cannot be used to identify the relationships among the ancient sharks because the soft tissues, on which the molecular analyses are based, are absent from the fossil record. However, molecular techniques are currently being used in conjunction with cladistic analysis and fossil remains to reveal the phylogenetic relationships between the modern sharks, rays, and chimaeras.

A dendrogram can be used to illustrate the sequence of evolution. This diagram can be likened to a tree, which is not upright but lying on its side. It consists of a trunk with a series of diverging limbs and branches. The species on each bifurcation or trifurcation are related by the possession of at least one characteristic shared with a common ancestor on the branch serving as its origin. The fishes can be related to each other based on the similarity of their sequences of genes (see left tree in fig. 2.13)[15] and common derived anatomical features (see right tree).[16] All comparisons of evolutionarily related taxa must be placed in the context of an unrelated outgroup to demonstrate that they are more related to each other than to an evolutionarily divergent group. It is not

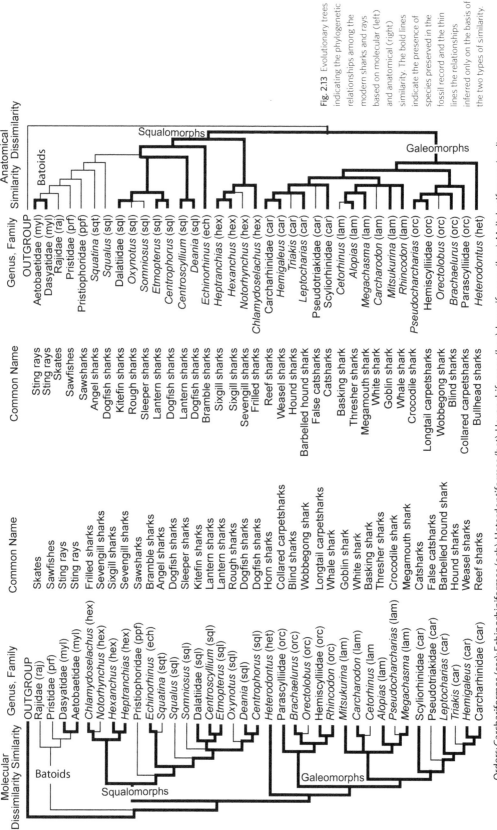

Fig. 2.13 Evolutionary trees indicating the phylogenetic relationships among the modern sharks and rays based on molecular (left) and anatomical (right) similarity. The bold lines indicate the presence of species preserved in the fossil record and the thin lines the relationships inferred only on the basis of the two types of similarity.

Orders: Carchariformes (car); Echinorhiniformes (ech); Heterdontiformes (het); Hexanchiformes (hex); Lamniformes (lam); Myliobatiformes (myl); Orectolobiformes (orc); Pristiophoriformes (prf); Rajiformes (raj); Squaliformes (sql); Squatiformes (sqt).

possible to identify just when each order appeared using this method of analysis; that must be based on the discovery of fossils in sedimentary deposits dated to a particular age.

One of the ctenacanths, which survived the inclement conditions of the Permian, was likely the common ancestor of the modern sharks. It constitutes the base of this phylogenetic tree of chondrichthyan fishes. Notice the second limb from the trunk of the tree based on similarity in genetic code leads to the Batoidea, the subclass of the rays and their relatives. This tree limb bifurcates with the first branch leading to the Rajiformes, the skates, and the second to the Pristiformes, the sawfishes, and the Myliobatiformes, the stingrays. The limb structure of this phylogenetic tree indicates that the rays are only distantly related to the modern sharks. Yet this conclusion is inconsistent with the branching of the evolutionary tree based on shared skeletal and muscular characters.[16] The first limb off the trunk of this tree leads to the galeomorphs and squalomorphs, with a limb from the latter leading to a series of bifurcations with the first leading to the Hexanchiformes. The second branch leads to three groups, the Echinorhiniformes, the majority of the Squaliformes, and a third group. This group consists of the dogfish sharks of the genus *Squalus*, the angelsharks of the genus *Squatina*, and the skates and stingrays. These latter groups have been combined into a group called the hypnosqualeans, based the possession of a cartilaginous process extending from the upper jaw into the eye socket, the angle of the braincase, and the wide separation of the eyes in the chondrocranium.[17] It is more likely that the molecular tree indicates the correct evolutionary relationship between the sharks and rays. Two genetic studies, one based on the similarity of sequences of mitochondrial genes[18] and the other based on the similarity of sequences of nuclear genes,[19] indicate that the rays diverged early from the sharks before the evolution of the squalomorphs. Fossils in sedimentary deposits in Europe indicate that the rays were present in the oceans in the early Jurassic when the earliest modern sharks began to flourish.[3]

Both cladistic and genetic analyses indicate that the modern sharks are divided into two lineages, the squalomorphs and galeomorphs. Notice the separation of the trunks of the two trees near the base to form two main limbs, each leading to the different families and genera. The Squalomorphi are considered more primitive in their anatomy, possessing smaller brains, and having evolved in deepwater environments, where they are currently more abundant.[2] The Galeomorphi are more advanced in their anatomy, have larger brains, and are likely to have evolved in shallow water tropical environments, where they are currently most abundant. Genetic studies indicate that rays evolved separately from the two modern lineages of sharks, unlike the latest cladistic study based on anatomical similarities. Both the genetic and anatomical studies emphasize the distinct ancestry of the elasmobranchs and holocephalans. The latter are not shown in the phylogenetic tree.

Fossil Record

The similarities between the evolutionary trees based on molecular and morphological similarities can be combined with fossil evidence to provide insights into the evolutionary history of the modern sharks.[20] The bold lines in fig. 2.13 indicate the presence of groups preserved in the fossil record and the thin lines the relationships inferred only on the basis of cladistic and molecular evidence. The squalomorph and galeomorph sharks and batoids began to radiate during the early Jurassic.

The first squalomorph sharks to appear were the six- and sevengill sharks of the order Hexanchiformes.[21] These sharks are easily distinguished from other sharks by the additional one or two gill slits and their single dorsal fin that originates far back on the body near the caudal fin. The sixgill sharks are most common in the cool upper waters of the temperate latitudes and occupy the cool deepwaters in the subtropical seas. Extensive fossil evidence at the end of the Cretaceous indicates that a large predatory sixgill shark, *Notianodon*, inhabited the cold temperate waters on either side of the equator.[22] The modern sixgill shark *Notorynchus* has a similar bipolar distribution, inhabiting the cool temperate latitudes in both the northern and southern hemispheres in the Pacific Ocean. Sixgill sharks in *Hexanchus* and sevengills in *Heptranchias* live in the same cold coastal waters in the temperate latitudes as their ancestors, but they also inhabit the cold deepwaters of the subtropics and tropics.

The Squatiniformes appeared later in the seas at the beginning of Cretaceous and became widespread by the middle Cretaceous (fig. 2.9). The modern species are restricted to temperate coasts due to their benthic lifestyle and limited mobility and thus are absent from offshore islands. They dispersed along the continental shelf prior to the separation of Pangea into many smaller continents during the late Cretaceous; hence, their distribution today is worldwide.

The Squaliformes appeared after the Hexanchiformes and Squatiniformes. *Protosqualeus* is present in marine deposits in northern Germany and northern France during the early Cretaceous.[23] The molecular evidence supports the evidence from the fossil record. *Squalus* is on a branch after the four genera of hexanchiform sharks and after a branch leading to two genera, *Echinorhinus* and *Squatina*, on the tree based on molecular similarity (see left tree in fig. 2.13). The genus *Squalus* first appeared in the upper Cretaceous with a plethora of fossils preserved in deposits in Asia, northern Africa, and North America.[23] This broad geographic distribution of the fossils is consistent with the long-distance migrations of modern dogfishes. Some taxonomists include the bramble sharks of the family Echinorhinidae within the Squaliformes,[24] others separate them into their own order, the Echinorhiniformes.[1] We will follow the former convention because of their many similarities with members of the Squaliformes.

The sawsharks of the Pristiophoriformes were the last to diversify of the squalomorphs. They were first found in deposits surrounding the eastern Pacific Ocean 33.4 mya, toward the middle of the Paleogene.[23] The sawsharks were widely distributed worldwide by the end of the Paleogene, as fossils have been found in sedimentary deposits from this epoch in Japan, Morocco, Oregon, Belgium, Holland, and California.[25] The living sawsharks are now restricted to continental and insular shelves, living in shallow waters in the temperate latitudes and deeper waters in the tropics.

The first galeomorphs to inhabit the seas were the Heterodontiformes. The earliest fossils are from Germany dated from the late Triassic and early Jurassic. Hence, the heterodontids shared the seas with the hexanchids during this geological period. The members of this order are either called bullhead sharks due to their elevated head with bulbous crests above they eyes or horn sharks due to the presence of spines at the front of their two dorsal fins. The ancient bullhead sharks resemble their descendents in skeletal and jaw anatomy and hence were also likely confined to the continental shelf. They inhabited the extensive continuous shelf surrounding Pangea.[18] Fossil remains are present in sedimentary deposits of Patagonia, South America in the Cretaceous Period. The ancient *Heterodontus* living in the Atlantic became extinct, whereas a similar species survived in the Pacific Ocean.

The Orectolobiformes were second order of galeomorphs to appear in the fossil record. *Brachaelurus* was present in the Early Jurassic and *Orectolobus* first appeared in the middle Jurassic.[20, 21] This is consistent with the molecular and anatomical evidence because the next branch after the one leading to the horn sharks on both phylogenetic trees leads to the carpet sharks with the adjacent limb leading to the Lamniformes and Carcharhiniformes.

The Lamniformes appeared next at the end of the Jurassic. *Palaeocarcharias* was present in deposits 146 mya in Europe.[25] This fossil shark had the sharp pointed teeth for seizing and swallowing fish as do the modern mackerel sharks. The goblin shark of the genus *Mitskurina* appeared later 115 mya during the early Cretaceous. The white sharks of the genus *Carcharodon*, the thresher sharks of *Alopias*, and the basking shark appeared even more recently, being preserved between 65.5 million and 23.0 mya in deposits from the Paleogene.

The sister group of the mackerel sharks is the Carcharhiniformes. The ancestral family in this order is the Scyliorhinidae, which includes the catsharks. These sharks mainly inhabit the deep ocean but also are found in cool water

The giant-tooth shark (*Carcharodon megalodon*) is the most imposing predatory shark of all time. It thrived only briefly on the geological time scale. It grew to over 15 m long, dwarfing the size of its extant relative, the white shark (see size comparison, insert in fig. 2.14a), and was first present in the seas in the second half of the Paleogene.[26] The giant-tooth (also referred to as *Megalodon*) had a blunt snout with very stout jaw. Its razor-sharp serrated teeth 8–10 cm high were capable of cutting through flesh and severing even the hardest skeletons of massive prey. At this time, the waters in all of the oceans were warm and subtropical (fig. 2.14a). Teeth of adults from both species are found in sedimentary deposits on the southeastern coast of North America and at higher latitudes along the southern coast of England and northern coast of Europe. The teeth of the giant-tooth appear at the same fossil deposits as the archeocetes, the ancestors of the baleen whales, and myctocetes at the end of the Paleogene. The teeth of smaller toothed white shark are associated with the fossil remains of the first seals and sea lions (pinnipeds) and dolphins (odontocetes) 23.3 mya at the very end of the Paleogene.

The waters of the North Atlantic began to cool from the middle Miocene through the early Pliocene, which was accompanied by a general increase in oceanic productivity. In response to this, the marine mammals increased in diversity from the middle Miocene through the early Pliocene. This explains the occurrence of teeth of the giant-tooth with pinnipeds, odontocetes, and mysticetes at multiple sites along the southeastern and gulf coasts of North America during this period of time (fig. 2.14b). Similar fossil remains of the giant-tooth occur with the remains of the same marine mammals along the northern coast of Europe. The predator-prey relationship between the giant-tooth and whales is evident from bite marks made by very large serrated teeth on tail vertebrae and flipper bones preserved in these deposits. It is likely that the rapid diver-sification of marine mammals was driven by an increase in ocean productivity at this time; the giant-tooth shark responded by seeking out areas frequented by whales.

The ocean waters cooled even further 5.2–1.6 mya, during the Pliocene, with a distinct zonation arising in the seas with very warm waters at the low latitudes, moderate water temperatures at the mid latitudes, and cool waters at the high latitudes (see fig. 2.14c). Only the smaller white shark was able to inhabit the higher latitudes along the northern coast of Europe, where it fed on phocine seals in the cool waters, as is evident from the common occurrence of fossils of both in sedimentary deposits in Europe. The giant-tooth could not penetrate these cooler waters, but was restricted to living at the mid latitudes.

This appears to be a classic case of coevolution between predator and prey in response to the cooling of the oceans at the end of the Eocene. The seals and sea lions were able to colonize the regions of high productivity during the Miocene and Pliocene because their fatty exterior blubber layer kept them warm in the cold waters. The circulatory system of the extant white shark may have evolved in response to this environmental cooling to conserve heat and maintain a warm body. This would certainly enable it to actively pursue and capture pinnipeds in the cooler water of the upper latitudes. The giant-tooth shark could inhabit the upper latitudes during the Paleogene and Miocene, when the waters off the northern coast of Europe were warm. However, it may have been unable to tolerate the cool temperate waters off the northern coast of Europe during the Pliocene as did the small-tooth white shark. It is not known why the giant-tooth became extinct 1.6 mya during the Pleistocene, while its closely related sister species survived to be a dominant predator today in the world's oceans. However, one explanation would be that the giant-tooth was unable to evolve endothermy, the ability to maintain an elevated body temperature, and absence of this physiology doomed it to extinction.

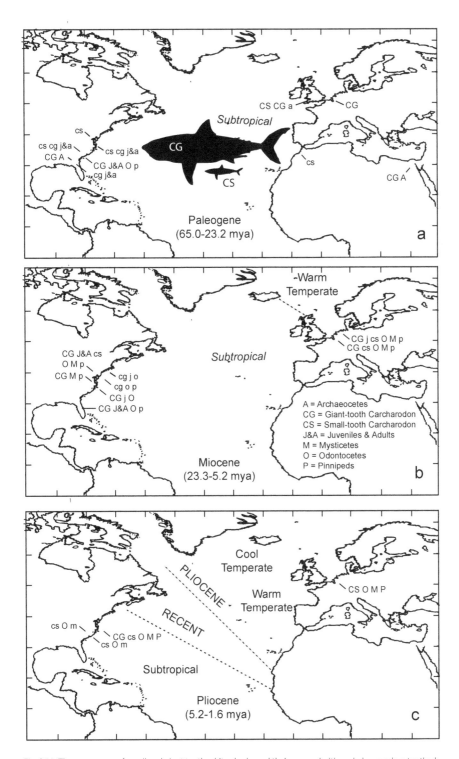

Fig. 2.14 The occurrence of small and giant tooth white sharks and their prey, primitive whales, modern toothed whales, baleen whales, and pinnipeds indicated together with the location of subtropical, warm temperate, and cool temperate waters on maps of the seas during the (a) Paleogene, (b) Miocene, and (c) Pliocene. Uppercase abbreviations indicate the fossils are common, lowercase that they are uncommon.

habitats. The oldest fossils of this family are from the upper Jurassic Period from deposits in Germany.[27, 28] The other carcharhinids such as the hound-sharks of the Triakidae appeared later in the Paleogene and Neogene. They are found in fossil deposits in Trinidad and western Europe.[25] The members of *Carcharhinus*, characterized by a large brain, complex social behavior, and the ability to migrate over long distances, appeared 53 mya, in the Paleogene. The carcharhinids expanded their range throughout the tropical seas with some species even moving into freshwater habitats. The rise of this group parallels the radiation of the bony fishes. The diversification of the bony fishes, which are the main prey of carcharhinids,[21] may have led to the great diversity in the size and shape of their teeth and the distinct patterns of serrations.

The first batoid fossils were found in deposits from the early Jurassic in Europe.[6] *Protospinax*, which was present at the same time, had a less rounded, flattened shape and thus resembled the angelsharks of the Squatiniformes (fig. 2.11a).[11] This group contains the modern family Rhinobatidae, whose members are shaped like a guitar. *Spathobatis*, an ancestral guitarfish and one of the first batoids, was present in the upper Jurassic (fig. 2.11b).[15] The majority of species in this order are in the family Rajidae, the skates, which have a round or triangular disk. They became common in the late Cretaceous.[3] *Cyclobatis* and *Heliobatis* were some of the earliest skates present at this time (figs. 2.11c & 2.12b).[11] The Torpediniformes, or electric rays, and Pristiformes, or saw-fishes, were present with the skates in the late Cretaceous. The fourth order of rays, the Myliobatiformes, became common 56.5 mya, at the beginning of the Eocene.

The ancestors of the chimaeras disappeared in the Carboniferous and Permian. Only *Myriacanthus* survived into the Triassic and Jurassic.[3] *Ischyrodus*, which occurs in fossil deposits in the late Jurassic, is the likely ancestor of the modern chimaeras.[3] It shares many characteristics with them, such as the long rat-like tail, the large erectile dorsal fin, and the protuberance from the head. Although holocephalans of many diverse body shapes were common during the Carboniferous, only a few survived the Permian to live in the Mesozoic together with the euselachians, and even fewer existed in the Cenozoic with the modern sharks.

ORDERS OF EXTANT CARTILAGINOUS FISHES

The class Chondrichthyes is composed of thirteen orders of sharks, rays, and chimaeras (fig. 2.15).[1] This class is separated into the subclass Elasmobranchii, the sharks and rays, and the subclass Holocephali, the chimaeras. The former group is separated into the subdivision Selachii, the sharks, and subdivision Batoidea, the rays. There are eight orders of sharks in the Subdivision Selachii,

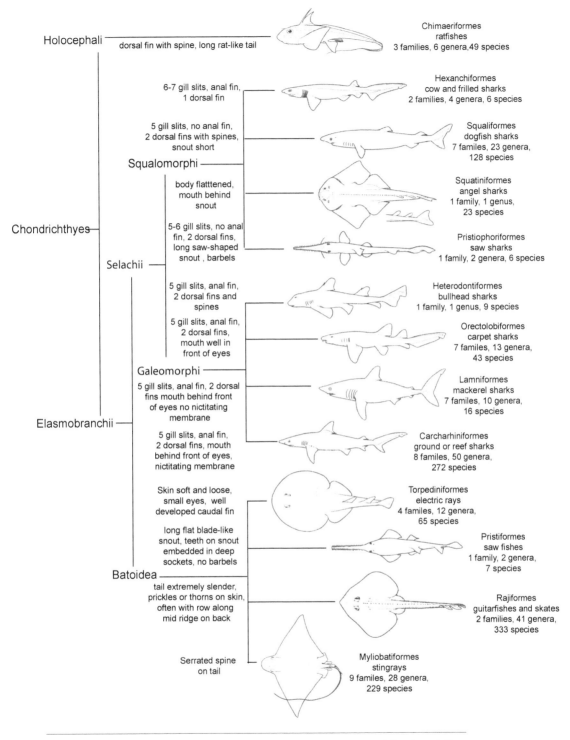

Holocephali — dorsal fin with spine, long rat-like tail

Chimaeriformes
ratfishes
3 families, 6 genera, 49 species

6-7 gill slits, anal fin, 1 dorsal fin

Hexanchiformes
cow and frilled sharks
2 families, 4 genera, 6 species

5 gill slits, no anal fin, 2 dorsal fins with spines, snout short

Squaliformes
dogfish sharks
7 families, 23 genera, 128 species

Squalomorphi

body flatttened, mouth behind snout

Squatiniformes
angel sharks
1 family, 1 genus, 23 species

5-6 gill slits, no anal fin, 2 dorsal fins, long saw-shaped snout, barbels

Pristiophoriformes
saw sharks
1 family, 2 genera, 6 species

Selachii

5 gill slits, anal fin, 2 dorsal fins and spines

Heterodontiformes
bullhead sharks
1 family, 1 genus, 9 species

5 gill slits, anal fin, 2 dorsal fins, mouth well in front of eyes

Orectolobiformes
carpet sharks
7 familes, 13 genera, 43 species

Galeomorphi

5 gill slits, anal fin, 2 dorsal fins mouth behind front of eyes no nictitating membrane

Lamniformes
mackerel sharks
7 families, 10 genera, 16 species

5 gill slits, anal fin, 2 dorsal fins, mouth behind front of eyes, nictitating membrane

Carcharhiniformes
ground or reef sharks
8 familes, 50 genera, 272 species

Chondrichthyes

Elasmobranchii

Skin soft and loose, small eyes, well developed caudal fin

Torpediniformes
electric rays
4 families, 12 genera, 65 species

long flat blade-like snout, teeth on snout embedded in deep sockets, no barbels

Pristiformes
saw fishes
1 family, 2 genera, 7 species

Batoidea

tail extremely slender, prickles or thorns on skin, often with row along mid ridge on back

Rajiformes
guitarfishes and skates
2 families, 41 genera, 333 species

Serrated spine on tail

Myliobatiformes
stingrays
9 families, 28 genera, 229 species

Fig. 2.15 Representatives of the thirteen orders of sharks, rays, and chimaeras, together with their distinguishing features as well as the numbers families, genera, and species in each order.

which differ most prominently from the dorsal-ventrally flattened Batoidea in their lateral gill openings and lack of the fusion of their pectoral fin to the side of the head.

Subclass Elasmobranchii

Subdivision Selachii

Superorder Squalomorphi

Four of the orders have a common form of jaw attachment to the braincase and are placed in the Squalomorphi. The **Hexanchiformes** are easily distinguished from other sharks, which have five gill slits, by their additional one or two gill slits and their single dorsal fin that originates far back on the body near the caudal fin. The cow and frilled sharks are most common near the surface in cool waters of the temperate latitudes and are found deeper in subtropical waters. Probably, the most unusual of these sharks is the frilled shark, *Chlamydoselachus anguineus*—so named for the many protruding gills that extend outside its gill openings. Also a member of this group is the bramble shark, which is full bodied, sluggish in nature and has large dermal denticles with sharp ridges widely separated over their bodies. This roughness to its skin led fishermen to call it the bramble shark, *Echinorhinus brucus*. The **Squaliformes**, or dogfish sharks, have two dorsal fins, some with spines, and have no anal fin as do all other sharks. The members of this group vary greatly in size. It includes the dwarf lantern shark, *Etmopterus perryi*, which reaches a maximum length of 21 cm, hardly longer than one's foot, and the Greenland shark, *Somniosus microcephalus*, which may reach a length of 730 cm, many body lengths long. Members of this group also differ greatly in their predatory tactics. The cookie cutter shark, *Isistius brasiliensis*, has an unusual circular jaw, which is rotated in a circular manner to remove cookie-shaped plugs of fat from tuna, dolphins, seals, and whales swimming at great depths in the ocean. The Greenland shark is sluggish and stays near the bottom. However, it suddenly opens its cavernous mouth to create a vacuum to draw a seal into its gullet. Adam Ravich, a cinematographer, braved the cold water of the arctic to film a large Greenland shark remove the energy rich, fatty layer from the seal tethered to the bottom. The shark opened its jaws wide, swam forward so the seal was within its jaws, and first rotated its massive body in one direction and then in the opposite direction. Seal carcasses, comprised solely of bones and skin, have been found by naturalists walking along the beaches of Sable Island, Newfoundland.[35] Greenland sharks are abundant in the shallow water around the sandy island during the winter months and feed upon the seals as they move to and from their colonies on the beaches.

The **Squatiniformes**, or angelsharks, have a flattened, ray-like body, with eyes looking upward and a mouth at the end of the snout. The Pacific angelshark, *Squatina californica*, lies immobile on the bottom partially buried in sand with only its mouth visible. When a crab walks or fish swims near the bottom close to the angelshark's mouth, it opens it with explosive speed, drawing the prey into its throat by the vacuum created by inflating its gill chamber. The **Pristiophoriformes**, saw sharks, have a cylindrical body and snout forming a long blade. Teeth protrude from either side of the rostrum with large teeth alternating with smaller teeth. These sharks also have a pair of fleshy extensions, or barbels, near the mouth. The blade is believed to be slashed back and forth to wound their prey.

Superorder Galeomorphi

There are four orders in the Galeomorphi. The **Heterdontiformes**, popularly called the bull or horn sharks, have an elevated head with bulbous crests above their eyes and blunt spines, from which their common names derive. They have spines at the front of their two dorsal fins. Members of the group are small, being at most one and a half meters long. These species live in shallow shelf habitats, where they can find bivalves and crustaceans that they crush with the pavement-like teeth at the back of their jaws. The Port Jackson shark, *Heterodontus portusjacksoni*, which lives in the cool waters off southern Australia, forms aggregations in caves during the day. Members of these groups forage in the surrounding waters at night before returning to their home cave the following day.

The **Orectolobiformes,** or carpet sharks, have two dorsal fins, but spines are not present on their fins as on the horn sharks. They have large gill slits with the last two to four relatively posterior on the body above or behind the pectoral fin. Their eyes, similar to the horn sharks, are on the side of the head. Many of these species have small mouths with barbels. They open their mouths to create a suction that draws invertebrates such as sea urchins and crabs into their mouths. The nurse shark, *Ginglymostoma cirratum*, is a common carpet shark that lives in the southeastern Atlantic. There are different explanations for the name nurse shark. One is based on their resting behavior that consists of one shark lying on the bottom surrounded by others with their heads touching its mid body as if it were a mother nursing calves.[29] The whale shark (*Rhincodon typus*) reaches a size of 18 m and is the largest fish in the ocean. It feeds usually by swimming slowly on a horizontal plane, swallowing plankton, and expelling the water through its gills. However, whale sharks also have been observed to bob up and down in a vertical orientation in order to swallow small fishes schooling at the surface, as do humpback whales.

The **Lamniformes**, or mackerel sharks, have their mouth posterior to their eyes unlike the carpet sharks. Furthermore, they have a spiral intestinal valve consisting of many rings stacked in appearance. This group includes the megamouth shark (*Megachasma pelagios*), which spends the day in midwater and swims at night near the surface. It follows the distribution of krill, small shrimp-like crustaceans that live in deepwater during the day and migrate to the surface at night. The shark cruises, mouth wide, through these schools, catching the krill in long finger-shaped protrusions from their gill rakers. Another planktivorous shark, the basking shark, *Cetorhinus maximus*, is also a member of this order. It also has elongated protrusions from its gills, but they are thinner and hair-like and catch prey as they pass into its large mouth and out its large gill slits.

The **Carcharhiniformes**, or ground sharks, differ from the mackerel sharks in that they possess a nictitating eyelid. This is unlike the human eyelid because it moves upward not downward to cover the shark's eye and protect it from harm. The majority of ground sharks have an intestine with a spiral valve, but the hammerhead sharks have a scrolled intestinal value. The hammerhead sharks with their laterally enlarged rostrum are members of this order. Many functions have been proposed for this lateral expansion of their heads such as providing greater space for larger nasal capsules better at chemical detection, separating the nares further for more accurate chemical localization, expanding the surface over which the ampullae of Lorenzini are distributed for easier electromagnetic prey detection and utilizing the local pattern of magnetic fields to guide them during their migrations. The results of studies supporting these very different functions will be presented in the chapters on the sensory capabilities of the cartilaginous fishes.

Subdivision Batoidea

There are four orders of rays in the Batoidea. The rays differ from the sharks in that they are dorsally and ventrally flattened with their greatly enlarged pectoral fin attached to the head. Their gill openings are underneath their head, but their spiracles are on the dorsal surface. Those that remain on the bottom often draw water through their spiracle rather than their mouth for the purpose of respiration.

The **Torpediniformes**, or electric rays, have powerful electric organs that are used to stun their prey. The Pacific electric ray (*Torpedo californica*) drifts passively with the current at night until it passes close to a small fish. It then bends its disk into a glove-like shape oriented toward its prey and discharges its electrical organs while quickly moving toward it and swallowing it.

The **Pristiformes**, or sawfishes, are larger than the saw sharks, reaching a maximum length of 6 m. Their snout is tapered into a long flat blade with teeth on either side. Unlike the saw sharks, the teeth are all equal in size and embedded in deep sockets. This is a formidable sword-like instrument for slashing back and forth and incapacitating schooling fish.

The **Rajiformes**, called skates, usually have a very long and slender tail. Most of them have large and sharp ridges on their skin, often in a line in the middle of their backs. This group contains the shovelnose guitarfish (*Rhinobatos productus*), which is abundant along the western coast of North America. Its body shape resembles that musical instrument. Yet the majority of species in the order are in the family Rajidae, the members of which have a more rounded disk. They are ubiquitous on sandy bottoms, where they ambush crabs that move on the bottom or small fishes that swim close by.

The fourth order of rays is the **Myliobatiformes**, or stingrays. They have a spine with serrations on either side, with which are associated poison glands. The butterfly rays of the family Gymnuridae, whose disk is extremely broad and tail short, are members of this order. These rays somewhat resemble stealth bombers. Also members of this order are the eagle rays of the family Myliobatidae. They have a more triangular disk, a longer tail, and can have more than one long poisonous spine on their tail. The large manta rays and smaller mobula rays are both members of this family. They are plankton feeders, funneling water into their mouths with their cephalic appendages and catching their prey in their gill rakers as water is expelled from their gills. The freshwater rays of the family Potomotrygonidae are abundant in the tributaries of the Amazon River over four thousand kilometers from the Atlantic Ocean on the eastern slopes of the Andes Mountains.

Subclass Holocephali

The chimaeras share many features with the elasmobranchs but have lost others in their long evolutionary history. They have multiple gills as do the other cartilaginous fishes, but the water passing through the mouth and over the gills exits through only one opening in their gill cover. They do not have a spiracle, a small opening posterior to the eye through which water flows from the mouth cavity into the external environment, as do most sharks and rays. The skin of the modern ratfishes is naked without sandpaper-like dermal denticles, although the ancestral species did have them covering their bodies. The chimaeras have a spiral intestine like their relatives, the sharks and rays. Males have claspers for insertion into the female's cloaca, and some of these are even branched into two or three appendages. In addition, they possess a tentaculum, a small pad at the end of a stalk extending from the forehead covered with sharp scales for clasping the female. The holocephalans were a very common and diverse group in the Paleozoic Era, over 350 mya, occupying some highly specialized niches similar to those of modern fishes. These species were dominant predators in this ancient ecosystem. However, there was a major reduction in the diversity of this group 250 mya, at the end of the Permian Period.

The members of the order Chimaeriformes are classified into three families, based the shape of their rostrum. The plownose chimaeras of the family Callorhinchidae have an elongate, flexible, and hook-like appendage extending from the snout. The snout of longnose chimaeras of the family Rhinochimaeridae tapers into a long, fleshy protruberance not unlike the jost or spear of a knight from the Middle Ages. The shortnose chimaeras of the family Chimaeridae have a short, rounded protruberance. The members of all three species have spines at the front of their dorsal fin, but the spine of the shortnose chimaeras has a poison gland associated with it.

SUMMARY

The first cartilaginous fishes evolved over 455 mya. The evolution of these fishes has been of a punctuated nature with rapid speciation occurring twice, once 380 mya in the second half of the Devonian and again 200 mya during the early Jurassic (see fig. 2.9). The sharks and chimaeras were ubiquitous during the Carboniferous and first two-thirds of the Permian before their precipitous decline toward the end of the Permian. While the sharks were largely predatory by nature, the ancient chimaeras diversified into more ecological niches during this period. In fact, some even had conical teeth that enabled them to seize and swallow their prey like the first sharks. The Paleozoic sharks such as *Cladoselache* swam somewhat stiffly because the bases of their fins were broad, restricting their ability to change direction quickly. They could not open their mouths very wide because their upper jaw was closely articulated with their brain. Finally, they were able only to seize prey with their multicusp teeth and swallow them whole. Yet the Paleozoic sharks resembled the modern sharks. The chimaeras became largely extinct at the end of the Permian. *Ctenacanthus* survived the Permian extinctions and is likely the direct ancestors of the euselachians, or modern sharks.

It was modern sharks and rays that evolved highly moveable pairs of jaws that could protrude outward and open wider and serrated teeth for sawing large chunks of flesh from their prey. The upper jaw is loosely connected by ligaments to the anterior chondrocranium and the lower jaw attached to the hyomandibula, a modified gill arch. The vertebral column of the modern sharks and rays became more calcified than that of the ancient sharks and less rod-like, with ligaments attaching separate centra to provide greater flexibil-

1. Why did the holocephalans not survive the Permian extinctions as did the elasmobranchs, whose descendents, the euselachians, radiated widely in the Mesozoic and Cenozoic Era?

2. Given your familiarity with the extinctions and subsequent radiations of the cartilaginous fishes, what do you believe would happen to the abundance and diversity of the sharks, rays, and chimaeras if the climate warms in the future?

3. Paleontologists have looked for the absence growth rings, alternating rings of more dense and less dense calcification, in the fossilized skeletal remains of dinosaurs. This has been given as evidence that they were warm-bodied. Could this same technique be used to determine whether the ancient small- and giant-tooth sharks were cold- or warm-bodied?

4. Has the anatomy of the sharks really remained constant through evolutionary time relative to that of other taxa such as the bony fishes?

ity to the torso while swimming. It is truly extraordinary that this group has maintained its role as a top predator in the ocean while maintaining a relatively unchanged form over such a long period of geological time.

* * *

KEY TO COMMON AND SCIENTIFIC NAMES

Basking shark = *Cetorhinus maximus*; bramble shark = *Echinorhinus brucus*; bull shark = *Carcharhinus leucas*; cookie cutter shark = *Isistius brasiliensis*; dwarf lantern shark = *Etmopterus perryi*; frilled shark = *Chlamydoselachus anguineus*; giant-tooth white shark = *Carchardon megalodon*; Greenland shark = *Somniosus microcephalus*; megamouth shark = *Megachasma pelagios*; nurse shark = *Ginglymostoma cirratum*; Pacific angelshark = *Squatina californica*; Pacific electric ray = *Torpedo californica*; Port Jackson shark = *Heterodontus portusjacksoni*; salmon shark = *Lamna ditropis*; sandtiger shark = *Carcharias taurus*; shortfin mako shark = *Isurus oxyrinchus*; shovelnose guitarfish = *Rhinobatos productus*; tiger shark = *Galeocerdo cuvier*; whale shark = *Rhincodon typus*; white shark = *Carcharodon carcharias*.

LITERATURE CITED

1. Nelson, 2006; 2. Grogan and Lund, 2004a; 3. Maisey, 1996; 4. Benton, 2005; 5. Grogan and Lund, 2004b; 6. Applegate, 1967; 7. Ahlberg *et al.*, 2009; 8. Goodwin *et al.*, 2001; 9. Scotese, 2006; 10. Miller *et al.*, 2003; 11. Carroll, 1988; 12. Klimley, 1987; 13. Dick,

1981; 14. Stahl, 1989; 15. Maisey *et al.*, 2004; 16. Shirai, 1996; 17. Musick *et al.*, 2004; 18. Douady *et al.*, 2003; 19. Winchell *et al.*, 2004; 20. Thies, 1983; 21. Capetta *et al.*, 1993; 22. Cioni, 1996; 23. Capetta, 1987; 24. Compagno *et al.*, 2005; 25. Duffin, 1988; 26. Purdy, 1996; 27. Gottfried *et al.*, 1996; 28. Lucas and Stobo, 2000, 29. Klimley, 1978.

RECOMMENDED FURTHER READING

Benton, M. J. 2005. Vertebrate Paleontology. Blackwell Publishing, Oxford.

Carroll, R. L. 1988. Vertebrate Paleontology and Evolution. W. H. Freeman and Company, New York.

Gottfried, M. D., L. J. V. Compagno, and S. C. Bowman. 1996. Size and skeletal anatomy of the giant "megatooth" shark *Carcharodon carcharias*. Pp 55–66 *in* Klimley, A. P., and D. G. Ainley (Eds.), Great White Sharks: The Biology of *Carcharodon carcharias*. Academic Press, San Diego.

Grogan, E. D. and R. Lund. 2004. The origin and relationships of early Chondrichthyes. Pp. 3–31 *in* Carrier, J. C., J. A. Musick, and M. R. Heithaus (Eds.), Biology of Sharks and Their Relatives. CRC Press, Boca Raton.

Maisey, J. G. 1996. Discovering Fossil Fishes. Westview Press, Oxford.

Musick, J. A., M. M. Harbin, and L. J. V. Compagno. 2004. Historical zoogeography of the Selachii. Pp. 33–78 *in* Carrier, J. C., J. A. Musick, and M. R. Heithaus (Eds.), Biology of Sharks and Relatives. CRC Press, Boca Raton.

Purdy, R. W. 1996. Paleoecology of fossil white sharks. Pp. 67–78 *in* Klimley, A. P. and D. G. Ainley (Eds.), Great White Sharks: The Biology of *Carcharodon carcharias*. Academic Press, San Diego.

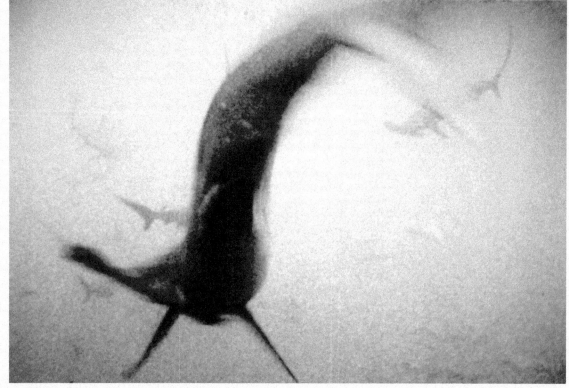

CHAPTER 3

Body Design and
Swimming Modes

When first diving in schools of hammerhead sharks, I was amazed at how one would disappear so quickly if approached within an arm's length, seemingly with a single flick of the tail. They could do this because they were in fact swimming with their whole bodies, due to the attachment of muscles to many vertebrae in the spine. The strength of these muscles was illustrated to me once with a potentially dangerous outcome when diving into a school of scalloped hammerheads to place an electronic tag on a shark at Malpelo Island, a remote island off the coast of Colombia. I passed so close to a large female that my shoulder brushed her side and frightened her. She explosively accelerated forward, her tail whipping to one side and slamming against my neck. Rising to the surface, my neck hurt terribly and I could not swallow my food during dinner that night without great effort. The shark might in a freakish accident have broken my neck. Shortfin mako sharks have also impressed me but in a different way. Keeping their bodies relatively stiff and mainly moving their tails, they ceaselessly swam while we labored to keep within range of their transmitters while tracking them from within a boat. This is because their muscles extend deep within their body and are attached to the base of the tail, which then can be moved from side to side in a swinging motion. Likewise, the

Sudden acceleration of scalloped hammerhead shown from two successive frames of 16-mm film taken at a seamount in the Gulf of California.

graceful swimming of the manta rays, seemingly flying like birds through the water using their extraordinary flexible and large pectoral fins for this purpose have impressed me. In contrast, the peculiar chimaeras swim in their own special way, paddling gracefully with outsized fins. These methods of swimming reflect skeletal and muscular adaptations for different ways of life. This chapter is an introduction to the remarkably different ways in which the sharks, rays, and chimaeras have figured out how to move through the water. You will begin by learning about differences in their skeletal and muscular anatomy and then move on to a description of their skin and the dermal denticles that cover their bodies. We will next examine hypotheses in the scientific literature explaining how they use their tail fins to propel themselves forward and the empirical evidence supporting them. With a familiarity of the building blocks of locomotion, you will at the end of the chapter be introduced to a commonly used classification of the diverse modes of swimming exhibited by the cartilaginous fishes.

SKELETON AND FINS

Let us first become acquainted with the body plan of the sharks and then proceed to those of the rays and chimaeras, whose plans are variants of the former with some conspicuous differences. The basic shark has a torpedo-shaped body that is pushed forward by a two-lobed tail at the posterior end and a mouth at the anterior end to seize its prey. Yaw, the tendency of the body to sway from side to side as the caudal fin propels it forward, is minimized by thin and flat appendages projecting upward and downward from the body, the dorsal and anal fins. Roll, the tendency to rotate, is reduced by paired flat horizontal appendages extending outward from the sides of the body, the anterior pectoral fins and posterior pelvic fins. These fins require supporting hard elements, which are less flexible at the base and more flexible at the distal edge of the fin, to keep the stabilizers stiff as they move through the viscous medium of water. These same fins, although different in size and shape, are necessary to reduce yaw and roll in the rays with their flattened bodies. The endoskeleton, or internal skeleton, can be separated into the axial and appendicular skeletons (fig. 3.1a).[1] The former consists of the chondrocranium, or braincase, and the vertebral column that eventually bends upward and extends into the upper lobe of the caudal fin. The latter comprises those elements that support the fins.

The skeletal elements of the sharks, rays, and chimaeras are composed of cartilage, a substance more flexible than bone. You can feel this substance by holding your nose with your thumb and forefinger and moving it from one side to the other two or three times. The cartilaginous septum easily bends from side to side yet still provides support that keeps the nostrils open in the

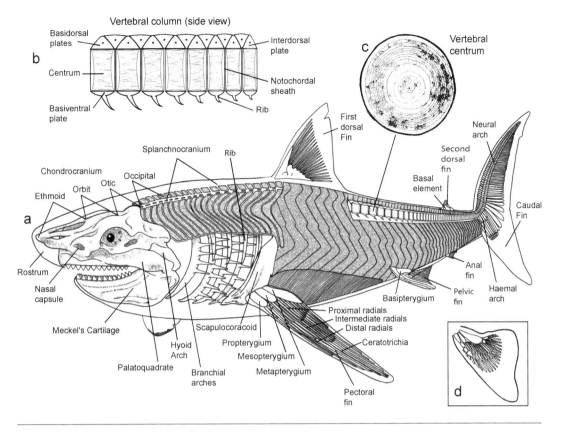

Fig. 3.1 White shark: (a) musculature and axial and appendicular skeletons of white shark; (b) longitudinal section of vertebral column; (c) cross-section of vertebral centrum with annular rings; and (d) the skeletal elements in aplesodic fin.

nose. The skeletal elements of the cartilaginous fishes are composed mainly of hyaline cartilage, which has a pearly bluish color and is considerably elastic in nature, covered with a thin layer of denser and stiffer prismatic cartilage. This is arranged into tesserae, blocks of calcified cartilage that fit together as tiles on a floor form a mosaic (figs. 3.2a & 3.2b).[2, 3] This cartilaginous tissue possesses much of the strength of denser bone without the added weight. Hyaline is an amorphous material that contains no nerves or blood vessels. Its structure is relatively simple, consisting of cells of a rounded or bluntly angular form. The tesserae, on the other hand, are more complex in structure. They are covered by a fibrous perichondrium (fig. 3.2c). This is composed of proteoglycans and collagen.[4] Proteoglycans have the ability to absorb water and create a large swelling pressure, while collagen serves as a reinforcing tensile material. Beneath the perichondrium is very dense prismatic cartilage impregnated with many minute spherules of hydroxyapatite, mineralized calcium phosphate, that are fused together (fig. 3.2d). The spherules are smaller and tightly packed on the superficial surface of the tesserae (fig. 3.2f). An elemental analysis indicates that the calcium and phosphate are most concentrated near the boundary between the tesserae and the hyaline (see insert in fig. 3.2d). There is a collagen fiber network here that anchors the tesserae to the underlying hyaline cartilage. The collagen fibers may contribute to the formation of tesserae as individual hydroxyapatite crystals, which appear white, seem to extend outward from the mineralized boarder on the lower right side of the micrograph (fig. 3.3e). The majority of vertebral elements in the cartilaginous fishes are composed of prismatic cartilage. It is the main constituent in the chondrocranium and jaws of the axial skeleton and the fin supports that make up the appendicular skeleton.

The vertebral column is mainly composed of areolar cartilage (see fig. 3.1a).[3] This tissue consists of chondrocytes, cells containing a single nucleus surrounded by a relatively large amount of clear cytoplasm, embedded within an extracellular matrix hardened by mineralization with crystals of calcium phosphate impregnated hydroxyapatite. This cartilage is also surrounded by an external fibrous perichondrium. The interior of the vertebral column is composed of many cartilaginous disks with an hourglass shape that fit next to each other, each being separated by a notochordal sheath (fig. 3.1b). Each disk, or centrum, is composed of areolar cartilage that is deposited often on an annual cycle in two concentric rings, one with less mineralization alternating with another with more mineralization (fig. 3.1c).[5] Each centrum in the vertebral column is adjacent to a neural arch above that protects the nerve chord and a haemal arch below that affords protection to the blood vessels. These elements are composed of prismatic and hyaline cartilage with tesserae covering their surface.

Let us now become acquainted with the cartilaginous elements in the axial skeleton of a shark. The axial skeleton, starting at the shark's anterior and moving toward the posterior, includes the chondrocranium, which supports the head and provides protection to the brain and sensory organs, and the vertebral column, which supports the body, tail, and caudal fin. The chondrocranium can be divided into four regions from front to back: (1) the ethmoidal region, which consists of the rostrum and nasal capsule, (2) the orbital region, consisting of the circular orbits that surround the eyes, (3) the otic region, which contains the space for the inner ear, and (4) the occipital region at the end of the chondrocranium, which articulates with the vertebral column (fig. 3.1a). The palatoquadrate, or upper jaw, is attached to the chondrocranium. yet it also articulates below with the Meckel's cartilage, or lower jaw, just behind the skull. The chondrocranium and jaws of the axial skeleton are covered with tesserae to fortify them. Most sharks have a single layer of tesserae on their upper and lower jaws, but the largest predators, the bull shark (*Carcharhinus leucas*), tiger shark (*Galeocerdo cuvier*), and white shark (*Carcharodon*

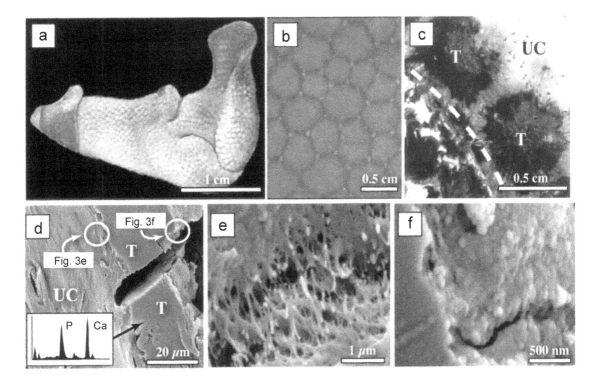

Fig. 3.2 Mineralized tesserae covering the surface of (a) the lower jaw and (b) the chondrocranium, of an elasmobranch together with (c) a cross-section of jaw cartilage showing two tesserae (T) and uncalcified hyaline (UC). Dashed white line = perichondrium. Micrographs showing a cross-section of the pelvic girdle showing tesserae with spectrograph indicating high concentrations of calcium and phosphorous present in (d) the cartilage near the boundary with the uncalcified hyaline with expanded views of the (e) internal and (f) external prismatic cartilage.

carcharias), have three to five layers.[6] A maximum of five layers were found in the largest white sharks; and these are likely needed to provide added support when biting through the dense bones of seals and sea lions.

Just behind the braincase is the splanchnocranium that provides support for the jaws, tongue, gills, and pharynx. The branchial arches support the gills; rays project outward from this part of the vertebral column. Each arch consists of four branchial elements that extend downward from the vertebral column and support the rays. The arches are joined at the base by a basal element. The majority of sharks and rays have five branchial arches. However, the frilled shark (*Chlamydoselachus anguineus*), cow sharks (in the genera *Heptranchias* and *Hexanchus* of the family Hexanchidae), the sixgill sawshark (*Pliotrema warreni*), and the sixgill stingray (*Hexatrygon bickelli*) each have six arches. There is one genus of cow shark, *Notorynchus*, whose species have seven arches.

The vertebral column is composed of a series of centra, calcified elements that are hourglass shaped (figs. 3.1b & 3.1c). They lead from the occiput, or attachment point to the chondrocranium, to the end of the upper lobe of the caudal fin. A soft notochordal sheath lies in the space between one centrum and another. The side of each is cone-shaped with the apices of the two cones of either side nearly connected at its center. Each centrum is composed of alternating dense and less dense bands of cartilage, and like tree rings one pair is generally deposited per year, as will be described in more detail in chapter 13. The number of centra within the vertebral column ranges from 60 to 477 in the living cartilaginous fishes.[1] Above each centrum is an arch-like structure, the neural arch, which has within its central cavity the spinal cord. It consists of a wedge-shaped basidorsal plate separated by an interdorsal plate. There is a circular opening in each plate, either the dorsal root foramen or ventral root foramen. These arched plates make up the neural arch, which has within its central cavity the notochord or spinal cord. Below each centrum is a similarly shaped basiventral plate separated by an interventral plate. These arched plates make up the haemal arch, which provides a protective space for veins and arteries to pass through. Attached to each basiventral plate from the branchial to pelvic region is a rib extending downward. The centra from the pelvic fin anterior are smaller, and the vertebral column angles upward in the caudal fin and extends to the upper hypochordal lobe of the fin. Short epichordal rays extend upward and long hypochordal rays extend downward in the caudal fin.

The appendicular skeleton consists of the paired and unpaired fins and their supporting elements. The paired fins consist of the pectoral fins, which are located between the head and the trunk, and the pelvic fins, which are situated between the trunk and the tail. These fins are supported by the pectoral and pelvic girdles. The largest element in the pectoral girdle is the scapulocoracoid. This element curves upward and laterally to articulate with the vertebral column. It consists of two scapular processes, which are joined ventrally by a small element, the sternal cartilage. Three cartilaginous elements at the base of the pectoral fin, from anterior to posterior the propterygium, mesopterygium, and metapterygium, are attached by ligaments to the scapulocoracoid. The pectoral basals are joined at their ends to slender and diagonal cartilaginous elements, the radials, that line up with each other and fan outward to support the base of the fin. In most species of sharks, there are three rows of radials, the proximal, intermediate, and distal radials. Attached to the radials are ceratotrichia, thin and flexible rays made of keratin. These are long bundles of proteinaceous filaments that lead to the end of the fin and provide further support. The pelvic girdle consists of the pubic bar, a flat plate that extends across the ventrum of the shark and has paired lateral nodes where each pelvic fin attaches to the skeleton. There is a longitudinal plate, the basipterygium, to which the radials articulate and extend outward within the fin. Ceratotrichia are attached to the radials and extend to the distal web of the fin. The unpaired fins of the appendicular skeleton consist of the two dorsal fins and the anal fin. These fins have three basal cartilaginous elements that anchor them to the trunk of the shark or ray. There are three rows of radials projecting outward from these elements that line up with the ceratotrichia. The dorsal fin spine is attached to the anterior basal element.

There are two different types of pectoral fins in the sharks and rays, aplesodic and plesodic fins.[7] Aplesodic fins are composed of cartilaginous radials that extend less than 60% outward into the fin, with the distal fin supported only by ceratotrichia (fig. 3.1d). The plesodic fins have radials that extend farther outward into the fin to provide further support to the fin. The distal radials in the third row are longer than those anterior or posterior to them toward the edges of the fin. The muscles in the aplesodic fins extend out to the third row of radials, whereas the muscles in the plesodic fins extend out to only the second row of radials. The squalomorph sharks, some inshore galeomorph sharks, and the rays have the former fins, whereas the oceanic galeomorphs such as some carcharhiniforms and lamniforms have the latter fins. The great white shark has a plesodic fin (fig. 3.1a). The extension of the radials over the fin's length makes the pleosodic pectoral fin stiffer and more rigid at greater swimming speeds. These rigid fins provide the oceanic members of the family Carcharhinidae such as the silky shark (*Carcharhinus falciformis*), blue shark (*Prionace glauca*), and oceanic whitetip shark (*Carcharhinus longimanus*) and the members of the family Lamnidae such as the white, shortfin mako (*Isurus oxyrinchus*) and salmon shark (*Lamna ditropis*) with a hydrodynamic advantage when cruising over long distances. The aplesodic pectoral fin (fig. 3.1d), well supported only at the base, permits greater freedom of movement to the distal web of the fin and thus increases the maneuverability of a shark in confined spaces. For example, some of the orectolobiform sharks such as the bamboosharks of the genus *Chiloscyllium* may bend the outer edges of their pectoral and pelvic fins downward to enable them to walk on the substrate similar to the locomotion of salamanders.[8] Other sharks with aplesodic fins appear well suited for accelerating quickly and maneuvering in tight spaces in the coastal waters.

The rays differ from the sharks mainly due to their flattened bodies (fig. 3.3a). The pectoral girdle is massive in order to support the many radial elements of their disk-shaped pectoral fin. It is proportionally larger than in the sharks and has a triangular shape with the vertex of the triangle articulating with a long and robust element of the vertebral column, the cervicothoracic synarcual. The anterior part of this element is attached to the posterior of the chondrocranium and the posterior portion to the intermediate vertebrae. This provides added support for the well-buttressed girdle, which supports the elongated basal elements of the disk-shaped pectoral fin. The propterygium,

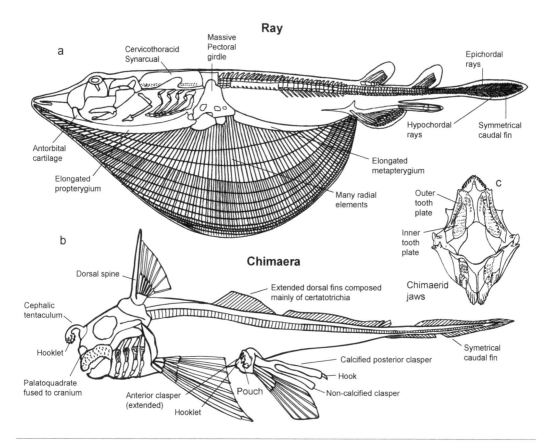

Fig. 3.3 Axial and appendicular skeleton of (a) a generalized ray and (b) the spotted ratfish (*Hydrolagus colliei*). (c) The upper jaw with two dental plates and lower jaw with one dental plate of the Pacific longnose chimaera (*Harriotta raleighana*).

the most anterior of the elements at the base of the pectoral fin, extends forward to the tip of the snout. Here it meets the antorbital cartilage, a conical extension of the chondrocranium. The metapterygium, the most posterior basal element, extends backward to the base of the pelvic fin. Both basal elements provide a wide base, to which are attached many radial elements, which extend outward into the disk-shaped pectoral fin. The caudal fin differs from that of the sharks in that the fin is often but not always symmetrical with the same number of upper epichordal and lower hypochordal rays.

The skeleton of a chimaera differs from those of sharks and rays in important respects. The palatoquadrate is fused tightly to the chondrocranium (fig. 3.3b). The upper jaw has two parallel tooth plates, and the lower has one plate for crushing the hard exoskeletons of invertebrates (fig. 3.3c). The lower jaw is articulated by ligaments to the hyoid arch, as is characteristic of the sharks and rays. Attached to the head and lying in a groove on the forehead of males is the cephalic tentaculum, a clublike organ with a fleshy pad on its end covered with small, sharp hooklets. This organ is distended during courtship and brought down on the dorsum of the female to induce her to let the male copulate with her. The first dorsal fin of the chimaeras is farther forward on the body than that of the sharks. It is short and triangular with a large spine immediately in front—both are attached to a basal plate bound by ligaments to the chondro-cranium. The second dorsal fin is low on the body with short radials at the base and ceratotrichia leading to the distal margin of the fin. The second dorsal of the spotted ratfish (*Hydrolagus colliei*) is divided, with a triangular anterior section and a ribbon-like posterior section. The pectoral fin of the chimeras is proportionately larger than that of the sharks yet is supported in the same manner by the scapulocoracoid with three basal elements attached to short radials that support ceratotrichia leading to the edge of the fin. The pelvic fin is supported by a proportionately larger basipterygium that is needed to pro-vide ample support for the pelvic fin as well as a single anterior and bifurcate posterior clasper. The former has a narrow base that leads to a flat and wide end with hooklets along the distal margin. The clasper is kept within a pouch most of the time but rotated outward during courtship to affix to the body of the female. The posterior clasper bifurcates into a soft external radius and a hard, mineralized internal radius with a single hook on its end. This latter or-gan is inserted into the cloaca of the female during copulation. The vertebral column extends along the entire length of the chimaera to the end of the cau-dal fin, which is symmetrical with ceratotrichia extending upward and down-ward along its entire length.

WHITE AND RED MUSCLE

The muscles of the cartilaginous fishes are composed of tissues that alternately contract and relax. These muscles, attached to a flexible vertebral column, are present on either side of sharks and above and below on the rays. The body is propelled forward by sequentially flexing and relaxing these stacks of muscles, which are located along the entire length of the body. Myomeres, small muscle segments on one side of the body, contract in a wave passing down the body while the myomeres on the opposite flank of the body relax si-multaneously. Each segment consists of muscle fibers stacked on top of each other within a zig-zag shaped segment that is folded around the backbone with its forward points nested together (fig. 3.4).[9] The W-shaped stack of epaxial muscles above the lateral line of the piked dogfish (*Squalus acanthias*) is inner-vated by a dorsal extension of the spinal nerve; the curved stack of hypaxial muscles below the lateral line is innervated by a ventral extension. The hori-zontal skeletogenous septum, a thin longitudinal sheet of connective tissue, at the lateral line separates the dorsal and ventral sections of each myomere. The

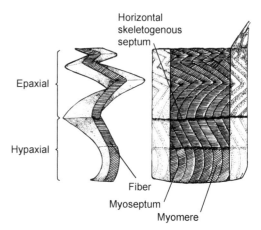

Fig. 3.4 Segment of the muscle from a chondrichthyan fish with an expanded myomere.

Epaxial

Hypaxial

Horizontal skeletogenous septum

Fiber

Myoseptum

Myomere

V-shaped folds are seen to overlap when viewed in a cross-section of the body. Each myomere is separated from its two nearest neighbors by a myoseptum, a sheet of connective tissue attached to the dermis of the shark that transmits the force of muscle contraction necessary to bend the trunk of the body to one side.

The arrangement of the myomeres produces the seemingly effortless undulation of the body and tail of sharks and rays. In the former, the undulations move rearward along the flanks; in the latter, along the top and bottom of the body. The contraction of myomeres on one part of the body draws the myosepta toward each other on that side causing a curvature, while the vertebral column prevents the body from shortening.[9] The zig-zag foldings of the overlapping myomeres permit force to be generated over a greater length than it could if the muscle segments were vertical in their alignment. This overlap produces a smooth generation of force. The direction of the muscle fibers differs over the length of the myomere. The fibers near the horizontal skeletogenous septum are longitudinal (oriented lengthwise) and are longer than obliquely oriented fibers to either side of the midline. This is because the flank of a shark or ray is wider in the middle than to either side of the body, and thus the longer fibers contract more than the shorter fibers to compensate for the greater distance over which they must move. The shorter obliquely oriented fibers contract little or not at all. The most superficial fibers attach directly to the skin, or dermis, which acts as an external tendon, transferring force along the body. As the wave of curvature moves down the body in the shark and ray, the skin on the opposite convex side is stretched and stores energy. This energy is released in the form of bending as a subsequent wave of contraction passes down the previously convex side of the body.

The forward thrust of the body is generated by the serial contractions of the trunk myomeres, resulting in the undulatory wave propagated posteriorly over some portion of the body. The length of the body thrown into waves by these muscular contractions varies from the entire body in the flexible, eel-like, anguilliform locomotion of the Paleozoic *Orthacanthus* and the extant frilled shark to the posterior quarter of the body and tail in the stiff thunniform (tuna-like) locomotion of the white, mako, and salmon sharks. The efficiency of locomotion increases as less of the body participates in these waves, yet at the same time the ability to accelerate is reduced.[10] The unique composition of the skin in some sharks enables them to avoid this conflict and become good accelerators as well as efficient cruisers. The chimaeras, on the other hand, keep their torso relatively stiff, and only rarely accelerate themselves forward by the sudden lateral back-and-forth movement of the tail. Most of the time, they move slowly forward or change their height in the water column by flapping their large pectoral fins. Their pectoral fins are stiff and have a relatively high height-to-length ratio that generates considerable thrust when the fin is flapped up and down (see fig. 3.3b).[11]

There are two types of muscle fibers, white and red muscle. The contractile proteins in exterior white muscle cells gain their energy molecules of adenosine triphosphate (ATP) by breaking down glycogen present in the muscle in the absence of a continuous supply of oxygen using a metabolic process termed anaerobic glycolosis. This enables sharks with predominantly white muscle such as the dogfish to accelerate quickly for short periods, but their muscle contractions slow down with the buildup of lactic acid—the end product of anaerobic metabolism. The buildup of lactic acid produces a burning sensation in human muscle, which can be experienced by exerting oneself vigorously on an exercise machine. Red muscle differs from white muscle. Its red hue is due to its greater vascularity (concentration of blood vessels), higher density of mitochondria, and greater amount of myoglobin than in white muscle. Red blood cells carry oxygen bound to hemoglobin through the circulatory system to the red muscle fibers, where the oxygen then binds to myoglobin and is transported to the mitochondria. The oxygen combines with fatty tissue within the mitochondria in the Krebs cycle, through energy-generating metabolism, to produce the ATP that supplies the energy for the contraction of this muscle. Oxygen is delivered at a slow but continuous rate so that red muscle contraction is tonic, and this is better for swimming continuously for long periods.

It is essential to know the location within the body of the two types of muscle and how they are attached to the vertebral column to understand how fish swim with flexible jack-like (carangiform) or stiff tuna-like motion. A large mass of white muscle is located in the interior and a small mass of red muscles in the exterior of the bodies of the majority of bony and cartilaginous fishes that swim with carangiform motion (fig. 3.5a).[12] For example, species that inhabit rocky and coral reefs on the shelves surrounding continents have this muscle arrangement, and it permits them to capture their prey with brief, rapid bursts of swimming. Each myomere is short and connected to an adjacent vertebral centrum within the torso. The myomeres extend only over six to eight percent of the body length. The red muscle, indicated on the illustration by solid ovals, form thin wedges just under the skin and exert force through the muscles and tendons to the vertebrae. The myomere labeled 5 is connected to vertebral segment 5 (fig. 3.5d). Thus, that particular segment, consisting of white muscle, indicated by horizontal stripes, and red muscle, denoted by vertical bars, will contract almost simultaneously pulling the vertebrae labeled 5 toward it to bend the spinal column. The arrow indicates the direction of the force exerted by the tendons attaching the muscle to the vertebrae. The serial contraction of myomeres thus results in a sinuous waveform passing down the length of the body, forcing water backward and producing forward thrust. This sinuous motion is apparent in the sudden acceleration of a scalloped hammerhead shark (see photograph on first page of chapter).

The red muscles are located toward the center of the body of the powerful tunas and mackerel sharks that swim with thunniform motion (figs. 3.5b & 3.5c).[12] These species are able to swim continuously for long periods over great distances at constant but slower speeds. The mass of red muscle of the shortfin mako is tear-shaped with its greatest mass in the center of the body. Here in the interior of the body the fibers retain the heat generated during metabolism, and this warmth enables the muscles to contract more quickly and forcefully. The muscle fibers are elongated, spanning 20 to 25% of the body length, and this shifts mass anteriorly and aids in streamlining. The white muscle fibers, indicated by horizontal stripes, pass anteriorly along the length of the body with red muscle, indicated by solid wedges, present in the anterior cones of these fibers (fig. 3.5e). Through long tendons, the force of many of these myomeres is projected backwards toward the caudal peduncle at the base of the tail. The anterior arrow indicates the direction of the force of the muscles; the posterior arrow denotes the direction of the force of the tendons attached to them. For example, the myomere 5 extends forward beyond vertebral centrum 1, the anterior location of its red muscle indicated by bars, yet joins posteriorly with a ligament to the vertebral column at centrum 10 near the caudal peduncle. The torso of these species is kept rigid while alternating contractions and relaxations of these long muscles on either side of the body to move the tail fin back and forth with high-amplitude lateral movements. This musculature enables mackerel sharks to swim very efficiently by keeping their forward torso stiff and streamlined.

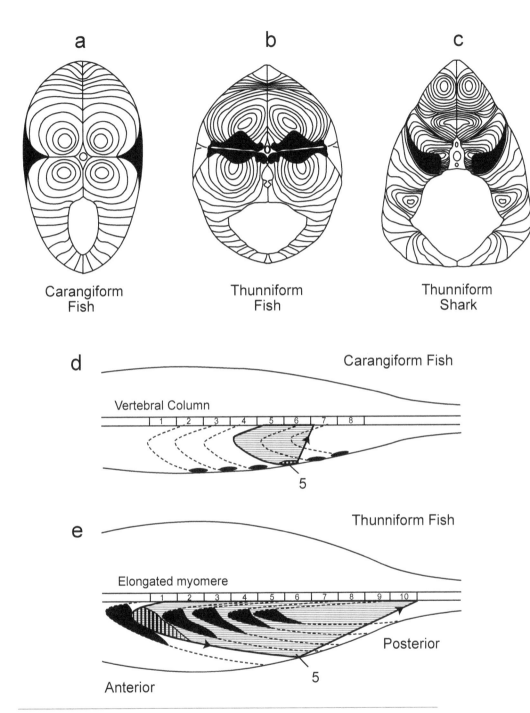

Fig. 3.5 Cross-sections of (a) a jack-like (carangiform) and (b) tuna-like (thunniform) swimming bony fish and (c) a thunniform swimming shark with an indication of the locations of white (clear) and red (solid) muscle. Longitudinal sections of the (d) carangiform and (e) thunniform swimming fish showing the attachment of myomeres to the vertebral centra.

The skin of the cartilaginous fishes is composed of a thin epidermis and a thicker underlying dermis. The former contains mucus-secreting cells that produce proteoglycans that spread over the body to reduce the drag caused by water flowing past.[13] The latter plays the major role in the stiffness and strength of the skin. The dermis is a white sheath of fabric composed of many layers of collagen, a very tough and elastic protein that is composed of triple-stranded and helically coiled fibers bound together to form fibrils. The fibrils within each layer are oriented in the same direction, but the fibrils in the next layer underneath are in a complementary direction.[3] The fibrils encircle the shark's body in a helical manner in these two complementary directions and can be described as right- and left-handed helices. Note their cross-ply orientation in a blacktip shark (*Carcharhinus limbatus*) (fig. 3.6a). The angles these helices make with the long axis of the shark vary between 50° and 70° in the forward torso between the pectoral and anal fins and between 45° and 50° in the thin caudal peduncle just anterior to the tail (fig. 3.6b). This collaginous corset constitutes a firm anchor for the swimming muscles, and acts in a way like an external skeleton (exoskeleton).

The skin of the sharks and rays is supported during swimming by hydrostatic pressure.[14] The internal pressure in the body of sharks increases more than tenfold from slow to fast swimming. The stiff skin acts as a tendon, transmitting the forces of the alternately contracting and relaxing muscles on either the shark's two flanks or the top or bottom of the ray's body to the caudal fin. During acceleration, the energy stored elastically in the stretched skin on the convex side is released when the muscles are relaxed on the concave side. This provides an elastic recoil, or whiplash, that accelerates the unbending of the body while the muscles on the heretofore convex side begin their contraction phase. For the skin to transmit this force to the tail, the skin must remain stiff for the duration of the muscle contraction. When the shark or ray is at rest, the muscles on both sides are the same length and cross-sectional area. At this time, the fibers in the skin conform to an angle of 60° with the longitudinal axis of the shark. The internal pressure within the shark is low and the skin is not stiff. However, the muscle on one flank shortens and increases in girth to bend the torso sharply during fast swimming. This causes the fibers in the skin overlying the contracting muscles to increase their angle relative to longitudinal axis of the shark. The change in fiber angle causes the skin to remain taut and avoid the loss of the tension or elastic energy on the concave side. This ensures that the force of the bending body is transmitted to the tail. Thus, it is both the vertebral column and skin that provide the support for the muscles to transmit force resulting in the back-and-forth movement of the tail.

Fig. 3.6 (a) Polarized light micrograph of the dermis of a blacktip shark, showing three layers of fibers. (b) Outline of a lemon shark, showing some of the helical angles to the fibers of collagen that surround the body in the dermis.

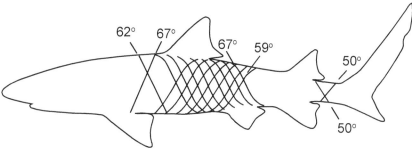

Toothlike scales cover the skin of most cartilaginous fishes. They function to reduce drag (the resistance to forward movement) caused by friction as the viscous medium of water passes in a posterior direction along the body during forward movement. These denticles, so named from the Latin word *dent* for tooth, cover the entire body of most sharks yet are distributed only on part of the bodies of rays and are uncommon on chimaeras. There are three elevated keels on the external surface of the denticles on the skin of most of the modern galeomorph sharks with each keel terminating in a pointed cusp (fig. 3.7a).[15] The central longitudinal ridge is always longer than the two lateral ridges lying on either side (fig. 3.7c). Each dermal denticle consists of a basal plate embedded under the skin and a crown that rises from that plate. There is a single pulp cavity with an elongated dental papilla extending from it base upward toward

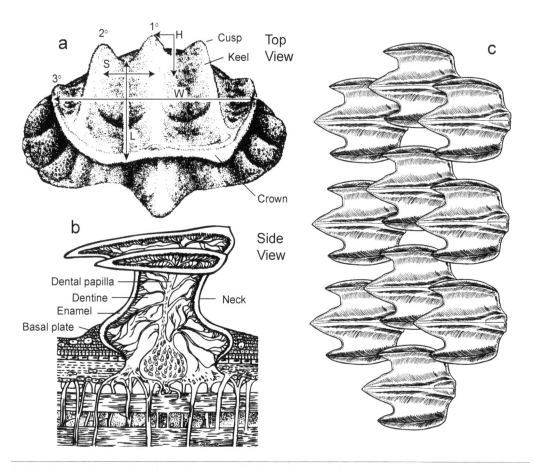

Fig. 3.7 (a) Top view of a dermal denticle with the three keels. H = keel height; L = crown length; S = inter-keel separation; W = crown width; 1°, 2°, and 3° = primary, secondary, and tertiary keels. (b) Side view of the denticle, showing the internal canal, or dermal papilla, surrounding dentine, and a thin layer of hard enamel. (c) The dermal denticles with three longitudinal ridges of the white shark. Notice that the rounded valleys between the keels on successive denticles line up with each other for maximum hydrodynamic efficiency.

the cusps (fig. 3.7b).[16] Offshoots from the papilla (odontoblasts) spread out into the three cusps and lead to the exterior of the denticle, where they secrete a matrix of dentine to form the crown and an outer covering of hard enamel. An outgrowth of connective tissue spreads out at the base to form the basal plate, which becomes calcified, similar to bone. The scales are formed continuously during the life of a cartilaginous fish. They are shed and replaced by larger denticles as an individual grows—although there is no synchronous pattern of molting as in birds and marine mammals. Close examination of the skin of sharks and rays reveals that some of the denticles are white, evidence that they are in the process of aging.[17]

Water is much denser than air. This has its advantages and disadvantages when a cartilaginous fish is swimming within the former medium. On the positive side, water provides greater buoyancy than air so that the fish needs to exert less effort to prevent it from sinking in the water column. On the negative side, it takes powerful tail beats to propel itself through this more heavy medium that is also much more viscous than air, or resistant to the passage of the medium past its body. Water does not pass easily around the shark, ray, or chimaera as it swims through it. The passage of the water creates a frictional drag as the boundary layer of water passes by the body and tends to stick to it. This drag increases with faster swimming, and its magnitude is related to the density of the fluid through which the fish swims and the size and shape of the fish. As an object moves faster, the passage of the water changes from a laminar flow, during which water passes in an undisturbed flat layer with molecules moving at the same velocity, to a turbulent flow, during which the molecules move in irregular paths with different velocities. At greater speeds, the passing water may form eddies with water reversing its direction 360° and passing again over the forward-moving fish, creating additional drag.

The flow of water over the surface of an object even at moderate speeds underwater changes the flow from laminar to turbulent, generating numerous small eddies or reverse flows that create drag, impeding forward motion.[18] Both empirical and experimental studies have shown that longitudinal grooves with V-shaped ridges[19, 20] and U-shaped valleys[21, 22] reduce the frictional drag up to 8%. For example, a 7% decrease in frictional drag is achieved by forcing wind over flat plates with longitudinal ridges separated by rounded valleys in comparison to smooth plates in a wind tunnel.[23] The researchers artificially created a checkerboard surface of many miniature plates, each with two rounded grooves separated by three ridges, with the ridges and grooves of each successively posterior plate lined up with those of the proceeding plate as on the skin of the white shark (fig. 3.7c). This drag reduction is accomplished largely by impeding the production of micro-turbulence, or minute eddies created as the water travels over the ridged surface. The overall dimensions of the ridges and valleys are critical to produce this effect. The tiny ridges must be high enough to deviate from the flat contour necessary to generate minute vortices, and the valleys separating the ridges must be narrow enough to prevent a significant increase in the rounded surface area exposed to frictional drag. The optimal ratio of the height of the keel to the inter-keel separation to optimize the reduction of drag is 1:2.[15]

Insights into the value of the dermal covering to the modern galeoform sharks in drag reduction have been learned from close examination of the gross anatomy of the denticles. Each particular feature on a denticle was carefully measured using a micrometer for species from three groups of sharks with very different lifestyles. A measurement was made of the width and length of the crown, the mass of the denticle, the density or concentration of scales over a centimeter of surface, the height of the primary keel, and the distance between this and a secondary keel (table. 3.1). The sharks were separated into three classes based on knowledge of their life histories. The first group of sharks included fast and active swimmers, many of which live in the open ocean, such as the silky, shortfin mako, and scalloped hammerhead sharks. The second group contained three sharks, judged intermediate in their swimming abilities. The third group was composed of sluggish and slow moving sharks that are epibenthic, living near the bottom, such as the nurse (*Ginglymostoma cirratum*), tiger, and sandtiger shark (*Carcharias taurus*). The constantly swimming pelagic sharks had denticles that were distinctly smaller and lighter than the sedentary sharks that spend much of their time on or near the bottom. For example, scalloped hammerhead shark, which makes extensive nightly migrations away from its daytime seamount abode searching the pelagic environment for squids, has crown widths and lengths of a maximum of 297 microns wide and 319 microns long, respectively. Its denticles weigh up to 7 micrograms with an average of 11.5 denticles per square centimeter of epidermal surface. These measurements can be compared to those from the sluggish nurse shark, which spends most of the day lying on the bottom but cruises slowly over the bottom at night searching for shrimps, crabs, and bivalves. The nurse shark's crown is as wide as 566 microns and as long as 599 microns—being over twice the width and length of the hammerhead denticles. Those of the nurse shark weigh as much as 107 micrograms, over ten times the weight of the hammerhead denticle. Furthermore, there are a maximum of only 2.4 denticles over a square centimeter of epidermal surface—a density a tenth that of the denticles on the scalloped hammerhead. This relationship appeared to remain constant between the two groups. The more active sharks had smaller, thinner, and lighter denticles, densely arranged in a honeycomb fashion. The less active sharks had fewer scales distributed farther apart that were larger, thicker, and heavier. The members of the intermediate group of species, including the bull, sandbar, and blue shark, had denticles with dimensions between those of the fast and slow swimmers.

Further support for the hydrodynamic value of the alternating V-shaped keels with U-shaped valleys comes from measurements of the height of the keels and the separation distance between them.[15] The silky shark (*Carcharhinus falciformis*) provides the best example of this. As silky sharks grow larger, so do the widths and lengths of their denticles. This is apparent from the two rising curves, one of denticle width and the other of length, on a plot of these measurements from silky sharks of increasing sizes (fig. 3.8).[24] Furthermore, the number of keels increases from three to six as the shark's length increases from 120 cm to 210 cm long (see insert). On the other hand, the heights of the keels and separation distances between the keels do not increase as the sharks continue to grow; the curves for keel height and keel separation do not

TABLE 3.1 Measurements of crown width and length; denticle mass, density, and mass/area; and keel height and spacing for nine fast-swimming, three intermediate-speed-swimming, and three sluggish, slow-swimming species of sharks.

Swimming Mode Species	Common name	Crown width (microns)	Crown length (microns)	Denticle mass (µg)	Denticle density (N/cm)	Denticle mass/ area (µg/cm²)	Keel height (microns)	Keel spacing (microns)
Fast-swimming, neritic								
Carcharhinus falciformis	Silky shark	150–300	160–290	6–12	6.9–15.5	39.9–137.6	9–19	40–57
C. limbatus	Blacktip shark	311–396	254–344	11–20	6.2–8.0	77.3–123.8	13–14	51–57
C. obscures	Dusky shark	180–500	170–560	4–41	5.6–30.3	46.7–302.8	8–26	41–76
C. signatus	Night shark	190–480	200–390	3–21			19–30	58–93
Isurus oxyrinchus	Shortfin mako shark	137–143	169–183	1–5			7–8	
Mustelus canis	Dusky smooth-hound shark	155–403	195–585	5–27	3.9–9.6	31.4–125.3	8–26	35–73
Rhizoprionodon terraenovae	Atlantic sharpnose shark	170–355	225–350	6–11	6.3–7.2	40.7–69.4	14–23	50–70
Sphyrna lewini	Scalloped hammerhead shark	137–297	173–319	7	11.0–11.5	76.7–80.6	8–12	38–44
S. mokarran	Great hammerhead shark	308	353	11	11.4	125.2	18	62
Intermediate-speed-swimming								
C leucas	Bull shark	467–896	435–738	156–194	1.2–2.5	238.2–390.0	49–54	117–143
C. plumbeus	Sandbar shark	170–789	125–625	3–169	1.7–14.1	38.8–396.1	18–52	55–125
Prionace glauca	Blue shark	205–560	235–455	5–21	1.5–8.6	25.2–97.3	14–45	73–128
Slow-swimming, epibenthic								
Ginglymostoma cirratum	Nurse shark	531–566	539–599	94–107	1.1–2.4	105.7–254.4	26–39	
Galeocerdo cuvier	Tiger shark	120–945	190–810	10–177	1.8–8.0	43.0–345.6	35–153	63–275
Carcharias taurus	Sandtiger shark	349–565	265–615	34–90	1.2–2.5	65.5–200.4	32–160	130–225

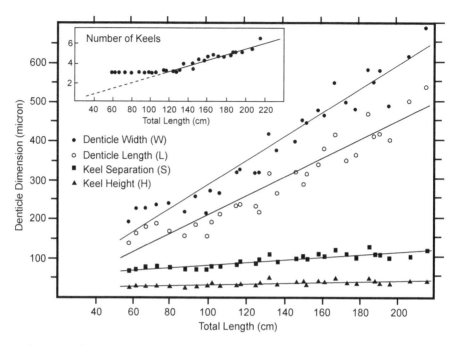

Fig. 3.8 Plot of denticle width and length and keel separation and height for silky sharks of increasing total lengths. It is significant that keel height and separation do not increase as the shark grows. This indicates that their dimensions are critical to their function in reducing drag during swimming.

rise much on the graph. There is something special about keel heights from 9–19 microns and separation distances from 40–57 microns. They may prevent the generation of small vortices as water passes over the ridged surface of the silky's skin with a concomitant increase in swimming efficiency. Hydrodynamic equations predict that a keel height-to-separation ratio of 1:2 is optimal for all sizes of sharks.[15]

The reef and mackerel sharks, which are active swimmers that inhabit the coastal and pelagic waters, are covered completely with closely interlocked dermal denticles with three or more keels. However, the basking shark (*Cetorhinus maximus*) also a galeoform shark but a slow-moving planktivore, has widely separated single-cusped denticles (fig. 3.9a).[26] Its different structure provides support that the three-ridge shape and close spacing has evolved in sharks for hydrodynamic efficiency. The dermal denticles of the bramble shark (*Echinorhinus brucus*; fig. 3.9b) and the Greenland shark (*Somniosus microcephalus*; fig. 3.9c), which are squalomorph species, also have widely spaced dermal denticles. The former is a sluggish shark that spends much of the time near the bottom and likely feeds on slow-moving fishes and crustaceans. The latter is also sedentary, surprising its prey by opening its cavernous mouth to create a vacuum that draws in its prey like the slurp gun used to collect tropical

Bionics is the creative technique that strives to produce innovative engineering solutions using living biological phenomena as prototypes. This field is based on the fundamental idea that biological organisms and their organs have been optimized by natural selection over time through the process of evolution. The structure of a shark's dermis and dermal denticles has served as a basis for the design of modern fast-performing swim suits.[25] The engineers for the swimsuit company Speedo examined the skin of many species of sharks at the New Zealand Museum of Natural History. They then constructed a Fastskin fabric with built-in ridges and valleys emulating natural skin of the shark. Furthermore, the fabric also was made of a super-stretch material made to compress the muscles and improve the suit's fit on the athlete's body. The result was been a reduction of drag, believed to be on the order of 3%, as well as muscle vibration that has led to an increase in swimming performance.

The engineers also performed 3-D scans on elite swimmers in eight specific postures to better contour the suit to the shape of the competitive swimmer. Laminated polyurethane panels were inserted within the suit at specific points to smooth out the body's silhouette like a corset. Hence, the engineers were providing a human with the collagen-based dermis and dermal denticles of sharks. This suit likely contributed to the fact that 80% of the swimming medals won in the 2000 Summer Olympics were won by athletes wearing Speedo's Fastskin suits. Swimmers wearing these suits also broke thirteen of fifteen world records. Michael Phelps, who has set thirty-seven world records and won fourteen gold medals at the Olympics, has worn this suit throughout much of his career. Swimmers were prohibited from wearing these suits in 2010, and they have only begun to set world records again—two being broken in the breaststroke in the 2012 Olympics in London.

fishes. The sand devil (*Squatina dumeril*; fig. 3.9d) has large denticles with a single cusp, and the cookie cutter shark (*Isistius brasiliensis*), large pavement-like denticles ideal for protection. This may also be true for the Caribbean lantern shark (*Etmopterus hillianus*) (fig. 3.9e), a deepwater species with numerous denticles with a single long spike-like cusp on them.

BIOMECHANICS OF SWIMMING

Morphologists and hydrodynamicists throughout the twentieth century sought to understand the function of the tail in the locomotion of fishes.[27] The tails of the mackerel sharks of the family Lamnidae and the most diverse group of bony fishes, the Teleostei, are generally symmetrical with the upper and lower lobes similar in shape and size. In contrast, the majority of sharks possess an asymmetrical tail with the upper lobe larger than the lower lobe. Although it is generally agreed that a symmetrical tail generates forward thrust, there has been considerable debate over whether the asymmetrical tail with its large upper lobe produces upward and downward force in addition to forward thrust.

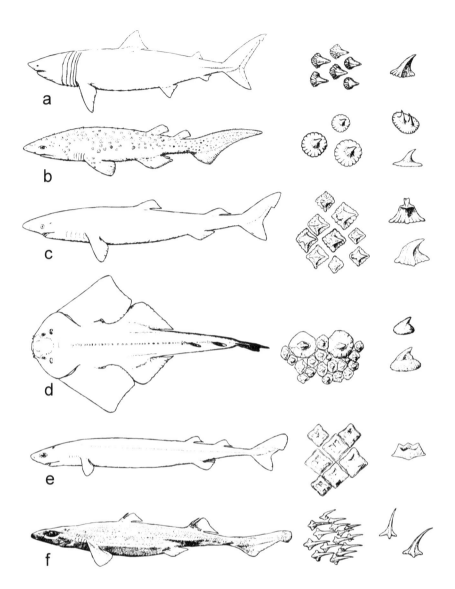

Fig. 3.9 Sharks with atypical dermal denticle shapes and spacing on the body. Shown are the dermal denticles of the (a) basking shark, (b) bramble shark, (c) Greenland shark, (d) sand devil, (e) cookie cutter shark, and (f) lantern shark.

Early studies led to the formulation of the classical hypothesis for the functioning of the heterocercal tail.[28, 29, 30] The dorsal edge of the caudal fin was believed to be relatively stiff due to its being supported by cartilaginous elements extending all the way to the distal end of the upper lobe. This rigid edge is thought to move before the flexible ventral region as the entire tail beats from side to side. As a result of the ventral lobe lagging behind the stiff dorsal edge of the tail, the caudal fin passes through the water at an angle to the horizontal and this provides both an upward force, or lift, and a forward force, or thrust (see thick solid arrow in fig. 3.10a). The lift produced by the asymmetric tail was observed to increase with a larger tail and greater frequency of tail beats

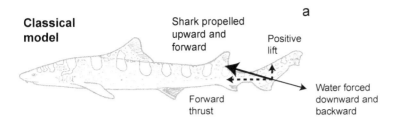

Classical model

a

Shark propelled upward and forward

Positive lift

Forward thrust

Water forced downward and backward

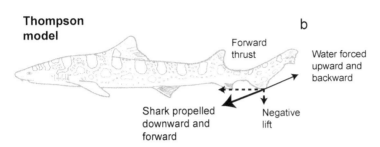

Thompson model

b

Forward thrust

Water forced upward and backward

Shark propelled downward and forward

Negative lift

Fig. 3.10 Schematic diagrams of the leopard shark illustrating (a) the classical and (b) Thompson's hypotheses explaining the forces generated by a shark during locomotion. (c) Evidence for the former model in the form of dye injected posterior to the pectoral fins and being propelled downward and backward by the caudal fin. One must distinguish the dye stream below the shark from the two shadows of the shark cast on the back of the apparatus.

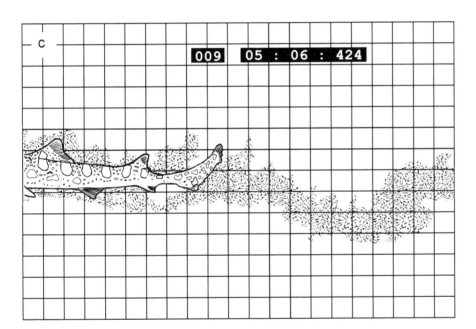

c

by moving the severed tails of sharks back and forth and observing the water movements.[27] Keith Thompson of the Peabody Museum of Natural History at Yale University presented a contrasting model of how the asymmetrical tail functioned in 1976 based upon his observations of motion pictures taken from behind of sharks of different species swimming in an aquarium.[32] He drew a time series of pictures depicting the movements of the caudal fin of the leopard, sandtiger, sandbar, and nurse sharks. The upper lobe of their caudal fin did not move before the lower lobe, but the two either moved simultaneously or the lower lobe actually led the upper lobe through much of the tail beat cycle. Because the upper lobe had a larger surface area than the lower lobe, his model predicted that that the forward force resulting from tail beats would be slightly downward (see thick solid arrow in fig. 3.10b).[31, 32] While the classical theory predicted that the tail pushes water backward and downward (see thin arrow, fig. 3.10a). Thompson argued that the tail pushes water backward and upward (see thin solid arrow, fig. 3.10b). Which of the two models most accurately explains the mechanics of swimming in sharks? The answer to this question would ultimately have to wait until the development of a new and innovative technique for studying the kinematics, or motions of the body, during swimming as shown here in Spotlight 3.2.

The pectoral fins are bent to produce either ascending or descending movements.[34] The shark initiates a sinking event by rotating the posterior margin of the pectoral fin slightly upward so that water contacts the upper surface of the fin at an angle closer to 90° to force water upward and produce greater negative lift that propels the shark forward but in a downward direction. The shark initiates an upward rise by rotating the posterior plane of the pectoral fin downward so that the water presses against the lower surface at an angle closer to perpendicular to force more water downward resulting in greater positive lift propelling the shark forward but in an upward direction. These angular adjustments of the pectoral fin function to maneuver the shark vertically in the water column. The real value of an asymmetric tail present in the majority of the sharks in comparison to the symmetrical tail possessed by the mackerel sharks may be the maneuverability achieved by modulating the upward trajectory from the beating tail by the upward and downward movements of the pectoral fins.

CLASSIFICATION OF SWIMMING MODES

You are now familiar with the composition of cartilage, the axial and appendicular skeletons, two muscle types and divergent modes of skeletal attachment, and biomechanics of fin propulsion of the cartilaginous fishes. It is time to present a generally accepted classification of the styles of swimming exhibited by the sharks, rays, and chimaeras. Each of the three lineages has one or more separate styles of locomotion. The sharks mainly exhibit three types of locomotion: anguilliform, carangiform, and thunniform swimming. The names of the motions refer to families of bony fishes that exhibit this form of swimming: the eels of the family Anguillidae, the jacks of the Carangidae, and the tunas of the Thunnidae. The shark's torso and tail in anguilliform swimming conform to a sinusoidal wave with a positive and negative peak (see position 5 in fig. 3.12a). The muscles on one side of the anterior torso contract so that the

Lara Ferry and Cheryl Wilga, both postdoctoral fellows, evaluated these two competing hypotheses while in the laboratory of George Lauder of the University of California–Irvine in the 1990s. Ferry and Lauder recorded high-speed video in three dimensions of the movements of the lateral and posterior views of the asymmetrically lobed heterocercal tail of the leopard shark that was swimming in a flow chamber.[27] Their apparatus could simultaneously record a lateral view and a posterior view of the shark's body. This was recorded by directing the camera at a mirror angled at 45° that was situated behind the shark, which, despite beating its tail, remained in place due to the continuous flow of water. In this respect, a flow chamber is like an exercise treadmill, on which a person can walk or run without moving forward. The two images were scaled by marked grids situated to the side and in front of the shark.

They viewed images of the tail beating both from the side and from the rear of the shark. They discovered that the upper lobe extended farther to the left than the lower lobe when the tail was at the extreme left, the trailing edge of the caudal fin was vertical at the midline of the beat, and the upper lobe moved farther to the right when the tip of the tail reached the extreme right. The lower lobe moved first, consistent with Thompson's observations. However, the lower lobe did not very move far and the upper lobe passed over it farther to either side. This additional movement of the terminal lobe forms an obtuse angle exceeding a vertical orientation of 90° as the tail swings back and forth. The tail thus forces water downward and backward as predicted by the classical model, not upward and backward as predicted by Thompson's model. This conclusion was corroborated by releasing dye in the vicinity of the caudal fin. The dark blue colored water was deflected backward and downward at an angle of approximately 20° from the horizontal (fig. 3.10c). This

angle is consistent with an analysis of the relative movement measured at various points on the caudal fin.

Wilga and Lauder found further support for the classical theory by recording the movement of water as it passed off the caudal fin of a shark in the flow chamber.[33] In this innovative experimental apparatus, one video camera recorded images of the position of shark's caudal fin and a second took images of the positions of particles suspended in the water forced in the opposite direction as the shark swam forward. The water movements were indicated by the reflection of light off small hollow glass beads of neutral buoyancy that were suspended throughout the flow chamber. A matrix of vectors, illustrating the direction and magnitude of flow, was formed from the successive positions of the beads over time recorded from the video record. Supporting the hypothesis that water was forced backward and downward by tail beats was the strong jet of water apparent posterior and ventral to the tail at an angle of −45° for the leopard shark (fig. 3.11b) and at an angle of −42° for the bamboo shark (fig. 3.11c). The general direction of jet angle (see dashed line, fig. 3.11a) is determined in the two images from the concentrations of black-colored vectors indicating water movement that are oriented backwards and downward to the right of the drawing of the tails of the two sharks. The specific angle is measured from the horizontal axis and a dashed line drawn perpendicular to an imaginary line drawn between clockwise rotating fluid (black shading) and the counterclockwise rotating fluid (gray shading). The body angle, drawn between the horizontal and a line normal to the ventrum of the shark, increases as the ring jet angle increases for both species. This relationship provides unequivocal evidence that when the asymmetrical tail accelerates water backwards and downwards there is an opposite reaction force propelling the shark both forward and upward.

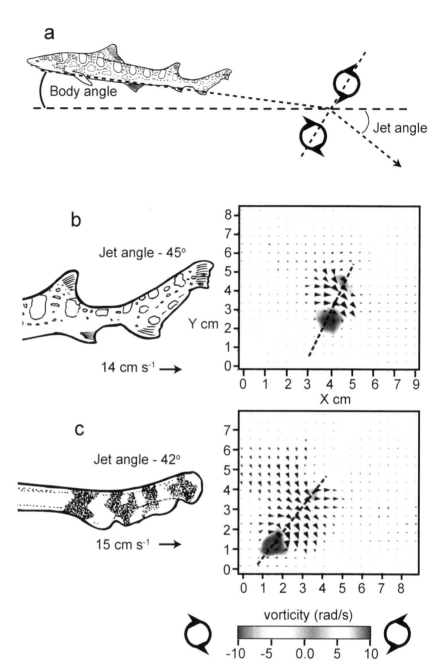

Fig. 3.11 (a) The posture of a swimming shark showing the jet and ring axes relative to its body angle and path of motion. Position of caudal fin (left) and matrix of vectors indicating the direction and speed of water movement (see arrows) and areas of strong clockwise water rotation (black) and similar counterclockwise water rotation (gray) for the (b) leopard shark and (c) bamboo shark.

forward body bends to that side—upward in the illustration. This loop passes down the length of the shark's body with the posterior surface of the loop pushing water backward, resulting in forward movement. When the midbody is bent most (see position 3), muscles on the other side of the anterior torso begin to contract while bending the body to the other side with the posterior surface of this new loop pushing more water backwards and producing further

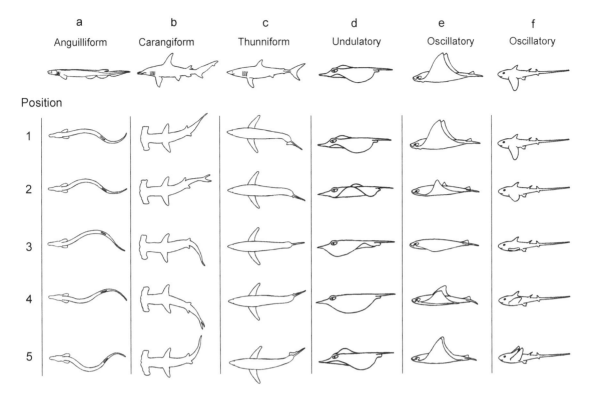

Fig. 3.12 Modes of propulsion in the cartilaginous fishes. The fusiform sharks propel themselves forward by undulating their body along its longitudinal axis, and their modes of swimming are (a) anguilliform, (b) carangiform, and (c) thunniform motion. The dorsally and ventrally flattened rays move themselves forward by passing a wave along their body or moving their appendages, modes of locomotion termed (d) undulatory and (e) oscillatory. The chimaeras propel themselves forward by (f) oscillating their fins. The successive bending of the torso over one swimming cycle is shown below on a horizontal axis.

forward movement. The nurse, cat, and frilled sharks use this form of loco-motion. The shark shown in fig. 3.12a is the frilled shark (*Chlamydoselachus anguineus*). It is elongated and eel-like with a flattened snake-like head and has a terminal mouth full of widely spaced, needle-sharp, three-cusped teeth. Frilled sharks live in the deep sea, where they likely do not have to move very quickly to feed on soft-bodied deepwater squid and fishes.[35] A few have been observed swimming at the surface but most specimens have been collected from the bycatch of fisherman deploying deep bottom trawls and gill nets. This was the mode of swimming used by *Orthacanthus*, a shark common during the Paleozoic. This eel-like predator reached a length of three meters. It had massive jaws and double-pronged teeth for seizing unsuspecting prey hiding within the dense vegetation in the extensive freshwater swamps of that time.

The majority of sharks propel themselves forward with fewer undulations of the body using the carangiform mode of swimming. Bull sharks, six- and sevengill sharks, saw sharks, reef sharks, and hammerhead sharks swim in this manner. These sharks bend only the posterior half of the torso from side to side. A scalloped hammerhead shark is shown first bending its posterior torso and tail downward toward the rear and then upward and backward to propel itself forward (fig. 3.12b). These swimmers still move in a relatively sinuous manner, although they bend their body less than the anguilliform swimmers. The mako, white, and salmon sharks are thunniform swimmers. They bend only the posterior quarter of their body. A shortfin mako shark is shown moving its tail back and forth from side to side (fig 3.12c). Notice that only its caudal fin moves back and forth while the rest of its body is held rigid. This makes the body appear stiffer when swimming than do the sharks using the anguilliform and carangiform modes of locomotion.

The dorsally and ventrally flattened rays propel themselves forward in either of two manners. Most rays pass a propulsive wave down the length of their flat and enlarged pectoral disk to force water backward and propel themselves forward—a form of locomotion described as undulatory (fig. 3.12d). The electric rays, skates, and most stingrays use this mode of locomotion.[36] Notice the southern stingray (*Dasyatis americana*) initially lifts its pectoral fin in a small upward loop that first increases and then decreases in size as the loop passes toward the rear pushing water backward (see positions 1–3). Many of the stingrays of the order Myliobatiformes exhibit oscillatory swimming. They flap their pectoral fins up and down with the posterior edge of the fin lagging behind the anterior part, thus pushing the water backwards. Shown here is the up-and-down flapping of the cownose ray (*Rhinoptera bonasus*), which migrates each year in massive schools up and down the eastern coast of North America (fig. 3.12e). This same mechanism is used by the chimaeras, which also oscillate their pectoral fins upwards and downwards, to propel slowly forward or upward and downward (fig. 3.12f).

SUMMARY

The purpose of this chapter has been to demonstrate how the cartilaginous fishes swim. There are five generally accepted locomotory modes of swimming for the sharks, rays, and chimaeras: anguilliform, carangiform, thunniform, undulatory, and oscillatory. The modes are not distinct but are graded among species. The sharks mainly exhibit the first three modes, while the rays and chimaeras the last two modes. The sharks, rays, and chimaeras all possess axial and appendicular skeletons composed of cartilage. The skeletal elements are composed of a thin, hard external covering of tesserae, plates of cartilage fortified with mineralized calcium phosphate, and an interior of softer hyaline cartilage. There are two types of muscle in these fishes, white and red, energized by two separate metabolic processes, one anaerobic and the other aerobic. The latter muscles are located more externally in the carangiform-swimming sharks and rays and situated more internally in the thunniform-swimming sharks. The muscular attachments to the vertebral elements differ, and these produce very different modes of locomotion. The cartilaginous fishes have a pressurized elastic skin that provides support for the muscles together with the vertebral column. Through alternating cycles of contraction and relaxation, they transmit force to the caudal fin to propel them forward. The dermal denticles reduce drag and increase swimming efficiency. Most sharks, rays, and chimaeras swim slowly but are capable of short bursts of swimming activity to capture their prey. The mackerel sharks, on the other hand, are adapted to continuous swimming and sustained cruising speeds over long distances. In conclusion, the muscles and skeletal elements of the cartilaginous fishes have evolved to enable the different members of the group have very different modes of locomotion.

⁎⁎ ⁎⁎ ⁎⁎

KEY TO COMMON AND SCIENTIFIC NAMES

Atlantic sharpnose shark = *Rhizoprionodon terraenovae*; basking shark = *Cetorhinus maximus*; blacktip shark = *Carcharhinus limbatus*; blue shark = *Prionace glauca*; bramble shark = *Echinorhinus brucus*; brownbanded bambooshark = *Chiloscyllium punctatum*; bull shark = *Carcharhinus leucas*; Caribbean lantern shark = *Etmopterus hillianus*; cookie cutter shark = *Isistius brasiliensis*; cownose ray = *Rhinoptera bonasus*; dusky shark = *Carcharhinus obscurus*; dusky smooth-hound shark = *Mustelus canis*; frilled shark = *Chlamydoselachus anguineus*; great hammerhead shark = *Sphyrna mokarran*; Greenland shark = *Somniosus microcephalus*; lemon shark = *Negaprion brevirostris*; leopard shark = *Triakis semifasciata*; night shark = *Carcharhinus signatus*; nurse shark = *Ginglymostoma cirratum*; oceanic whitetip shark = *Carcharhinus longimanus*; Pacific longnose chimaera = *Harriotta raleighana*; piked dogfish = *Squalus acanthias*; salmon shark = *Lamna ditropis*; sandbar shark = *Carcharhinus plumbeus*; sand devil = *Squatina dumeril*; sandtiger shark = *Carcharias taurus*; shortfin mako shark = *Isurus oxyrinchus*; silky shark = *Carcharhinus falciformis*; sixgill sawshark = *Pliotrema warreni*; sixgill stingray = *Hexatrygon bickelli*; small-spotted catshark = *Scyliorhinus canicula*; southern stingray = *Dasyatis americana*; tiger shark = *Galeocerdo cuvier*; white shark = *Carcharodon carcharias*.

LITERATURE CITED

1. Compagno, 1999; 2. Dean *et al.*, 2005; 3. Dean and Summers, 2006; 4. Summers and Long, 2006; 5. Francis, 1996. 6. Dingerkus *et al.*, 1991; 7. Wilga and Lauder, 2004; 8. Pridmore, 1995; 9 Liem and Summers, 1999; 10. Webb, 1984; 11. Combes and Daniel, 2001; 12. Shadwick, 2005; 13. Shepherd, 1994; 14. Wainwright *et al.*, 1978; 15. Raschi and

1. What species of sharks have uniquely shaped dermal denticles? How do their shapes and spacing differ between species? You can find drawings of the denticles of a large diversity of sharks in Bigelow and Schroeder's monograph, "Fishes of the Western North Atlantic: Lancelets, Cyclostomes, Sharks." Can you relate these anatomical differences to distinct swimming behaviors and habitat preferences?

2. Can you explain the biomechanics of more complicated movements such as rolling and relate these to the pilot's control of a fighter airplane? This topic is covered in an article titled "Three-dimensional kinematics and wake structure of the pectoral fins during locomotion in leopard sharks *Triakis semifasciata*" (Wilga and Lauder, 2000).

3. What species make the longest migrations? This information is available in the recent scientific articles on the long-distance movements of fish of various species that have sophisticated satellite transmitters placed on them (see chapter 14). Their movements are overground and should be distinguished from the speed at which they swim. Finding out what species swim the fastest over short distances is more challenging because the transmitter must be outfitted with a paddlewheel that rotates as the shark passes through the water to record its speed.

4. As is apparent from the material presented in this chapter, the bulk of the studies of swimming biomechanics of the cartilaginous fishes have focused on the sharks. There are fewer scientific articles directed at understanding the swimming behavior of the rays and almost none on the chimaeras. What kind of studies could be performed to learn more about the biomechanics of swimming of the members of these other two taxonomic groups? Conduct a library search and read papers in the scientific literature before making your recommendations. This paucity of information makes the subject ripe for studies leading to an advanced graduate degree.

Elsom, 1986; 16. Bertin, 1985; 17. Budker, 1938; 18. Reif, 1985; 19. Walsh and Weinstein, 1978; 20. Walsh, 1982; 21. Bechert *et al.*, 1985; 22. Bechert *et al.*, 1986; 23. Walsh, 1980; 24. Raschi and Tabit, 1992; 25. Skulberg, 2009; 26. Bigelow and Schroeder, 1948; 27. Ferry and Lauder, 1996; 28. Grove and Newell, 1936; 29. Alexander, 1965; 30. Simons, 1970; 31. Thompson, 1976; 32. Thompson and Simanek, 1977; 33. Wilga and Lauder, 2002; 34. Wilga and Lauder, 2000; 35. Compagno *et al.*, 2005; 36. Rosenberger, 2001.

RECOMMENDED FURTHER READING

Compagno, L. J. V. 1999. Endoskeleton. Pp. 69–92 in Hamlett, W. C. (Ed.). Sharks, Skates, and Rays: The Biology of Elasmobranch Fishes. Johns Hopkins University Press, Baltimore.

Combes, S. A., and T. L. Daniel. 2001. Shape, flapping and flexion; wing and fin design for forward flight. *J. Exp. Biol.*, 204: 2073–2085.

Dean, M. N., and A. P. Summers. 2006. Mineralized cartilage in the skeleton of chondrichthyan fishes. *Zoology*, 109: 164–168.

Ferry, L. A., and G. V. Lauder. 1996. Heterocercal function in leopard sharks: a three-dimensional kinematic analysis of two models. *J. Exp. Biol.*, 199: 2253–2268.

Liem, K. F., and A. P. Summers. 1999. Muscular system: gross anatomy and functional morphology of muscles. Pp. 93–114 *in* Hamlett, W. C. (Ed.), Sharks, Skates, and Rays. Johns Hopkins University Press, Baltimore.

Raschi, W., and J. Elsom. 1986. Comments on the structure and development of the drag reduction-type placoid scale. Pp. 409–436 *in* Uyeno, T., R. Arai, T. Taniuchi, and

K. Matsuura (Eds.), Indo-Pacific Fish Biology: Proceeding of the Second International Conference on Indo-Pacific Fishes. Ichthyological Society of Japan, Tokyo.

Rosenberger, L. J. 2001. Pectoral fin locomotion in batoid fishes: undulation versus oscillation. *Jour. Exp. Biol.*, 204: 379–394.

Shadwick, R. E. 2005. How tunas and lamnid sharks swim: an evolutionary convergence. *Am. Sci.*, 93: 524–531.

Thompson, K. S. 1976. On the heterocercal tail in sharks. *Palaeobiol.*, 2: 19–38.

Wainwright, S. A., F. Vosburgh, J. H. Hebrank. 1978. Shark skin: function in locomotion. *Science*, 202: 747–749.

Wilga, C. A. D. and G. V. Lauder. 2004. Biomechanics of locomotion in sharks, rays, and chimeras. Pp. 139–164 in Carrier, J. C., J. A. Musick, and M. R. Heithaus (Eds.), Biology of Sharks and Their Relatives. CRC Press, Boca Raton.

Wilga, C. D., and G. V. Lauder. 2002. Function of the heterocercal tail in sharks: quantitative wake dynamics during steady horizontal swimming and vertical maneuvering. *J. Exp. Biol.*, 2005: 2365–2374.

CHAPTER 4

Water and Ionic Regulation

The cartilaginous fishes largely reside in marine environments. Yet in 1937, a small 1.5 m–long bull shark (*Carcharhinus leucas*) was caught in a fishing trap 2,800 km up the Mississippi River by a surprised fisherman expecting to catch a sturgeon, a common inhabitant of the river.[1] A picture of the shark appeared in the *Alton Evening Telegraph* with the following caption: "Far from home was this smooth-hound deep-sea shark—and maybe just a bit homesick—the other day." The fisherman took the shark to the local fishing market, where it was the object of great interest; several hundred people came to see it. This caused fear among the local populace, who commonly swam in the river, because the largest bull sharks were known to attack humans. In fact, two swimmers were killed by what was likely a bull shark that migrated into the freshwater in Matawan Creek in New Jersey on July 6, 1916. A second attack occurred, possibly by a white shark (*Carcharodon carcharias*), less than a week later near the recreational coastline. These attacks, highly publicized by local newspapers, led to the local hysteria on which were based the novel and movie *Jaws*.[2, 3] The longest freshwater migration of a bull shark reported so far is to the Rio Ucayali, a tributary of the Amazon River 4,200 km from the Atlantic Ocean.[4] Freshwater rays within the family Potamotrygonidae

South American freshwater stingray (*Potamotrygon motoro*), common in the Amazon watershed including the streams on the eastern slope of the Andes.

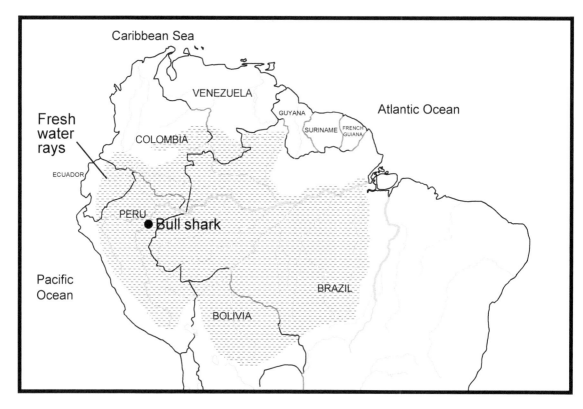

Fig. 4.1 Map of the Amazon River showing the farthest upstream migration of a bull shark on the Rio Ucayali above the junction with the Rio Marañon 4,200 km from the sea and the distribution of the freshwater rays in the tributaries on the eastern slope of the Andes Mountains (see cross-hatching).

are very common throughout the Amazon watershed, including the streams flowing down the eastern slopes of Andes Mountains, 4,500 km from the sea in the same region (fig. 4.1).[4,5] Yet these sharks and rays are exceptions to the general rule, as the majority of these fishes live in saltwater.

The concentrations of salts and other chemicals in the bodies of chondrichthyan fishes differ from those present in salt- or freshwater. These fishes maintain a constant ionic concentration within their blood and tissues. In the sea, they excrete sodium and chloride ions that diffuse into their bodies because the concentration of these ions in the ocean exceeds that within their bodies. On the other hand, they gain water because the concentration of solutes, salts, and other chemicals in their tissues is greater than in the surrounding ocean. However, the bull shark and stingrays of the genus *Dasyatis* enter estuaries and rivers with lower salinities than in the oceans. The relationship between their ionic and solute composition and that of the environment changes once they move into the freshwater. This chapter is devoted to a description of the internal hydro-mineral balance of the sharks, rays, and chimaeras. Particular attention will be given to how species regulate the salt composition of their tissues in order to temporarily inhabit estuarine and freshwater environments or reside wholly in the freshwater environment.

OSMOREGULATION

Osmoregulation, or the maintenance of an internal balance between water and dissolved solutes, involves the physical relationship between the body fluids and the surrounding medium.[6] Blood and other body fluids are always separated from the external environment by permeable surfaces, particularly the gills. Unless the internal and external fluids have the same solute concentration measured in osmolarity, water will pass from the region of less solute concentration to that with the greater concentration. Water will pass through the gills of cartilaginous fishes into the body when its body fluids contain a higher osmolarity than in the surrounding aqueous environment, and conversely water will leave the body when the surrounding fluids have a lower osmolarity then the internal tissues. Salts (electrically charged ions such as Na^+ and Cl^-), will also either enter or leave the body based on their concentration and the electrical gradient between the internal and external environments.

The cartilaginous fishes that live only in the oceans are stenohaline, capable of adapting only to the narrow range of salinities that occur there. They maintain an osmolarity in their blood near that of their surrounding marine environment. Their internal osmolarity is slightly higher than that of saltwater, which is 935 millimoles per liter of water ($mmol \cdot l^{-1}$). For example, the osmolarity of the piked dogfish (*Squalus acanthias*) is either 948 or 1018 $mmol \cdot l^{-1}$, depending on the study, that of the Atlantic stingray (*Dasyatis sabina*) is 1034 $mmol \cdot l^{-1}$, and that of the rabbit fish (*Chimaera monstrosa*) is 1046 $mmol \cdot l^{-1}$ (table 4.1). Hence, water enters their bodies at a slow rate. They eliminate from their body this excess water by producing modest amounts of urine (fig. 4.2a).

The similarity of the osmolarities of the blood and seawater does not indicate that the solutes in their tissues are necessarily the same. Sodium and chloride salts in the ocean exist in concentrations 440 and 495 $mmol \cdot l^{-1}$, adding up to an osmolarity of 935 $mmol \cdot l^{-1}$. The concentrations of the two salts in the plasma of the piked dogfish,[8] in which the most solutes were measured, were 296 and 276 $mmol \cdot l^{-1}$. These concentrations are much lower than those in saltwater (see stippled bars in histograms). Hence, the inorganic salts that diffuse through the gills into the body tissues of the dogfish must be excreted by a specialized organ, the salt gland. Other organic solutes exist in the plasma of elasmobranchs, contributing to its high osmolarity. These are two nitrogenous compounds, urea and trimethylamine oxide (TMAO). The levels of urea and TMAO measured within the plasma of the piked dogfish in the same study

TABLE 4.1. Blood osmolarity, and sodium and chloride ion, urea, and TMAO concentrations in plasma of sharks, rays, and chimaeras in seawater and freshwater

Medium Species	Common Name	Osmolarity (mmol·l⁻¹)	Sodium (mmol·l⁻¹)	Chloride (mmol·l⁻¹)	Urea (mmol·l⁻¹)	TMAO (mmol·l⁻¹)
Seawater[7]		935	440	495	0	
Squalus acanthias[7]	Piked dogfish	1018	286	246	351	
Squalus acanthius[8]		948	296	276	314	76
Dasyatis sabina[9]	Atlantic stingray	1034	310	300	394	
Chimaera monstrosa[10]	Rabbit fish	1046	338	353	332	
Freshwater[9]		38	3.0	3.7		
Carcharhinus leucas[5]	Bull shark		200	180.5	132	
Carcharhinus leucas[14]			245	219	169	
Dasyatis sabina[9]	Atlantic stingray	621	212	208	196	
Potamotrygon sp.[14]	Freshwater stingray	308	150	150	1.3	
Potamotrygon sp.[15]		282	164	152	1.1	
Potamotrygon sp.[16]		320	178	146	1.2	

were 314 and 76 mmol·l⁻¹, and they add up to a solute total of 962 mmol·l⁻¹—a level close to the plasma's overall osmolarity of 948 mmol·l⁻¹. There are also small concentrations of potassium and calcium in the tissues of the dogfish. The concentrations of the sodium and chloride ions in the plasma of the Atlantic stingray are similar to those of the dogfish.[9] However, the concentration of urea is higher, giving an overall osmolarity of 1014 mmol·l⁻¹, near the 1034 mmol·l⁻¹ concentration in its plasma. The plasma of the rabbit fish has even greater concentrations of sodium and chloride ions, a similar urea concentration, but no detectable TMAO.[10]

Urea is a destabilizer of proteins. The occurrence of such high concentrations of this molecule in the muscles of cartilaginous fishes might be expected to adversely affect their metabolic processes. The deleterious effect of urea is counteracted by the accumulation of TMAO, a protein stabilizer. Urea's tendency to depress the binding affinity of proteins with other molecules is reduced by combining urea and TMAO in a ratio of 2:1.[11] This combination also increases the contractile properties of muscle fibers.[12] The concentration of TMAO in the muscle of the dogfish is roughly half that of urea.[8] A similar relationship exists between the amounts of TMAO and urea in the muscles

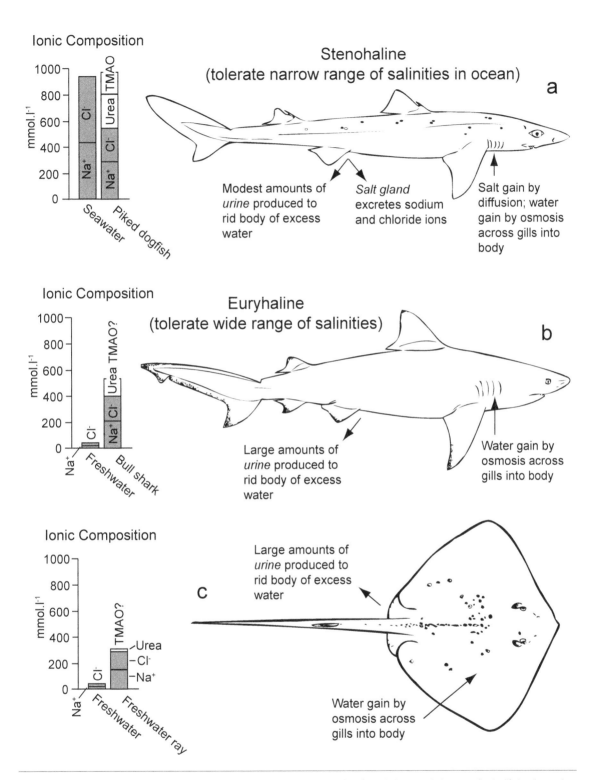

Fig. 4.2 Diagram of the inward and outward transport of water and salts in (a) the piked dogfish, which is stenohaline, spending its life in saltwater in a narrow range of salinities; (b) the bull shark, which is euryhaline, living most of the time in saltwater but temporarily entering rivers and lakes with low salt content; and (c) the freshwater stingray, which spends its entire life in freshwater. The concentrations of inorganic and organic solutes in the tissues of these three species are compared in the histograms.

A few sharks and rays are able to inhabit the waters, such as estuaries, rivers, and lakes, with salinities lower than seawater. These species are euryhaline, or broadly tolerant to variable saline environments. The bull shark and many species in the genus *Dasyatis* at times do enter these less saline environments. In contrast, the chimaeras live exclusively in the marine environment. The osmolarity of freshwater is very low, only 38 mmol·l⁻¹, and it contains concentrations of sodium and chloride ions of only 3.0 and 3.7 mmol·l⁻¹, respectively (table 4.1, histogram in fig. 4.2b). Bull sharks leave the salty waters of the Caribbean Sea, swim upstream in the San Juan River, and enter Lake Nicaragua, where they may reside for more than a year. The bull shark maintains internal concentrations of sodium, chloride, and urea considerably less than those of the stenohaline piked dogfish.[5,14] Atlantic stingrays frequently enter the St. Johns River, which is largely estuarine, and then migrate into Lake Jesup, which is a freshwater environment. These euryhaline rays also have considerably lower concentrations of sodium, chloride, and urea than those living in the more saline ocean.[9] The concentrations of the ions in these euryhaline species are lower than those of the stenohaline species, but their osmolarities are still considerably higher than that of freshwater. Hence, they have to produce large amounts of urine to rid their body of the water gained through osmosis across their gills.

The rays in the family Potamotrygonidae are unique among the cartilaginous fishes because they reside in freshwater during their entire life. The concentrations of sodium and chloride ions in their plasma are half those of the marine cartilaginous fishes. Urea is almost nonexistent in their plasma, ranging in concentration from 1.1 to 1.3 mmol·l⁻¹. Hence, the overall solute concentration within their blood only ranges from 282 to 320 mmol·l⁻¹. These species have a different osmoregulatory mechanism than the marine species. They must rid themselves of the water, which passes through the permeable surface of their gills into their circulatory system, as well as retain concentrations of sodium and chloride ions (fig. 4.2c). They maintain their higher osmolarity by producing copious amounts of urine.

of the ratfish. Thus, in the cartilaginous fishes the accumulation of these two compounds in the appropriate ratio prevents urea from having a deleterious effect on proteins.

OSMOREGULATORY ORGANS

The chondrichthyan fishes have three organs that are involved in osmoregulation: the rectal gland, the kidneys, and the gills.[17] Located above the posterior intestinal track is the rectal gland, a tubular organ containing many thousands of tubules that extend outward in a radial arrangement from a central lumen that ends in a short duct through which its secretions are expelled into the intestine. The rectal gland excretes sodium and chloride salts through the anus along with feces. These excess salts enter the fish by diffusion through the permeable gills and are transported throughout the body by the circulatory system. Excess water that enters through diffusion is excreted through the kidney.

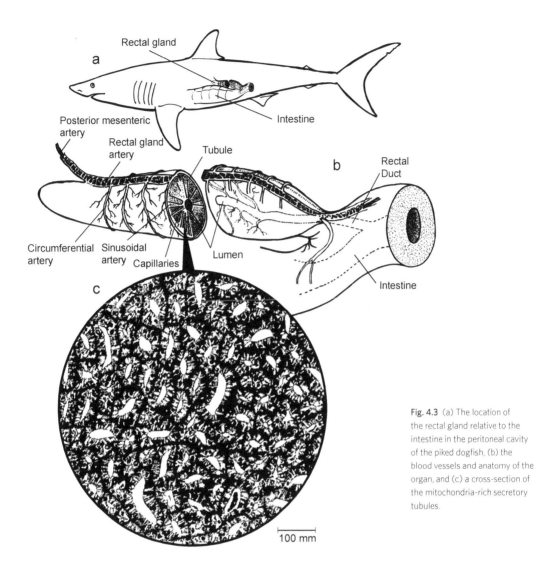

Fig. 4.3 (a) The location of the rectal gland relative to the intestine in the peritoneal cavity of the piked dogfish, (b) the blood vessels and anatomy of the organ, and (c) a cross-section of the mitochondria-rich secretory tubules.

The concentrations of both urea and TMAO in the blood of the cartilaginous fishes exceed those in the marine environment. For this reason, they must be retained within their bodies. The tubules within the kidney reabsorb both nitrogenous compounds after they are filtered out of the blood in the glomeruli. These compounds are then maintained at a high relative concentration within the blood because of the gill tissues' low permeability to them.[6]

Rectal Gland

The tubular rectal gland is suspended from the mesentery above the small intestine, just behind the dorsal fin (fig. 4.3a). In the dogfish, the gland is long and slender, whereas in the round stingray, *Urobatis jamaicensis*, it is S-shaped and formed of multiple lobes.[18] The mass of the dogfish's rectal gland ranges from 444 to 600 milligrams per kilogram (mg/kg) of body mass.[19, 20] The rectal gland is reduced in size roughly tenfold from 20 to 60 mg/kg in the freshwater-tolerant bull shark and further still to 15 mg/kg in the freshwater stingray.[21] The smaller size of the rectal gland of the euryhaline bull shark and freshwater ray is consistent with their having little need to excrete sodium and chloride ions. These species need rather to counteract the osmotic gain of water and the loss of ions through their gills in order to maintain their higher solute concentration in the dilute freshwater environment.

The rectal gland has three concentric layers of tissue that surround its hollow center, or lumen. These layers are an outer capsule with an interior lining of smooth muscle, a middle matrix of arteries and tubules, and an inner membrane. The capsule is covered by a peritoneal membrane with arteries embedded in connective tissue with an associated network of nerves (fig. 4.3b). The inner matrix consists of many sinusoidal arteries oriented inward in a radial direction that are embedded in connective tissue with capillaries and nerves associated with the arteries. There are also tubules that alternately radiate inward and outward that can extend along the entire length of the gland. The posterior mesenteric artery brings blood to the rectal gland artery, which extends along the longitudinal axis of the gland on its dorsal surface. From it originate many circumferential arteries, which form progressively smaller arterioles that become capillaries as they reach the lumen. Veins near the lumen collect the blood and drain into larger veins, which eventually join with a central vein toward the rear of the gland, which flows into the posterior intestinal vein. The secretory tubules consist of epithelial cells, which have many mitochondria evident by their dark shade and circular shape, arranged in a circumferential manner (fig. 4.3c).[24, 25] Each cell also has many invaginations in its outer membranes to provide a large surface for transport of sodium and chloride ions from the blood into a secretory tubule.

The concentrations of sodium and chloride in the lumen exceed those in the plasma as well as those in the external marine environment. For example, the concentrations of sodium and chloride ions of 540 and 533 mmol·l^{-1} in the lumen of the rectal gland of the piked dogfish (table 4.2) exceed those in its plasma of 286 to 296 and 246 to 276 mmol·l^{-1} (table 4.1). These unequal concentrations indicate active transport of ions from the capillaries into the tubules in the rectal gland. Moreover, the concentrations of sodium and chloride ions in the tubules of the dogfish exceed those of the surrounding marine environment (440 and 495 mmol·l^{-1}), which in turn are considerably greater than the concentrations of these two electrolytes in the plasma of the piked dogfish. Similar high concentrations of sodium ions and chloride ions are present in the rectal glands of the small-spotted catshark (*Scyliorhinus canicula*), white-

Medium Species	Common Name	Sodium (mmol·l⁻¹)	Chloride (mmol·l⁻¹)
Seawater[7]		440	495
Squalus acanthias[7]	Piked dogfish	540	533
Scyliorhinus canicula[23]	Small-spotted catshark	554	
Hemiscyllium plagiosum[22]	Whitespotted bambooshark	535	
[a]*Dasyatis sabina*[24]	Atlantic stingray		583
Leucoraja ocellata[25]	Winter skate	490	

[a]*D. Sabina* present in both salt- and freshwater environments.

spotted bambooshark (*Hemiscyllium plagiosum*), Atlantic stingray, and winter skate (*Leucoraja ocellata*) (table 4.2).

Kidney

The kidneys of chondrichthyans are located on each side of the vertebral column on the top wall of the abdominal cavity. The most anterior section of the kidney is threadlike, often intermingling with the gonads in males, but gradually it becomes more bulky toward the rear of the body cavity. The kidneys of the sharks are long and narrow and are present along most of the body cavity; those of the rays are short, wide, and lobate and are confined to the rear of the body cavity (fig. 4.4a).[26] Examination of a cross-section of little skate's kidney, which appears oblong with flattened dorsal and ventral sides, reveals that there are two distinct zones, the bundle zone on the top and side of the kidney and the sinus zone in the bottom and center of the kidney (figs. 4.4b & 4.4c).[27] The former is thinner than the latter, and consists of closely intertwined tubules covered with a sheath with little vascular tissue, while the second is wider, and the tubules are separated from each other by large vascular spaces.

Within the kidney is a solid mass of nephrons, each consisting of tubules with associated arterioles, veins, and capillaries that regulate the water content and concentrations of the solutes in the blood (fig. 4.4c).[28] In the stenohaline cartilaginous fishes, which have slightly higher solute concentrations than that of saltwater, some water passes from the blood into the tubules and eventually flows through the ureter to be excreted from the body as urine. Urea

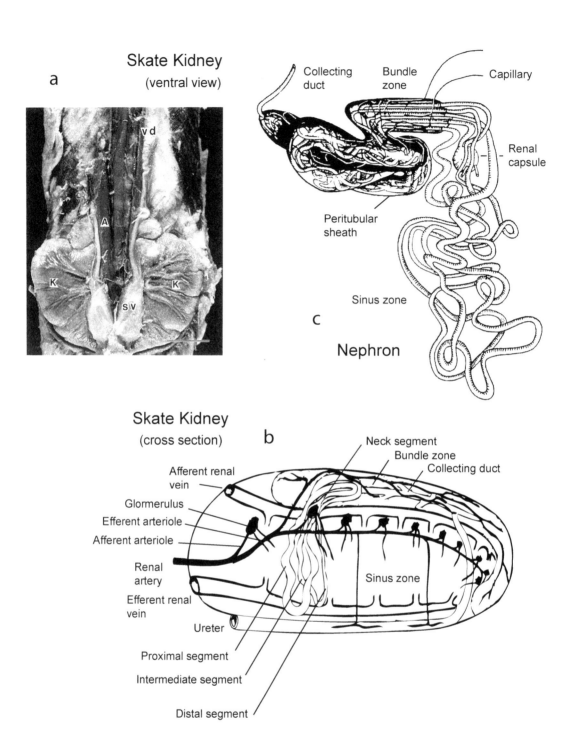

Skate Kidney
a (ventral view)

v d

A

K   K

s v

Collecting duct   Bundle zone   Capillary

Renal capsule

Peritubular sheath

Sinus zone

c

Nephron

Skate Kidney
(cross section)   b

Neck segment
Bundle zone
Collecting duct

Afferent renal vein

Glormerulus
Efferent arteriole
Afferent arteriole

Renal artery

Sinus zone

Efferent renal vein

Ureter

Proximal segment

Intermediate segment

Distal segment

Fig. 4.4 (a) Low-power photomicrography of the skate's kidney. A = aorta; K = kidney; sv = seminal vesicle; vd = vas deferens. (b) Cross-section of a little skate kidney showing the general arrangement of the blood vessels and a single nephron. (c) Drawing of a nephron of skate passing through the bundle and sinus zones.

and TMAO must also be retained within the blood at higher concentrations than in saltwater; this is accomplished in the kidney. The nephron consists of a glomerulus, or tight bundle of capillaries, surrounded by Bowman's capsule. This capsule is the beginning of an excretory tubule that encloses the glomerulus and makes a series of loops back and forth through the sinus and bundle zones.

The kidney is a complicated organ, and its structure and function can best be understood by following a generalized diagram (fig. 4.5).[16] Arteries branch off from the dorsal aorta at regular distances, and bifurcate into dorsal and ventral arteries. These become the afferent, or inward-flowing, arterioles, which supply blood to small capillaries in the bundle zone. They also become intertwined tightly within multiple glomeruli within the sinus zone near the bundle zone. Short efferent, or outward-flowing, arterioles leave each glomerulus and drain into the sinus zone. At the other end of the Bowman's capsule, the neck segment of the tubule passes into the bundle zone of the kidney, where it bends 90° upward and continues upward through the bundle zone, then reverses directions in loop 1 and continues downward as the proximal tubule into the sinus zone, where it reverses direction again in loop 2, and now leads upward through the sinus zone, and becomes the intermediate segment of the tubule as it passes into the bundle zone. As the tubule continues farther into this zone, it becomes the distal tubule before making another abrupt turn in loop 3 and moves back into the sinus zone. Here it makes a final turn, or loop 4, and continues upward through the sinus zone and becomes the collecting tubule. It proceeds upward through the bundle zone until it empties into the collector duct that leads to the ventral ureter (fig. 4.4b). This carries urine to the cloaca, where it is excreted from the body.

Water as well as solutes pass from the blood into arterioles in the glomerulus of the Bowman's capsule and leave as urine in the neck segment of the tubule. This begins the first of four loops, the first and third within the bundle zone and the second and fourth within the sinus zone. The fluid flows in opposite directions within two successive tubules within the bundle zone. This countercurrent flow in tubules in close proximity enables urea to be reabsorbed and retained within the body.[27] Evidence for the reabsorption of urea is the lower concentration of urea in the urine of the saltwater-inhabiting piked dogfish and the freshwater-inhabiting Atlantic stingray, 100 and 20 mmol·l^{-1}, respectively (table 4.3) than in their blood, 314 to 351 and 196 mmol·l^{-1}, respectively (see table 4.1). Similarly, the reabsorption of TMAO is apparent from the lower concentration of 10 mmol·l^{-1} (see table 4.3) in the urine of the piked dogfish than its concentration in its blood, 76 mmol·l^{-1} (see table 4.1). Also, there is some reabsorption of sodium and chloride ions because the concentrations of the two ions are slightly lower in the urine than in the blood. For

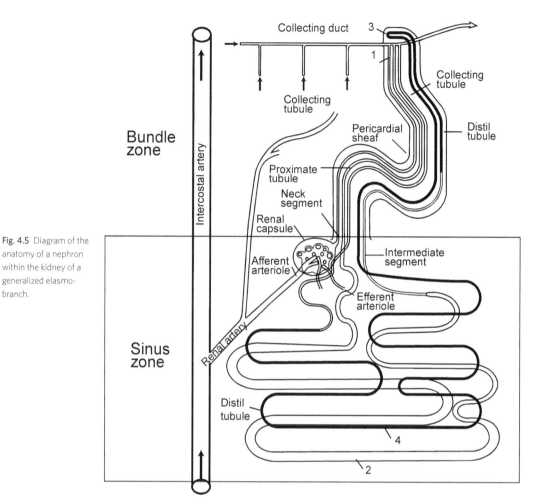

Fig. 4.5 Diagram of the anatomy of a nephron within the kidney of a generalized elasmobranch.

example, the concentrations of both ions in the urine of the dogfish are 240 mmol·l^{-1} and in the blood 286 to 296 and 246 to 276, depending on the study. The tubules are loosely packed in the sinus zone. This may facilitate diffusion of water and solutes between the blood within the sinus and the urine within the tubules.[35] This is likely what enables stenohaline sharks and rays to produce a dilute urine, and this compensates for their tendency to absorb water through their gills in saltwater.

An average of 10,354 glomeruli are in each kidney of the female piked dogfish.[36] There are fewer glomeruli in the kidneys of male and female small-spotted catshark, 1,140 and 1,474 respectively.[37] The kidney of the little skate has roughly 2,240 glomeruli.[36] The nephrons of these stenohaline sharks and ray are very long. For example, the nephrons of the dogfish have a mean length of 3.3 cm,[38] those of the winter skate 9.0 cm.[39]

The nephrons of the freshwater rays of the family Potamotrygonidae are

TABLE 4.3.   Renal osmolarity, and sodium and chloride ion, urea, and TMAO concentrations for sharks, rays, and chimaeras in seawater and freshwater.

Medium Species	Common Name	Osmolarity (mOsm.l⁻¹)	Sodium (mmol.l⁻¹)	Chloride (mmol.l⁻¹)	Urea (mmol.l⁻¹)	TMAO (mmol.l⁻¹)
Seawater[7]		930	440	495	0	
Squalus acanthias[29,30]	Piked dogfish	800	240	240	100	10
Scyliorhinus canicula[31]	Small-spotted catshark	960	238	361	124	
Hemiscyllium plagiosum[22]	Whitespotted bambooshark	797	249	225	248	
Leucoraja erinacea[32]	Little skate	967	179.8	208.9		
Hydrolagus colliei[33]	Spotted ratfish	820	162	268		
Freshwater[9]		38	3.0	3.7	Unknown	
Dasyatis sabina[34]	Atlantic stingray	53	10	8	20	

shorter and less developed than those of the stenohaline sharks. This reduction in length is likely related to the absence of a need to retain urea in their blood to adjust their osmolarity to that of the surrounding environment as in marine cartilaginous fishes. These rays have low blood concentrations of urea, 1.1 to 1.3 mmol·l⁻¹ in contrast to the 314 to 394 mmol·l⁻¹ in stenohaline sharks and rays. The urea concentrations in the blood of the freshwater rays are also considerably lower than those of euryhaline sharks and rays, which enter rivers and lakes for part of the year. The bull sharks that enter Lake Nicaragua have blood concentrations of urea ranging from 132 to 169 mmol·l⁻¹, and Atlantic rays inhabiting Lake Jesup have a concentration of 169 mmol·l⁻¹. The tubule in a nephron of a freshwater ray passes through only one loop in a sinus zone and another loop in the bundle zone. Furthermore, the tubules in the latter zone are not held tightly together within a peritubular sheath. This simpler nephron is consistent with their having little need to reabsorb urea into the blood.

Gills

There are ten gills in most sharks and rays, five on either side of the pharynx. These consist of cartilaginous rods with gill rays radiating outward from each arch (fig. 4.6a). Each arch is separated from the next by sheets of connective tissue that make up the interbranchial septum. Located on the gill rays are the gill filaments, which are supplied with blood through an afferent filamental

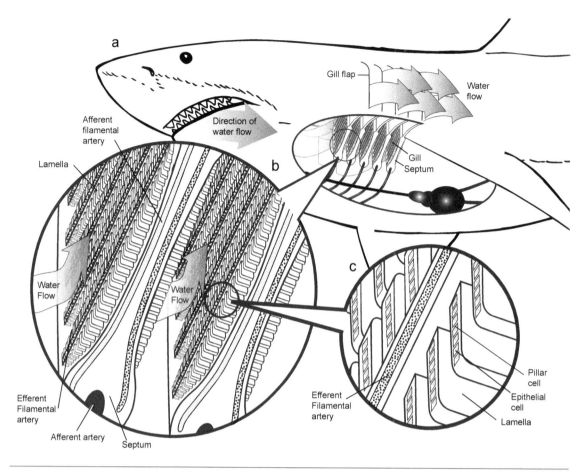

Fig. 4.6 (a) Diagram of a shark, showing pathway of water through mouth, across the gills, and out the gill slits. (a) Magnified section of gill arch, showing the septum, filamental arteries, filaments, and their lamellae. (c) Close-up of filament showing the epithelial and pillar cells that make up a lamella. The gill resembles in a crude sense the radiator of a car, but instead of cooling it optimizes the diffusion of oxygen into the bloodstream.

artery on one side and drained of blood through an efferent filamental artery on the other (fig. 4.6b). Within the filaments are afferent lamellar arterioles, which perfuse lamellae, minute protrusions that run perpendicular to the filament's longitudinal axis, and efferent lamellar arterioles that collect the blood after perfusion. Each lamella is composed of two flat sheets of epithelial cells held apart by intermittent pillar cells (fig. 4.6c). Blood collects in the space between the epithelial cells and pillar cells. It is here, where there is a short diffusion distance, that oxygen is absorbed into the blood from the water passing by the gill. Water also is likely absorbed within the blood at this site. However, the epithelial cells in the sharks and rays are spaced together so closely that the two membranes are likely impermeable to ions.[40] Hence, the gills of the stenohaline sharks and rays are thought not to function in the uptake of sodium and chloride ions, but rather that this is done by the rectal gland. However, there is some evidence that the gills of the bull shark and freshwater rays are involved in the uptake of these ions during acid-base regulation.[16]

SUMMARY

It should be evident that chondrichthyan fishes have successfully adapted to the ionic composition and thermal nature of the world's oceans. They are able to control the influx of ions and efflux of water using their rectal gland, kidney, and gills. They remain nearly isosmotic with saltwater because in addition to sodium and chloride ions their tissues contain two other solutes in large concentrations, urea and TMAO. These two compounds occur in a concentration ratio of 2:1—at this mixture, the latter compound nullifies the toxicity of the former. Yet the concentrations of the sodium and chloride ions in sharks, rays, and chimaeras are less than those of the two ions in saltwater. Hence, they must be excreted by the salt gland. Urea and TMAO are retained within the body in their high concentrations by the countercurrent absorption mechanism of the kidneys. The rectal glands in the freshwater-tolerant bull shark and freshwater stingray are very small. This is consistent with there being little need to excrete sodium and chloride ions because their osmolarities are greater than those of the surrounding freshwater. These species need rather to counteract the osmotic gain of water and do this by producing copious urine. As a group, the sharks, rays, and chimaeras are most suited for life in saltwater.

* * *

KEY TO COMMON AND SCIENTIFIC NAMES

Atlantic stingray = *Dasyatis sabina*; bull shark = *Carcharhinus leucas*; Haller's round ray = *Urolophus halleri*; little skate = *Leucoraja erinacea*; piked dogfish = *Squalus acanthias*; rabbit fish = *Chimaera monstrosa*; small-spotted catshark = *Scyliorhinus canicula*; spotted ratfish = *Hydrolagus colliei*; white shark = *Carcharodon carcharias*; whitespotted bambooshark = *Hemiscyllium plagiosum*; winter skate = *Leucoraja ocellata*.

LITERATURE CITED

1. Thomerson, 1977; 2. Schulz and Mallin, 1975; 3. Fernicola, 2001; 4. Thorson, 1972; 5. Thorson *et al.*, 1967; 6. Pang *et al.*, 1977; 7. Burger and Hess, 1960; 8. Robertson, 1975; 9. Piermarini and Evans, 1998; 10. Robertson, 1976; 11. Yancey and Somero, 1980; 12. Altringham *et al.*, 1982; 13. Urist, 1962; 14. Griffith *et al.*, 1973; 15. Wood *et al.*, 2002; 16. Evans *et al.*, 2004; 17. Olson, 1999; 18. Bonting, 1966; 19. Shuttleworth, 1988; 20.

1. Why do sharks, rays, and chimaeras possess specialized organs such as the salt gland, kidney, and gills to maintain their inner ionic composition?

2. How do they maintain the proper functioning of their enzymatic reactions to counteract the denaturing effect of urea on intracellular proteins?

3. Only a few sharks and rays inhabit freshwaters. Why do you think this is so? Perhaps this has less to do with osmoregulation than with the limited value of electroreception in this nonconductive medium (see chapter 9).

Burger, 1972; 21; Thorson *et al.*, 1978; 22. Wong and Chan, 1977; 23. Payan and Maetz, 1970; 24. Beit, 1977; 25. Holt and Idler, 1975; 26. Lacy and Reale, 1999; 27. Hentschel, 1988; 28. Lacy *et al.*, 1985; 29. Burger, 1967; 30. Shannon, 1940; 31. Henderson *et al.*, 1988; 32. Stolte *et al.*, 1977; 33. Read, 1971; 34. Janech and Piermarini, 2002; 35. Friedman and Hebert, 1990; 36. Antkowiak and Boylan, 1974; 37. Green, 1986; 38. Ghouse *et al.*, 1968; 39. Nash, 1931; 40. Wilson and Laurent, 2002.

RECOMMENDED FURTHER READING

Carrier, J. C., M. R. Heithaus, and J. A. Musick. 2009. Sharks and Their Relatives: Physiological Adaptations, Behavior, Ecology, Conservation, and Management. CRC Press, Boca Raton.

Evans, D. H., P. M. Piermarini, and K. P. Choe. 2004. Homeostasis, osmoregulation, pH regulation, and nitrogen excretion. Pp. 247–268 *in* Carrier, J. C., J. A. Musick, and M. R. Heithaus (Eds.), Biology of Sharks and Their Relatives. CRC Press, Boca Raton.

Piermarini, P. M., and D. H. Evans. 1998. Osmoregulation of the Atlantic stingray (*Dasyatis sabina*) from the freshwater Lake Jesup of the St. Johns River, Florida. *Physiol. Zool.*, 71: 553–560.

Robertson, J. D. 1976. Chemical composition of the body fluids and muscle of the hagfish *Myxine glutinosa* and the rabbit-fish *Chimaera monstrosa*. *Jour Zool.*, 178: 261–277.

Thorson, T. B., C. M. Cowan, D. E. Watson. 1967. Body fluid solutes of juveniles and adults of the euryhaline bull shark *Carcharhinus leucas* from freshwater and saline environments. *Physiol. Zool.*, 46: 29–42.

Thorson, T. B., R. M. Wotton, and T. A. Gorgi. 1978. Rectal gland of freshwater stingrays, *Potamotrygon* spp. (Chondrichthyes: Potamotrygonidae). *Bio. Bull.*, 154: 508–516.

Yancey, P. H., and G. N. Somero. 1980. Methylamine osmoregulatory solutes of elasmobranch fishes counteract urea inhibition of enzymes. *Jour. Exp. Zool.*, 212: 205–213.

CHAPTER 5

Warming of the Body

I remember my surprise when putting my hand on the back of a white shark, which was well out of the water as it swam beside my small boat at the Farallon Islands, in discovering that the body of the shark was really warm—much warmer than the cold water in which it was swimming. I placed my hand on the shark's back to find a soft muscular spot to insert the barb and line leading to an ultrasonic transmitter at the end of my pole spear. My hand would become numb in less than a minute or two when immersed in the cold waters off central California. The white shark can tolerate the cold water in temperate and subpolar latitudes because of its ability to keep warm, while its relative, the mako shark, is able to keep warm while swimming downward into the cold water present at great depths in the temperate and semitropical latitudes. It is likely that the white shark evolved this capability to exist in the cold waters where its prey, seals and sea lions, are common. Of course, their bodies are girdled by a thick layer of fat, which can be up to a half of the body mass in juvenile northern elephant seals; this insulates them from the cold. Sharks of the order Lamniformes elevate their body temperature by other means that will be described in this chapter.

The body temperature of the majority of cartilaginous fishes is the same as that of the surrounding marine environ-

The shortfin mako shark possesses a muscular rete that keeps its body warm when it descends into the cool deep waters of the ocean.

ment. This is to be expected of species that live in a liquid that has a higher specific heat than a gas such as air. Heat is lost more rapidly from a warmer body to a surrounding aqueous environment than from an equally warm body to the gaseous mixture of air because the former has many times the capacity of that latter to store heat.[1] Heat, which is produced during aerobic metabolism, is lost from the body of a cartilaginous fish when blood passes through its gills. As the cold water is drawn through the mouth, past the gills, and out the gill slits, heat diffuses outward across the gills to warm the passing water, and oxygen diffuses inward from the water during the process of respiration. Furthermore, heat is lost through the muscles and skin to a colder environment. Thus, it is not surprising that most chondrichthyans are ectothermic, or cold-bodied.

Despite these challenging obstacles to heat retention, members of two families of sharks, the Lamnidae and Alopiidae,[2] and one family of rays, the Mobulidae,[3, 4] have evolved anatomical and physiological adaptations that enable them to keep parts of their body warmer than the external environment. Endothermy, or the ability to maintain a temperature above that of the surrounding environment, is energetically demanding.[2] Heat is released during the oxidation of carbohydrates, fats, and proteins while creating the energy-rich adenosine triphosphate (ATP) molecules within red muscles with high densities of mitochondria. Specialized retes, bundles of blood vessels, retain this heat within parts of the body such as the muscles, stomach, brain, and eyes. There are benefits to being warm-bodied. Warm muscles contract faster and with greater force, and recover more quickly than cold muscles during burst swimming.[1] Prey items in a warmed stomach are digested more rapidly because of the increase in the enzymatic activity associated with an elevated temperature.[5] Finally, warming of the brain and eyes reduces the cooling of the cranial cavity when the elasmobranch descends into colder deepwaters in the ocean or migrates into cooler temperate waters. The warming of these organs ensures uniform and rapid processing of information from the central nervous system, enabling endothermic sharks to be more active and responsive filter feeders or predators.[6] Overall, the ability to remain warm enables these species to migrate from the warm tropical waters into colder temperate and boreal waters at higher latitudes. This chapter will focus on understanding the physiological processes and anatomical adaptations that enable some sharks and rays to elevate the temperature of key body parts over the ambient temperature of the ocean.

The temperature of the tissues and organs within the body of most cartilaginous fishes is the same as the water surrounding them. This is true for all of the chimaeras, the members of eight of the nine orders of sharks, and the members of three of the four orders of rays. The geographic extent of the movements of these sharks and rays is limited by their inability to warm their bodies. This is why many of them remain year round in warm tropical waters. Sharks such as the sandbar (*Carcharhinus plumbeus*) spend winter and spring in the warm waters off the southeastern coast of North America but migrate into the cooler temperate latitudes during summer. Here they inhabit the warm waters along the northeastern coast of North America until the fall, when they migrate southward toward the subtropics.[7] Similarly, rays such as the large, planktivorous giant manta (*Manta birostris*) and smaller smoothtail mobula (*Mobula thurstoni*) and Chilean devil ray (*Mobula tarapacana*), spend winter and spring in the warm waters of the eastern tropical Pacific Ocean off Central and South America. They then migrate in massive schools northward into the Gulf of California as the cold waters of the California Current withdraw from the gulf and are replaced by the warm waters from the Eastern Tropical Countercurrent. Only a few cold-bodied species such as the sluggish Greenland shark (*Somniosus microcephalus*) inhabit the cold waters of the polar regions of the North Atlantic Ocean. Its sister species, the Pacific sleeper shark (*Somniosus pacificus*), lives off the northwestern coast of North America in the cold waters of the Gulf of Alaska. Those cartilaginous fishes that are unable to regulate their body temperature either inhabit the tropics year round or migrate during summer into the temperate latitudes, where the seasonally warm waters elevate the temperature of their bodies and enable them to actively chase and capture prey. Frigid waters likely slow down their bodily functions, such as nervous signal transmissions, muscle contractions, and digestion. This probably results in a shift from an active predatory role to that of a scavenger or an ambusher predator that can capture its prey only once it comes within striking distance.

However, some ectothermic sharks may perform diel vertical migrations to conserve energy and increase bioenergetic efficiency.[8] These sharks rest in the cooler, deeper waters when not feeding and move upward toward the surface into warmer waters, where they can search for their prey while expending less energy. Such is the case with the small-spotted catshark (*Scyliorhinus canicula*) at Lough Hyne, a region of the Atlantic Ocean off the coast of Ireland. Male catsharks migrate up the submarine slopes from deeper, colder waters that they occupy during the day to the warmer, shallow waters above the thermocline where they spend the night feeding on small fish. The activity levels of individuals tracked on a twenty-four-hour basis were higher during the night than during the day, indicative of their nighttime foraging and their daytime return to the cooler depths to digest their meals and rest. A theoretical model of the energy budget of these small-spotted catsharks indicated that they would lower their daily energy consumption by just over 4% if they adopted this hunt warm–rest cool strategy.

The ectothermic sharks and rays utilize a central system for distributing their blood.[9, 10] The main supply of blood for their muscles flows posteriorly in the interior of the body within the dorsal aorta along its body length just below the vertebral column (fig. 5.1a). This artery, located immediately below the vertebral column above the postcardial vein, extends from behind the otic capsule to the caudal peduncle. Blood returns to the heart flowing anteriorly within this vein. From these two blood vessels deep within the body of the ectothermic shark, arteries and veins radiate outward to supply blood to and drain blood from the muscles, the majority of which are of white tissue (fig. 5.1b). These muscles can contract and relax only for short periods of time during burst swimming before the glucose stored within them is used up and lactic acid buildup changes the pH of the muscle and inhibits anaerobic metabolism.

WARM-BODIED CARTILAGINOUS FISHES

It is a physiological challenge for any animal to elevate its temperature in cold water. Heat is constantly lost not only through conduction from the skin to the surrounding water but also when cold water is drawn over the gills during respiration to extract oxygen for metabolism.[10] The acquisition of oxygen is essential for metabolism. This energy is used to contract muscles during locomotion. Terrestrial animals such as birds and mammals obtain their oxygen by breathing air, which is rich in oxygen and has a low capacity for absorbing heat. Marine animals, on the other hand, must extract oxygen from water, which contains oxygen at a much lower concentration. On the other hand, water's capacity for absorbing heat is roughly four times that of air. The temperature of the blood flowing into the gills is slightly higher than that of the surrounding water because of warming by the heat released during metabolism. However, heat diffuses faster than oxygen molecules, and hence, the temperature of the blood equilibrates with that of the water by the time the blood in the gills is oxygenated.

The amount of extractable oxygen in blood is small, and thus there is a limited supply of oxygen that becomes available for metabolism. Therefore, only a small amount of heat is released during this process—enough to heat the body of the fish roughly a degree Celsius. This small rise in temperature

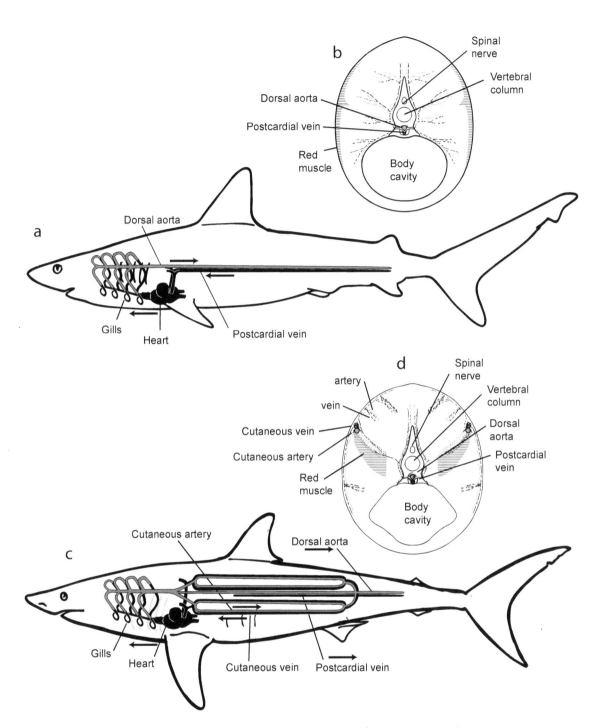

Fig. 5.1 The vascular system of an ectothermic shark viewed (a) from side and (b) from a cross-section, and an endothermic shark viewed (c) from side and (d) from a cross-section. The main vessels of ectothermic sharks run along the vertebral column with the small blood vessels radiating outward. The central vessels in endothermic sharks are smaller and most of the blood flows through a pair of large subcutaneous arteries and veins. Note that the cutaneous arteries and veins are visible from both sides of the body of the endothermic shark.

disappears when the blood next passes through the gills, and heat is again lost from the blood to the cooler surrounding waters. This cyclic process prevents the accumulation of heat within the body of a cartilaginous fish. An increase in its activity does not elevate its temperature further because any increase in the attendant oxygen consumption results in a simultaneous loss of heat during respiration. Indeed, it is impossible for a cartilaginous fish to elevate its temperature significantly by energy production alone.

Muscular, Visceral, and Orbital Retes

Despite these physiological obstacles, members of one order of sharks, the Lamniformes, are able to maintain the temperature of parts of their bodies elevated above that the external environment.[10] This order includes the shortfin mako (*Isurus oxyrinchus*) and longfin mako (*Isurus paucas*), which inhabit mainly the warm temperate, subtropical, and tropical latitudes; the white shark (*Carcharodon carcharias*), which lives mainly in the temperate latitudes; and salmon shark (*Lamna ditropis*) and porbeagle shark (*Lamna nasus*), which mainly inhabit the boreal latitudes. This order also includes the thresher (*Alopias vulpinus*), bigeye thresher (*Alopias superciliosus*), and pelagic thresher (*Alopias pelagicus*). Members of two genera of rays, *Manta* and *Mobula*, possess this ability too. Common to all of these species is their propensity to make long-distance migrations into waters greatly different in temperature. These species likely tolerate these fluctuating temperatures by regulating their own body temperatures.

These species possess a *rete mirabile* (Latin for wonderful net) in their circulatory system.[1] This structure provides a barrier against the loss of heat produced during metabolism. The rete is supplied with cool, oxygen-rich blood from the gills by a thick-walled, lateral cutaneous artery located just under the skin near the midline of the flank (fig. 5.1c). Many retial arterioles extend from the artery inward leading to the large mass of red muscle located near the vertebral column (fig. 5.1d). They are intermingled among an equal number of thin-walled retial venous vessels, carrying the oxygen-depleted blood from the red muscles back to the lateral cutaneous vein just under the skin. This blood vessel returns the oxygen-deficient blood to the gills. The intermingling of arterial and venous vessels leading to interior red muscle is more apparent in a generalized diagram of a muscular rete (fig. 5.2).[11] This internal network of closely associated arteries and veins is called a countercurrent heat exchanger. Blood is said to flow in a countercurrent manner because it flows through arteries in one direction into the muscles and through veins in the opposite direction out of the muscles. The directions of flow are indicated by arrows in fig. 5.2. The arrangement of blood vessels is termed a heat exchanger because the

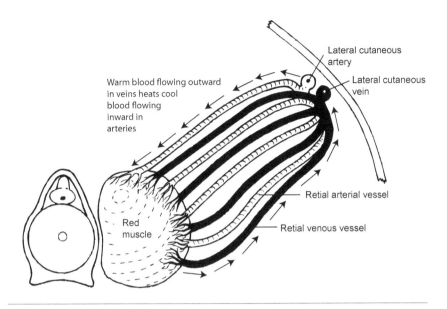

Fig. 5.2 A generalized diagram of a muscular rete with arterioles and venules leading from the lateral cutaneous artery and vein, respectively, to the red muscle located near the vertebral column. Arrows indicate the flow of blood.

heat produced from metabolism of proteins or lipids stored within the muscle tissues diffuses through the thin walls of veins carrying blood to the exterior surface of the fish and again through the thin walls of the arteries to warm the cool blood flowing toward the interior muscles of the fish. The temperature of the arterial blood arriving at the interior muscle mass is roughly similar to that of the warm core of the body.

The paired arteries and veins bring blood to and from slabs of vascular tissue on upper and lower surfaces of myotomes composed of red muscle fibers. These myotomes are capable of prolonged cycles of contraction and relaxation, and are used during continuous and sustained swimming. Their energy comes from the aerobic metabolism (oxygen-requiring metabolism) of glucose and lipids, which requires warm temperatures in addition to oxygen, which is carried by the hemoglobin in red blood cells. The warming of a muscle fiber by ten degrees Celsius results in a threefold increase in the ability to contract and relax, with no reduction in the power exerted during each cycle. This is likely accomplished by speeding up the delivery of oxygen to muscle mitochondria.[12] The warming of the muscles enables these species to swim continuously and rapidly through a viscous medium such as seawater. Continuous locomotion in water is more challenging than in air because the former medium is more dense and viscous than the latter. Hence, the endothermic sharks are fusiform in shape, and their pectoral, dorsal, and anal fins fit into grooves in the body to reduce drag and thus permit efficient long-distance swimming.

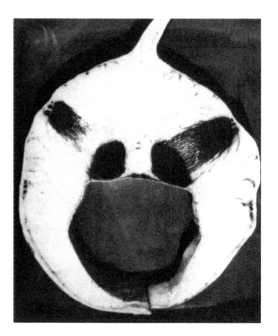
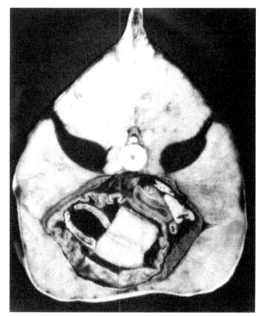
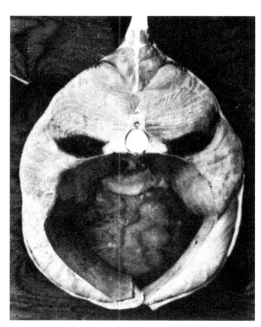

Fig. 5.3 Cross-sections showing the dark red muscle in the interior of body of three members of the family Lamnidae, the (a) porbeagle, (b) shortfin mako, and (c) white sharks.

The locations of the rete and red muscle are slightly different in the bodies of the different species of lamniform sharks; this affects their thermoregulatory capacity and swimming behavior.[13] In the porbeagle shark, the red muscle lies directly against the vertebral column (fig. 5.3a). Yet the red muscle is not attached to the vertebrae along much of its length and is free to slide during alternating cycles of contraction and relaxation. The red muscle is slightly separated from the vertebrae with some intervening white muscle fibers in the shortfin mako (fig. 5.3b). The muscle is not as free to slide back and forth as the red muscle in the porbeagle shark. The rete of the mako is composed of a homogenous mass of vascular tissue containing vessels crowded together. The red muscle in the white shark is separated with fewer intervening white muscle fibers from the vertebral column than in the mako shark (fig. 5.3c).

The distribution of the red muscle along the length of the body also differs among the three species. The red muscle is concentrated at the thickest section of the body in the porbeagle shark, while the muscle is more evenly distributed along the length of the body in the white shark. The location of the red muscle in the mako shark is intermediate between the two. The distribution of red muscle in these three species may provide insight into their swimming modes. The uncoupling of the muscle from the backbone of the mako shark would result in stiff swimming movements by the torso, in which only the caudal fin is moved back and forth. The extension of muscle farther toward the posterior of the white shark, and its attachment along part of its length, suggests that the white shark swims more sinuously than the mako shark. The porbeagle may swim in an intermediate manner, more stiffly than the mako, but less sinuously than the white shark.

The cutaneous blood vessels in the shortfin mako shark supply a teardrop-shaped rete,[13] which keeps a large slab of red muscle warm in the interior of the body. The distribution of temperature within the muscle mass has been measured immediately after the capture of an individual by inserting a thermistor, or temperature sensor, at the end of a probe various distances into the musculature. The distribution of temperature, across and along the length of the shortfin mako, can be described by superimposing contours of temperature on longitudinal and cross-sections of the shark's body.[10] The warmest red muscle is just behind the pectoral fins (see longitudinal section in fig. 5.4a) and lies just to the side of the vertebral column and peritoneum (see cross-section of body in fig. 5.4b). The temperature of the central muscle mass was 27.2° C for a shark that had previously been swimming in water of 21° C. The temperature of the internal muscle was thus elevated 6.2° C. The muscle mass is cooler closer to the surface of the body with a minimum temperature of 22° C measured under the skin of the shark, where the paired arteries and veins are located. The warmth of the muscles also decreases away from the midsection of the shark in the anterior and posterior directions. The muscle mass near the gill slits and near the caudal peduncle was 22° C, only one degree higher than the external aqueous environment.

Carey and his colleagues recorded the muscle temperature, the ambient water temperature, and the swimming depth of a free-swimming white shark that swam for three-and-a-half days from Montauk Point, New York, to Hud-

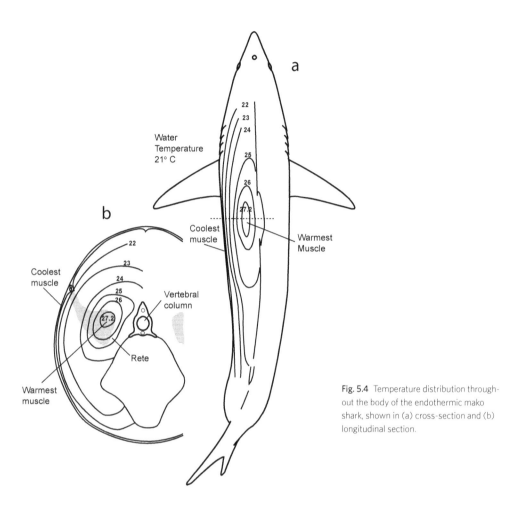

Water
Temperature
21° C

a

22
23
24
25
26
27.2

Coolest
muscle

Warmest
Muscle

b

22
23
24
25
26
27.2

Coolest
muscle

Warmest
muscle

Vertebral
column

Rete

Fig. 5.4 Temperature distribution throughout the body of the endothermic mako shark, shown in (a) cross-section and (b) longitudinal section.

son Canyon.[14] The shark made a number of ascents to the surface and descents to the bottom, but during the majority of the time swam within the thermocline. This behavior is apparent from small zig-zag-like excursions up and down across that region of the depth profile characterized by closely spaced isotherms, or temperature contours (fig. 5.5a). During this time, the muscle temperature of the shark exceeded that of the surrounding water as much as 5° C (fig. 5.5b). Note that the upper line, indicating body temperature, is consistently five degrees above the lower line, which shows the temperature of the surrounding water. An even greater elevation of muscle temperature above water 11.0° C has been recorded from the salmon shark.[15]

The lamniform sharks have a second rete, which warms their stomach. This may speed digestion by accelerating the activity of the enzymes that break down food within the stomach. The structure of the visceral rete can best be understood from a diagram in which the body wall and pectoral girdle are cut away, the ventricle has been lifted upward, and the two atria have been

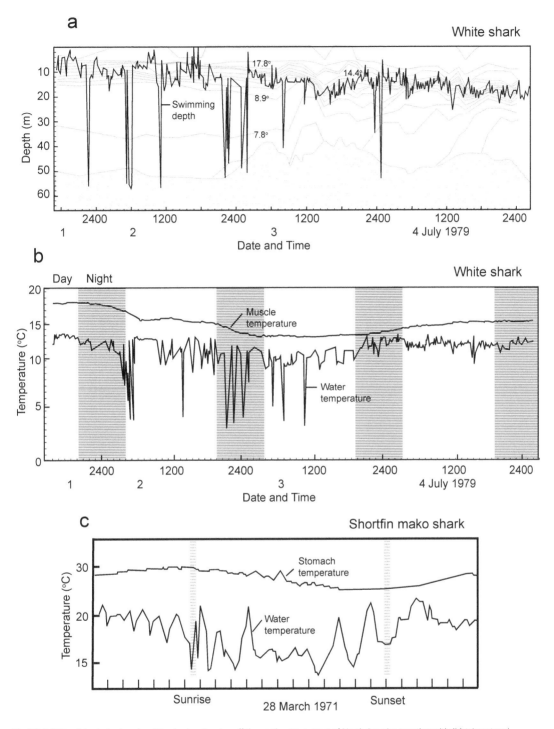

Fig. 5.5 (a) The diving behavior of a white shark swimming off the northeastern coast of North America together with (b) telemetered measurements of its muscle and surrounding water temperatures. Note that the shark's muscle temperature was always higher than that of the surrounding water temperature as it swam above the thermocline, indicated by the closely spaced isothermal contours. (c) Telemetered stomach and surrounding water temperature of shortfin mako shark over a twenty-four-hour period. Note that the stomach temperature exceeded that of the ambient water temperature yet unlike the muscle temperature remained relatively constant throughout this period.

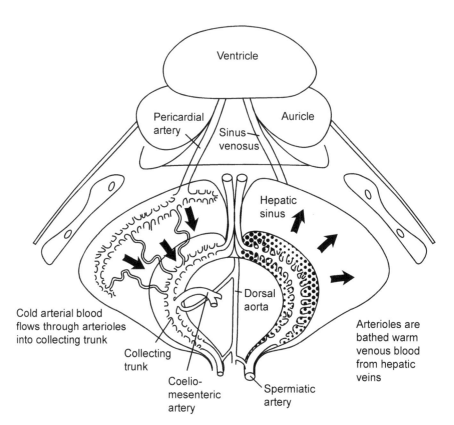

Fig. 5.6 Anatomy of the visceral rete that warms the stomach of the mako shark.

Labels in figure:
Ventricle
Pericardial artery
Sinus venosus
Auricle
Hepatic sinus
Dorsal aorta
Cold arterial blood flows through arterioles into collecting trunk
Arterioles are bathed warm venous blood from hepatic veins
Collecting trunk
Coelio-mesenteric artery
Spermiatic artery

pushed aside (fig. 5.6).[5] The rete is close to the esophagus. Two pericardial arteries enter the rete from above and branch into many small interconnecting arterioles that form the sponge-like mass of vessels filling the sinus that eventually join at its bottom with the collecting trunk which leads to the coeliomesenteric artery. Cold arterial blood (see arrows on left of diagram) entering the pericardial arteries passes through the many arterioles, which are bathed in warm venous blood (see arrows on right) flowing into the sinus in the reverse direction from the hepatic veins. Each collecting trunk, through which the warmed arterial blood passes from the arterioles, leads to a coelio-mesenteric and spermiatic artery. These are attached to the mesentery in the peritoneal cavity that surrounds the stomach and intestines.

Temperatures were first recorded by a thermistor implanted in the stomach of a female shortfin mako shark, caught and released near the boundary between the Gulf Stream and the continental shelf along the eastern coast of North America. The stomach of the female remained from 6 to 8° C warmer than the surrounding waters throughout five days that she was tracked (fig. 5.5c).[5] Unlike muscle temperature, the stomach temperature of the shortfin mako shark remains constant over time. The stomach temperatures of white sharks tracked off the South Farallon Islands, central California, exceeded the surrounding water temperatures by as much as 14.3° C and also remained relatively constant over time.[16] The stomachs of salmon sharks can be even warmer, ranging from 5 to 16° C above the temperature of the water in Prince William Sound on the coast of Alaska.[16]

The lamniform sharks have an orbital rete, located within the eye capsule, which warms the eyes and brain.[6, 17] The heating of the retina of the eye may speed the processing of visual information, thereby improving an individual's response to sudden visual stimuli in the cold, dimly lit marine environment. This vascular heater may also reduce severe temperature fluctuations, which may disrupt the uniform and rapid integrative processes in the central nervous system.[6] Two myliobatiform rays also have a cranial rete that heats the brain, but it warms the nares, not the eyes, and may enhance the processing of smells.[4]

The elevation of cranial temperatures requires a source of heat (aerobic metabolism) and a means for reducing heat loss to the environment (a countercurrent heat exchanger).[2] The main source of heat of the cranial rete in the endothermic sharks is the activity of the extraocular muscles. They are very red in color, indicating the presence of many oxidative fibers that generate heat at the same time that they produce energy in the form of ATPs. These muscles are disproportionately large relative to the extraoccular muscles of ecotothermic sharks.[15] The presence of so many red muscle fibers within extraoccular muscles that already have enough sarcoplasmic reticulum and proteins that cycle calcium for rapid muscle contraction implies that the eye muscle also has another thermogenic function.[6] Yet this muscle does not appear to produce enough heat to account for the elevated temperatures of the eye and brain. A small rete, located in the orbital cavity, provides a venous pathway from the red muscle, which is elevated 5° C above the background temperature, to the brain.

The small rete is located in the orbital cavity of the white shark and shortfin mako shark. Fig. 5.7a diagrams this orbital, or more specifically hyoidean, rete in the white shark; fig. 5.7b shows a ventral view of the rete, and fig 5.7c shows the uncoiled pseudobranchial artery and its two terminal branches.[18] The major blood vessels supplying the eye and brain with blood are the hyoidean efferent and pseudobranchial arteries. They are joined by the paired dorsal aortae, which are the end branches of the dorsal aorta. The hyoidean rete consists of many small arteries that arise from fusion of the hyoidean efferent artery and a branch of the dorsal aorta. The rete extends laterally from under the base of the eye socket through the orbital fenestra, and into the orbital cavity. The rete is more developed in the white shark than the shortfin mako shark. The pseudobranchial artery in both species is highly convoluted and surrounded by the warm blood of the venous sinus within the orbital cavity

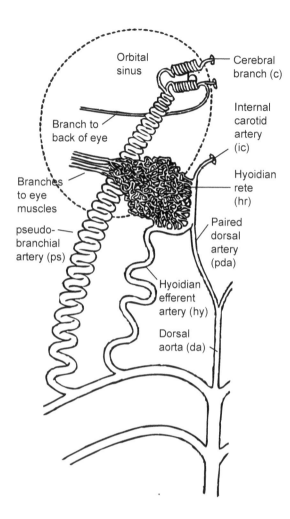

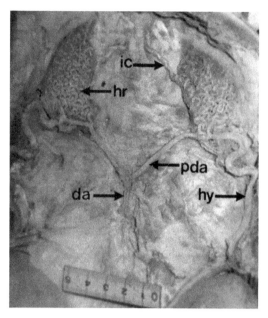

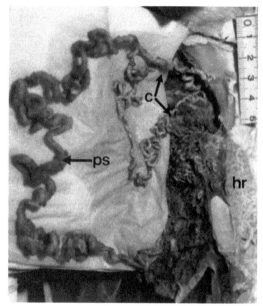

Fig. 5.7 (a) Anatomy of the orbital rete that warms the eyes and brain of white shark; (b) its hyoidean rete; (c) its pseudobranchial artery.

(see fig. 5.7c). As it proceeds through the orbit, the elaborate coilings and convolutions function to increase the surface area of the vessel. This results in the continuous warming of the blood as it passes into the two cerebral branches that supply blood to the brain.

This rete retains metabolic heat by transferring it from the warm venous blood to the cold arterial blood that both pass through the orbital sinus. As the cool arterial blood exiting the first gill arch passes though the orbital rete, it is warmed by the flow of blood in the reverse direction originating from the few veins flowing through the sinus. The large surface area of the many small and coiled arteries enhances the efficiency of heat transfer and enables the arterial blood to be warmed nearly to the temperature of the venous blood. This thermoregulatory process in the mako and white sharks results in an elevation of 3 to 6° C of the temperature difference between the eyes and brain and the surrounding seawater.[6]

The rete of the giant manta ray consists of a cerebral portion, which surrounds the brain, and a precerebral portion, which extends forward toward the ray's snout beyond its brain and to the side toward its two olfactory bulbs.[4] The rete appears as a mass of intertwined blood vessels, following the removal of the skin and connective tissue covering the braincase. The rete is located just above the mouth (see m in fig. 5.8a). The major arteries that give rise to the cerebral and precerebral retes are under the brain on the floor of the cranial cavity of the chondrocranium. They are apparent in a ventral view of this vascular plexus that covers the brain (fig. 5.8b). The arteries are shown in black, the veins in white. The profunda cerebri arteries are branches of the internal carotid arteries that enter the cranial cavity from either side of the telencephalon. An anterior communicating artery passes underneath the brain on the cranial shelf and connects the right and left internal carotid arteries. The profundae cerebri arteries converge in a posterior direction and join on the underside of the medulla to become the spinalis impar artery that continues on to the spinal cord. Branches from the profundae cerebri arteries supply blood to the fore- and midbrains of the manta ray. A mass of vessels extend upward and branch into arterioles that compose the caudal rete that exists on either side of the hindbrain. The mass of these vessels occupying the cranial cavity anterior to the brain is the precerebral rete, and these arterioles extend outward, gradually reducing in size until they terminate near the olfactory capsules. On each side of the head is a dorsal collector vein formed by the many tributaries passing forward from both the superficial and deep regions of the rete (see the many branching white veins). Many small venules join to form successively larger tributaries and eventually become the cerebral vein. They come in close contact with and run parallel to the many branches of the profundae cerebri arteries. A countercurrent system of heat exchange likely exists here, where heat passes from the warm venous blood carried in the dorsal conductor venules to the cool arterial blood in the profundae cerebri arterioles. Such a system requires a source of heat, which is likely the activity of eye muscles of the mobulid rays. This inference is based on the disproportionately large number of red fibers in these muscles.

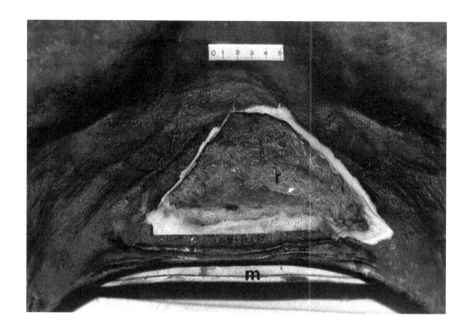

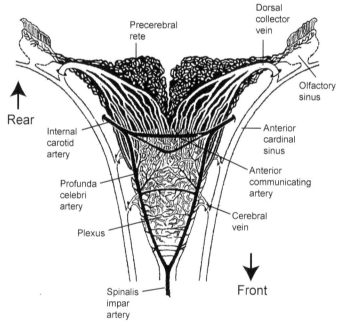

Fig. 5.8 (a) Top view of cranial rete in manta ray, revealing the plexus of blood vessels after the removal of skin and connective tissue. (b) Ventral view of the vascular plexus on the dorsal surface of the brain, the precranial rete, and the anterior plexus associated with the olfactory bulb.

Ectothermy versus Endothermy

The warm-bodied and cold-bodied sharks have very different lifestyles. The former live a constant fast, high-energy life whereas the latter have a slower approach to life punctuated by short bursts of activity.[20] So why are the majority of the chondrichthyan fishes ectothermic in nature? To start with, they have to expend relatively little energy on thermoregulation because their body temperatures remain close to that of the environment. This energy can alternatively be invested in growth and reproduction. Cold-bodied sharks require less food, and thus they can devote more time to courtship and other activities. Yet at the same time, there are costs to being cold-bodied. The small amount of red muscle in their bodies limits aerobic respiration, and thus they must swim slowly much of the time. They can only punctuate this inactivity by short bursts of swimming, during which they engage in social behavior and foraging. Before long, the product of anaerobic respiration, lactic acid, builds up in their muscles and causes fatigue.

For these reasons, ectothermic sharks may spend half the day either resting on the bottom or milling back and forth in a confined area before making a foraging migration during the other half of the day. Examples of species that rest on the bottom are the Port Jackson shark (*Heterodontus portusjacksoni*) and the whitetip reef shark (*Triaenodon obesus*). The former aggregate in small groups within rocky crevices during the day and disperse at night to search for benthic echinoderms and crustaceans in the cool, temperate waters of Southern Australia.[21] The latter rest in small groups within reef crevices or lie by themselves on sandy bottoms during daytime and actively forage for nocturnal fishes such as cardinal and parrot fishes at night in the tropics in both the eastern and western Pacific Ocean.[22] Other sharks such as the blacktail reef shark (*Carcharhinus amblyrhynchos*) and the scalloped hammerhead shark (*Sphyrna lewini*) congregate in large schools during daytime and disperse at nighttime to forage by themselves in the surrounding waters. Blacktip reef sharks form large aggregates at reefs during daytime in the protected, shallow waters within atolls in the western Pacific and move through channels leading them out of the circular atoll to forage in deeper water along the slopes of these coralline islands.[23] Scalloped hammerhead sharks form large schools during daytime within the Gulf of California and the eastern Pacific Ocean and disperse at nighttime from these schools to travel great distances into the surrounding oceanic waters to feed on fishes and cephalopods.[24] The social system adopted by these four species of sharks is classified as a central-place refuging system, common to birds and mammals that alternate periods of rest and foraging.[25] Male catsharks migrate along submarine slopes from deeper colder waters

The salmon shark has well-developed muscular retes. These enable it to maintain body temperatures as high as 21.2° C above the local water temperature.[19] However, the sarcoplasmic reticulum in this species' heart muscles also has an increased capacity to absorb calcium ions. Barbara Block, a professor at Stanford University based at the Hopkins Marine Laboratory in Monterey, and colleagues have found that this physiological process likely speeds up the excitation-contraction cycle of its muscles and thus improves heart function at low temperatures. The rate of uptake of Ca^{2+} ions in tissue taken from the atrium and ventricle of a salmon shark's heart (see solid and open circles in fig. 5.9) was observed to be higher than the uptake of the same ion in the atrium of the blue shark's heart (solid squares) over a 30° C range of temperatures from 5 to 35° C. The cardiac tissues of the salmon shark have more of the unique calcium-associated proteins, SERCA2 and RyR2, associated with the sacoplasmic reticulum than the blue shark. This is evident from the dark bands, indicating the separation of these two proteins with molecular weights of 115 kDa and 565 kDa, on an electrophoretic gell of tissue from the atrium (A) and ventricle (V) of the salmon shark and the absence of the same bands from the tissue from the atrium and ventricle of the blue shark (see insert in top left). The heart must vigorously pump red cells containing oxygen bound to hemoglobin to its muscle tissues in order to perform the burst swimming necessary to capture its prey, salmon, in the cold arctic waters. This medium, of course, has a very high specific heat, or ability to conduct heat away from the body of the shark. Both the muscular retes and enhanced delivery of calcium ions to the heart tissues have enabled the salmon shark to occupy an expanded geographic niche, in which it is a successful apex predator. It appears to be the species best adapted to migrating between subtropical and subarctic waters. Salmon sharks migrate widely, from the 4° C waters off the coast of the Aleutian Islands in the Gulf of Alaska to the 24° C waters west of the Baja Peninsula, a migration covering a latitudinal gradient of 30°.

they occupy during the day to the warm shallow waters above the thermocline, where they spend the night feeding on small fish.[8] Their activity levels are higher during nighttime than daytime, indicative of their foraging at this time and returning to the cooler depths during daytime to digest their meals and rest during the day. Even the blue shark (*Prionace glauca*), an ectotherm that moves over long distances, either swims at the surface or makes shallow dives during daytime and makes deep dives at nighttime to forage on midwater squid and fish.[26]

Warm-bodied sharks are capable of constant activity both during the day and night. Hence, shortfin mako sharks that have been tracked by boat for twenty-four hours make repeated dives both during daytime and nighttime to search for food in the waters of the Southern California Bight.[27, 28] Endothermic sharks such as the white and salmon sharks, carrying satellite transmitters that provide very accurate daily positions, have been shown to make

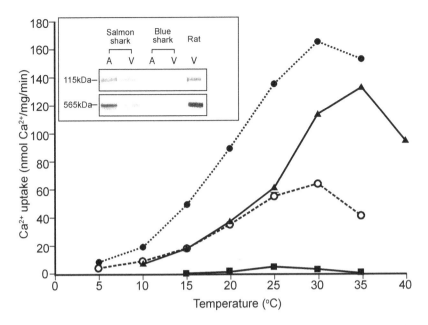

Fig. 5.9 The rate of uptake of Ca²⁺ over varying temperatures of tissue from the salmon shark's atrium and ventricle (black and open circles), the blue shark's atrium (black squares), and the rat's ventricle (black triangles). The insert in the upper left shows electrophoretic bands. The degree of darkness indicates the amounts of two proteins, SERCA and RyR, associated with calcium transfer to the sarcoplastic reticulum in tissues from the atrium (A) and ventricle (V) of the salmon shark, blue shark, and rat.

long-distance migrations in the North Pacific Ocean.[19, 29] The salmon sharks move between the cold waters of the boreal latitudes and the warm waters of the subtropics.[29] The endothermic warming of the muscles, viscera, eyes, and brain enables these sharks to be very active predators in cold-water environments. The few ectotherms that inhabit these waters are sluggish, and their method of prey capture is limited to being cryptic and ambushing their prey. Hence, a cost-benefit analysis of ectothermy relative to endothermy indicates that this metabolic dichotomy affects far more than just body temperature but also their behavior and ecology.

SUMMARY

As the cold water is drawn through the mouth, past the gills, and out the gill slits, heat diffuses outward across the gills. Furthermore, heat is lost through the muscles and skin to a colder environment. Thus, it is not surprising that most chondrichthyans are ectothermic, or cold-bodied. Despite these challenging obstacles to heat retention, members of two families of sharks, the

David Cairns of the Canadian Department of Fisheries and Oceans and his colleagues have argued that marine communities on a global latitudinal scale are governed by the temperature-dependent predatory and avoidance capabilities of predators and prey.[30] Birds and mammals are the major predators at high latitudes, while sharks and other large fishes are the dominant predators in the low latitudes. The ability of a marine predator to perform burst swimming to catch its prey and for its prey to swim in a similar manner to avoid being eaten is related to the temperature of the surrounding water. Burst-swimming ability increases with increasing water temperature in the ectotherms but is independent of temperature in the endotherms. If capture success depends on relative swimming speeds, at low temperatures ectothermic prey will be highly vulnerable to predation by endotherms. Conversely, high temperatures will increase the vulnerability of endothermic prey to ectothermic predators.

This prediction is supported by the greater abundance of endothermic pursuit-diving predators such as the penguins, auks, and some cormorants as well as seals, sea lions, and walruses in the cold temperate to arctic latitudes (where the summer surface temperatures are cooler than 15° C) than in the warmer temperate and tropical latitudes. This is apparent in fig. 5.10, where the preponderance of endothermic species is indicated by dark fills in the first two rows. The endothermic predators are ubiq-

uitous at 45° N and higher in latitude where the summer temperatures are 15° C in the cold temperate latitudes and 5° C in the Arctic waters. However, these same predators are absent at lower latitudes due to the difficulties they encounter in capturing prey, which are more active in the warmer waters and better able to escape without being eaten. The ectothermic fishes and sharks as well as the partially endothermic tunas and lamniform sharks, whose swimming performances are enhanced by increasing water temperature, are more successful feeding on ectothermic prey in the lower latitudes near the equator. According to the researchers, partial endotherms are fishes that can raise their temperature above the ambient by a certain offset but are not able to maintain a constant elevated temperature as marine mammals. In the figure, the partial endothermy of the tunas and lamniform sharks is indicated by their white and dark fills in the lower two rows, corresponding to latitudes below 35° N; these zones are warm temperate and tropical, with temperatures ranging from 15° to 25° C. The baleen whales are only seasonal visitors to these waters (as indicated by the two asterisks). This is an attempt to explain why pursuit-diving birds and pinnipeds are largely absent from these waters despite high levels of primary productivity and abundant fisheries that indicate there is ample local food supplies to support them—an incongruity that has puzzled marine biologists for many years.

Lamnidae and Alopiidae, and one family of rays, the Mobulidae, have evolved anatomical and physiological adaptations that enable them to maintain parts of their body warmer than the external environment. These species possess a rete mirabile that provides a barrier against the loss of metabolic heat. This internal network of closely associated arteries and veins is a countercurrent heat exchanger because the metabolic heat diffuses through the thin walls of veins carrying blood to the exterior surface of the fish and again through the thin walls of the arteries to warm the cool blood flowing toward the interior muscles of the fish. The lamniform sharks have two more retes, one that warms their stomach and another that warms the eyes and brain. The former may speed di-

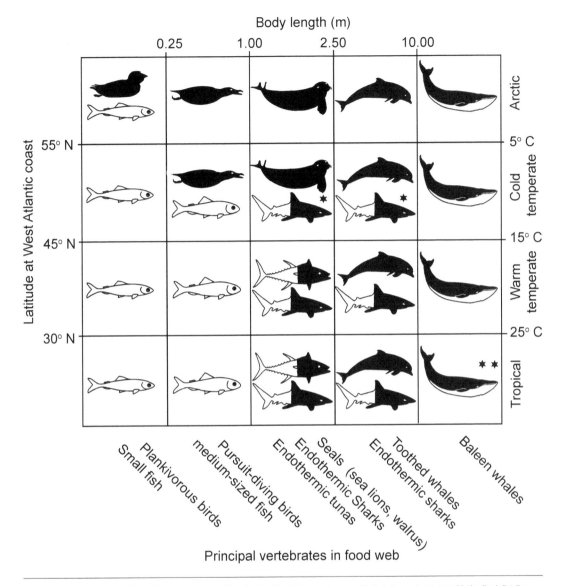

Fig. 5.10 The dominant vertebrates in the food webs of the North Atlantic Ocean shown with their thermal status and latitudinal distributions. The white symbols denote the ectotherms, the black symbols the endotherms, while the white-and-black symbols refer to partial endotherms that elevate their temperatures above the ambient water temperature but do not maintain a constant higher temperature. A single asterisk indicates that the group is at the northernmost extent of its range; double asterisks denote a group that do not feed in warm waters, but use them as a wintering ground for reproduction.

gestion by accelerating the activity of the enzymes that break down food in the stomach. The latter may speed the processing of visual information, thereby improving an individual's response to sudden visual stimuli in the cold, dimly lit marine environment. Two myliobatiform rays also have a cranial rete that heats the brain, but it warms the nares not the eyes, and may enhance the processing of smells. Hence, the warm-bodied and cold-bodied sharks have very different lifestyles. The former have a constant fast high-energy lifestyle while the latter have a slower approach to life punctuated by short bursts of activity.

⁂ ⁂ ⁂

KEY TO COMMON AND SCIENTIFIC NAMES

Bigeye thresher shark = *Alopias superciliosus*; blacktail reef shark = *Carcharhinus ambly-rhynchos*; blue shark = *Prionace glauca*; Chilean devil ray = *Mobula tarapacana*; Greenland shark = *Somniosus microcephalus*; giant manta ray = *Manta birostris*; longfin mako shark = *Isurus paucas*; Pacific sleeper shark = *Somniosus pacificus*; pelagic thresher shark = *Alopias pelagicus*; porbeagle shark = *Lamna nasus*; Port Jackson shark = *Heterodontus portusjacksoni*; salmon shark = *Lamna ditropis*; sandbar shark = *Carcharhinus plumbeus*; scalloped hammerhead shark = *Sphyrna lewini*; shortfin mako shark = *Isurus oxyrinchus*; small-spotted catshark = *Scyliorhinus canicula*; smoothtail mobula = *Mobula thurstoni*; thresher shark = *Alopias vulpinus*; white shark = *Carcharodon carcharias*; whitetip reef shark = *Triaenodon obesus*.

LITERATURE CITED

1. Carey, 1982; 2. Block and Finnerty, 1994; 3. Alexander, 1995; 4. Alexander, 1996; 5. Carey *et al.*, 1981; 6. Block and Carey, 1983; 7. Springer, 1960; 8. Sims *et al.*, 2006; 9. Muñoz-Chápuli, 1999; 10. Carey, 1973; 11. Satchell, 1999; 12. Stevens and Carey, 1981; 13. Carey *et al.*, 1985; 14. Carey *et al.*, 1982; 15. Rhodes and Smith, 1983; 16. Goldman, 1997; 17. Wolf *et al.*, 1988; 18. Alexander, 1998; 19. Weng *et al.*, 2005; 20. Randall *et al.*, 2002; 21. McLoughlin and O'Gower, 1971; 22. Randall, 1977; 23. Johnson, 1978; 24. Klimley, 1993; 25. Klimley and Nelson, 1984; 26. Carey and Scharold, 1990; 27. Klimley et al., 2002; 28. Sepulveda *et al.*, 2004; 29. Block *et al.*, 2011; 30. Cairns *et al.*, 2008.

RECOMMENDED FURTHER READING

Alexander, R. L. 1996. Evidence of brain-warming in the mobulid rays, *Mobula tarapacana* and *Manta birostris* (Chondrichthyes: Elasmobranchii: Batoidea; Myliobatiformes). *Zool. Jour. Linnean Soc.*, 118: 151–164.

Block, B. A., and F. G. Carey. 1983. Warm brain and eye temperatures in sharks. *Jour. Comp. Physiol. B*, 156: 229–236.

Cairns, D. K., A. J. Gaston, and F. Huettmann. 2008. Endothermy, ectothermy and the global structure of marine vertebrate communities. *Mar. Ecol. Progr. Ser.*, 356: 239–250.

Carey, F. G., J. G. Casey, H. L. Pratt, D. Urquhart, and J. E. McCosker. 1985. Temperature, heat production and heat exchange in lamnid sharks. *Memoirs, South. Calif. Acad. Sci.*, 9: 92–108.

Discussion Questions

1. What are the costs and benefits to endothermy versus ectothermy?

2. Provide examples of sharks with each energy economy and relate their anatomy, behavior, and movement patterns to their ectothermic and endothermic lifestyles.

3. The manta ray, which appears to possess a cranial rete, seems to be an exception to the rule. Mantas live in subtropical and tropical waters and rarely if ever visit the cold waters of the temperate and boreal latitudes. They are filter feeders that feed largely at the surface. Why do they have a cranial rete? Might one suspect that they make deep dives into cooler waters in search of dense aggregations of small mesopelagic fishes or crustaceans? Hence, they would need it to stay warm at this time.

4. Why are most sharks and rays ectotherms and not endotherms? There seem to be obvious advantages to being warm-bodied.

Carey, F. G., J. M. Teal, and J. W. Kanwisher. 1981. The visceral temperatures of mackerel sharks (Lamnidae). *Physiol. Zoology*, 54: 334–344.

Goldman, K. J. 1997. Regulation of body temperature in the white shark, *Carcharodon carcharias. Jour. Comp. Physiol., B*, 167: 423–429.

Sims, D. W., V. J. Wearmouth, E. J. Southall, J. M. Hill, P. Moore, K. Rawlingson, N. Hutchinson, G. D. Budd, D. Righton, J. D. Metcalfe, J. P. Nash, and D. Morritt. 2006. Hunt warm, rest cool; bioenergetic strategy underlying diel vertical migration of a benthic shark. *J. Anim. Ecol.*, 75: 176–190.

Weng, K. C., P. C. Castilho, J. M. Morrissette, A. M. Landeira-Fernandez, D. G. Holts, R. J. Schallert, K. J. Goldman, and B. A. Block. 2005. Satellite tagging and cardiac physiology reveal niche expansion in salmon sharks. *Science*, 310: 104–106.

Lemon shark attracted to chemical 'repellent'

CHAPTER 6

Sense of Smell: Chemoreception

The cartilaginous fishes, in particular the sharks, can detect their prey at great distances using their senses of smell and hearing. First, we will discuss smell, the sense with the greatest distance of detection. Because the olfactory receptor is sensitive to chemicals borne in water, the sense of smell is referred to by aquatic scientists as chemoreception. Everyone is familiar with the image of a white shark swimming in a pool of blood that was released into the water to attract the sharks to the vicinity of a shark cage. It is from here that the sharks can be safely filmed by cinematographers for nature documentaries. The white shark has the largest olfactory bulb relative to its body weight of all the chondrichthyan fishes,[1] and it is able to locate the carcasses of whales from a considerable distance using this sense. Yet you will also find out that lemon and nurse sharks are only mildly stimulated by the components or fractions of blood. They are more sensitive to the amino acids and amines in the body fluids of their prey. These species do not feed on seals with a high blood volume as does the white shark. Hence, they may not be so sensitive to blood in the water. Not only are amino acids and amines released by dead and decomposing prey, but they are released from live prey as they are consumed and can attract another predator to the location where a member of a school

Above. Shark Chaser, a mixture of copper acetate and nigrosine dye, was issued to all naval personnel and fliers to use as a shark repellent in the Pacific theater during World War II. *Below.* A researcher is shown using Shark Chaser to create a chemical corridor. Notice that the large lemon shark is not repelled but is turning into the path of the chemicals.

was eaten. Another chemical, which is extremely attractive to bony fishes, is dimethylsulfoniopropionate (DMSP). It is produced by phytoplankton and the algal symbionts in corals and released when they are consumed by zooplankton and herbivorous fishes, respectively. When released in a coral reef environment, DMSP has been shown to attract three species of planktivorous fishes[2] and seven species of predatory fishes.[3] It is possible that DMSP also serves as a feeding stimulus for cartilaginous fishes. The United States Navy developed a chemical repellent during World War II for seamen and pilots to use when cast into the sea where sharks are common. It was eventually found to be ineffective at repelling sharks. Ironically, the Moses sole (*Pardachirus marmoratus*) secretes a chemical from glands on its body that is distasteful to sharks and inhibits them from biting down on it.

Water is more viscous than air, flowing more slowly and changing direction less rapidly. Therefore, a parcel of water remains more intact as it moves over greater distances than does a parcel of air. Hence, the chemicals borne by ocean currents can disperse far from their source in a continuous gradient. White sharks scavenge on humpback and gray whales that die during their annual migrations along the western coast of North America.[4] They are attracted by the chemicals released from the decomposing body of the whale as it floats on the surface of the sea. White sharks are also attracted by artificial chum made up of macerated fish and blood mixed with water. The oils from the macerated fish disperse on the surface of the water in a long corridor in the direction that the wind is blowing (fig. 6.1a). The blood, water-soluble chemicals, and small particles of macerated tissue in the chum are transported downcurrent with increasingly smaller pieces of fish tissue settling to the bottom at greater distances from the source of the odor corridor. A mixture of these ingredients and fluorescent dye, released into the waters of Spencer Gulf off the southern coast of Australia, was visible for four-and-a-half hours as a chemical corridor from the boat to a distance of 4.1 km.[5] A large female white shark was tracked in the vicinity of the odor corridor. Upon swimming into the corridor, she abruptly turned 90° and headed southward in a straight line toward the boat, which was moored and dispensing chemical chum north of Dangerous Reef. The path of the shark, shown by white arrows (fig. 6.1b), was along the increasing concentration of chum, denoted by the gradual change in color from light to dark grey in an elongated oval indicating the odor corridor.

Molecules of various chemicals in continuous water flows create a gradient from a high to a low concentration. These chemicals can alert a cartilaginous fish of a mate or prey item at great distances and provide information that can be used to locate it. Cartilaginous fishes detect molecules of these chemicals as they pass through their nasal cavities. The source of an odor can be located in either of two ways. First, the fish can sense a chemical with its nares in flowing

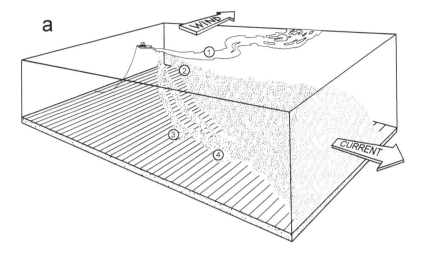

a

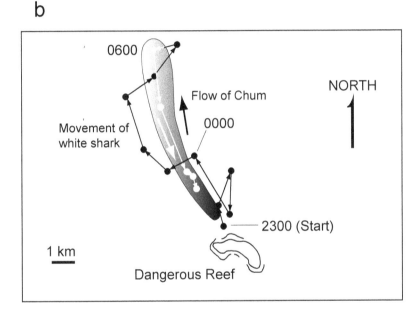

b

Fig. 6.1 (a) Hypothetical representation of the dispersal of a chemical odorant from a container aboard a boat. The chemical mixture consists of (1) the oil slick with its direction governed by the wind and surface currents, (2) the subsurface corridor with blood and small organic particles moving downcurrent, (3) large pieces of bait descending to the bottom, (4) smaller and smaller particles settling out of the water column at greater distances. (b) The path of a white shark, indicated by white arrows, up a concentration gradient of chum, denoted by an elongated oval with darker shades of gray indicating higher concentrations of blood, oils, and macerated fish.

water and use another sense to determine the direction of flow. Second, the fish can turn its head until one olfactory sac is more stimulated than the other, and then turn its head in the other direction until the olfactory sac on the other side becomes more stimulated, and so forth. This results in the individual approaching the odor source along a sinusoidal path.

OLFACTORY ORGAN

The cartilaginous fishes possess two nasal organs sensitive to water-borne chemicals on either side of the ventral surface of the head in front of the mouth (fig. 6.2a). These fleshy sacs are contained in elliptical cavities situated within the cartilaginous skull. Each organ has an opening, or incurrent nostril, facing toward the outside of the body, through which water flows into the chemical-sensing olfactory chamber, and an excurrent nostril oriented toward the inside of the body, through which water flows out of the chamber into the external environment. The nostrils are separated by a flap of skin (fig. 6.2b).[6] The two nasal openings are situated at an oblique angle to the long axis of the body. A shallow depression exists in front of the incurrent nostril; the rear margin of the opening is elevated and stiff. On the other hand, the front margin of the excurrent nostril is elevated and its rear margin is depressed. The forward movement of the shark creates a pressure differential between the two nostrils, resulting in water passing into the incurrent nostril, through the olfactory chamber, and out the excurrent nostril. It is not known for certain whether cilia within the olfactory chamber draw water through in the absence of external water movement. This does not appear to be so because dye placed immediately in front of an incurrent nostril of a restrained lemon shark (*Negaprion brevirostris*), was not drawn into the olfactory organ.

The anatomy of the nares varies greatly among species from different families and with different lifestyles.[7] In general, the more sedentary epibenthic species have wide, rounded nasal openings. These maximize the exposure of the nasal epithelium to the flow of water and dissolved chemicals. Examples of the species are the brownbanded bambooshark (*Chiloscyllium punctatum*) and the eastern shovelnose ray (*Aptychotrema punctatus*). The brownbanded bambooshark rests in crevices or under corals much of the day and at night searches for crustaceans and fishes on the bottom. This epibenthic species has large incurrent openings and even larger excurrent openings to its nares (fig. 6.3a). The shovelnose ray spends much of the time on the bottom lying cryptic against a sandy background, from which it can ambush a prey item that comes too close. It also swims slowly along the bottom at night searching for mollusks, crustaceans, and fishes on the bottom. This guitar-shaped ray has equally large nasal openings separated by a large nasal flap (fig. 6.3b). The neritic species of sharks and rays, which swim faster and higher in the water column, either have slit-like nasal openings or nares that are covered with large flaps to reduce the rate of flow over their nasal epithelium at high speeds. Examples of these are the nervous shark (*Carcharhinus cautus*) and the spotted eagle ray (*Aetobatus narinari*). The former, which is often observed swimming over reefs in the western Pacific Ocean, has slit-like nares similar to those of the lemon shark, common on reefs in the eastern Atlantic Ocean (fig. 6.3c). The spotted eagle ray is also an active swimmer, cruising close to the surface or near bottom when solitary. It joins large schools at certain times of the year. Its nares are slit-like and smaller than those of other benthic rays.

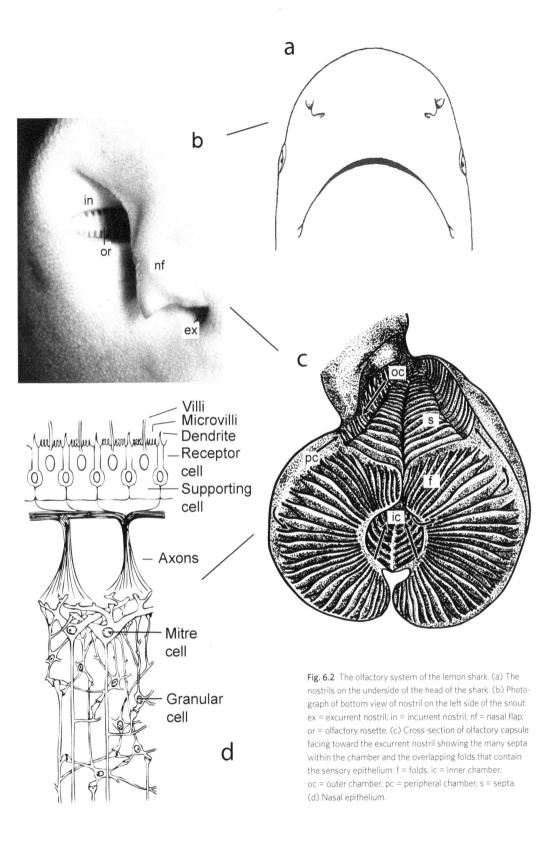

a

b

in

or

nf

ex

c

oc

s

pc

f

ic

Villi
Microvilli
Dendrite
Receptor
cell
Supporting
cell

Axons

Mitre
cell

Granular
cell

d

Fig. 6.2 The olfactory system of the lemon shark. (a) The nostrils on the underside of the head of the shark. (b) Photograph of bottom view of nostril on the left side of the snout: ex = excurrent nostril; in = incurrent nostril; nf = nasal flap; or = olfactory rosette. (c) Cross-section of olfactory capsule facing toward the excurrent nostril showing the many septa within the chamber and the overlapping folds that contain the sensory epithelium: f = folds; ic = inner chamber; oc = outer chamber; pc = peripheral chamber; s = septa. (d) Nasal epithelium.

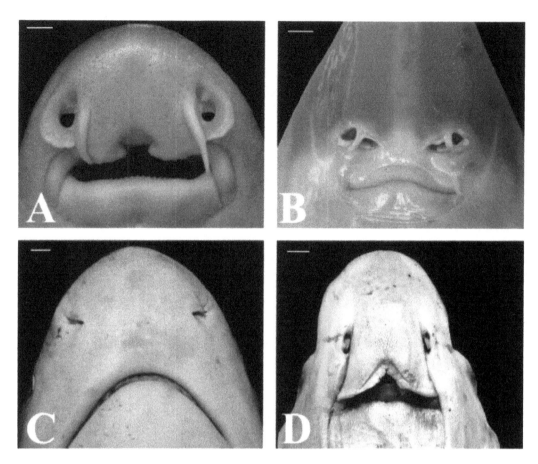

Fig. 6.3 The nares on the ventral side of the body for two benthic species, (A) the brownbanded bambooshark and (B) the eastern shovelnose ray, and two neritic species, (C) the nervous shark and (D) the spotted eagle ray. Notice that the benthic species have large circular nares and the neritic species slit-like nares.

The incurrent openings to its nares are oriented in front of the excurrent openings (fig. 6.3d).

The olfactory chamber is divided by many closely spaced circular partitions, or septa (fig. 6.2c). Water flows through the incurrent opening into an inner chamber, outward across many septa spaced next to each other, through the peripheral chamber external to the septa, and collects in the outer chamber before exiting through the excurrent opening into the surrounding environment.[8] Each septum consists of many overlapping flaps, or lamellae, that are covered with an epithelium that senses chemicals with receptor cells distributed on their outer and inner sides. The surface of the epithelium is entirely covered with receptor and supporting cells (fig. 6.2d).[9] A receptor cell has a single dendrite that extends into the nasal capsule and an axon that together

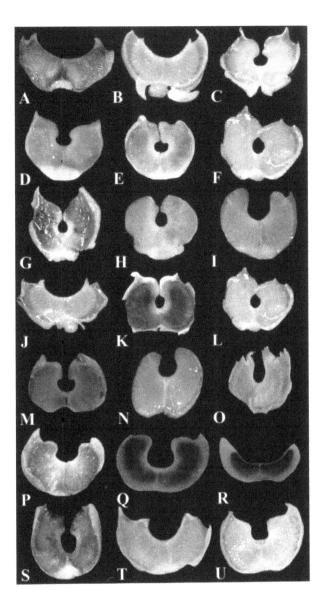

Fig. 6.4 The nasal capsules of sixteen species of sharks and five species of rays. A = spotted eagle ray; B = eastern shovelnose ray; C = blacktail reef shark; D = pigeye shark; E = nervous shark; F = bull shark; G = blackktip reef shark; H = spottail shark; I = brownbanded bambooshark; J = blue-spotted stingray; K = tiger shark; L = Australian weasel shark; M = snaggletooth shark; N = epaulette carpetshark; O = Port Jackson shark; P = sicklefin lemon shark; Q = tawny nurse shark; R = giant shovelnose ray; S = scalloped hammerhead shark; T = whitetip reef shark; and U = stingaree species.

with other axons forms a junction with a mitre cell within the olfactory lobe of the brain. Together with the mitre cells are inhibitory cells, called granular cells due to their appearance. Between the dendrites of the receptor cells are villi and microvilli (long and short cilia) extending from the supporting cells and projecting into the lumen, or cavity, of the olfactory sac.

Fig. 6.4 shows the nasal capsules of sixteen species of sharks and five species of rays. The inner chamber is visible as the small circular opening in the center of the capsule. Many septa with lamellae lie on the lower wall of the capsule on either side of the inner chamber. The radial arrangement of the septa with their attached lamellae is most apparent in the capsules of the eastern shovel-nose ray (see whiter tissue in b) and sicklefin lemon shark (see darker purple tissue in p). The peripheral canal is above the septa and is apparent in the former capsule by its purple color and in the latter by its white color. If both sides of these two capsules were pressed together, the outer chamber would become evident by its U shape. The outer chamber is most apparent from the capsule of the blacktail reef shark (c) and the tiger shark (k).[7]

There is considerable variation in the number of lamellae and the epithelial area in the olfactory organs of the elasmobranchs. This is apparent in fig. 6.4. There is particularly great variation in the number of septa with olfactory lamellae.[7] Counts of septa with lamellae range from 58 in the nasal capsule of the epibenthic eastern shovelnose ray to 231 in the capsule of the neritic scalloped hammerhead shark. On average, the neritic sharks and rays that are more active and swim up in the water column have more lamellae than epibenthic species. Each of the bars in fig. 6.5 denotes the average number of olfactory lamellae for each species. Note that the neritic species generally have more lamellae than the epibenthic species. Note too that eleven of the fifteen sharks, which have cylindrical shapes, thus adapted for swimming up in the water column, are in the neritic category with higher average numbers of olfactory lamellae. Four of the five rays, which have flat shapes adapting them for life near the bottom, are grouped in the epibenthic category with lower average numbers of olfactory lamellae. The majority of the neritic species feed primarily on fishes, as we see from the intermediate dark checkerboard shading of nine of the twelve bars on the left. Their greater numbers of chemically sensitive lamellae may enable them to better locate and capture fishes. On the other hand, none of the epibenthic species feed mainly on fishes. Four of the nine species feed on crustaceans that live on the bottom and three on polychaetes in tubes in the bottom substrate. Three of the four species that feed mainly on crustaceans have the fewest lamellae, indicating that they do not rely on olfaction to locate these species and feed upon them.

Not only do neritic species have higher average numbers of lamellae than epibenthic species, but they have a greater surface area to their lamellae. Fig. 6.6 shows the average surface area of epithelium on the lamellae expressed as a function of the average body length. The same neritic species as those in fig. 6.5 are indicated by boldface abbreviations, the epibenthic species by abbreviations in plain type. A regression line that expresses a linear relationship between lamellar surface and body length was fitted to the average epithelial areas for the sharks and rays with both lifestyles. The slope of the line for the neritic species is much steeper than that for the epibenthic species. That means that for a certain size, say a little over 100 cm in length, the average lamellae surface area of three neritic species, **CS**, the spottail shark (*Carcharhi-*

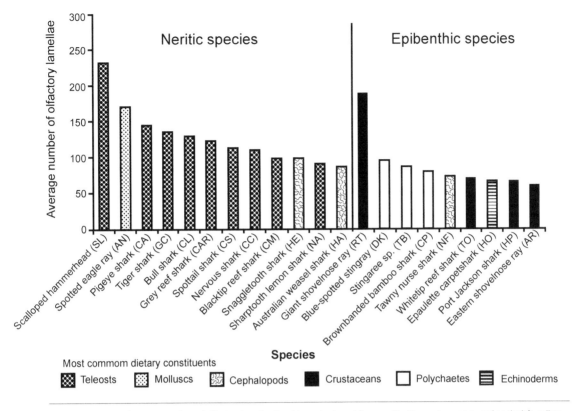

Fig. 6.5 Histogram of average numbers of olfactory lamellae for sixteen sharks and five rays. Neritic species are grouped on the left, epibenthic on the right. Dietary preferences are indicated by fill types shown in the legend below.

nus sorrah), **SL**, the scalloped hammerhead, and **HE**, the snaggletooth shark (*Hemipristis elongatus*), is much greater than the chemical-sensing area of the epibenthic species, **TO**, the whitetip reef shark (*Triaenodon obesus*).

The main conclusion here is that on the average neritic species appear to rely more on olfaction than epibenthic species. There are exceptions. For example, the giant shovelnose ray (*Glaucostegus typus*), an epibenthic species that feeds on crustaceans, possesses an average of 187 lamellae in its nasal capsule—the second highest count, exceeded only by the scalloped hammerhead shark. This outlier is particularly odd because it contrasts so much with the other crustacean feeders, which have the lowest counts of lamellae. One of them is a related species, the eastern shovelnose ray, with an average of 58 lamellae—the lowest average count of any of the sharks and rays examined. An answer to this incongruity may be that the giant shovelnose ray has a lamellar surface area only slightly larger than that of the eastern shovelnose ray (see RT and AR, fig. 6.6), and the two may not differ that much in their olfactory capabilities.

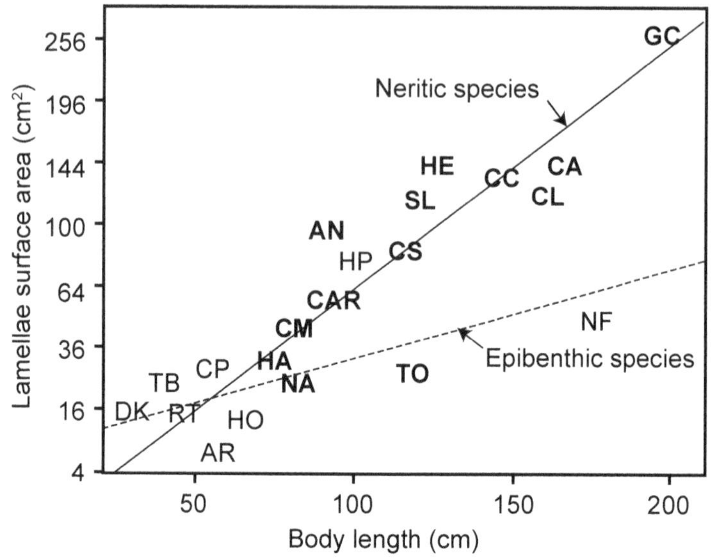

Fig. 6.6 The average surface area of the nasal epithelium in the lamellae of sixteen sharks and five rays expressed as a function of body length. The abbreviations (see fig. 6.5) are in bold type for the neritic species and in normal type for the epibenthic species. The functional relation for the former species is shown by a solid regression line and for the latter species by a dashed regression line.

CHEMICAL SENSITIVITY

The sensitivity to a wide spectrum of chemicals has been determined for two species of sharks, the lemon (*Negaprion brevirostris*) and the nurse (*Ginglymostoma cirratum*) (table 6.1).[10] Are sharks simply attracted by blood as one might conclude from the many films showing white sharks swimming within pools of blood, or are sharks attracted by wider range of chemicals? The olfactory capabilities of these two species was assessed by inserting two electrodes in the olfactory lobe (see locations 1 and 2 in fig. 6.7a), which receives input from the efferent nerves leading from the nasal capsule. These electrodes record the voltage, or potential for electrons to flow, between two locations in the brain with different concentrations. The oscillating voltage measured by these electrodes over time is measured by an electroencephalogram (EEG). It records the brain's electrical activity in response to chemical stimulation within the nares. The amplitude and frequency of the oscillations increase in the presence of some chemicals and decrease in the presence of others. In addition, an EEG can also be recorded from electrodes inserted into the shark's medulla (see locations 3 and 4), which controls gill movements, to determine whether they open or close in response to perception of a chemical. Finally, an electromyogram (EMG), recorded with two electrodes placed on the muscles of the shark, measures the degree of muscular activity associated with tail beats. The electrodes provided a record of the electrical activity within the brains of lemon and nurse sharks when minute concentrations of chemicals were introduced into the water while the sharks swam in a large flow chamber. The wires

TABLE 6.1. Electroencephalogram (EEG) and behavioral responses of lemon and nurse sharks to chemicals and extracts added to water at varying concentrations.

| Compound | Chemical | Concentration (molar) | | Response | |
		Highest	Lowest	EEG	Behavioral
Electrolytes					
	NaCl	10^{-1}	10^{-7}	+	+
	CuSO$_4$	10^{-1}	10^{-7}	−(?)	+
	Artificial seawater	NA	NA	−	+
	2 X seawater	NA	NA	+(?)	+
	0.5 X seawater	NA	NA	−	+
Lipids					
	Cod liver oil	NA	NA	−	−
	Tuna liver oil	NA	NA	−	−
Amino acids/amines					
	Glycine	10^{-3}	10^{-9}	+	++
	Glutamic acid	10^{-3}	10^{-9}	+	++
	Cysteine	10^{-3}	10^{-9}	+	+
	Betaine	10^{-3}	10^{-9}	++	++
	Trimethyl amine	10^{-3}	10^{-9}	++	++
	TMAO	10^{-3}	10^{-9}	++	++
Blood fractions					
	Hemoglobin-human	NA	NA	NT	+
	Hemoglobin-bovine	NA	NA	NT	+
	Albumin-human	NA	NA	NT	+
	Albumin-bovine	NA	NA	NT	+
Extracts					
	Sea cucumber	NA	NA	+	0 −
	Aplysia ink	NA	NA	−	0 −

[a] Sudden alterations in the EGG waveform indicated by +; major changes in the EEG by ++. Movement toward the chemical source followed by brief searching or approach behavior denoted by a +; persistence of this behavior for longer than 10 seconds, repeated biting of objects near the release point, or constant milling around that point and the exclusion of other sharks from the area, indicated by ++. If the shark turned away from the stimulus, but then seemed to ignore the stimulus, the response was designated 0. Not available = NA; not taken = NT.

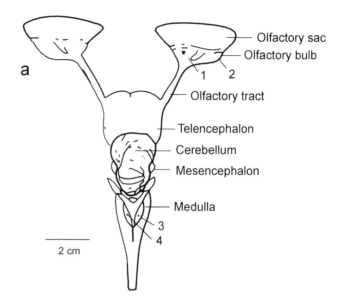

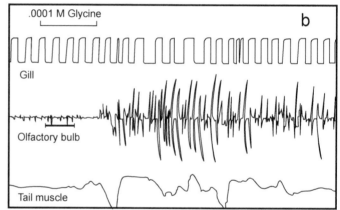

Fig. 6.7 (a) Brain of the lemon shark showing locations where electrodes were placed in the olfactory bulb and the medulla. (b) EEG waveform recorded from electrodes placed at positions 3 and 4 to record the electrical activity associated with gill movements (top), EEG waveform placed at positions 1 and 2 to record the brain activity associated with sensing chemicals (middle), and EMG recorded from electrodes placed in muscles to record beats of the tail during swimming (bottom) in response to the introduction of the chemical glycine into the tank. The scale on the middle trace indicates a period of three seconds.

from the implanted electrodes led through a swivel that led to an EEG recording apparatus beside the tank. The swivel rotated to permit the sharks freedom of motion while they swam around the large rectangular channel with rounded ends. Along with the EGGs, the movements of the sharks were recorded by a motion picture camera. The behavioral responses of the sharks were judged as positive when the shark accelerated upstream and moved toward the source of the stimulant and negative when the shark swam downstream and away from the source of the chemical stimulant.

A typical EEG is shown for a lemon shark when glycine was introduced into the water (fig. 6.7b). The line in the upper left corner of the record indicates the amount of time the chemical stimulus took to disperse within the center of the tank. The top trace denotes the neural activity associated with the rhythmical respiratory movements of the shark's gills; the middle trace shows the response of the olfactory bulb to the detection of the chemical by the receptors in the nasal capsule; and the bottom trace reveals the muscular activity associated with beating the tail. The positive and negative excursions of the waveform recorded from the olfactory bulbs increased at least tenfold after glycine was released into the channel, indicating that the shark perceived the chemical. The small truncated square wave of the uppermost waveform recorded from the medulla indicates that the shark suddenly closed its gill slits to ensure that it was streamlined as it accelerated forward toward the chemical source. Both changes in waveform coincided with more frequent and larger excursions in the trace on the electromyogram. This elevated electrical activity was associated with wider but less frequent sweeps of the tail as the shark swam toward the source of the stimulant.

Describing negative or aversive responses to chemicals was more challenging because it was difficult for the sharks to reverse direction and flee in the narrow confines of the experimental tank. However, the sharks at times did switch directions and swim in the direction opposite to the incoming current, particularly in response to the extract of sea cucumber, which is highly toxic. These sharks even hovered close to the outflow side of the tank. Hence, some tests were conducted in large enclosures, more similar to open ocean conditions.

The first chemicals tested were two series of electrolytes in what is a standard approach to analyzing the sensitivity of chemoreceptors. The olfactory organs of mammals and insects are particularly sensitive to these electrolytes, and their stimulating effect is proportional to their electrostatic field strength.[11] The sharks exhibited an awareness of these electrolytes, as evident from the + in the table denoting their EEG and behavioral responses, but they did not appear to discriminate between them by their ionic strength. The response of the EEG of the sharks to the toxic metallic ion in copper sulfate ($CuSO_4$) was weak and of short duration. The great sensitivity of the shark's olfactory system was evident from the discrimination of artificial seawater, containing the ten major electrolytes of seawater, from natural seawater. The sharks swam toward the source of the artificial seawater, although they did not exhibit a change in the waveform of their EEG. The sharks were also able to distinguish water with higher and lower salinities from natural seawater. The lemon and nurse sharks may sense hyposaline waters with their olfactory sense and avoid them while the bull shark may enter estuaries and rivers upon sensing the lower salinity. Only a mild response was exhibited by the sharks to lipids and their related compounds, despite the widely held belief that sharks prefer prey fish or meats with higher oil concentrations. Neither a distinguishable EEG nor even a mild behavioral response was exhibited in response to the presence cod and tuna liver oil despite the widespread use of the two in chumming sharks at Dangerous Reef in South Australia.

Contrary to what you might believe from the movies, the chemicals most attractive to the sharks were amino acids and amines. Conspicuous changes in the waveform of the sharks' EEGs and prolonged behavioral responses were exhibited in response to two amino acids, glycine and glutamic acid, as has been found in the rainbow trout[12] and other fishes.[13] Betaine, trimethylamine (TMA), and trimethylamine oxide (TMAO), which are all produced during the decomposition of the tissues of fish, also elicited very strong positive electrophysiological and behavioral responses from the sharks. Although sharks have often been observed to be attracted to fresh blood in the water, the results of the controlled tests with fractions of blood did not reveal very strong responses to these supposed attractants. Both species exhibited some awareness of all four purified fractions of human and bovine blood when they were released into the water. The human and bovine hemoglobin fractions elicited a clear orientation, with the lemon sharks accelerating their swimming toward the source of the blood fraction. The human and albumin fractions were less attractive. The response in the direction of the source of the chemicals was slower than in the presence of the former fractions. However, at no time did the sharks mill in the area for a period longer than fifteen seconds and bite the incurrent siphon as observed for amino acids and amines.

The sharks were attracted most strongly to chemical metabolites produced by their prey. The concentrations of the metabolites are known in the tissues of crab, shrimp, oyster, and bony fish (table 6.2).[14] Four types of common metabolites were extracted from these prey types because they had been observed in numerous studies to stimulate feeding in fishes and crustaceans: amino acids, nucleosides and nucleotides, quaternary ammonium compounds, and organic acids. Amino acids were the largest chemical component in the tissues of three of the four types of prey. The amino acid glycine, which produced slightly less activity in the brain but very strong behavioral responses in the two species of sharks, was present in very high concentrations in the four prey species. They were particularly high in the crab and shrimp, 68.00 and 93.70 mM, respectively. These concentrations were among the highest recorded for a metabolite in the tissues of all species from which extracts were taken. The two amines, betaine and TMAO, which when dispensed into the water elicited much higher electrical activity in the brain and stronger behavioral responses by the lemon and nurse, are also in high concentrations in the tissues of the oyster, crab, shrimp, and mullet. There are some metabolites to which the sharks were not exposed that are present in high concentrations in the four types of prey such as the organic acid L-lactic acid. It is likely that these chemicals also stimulate feeding responses in sharks.

Neither of the highly toxic natural chemicals emitted from sea cucumbers and sea slugs elicited strong electrophysiological and behavioral responses

TABLE 6.2. Concentrations of metabolites in the tissues of four common prey items, crab, shrimp, oyster, and mullet, of chondrichthyan fishes.

| Compound | Chemical | Concentrations (mM) in tissue[a] | | | |
		Crab	Shrimp	Oyster	Mullet
Amino Acids					
	Alanine	12.80	21.20	7.32	2.64
	Glutamic acid	1.70	1.65	1.01	0.42
	Glycine	68.00	93.70	4.79	3.05
	Histidine	0.83	0.32	0.10	6.72
	Lysine	1.34	0.49	0.33	1.18
	Total[b]	175.24	188.45	55.05	29.78
Nucleotides, nucleo-sides, and related					
	Inosine 5'-monophosphate	0.95	2.00	0.11	4.13
	Inosine	0.41	0.06	0.07	1.00
	Total	8.07	9.00	0.90	5.46
Quaternary ammoniums					
	Betaine	22.20	50.40	20.30	6.20
	Homarine	2.81	6.90	1.99	
	TMAO	20.30	54.00		5.90
	Total	45.31	111.30	22.29	12.10
Organic acids					
	L-Lactic acid	26.10	27.50	0.55	34.30
	Succinic acid			1.99	
	Total	26.10	27.50	2.78	34.30
	Grand total	254.72	336.25	81.03	81.64

[a]Metabolites with concentrations that were less than 1.0 mM were excluded from the table. A blank space indicates the substance was not detected in the extract.
[b]Chemicals with concentrations less than 1.0 mM other than glutamic acid, which was included in table 6.1, are not included in this table. Hence, the constituents under the headings do not add up exactly to the totals.

There was a major effort by researchers to develop a chemical shark repellent during World War II to protect the many wounded sailors and airmen who found themselves in shark-infested waters in the tropical Pacific Ocean.[17] The most infamous incident occurred on 30 July 1945, when the heavy cruiser U.S.S. *Indianapolis* sank. In addition to the many seamen who drowned, there were estimated to be 60–80 deaths due to shark attacks among the survivors, who were adrift for five days.[18] The publicity surrounding this tragedy led to widespread anxiety and fear among servicemen, and it created a significant morale problem. The U.S. Navy promptly initiated research to develop a chemical shark repellent to renew confidence among sailors and fliers in the odds of survival if cast adrift at sea. The result of this research was the development of a packet of chemicals, known as Shark Chaser, which was made up of 80% nigrosine-type black dye and 20% copper acetate (see photograph package at beginning of chapter). The packet was held together by a wax binder that controlled its release over a period of from three to four hours.[19]

The first tests to develop a shark repellent were carried out at the Naval Research Laboratory. The chemicals tested initially were chosen on the basis of an anecdote—commercial shark fisherman ceased fishing for sharks if one was left on a long line too long and began to de-compose before being removed from the water.[20] It was widely believed that rotting shark meat inhibited feeding by sharks. A quick chemical assay indicated that the most abundant by product of decomposition was ammonium acetate. Other tests screening a variety of chemicals at Woods Hole Oceanographic Institution provided evidence that copper sulfate almost always inhibited dogfish from feeding. The copper portion of the compound was combined with acetate because ammonium acetate was found be stable in saltwater for only a short period of time. Because copper acetate was invisible, a black pigment was later added to enable the sailors and airmen to see where the chemical diffused in the surrounding waters. This chemical mixture was observed by the researchers to inhibit some sharks from feeding, even when they were already scavenging on fish shoveled into the water from the deck of a shrimp trawler. Field tests were then conducted on Shark Chaser off northern Peru, Biloxi, Mississippi, and Mayport, Florida, on five species of sharks: blacktip (*Carcharhinus limbatus*), Atlantic sharpnose (*Rhizoprion-odon terraenova*), scalloped hammerhead, lemon, and tiger sharks. These tests concluded that Shark Chaser was effective only 67% of the time. Although the circulation of this information among members of the armed services increased morale, there were numerous reports during and after the war of the sharks passing through the black

from the sharks. However, these observations are inconsistent with those from other studies that indicate these compounds excreted by invertebrates and fishes deter sharks from feeding upon them. Holothurin is a neurotoxin, which sea cucumbers release into the surrounding water when approached by large predatory fish. An extract made from the Cuvierian gland from a sea cucumber resulted in a brief change in the waveform of the EEG when released into the water but at the same time triggered only a sudden and temporary movement away from the source of the extract. By contrast, a 1 µg ml^{-1} solution of holothurin, collected from the same species of sea cucumber and dispensed by another researcher into a 1520 liter tank in which a 22 kg lemon shark was swimming resulted in the shark frantically avoiding the source of the chemical.

cloud or seizing the cakes of chemical and swimming off with a cloud of the chemicals streaming out of its gills (see shark swimming into chemical corridor at beginning of chapter). This might be anticipated because Shark Chaser was never thoroughly tested on enough dangerous sharks under different environmental conditions. These reports led to further tests in the 1950s of the toxicity of many chemical substances. However, none of these chemicals ever achieved the desired repellent effect.[21] In fact, a mathematical analysis indicated that it was unlikely that a chemical would deter a highly motivated shark from attacking a human.[16] This analysis was based on the known toxicity of the chemicals, the time spent within the chemical field as the shark lunged at the subject, and the physiological reaction time necessary for the chemical to alter the shark's behavior.

Thus it was a surprise when Eugenie Clark, a biologist at the Mote Marine Laboratory, discovered that pardaxin, a milky exudates from pores at the base of the dorsal and anal fins of the Moses sole was highly repellent to sharks, and inhibits a predator from biting down (fig. 6.8a).[22,23] Live and dead specimens of the species that were tied to a set line interspersed with nontoxic reef fish were found to be uneaten by four different species of shark—the silvertip (*Carcharhinus albimarginatus*), blacktail reef, blacktip, and whitetip reef sharks—after periods of ten hours. This was in contrast to the other prey fish, which were consumed by the sharks within an hour. Captive whitetip reef sharks exhibited distressful responses after seizing and attempting to swallow the flatfish in their mouth (fig. 6.8b). They immediately expelled the sole from their mouth and violently shook their head. They then accelerated around the tank, swimming with jerky movements while holding open their mouth for up to a minute as if trying to rid their mouth of the noxious liquid. A chemical analysis of the secretion of Moses sole indicated that the toxic component is an acidic protein composed of 162 amino acids.[24] A dose of 25 µg ml^{-1} of this protein introduced into the water near the head of a restrained piked dogfish (*Squalus acanthias*) resulted in the shark immediately displaying an averse response. It struggled spasmodically while keeping its mouth wide and slowing its opercular rate to minimize the swallowing of water. The chemical is lethal to dogfish after a period of one hour if dispensed into water at a dosage of 5.1 µg ml^{-1} g^{-1} of body mass. The chemical likely irritates to some extent either the chemoreceptors in the olfactory capsule, free nerve endings in the oral cavity, or the taste buds. However, the likely site of its toxic and lethal action is the gill membrane, where osmoregulation is disrupted, resulting in a lethal imbalance in ionic composition of the body fluids.

The toxic compound resulted in the death of the shark after a period of fifty minutes.[15] This chemical is a powerful surfactant, a class of chemicals highly irritant to olfactory receptors; it is extremely irritating to the nasal passages of humans.[16] The black ink of large sea slug of the genus *Aplesia* is another reputed chemical toxin. However, it produced no apparent change in the EGG and only a brief alteration in the shark's swimming pattern.

CHEMICAL ORIENTATION

There are thought to be two methods by which cartilaginous fishes are capable of finding the source of a chemical that has dispersed over a distance.[25]

Releasing
sole

Fig. 6.8 (a) The Moses sole, shown
here in front of Eugenie Clark, secretes
a milky fluid from the base of the
dorsal and anal fins that is repellent to
sharks. (b) A whitetip reef shark exhibited a distress response to the toxin
secreted by Moses sole when seized
from the end of a long line.

Chemical dispersal will occur slowly in stagnant water as molecules of an attractant diffuse outward from an area of high concentration nearer to the source. However, the dispersal is more rapid if a current carries the chemical away from the source. An individual can find the source of a chemical attractant by turning in the direction from which the current is flowing upon detecting the presence of the attractant. It then can then move upcurrent toward the source as long as the chemical is perceptible. This ability is called rheotaxis.

The second way that a cartilaginous fish searches for the source of a chemical attractant does not require moving water but depends upon sensing the gradient between the concentrations detected by the olfactory organs on either side of the head. More of the chemical attractant will pass through one nostril than the other when the shark swims in a continuous chemical gradient. The shark turns its head slightly to the side of the more stimulated nostril and swims forward until less of the chemical attractant flows into the nostril on the other side of the head, and the fish then turns its head slightly to the other side and swims forward until more attractant flows into that nostril, and so forth. This action is repeated again and again so that shark approaches the source of the chemical attractant in a sinusoidal swimming path. The bilateral comparison is most effective in a steep chemical gradient close to the source of the attractant and less effective as the gradient rapidly flattens through dispersion and dilution of the chemical with water. This form of orientation is termed tropotaxis if the comparison of the strength of the stimulus is carried out roughly simultaneously between two bilaterally separated receptors.[26] It is termed klinotaxis if the comparisons of stimulus intensity are separated by a period of time. In this latter form of orientation, the stimulus comparisons may occur at a single receptor. Given the closeness of the nostrils in most sharks, tropotaxis is an unlikely mechanism of orientation, except in the presence of the steepest chemical gradients. However, this ability might be more likely in the hammerheads, whose olfactory organs are separated along either side of the front margin of their laterally elongated head.

The large outside enclosure used in the EGG and behavioral studies described earlier provided one of the first opportunities to understand the method by which sharks locate the source of chemical attractants.[10] Because the tide flowed at different rates at either side of the observational pen, it was possible to introduce the most attractive chemical compounds into either an area of slow or of rapid flow. The behaviors of the stimulated sharks were then studied relative to the rate and direction of flow. The tracks of the sharks were filmed as they located the source of the chemical attractant. The lemon sharks swam upstream immediately upon stimulation by the chemical regardless of whether this behavior led them closer to the source of the stimulant. The detection of the attractant triggered a rheotaxic response, in which these sharks

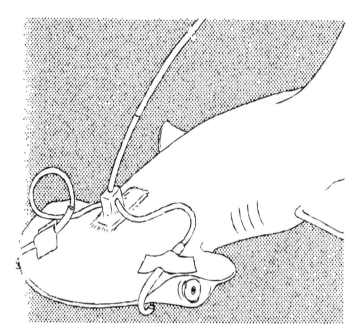

Fig. 6.9 Bonnethead shark with headmount glued on the dorsal surface of its head.

swam upstream while expecting to encounter the source of the chemical stimulant. On the other hand, the nurse sharks did not always move into the strongest currents after chemical stimulation. They swam toward the source of the chemical attractants in a sinusoidal path, with the extent of the side-to-side excursions decreasing as they neared the source. This was evidence for the use of the mechanism of tropotaxis. The speed and efficiency with which the nurse sharks located the source increased after successive tests. In another study, nurse sharks appeared to use rheotaxis because they were able to localize the source of a chemical attractant very accurately in flowing water but were only able to locate the approximate location of a chemical source in stagnant water.[27]

Further insight into how sharks localize the source of chemicals was achieved by delivering different concentrations of chemicals to the nares of bonnethead sharks while they swam in stagnant or flowing water.[26] A small mount was attached to the head of each bonnethead with two tubes leading away from the shark to stimulus reservoirs with different concentrations of attractants (fig. 6.9). The two solutions were passed through the tubes attached to the mount outward over the dorsal head of the shark. Here they were attached to two bent glass tubes with their tips immediately adjacent but not in contact with the incurrent openings of the shark's nostrils. A plunger on a syringe was pressed to force the chemical attractants out of each stimulus reservoir into each of the two delivery tubes leading to the right and left nostrils.

No current

a   b

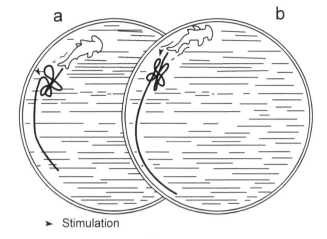

➤ Stimulation

Current

c   d

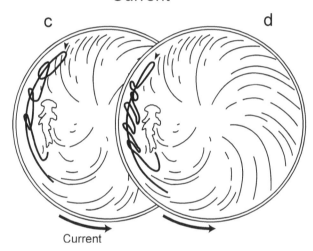

Current

Fig. 6.10 (a) Swimming paths of sharks in response to the introduction of crab homogenate in still water and (b) delivery of homogenate from head mount with bilateral stimulation in still water. (c) Paths of sharks in response to introduction of homogenate in water with a circular flow of 15 cm sec^{-1} and (d) delivery of homogenate from headmount in circular flow.

The stimulus solutions were composed of homogenates of macerated blue crab, a natural prey species of the bonnethead. The behavior of each subject was monitored by a video recorder when the attractants were inserted into their nares.

In stationary water the sharks would stop swimming continuously around the circular tank and begin swimming in tight circles upon encountering a small 10 ml bolus of crab extract poured into the tank (see arrow in tank in fig. 6.10a). In contrast, their behavior was different when the same amount of stimulus was added to the tank when there was a slow continuous flow of water around the tank. The introduction of a small amount of dyed water into the tank indicated that the chemical attractant remained localized in stagnant

Stephen Kajiura and Jessica Forney of Florida Atlantic University and Adam Summers of the University of California at Irvine first used comparative morphology to learn whether the laterally elongated rostrum enables hammerhead species to better perform olfactory klinotaxis. As explained earlier, this is the ability to compare the odor stimulus intensities between the left and right nasal capsules and orient toward the side of greatest chemical concentration. These researchers made numerous measurements of the dimensions of three anatomical features on specimens of the ten species in the family Sphyrnidae with cephalofoils of varying widths and one species, the sandbar shark, in the family Carcharhinidae with its pointed snout. They first determined the distance between the midpoints on either side of the head between the incurrent and excurrent openings to the nares and the prenarial groove (see measurement indicated in upper left in fig. 6.11a). The farther apart the two nasal capsules were, the greater the difference in the intensity of chemical stimuli perceived by the two bilaterally separated receptors, if the shark is swimming in a continuous chemical gradient. A larger intensity difference should make it easier for the shark to localize an odor source, assuming it utilizes olfactory klinotaxis. The mid-narial separation of the winghead shark (*Eusphyrna blochii*) and the scalloped hammerhead were statistically greater than that of the sandbar shark (*Carcharhinus plumbeus*). The separations of the nares of both sphyrnids are greater than the carcharhinid for all lengths—this is apparent from the regression lines for the former two species being higher than the line for the latter species on the graph. For example, the mid-nasal separation distances for winghead and scalloped hammerhead sharks that were 90 cm long averaged 25 and

17 cm, respectively, while the mid-nasal separation for sandbar sharks of the same length averaged only 7.5 cm (see arrows on graph). The next dimension measured was the effective sampling length, consisting of the prenarial groove plus the width of the nares (see upper left in fig. 6.11b). Water is sampled with odorant molecules from the inner edge of the former to the outer edge of the latter. The sampling lengths of scalloped hammerhead sharks of all sizes were statistically longer than those of the sandbar sharks. The winghead shark had by far the longest effective sampling length, but the few measurements precluded statistical comparison with the other species. For example, a winghead shark 90 cm long had an effective sampling length of 15 cm versus the 4 cm and 0.3 cm lengths, respectively, for the scalloped hammerhead and sandbar shark. The third feature measured was the area of the sensory epithelium within the lamellae (see upper left in fig. 6.11c). The lamellar surface could be determined only for the scalloped hammerhead, bonnethead (*Sphyrna tiburo*), and sandbar sharks. In this case, there was no statistical difference between the scalloped hammerhead and sandbar sharks, but the former two had significantly larger lamellar surfaces than the bonnethead shark.

In conclusion, the winghead has 49% greater separation between the nares than the scalloped hammerhead, and the latter has 79% greater separation than the sandbar shark. This supports the contention that lateral expansion of the head enables sphyrnid sharks to sample a wider swath of seawater and thus increases their probability of encountering odor molecules. The probability of an odor molecule landing upon the olfactory epithelium is related to the width of the nares and length of the prenarial groove. The scalloped hammerhead had a 1,300% greater

water but dispersed around the tank in a roughly discrete corridor in the moving water. Upon coming upon the chemical bolus, the shark reversed its direction and followed the bolus around the edge of the tank, swimming against the current in a path consisting of connected loops (see tank in fig. 6.10c).

The chemical was then injected in equal amounts into the two nares in stagnant water. The sharks again swam in tight circles in the immediate vi-

effective sampling length than a similarly sized sandbar shark, and the winghead shark had an even greater sampling length. The scalloped hammerhead shark did not have a significantly larger lamellar surface than the sandbar shark. This observation seems incompatible with the scalloped hammerhead having a longer effective sampling length than the sandbar shark. However, the lamellae of the former are smaller due to the dorsal ventral flattening of its head while those of the sandbar shark are larger due to its conical head.

Jayne Gardiner of the University of South Florida and Jelle Atema of Boston College conducted experiments to determine whether sharks can determine the direction of a chemical plume based on the differences in the timing between when their bilaterally separated nares encounter odor pulses.[29] This mechanism differs from chemical localization based upon detecting different concentrations of chemicals at bilaterally separated nares.[27] They presented carefully timed odor pulses directly into the nares of dusky smooth-hound sharks (*Mustelus canis*). The sharks turned toward the side that was stimulated first, even when presented delayed odor pulses of greater concentrations on the other side. This behavioral response could steer a shark into an odor patch, and hence enable it to stay within an odor plume. A wider separation to its nares would confer upon a shark an enhanced ability to discriminate smaller differences in the angle of the odor plume relative to the two nares when swimming at higher speeds. This is apparent in a diagram with lines drawn through one nasal sac at increasing angles relative to a line drawn through both nares. The distance to the second nasal sac from the line with an angle of 60° is greater than those distances from the lines with angles of 45° and

30° (see bold arrow in Fig. 6.12a). The difference between when a chemical impinges on the left nostril and that on the right increases from the dusky smooth-hound, sandbar, scalloped hammerhead, to a winghead hammerhead if swimming directly ahead (see increasing lengths of bold arrows in fig. 6.12b). Finally, this relationship is evident in a graph of the differences in time when the chemical pulse is perceived by bilaterally separated nares when chemical pulses reaches them different angles (fig. 6.12c). The time differences are shown for the dusky smooth-hound for different angles. The greatest difference would occur if chemical pulses arrived at an angle of 60°, and the time differences get smaller with decreasing angles of 45° and 30° until they are nonexistent at an angle of 0° (see dotted line in fig. 6.12c). The time difference for one angle, for example 60°, is greatest for the winghead with the largest nasal separation (see solid line), next largest for the scalloped hammerhead and sandbar sharks (see dashed and dotted and dashed lines), and least for the dusky smooth-hound (dotted line). Notice on the graph that the curve for the winghead is farther to the left and higher than the other three curves, with that of scalloped hammerhead next to the right followed by that of the sandbar shark, and finally farthest to the right that for the dusky smooth-hound. At any particular angle at which the chemical pulse approaches, the internarial timing difference of the winghead is the greatest, followed by those of the scalloped hammerhead, sandbar, and dusky smooth-hound sharks. Hence, Gardiner and Atema argue that this benefit to the laterally elongated rostrum and widely separated nares may have contributed to the evolution of the hammerhead sharks.

cinity of the bolus, as they had done when the chemical was poured into the tank at that site in still water (see tank in fig. 6.10b). Equal amounts of attractant were then simultaneously injected into both nostrils in the presence of a circular current. The sharks reversed direction and began to swim upcurrent in a sinusoidal pattern as they had done when the chemical was dispersed in moving water (see tank in fig. 6.10d). Then the stimulant was delivered solely

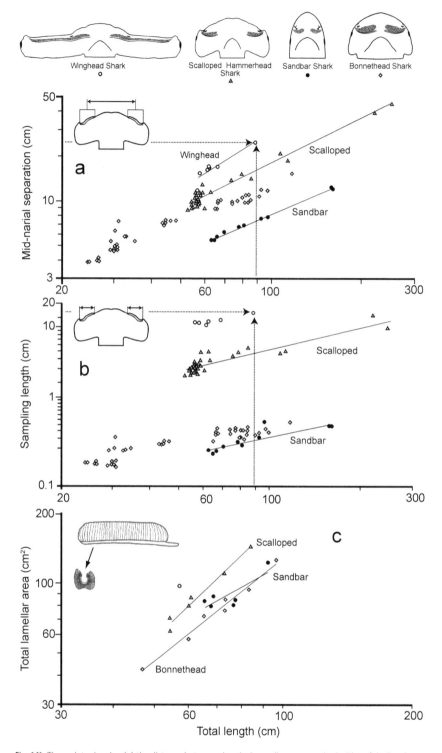

Fig. 6.11 Three plots showing (a) the distance between chemical sampling areas on both sides of the head, (b) the chemical sampling length from the inner margin of the prenarial groove to the outer margin of the nares, and (c) the total lamellar area for the winghead shark, scalloped hammerhead shark, bonnethead shark, and sandbar shark.

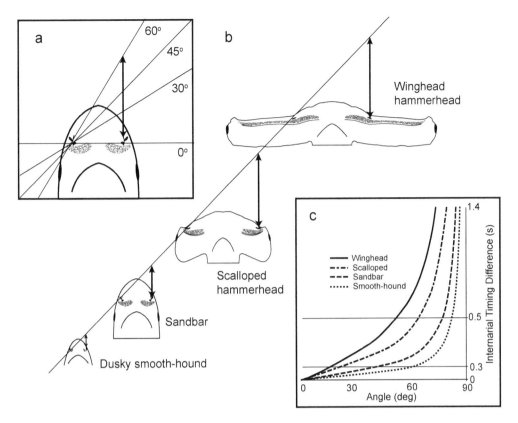

Fig. 6.12 (a) Differences in arrival time (arrow) between two bilaterally separated nares for chemical pulses arriving at different angles for the dusky smoothhound, (b) differences in arrival times of chemicals arriving at an angle of 60° for four different sharks, and (c) the internarial time differences for four different sharks occurring at increasing angles at which chemical pulses arrive at the nares.

to the left or right nostril after a bonnethead moved away from the walls of the tank. The shark always turned toward the side of the stimulated nostril. Furthermore, when the concentration of the stimulant injected into one nostril was greater than that simultaneously injected into the other, the bonnethead turned toward the side of the more stimulated nostril. This experiment is the best evidence to date that sharks are capable of tropotaxis or following a chemical gradient to its source by bilateral comparisons of stimulus strength between their two nostrils.

SUMMARY

Molecules of various chemicals in continuous water flows create a gradient from a high to a low concentration. This can alert a cartilaginous fish to a prey item at great distances and provide information that can be used to locate it.

Cartilaginous fishes detect these molecules, which are dissolved in water, as they pass through their nasal cavities; this process is called chemoreception. Each olfactory organ has an opening, or incurrent nostril, facing toward the outside of the body, through which water flows into the chemical-sensing olfactory chamber, and an excurrent nostril oriented toward the inside of the body, through which water flows out of the chamber into the external environment. The cartilaginous fishes are sensitive to a wide spectrum of chemicals. Those most attractive are amino acids and amines. High concentrations of amino acids are found in the tissues of prey species such as crabs, shrimp, oysters, and bony fish. A cartilaginous fish can locate the source of an odor in one of two ways. First, it can swim into the current by sensing the presence of a chemical with its nares and use another sense to determine the direction of flow. Second, the individual can turn its head to one side until one olfactory sac is more stimulated than the other, and then turn its head to the other side until the olfactory sac on the other side becomes more stimulated than the former, and so forth. Alternatively, the cartilaginous fish can turn in the direction of the nares stimulated first by a chemical pulse from a chemical plume. This results in the individual approaching the odor source in a sinusoidal path.

* * *
** ** **

KEY TO COMMON AND SCIENTIFIC NAMES

Atlantic sharpnose shark = *Rhizoprionodon terraenovae*; Australian sharpnose shark = *Rhizoprionodon taylori*; Australian weasel shark = *Hemigaleus australiensis*; blacktail reef shark = *Carcharhinus amblyrhynchos*; blacktip shark = *Carcharhinus limbatus*; blacktip reef shark = *Carcharhinus melanopterus*; blue crab = *Callinectes sapidus*; blue-spotted stingray = *Dasyatis* (= *Neotrygon*) *kuhlii*; bonnethead shark = *Sphyrna tiburo*; brownbanded bambooshark = *Chiloscyllium punctatum*; bull shark = *Carcharhinus leucas*; dusky smooth-hound shark = *Mustelus canis*; eastern shovelnose ray = *Aptychotrema rostrata*; epaulette shark = *Hemiscyllium ocellatum*; giant shovelnose ray = *Glaucostegus typus*; lemon shark = *Negaprion brevirostris*; Moses sole = *Pardachirus marmoratus*; nervous shark = *Carcharhinus cautus*; nurse shark = *Ginglymostoma cirratum*; pigeye shark = *Carcharhinus amboinensis*; piked dogfish = *Squalus acanthias*; Port Jackson shark = *Heterodontus portusjacksoni*; reef shark = *Carcharhinus wheeleri*; scalloped hammerhead shark = *Sphyrna lewini*; sea cucumber = *Actinopyga agassizi*; sea slug = *Aplysia* sp; sicklefin lemon shark = *Negaprion acutidens*; silvertip shark = *Carcharhinus albimarginatus*; snaggletooth shark = *Hemipristis elongatus*; spottail shark = *Carcharhinus sorrah*; spotted eagle ray = *Aetobatus narinari*; stingaree sp. = *Trygonoptera* sp.; tawny nurse shark = *Nebrius ferrugineus*; thresher shark = *Alopias vulpinus*; tiger shark = *Galeocerdo cuvier*; whitetip reef shark = *Triaenodon obesus*.

Discussion Questions

1. Compare the chemical sense of the cartilaginous fishes in the viscous medium of the ocean to the sense of smell of terrestrial species in the rarefied medium of air. Where is the distance of detection of a chemical source greatest, on land or in the sea?

2. Scientists have given considerable attention to determining whether the laterally elongated rostrum of the hammerhead aids in chemoreception. Are there any more experiments that should be conducted to provide insight into the function of the elongated rostrum?

3. Little research has been conducted on chemoreception in the rays and chimaeras. Can you design a research plan for learning about olfactory abilities of these species? It would certainly make a first-rate thesis project.

4. The white shark has the largest ratio of olfactory bulb body mass of all the sharks and rays. Can you provide a rationale why it has such a strong sense of smell?

LITERATURE CITED

1. Demski and Northcutt, 1996; 2. DeBose *et al.*, 2008; 3. DeBose and Nevitt, 2007; 4. Long and Jones, 1996; 5. Strong, Jr. *et al.*, 1996; 6. Zieske *et al.*, 1987; 7. Schluessel *et al.*, 2008; 8. Zeiske *et al.*, 1986; 9. Montgomery, 1988; 10. Hodgson and Mathewson, 1978; 11. Hodgson, 1974; 12. Hara, 1973; 13. Hara, 1975; 14. Carr, 1988; 15. Sobatka, 1965; 16. Nelson, 1983; 17. Sisneros and Nelson, 2001; 18. Brown, 1980; 19. Tuve, 1963; 20. Zahuranec, 1978; 21. Gilbert and Springer, 1963; 22. Clark, 1974; 23. Clark, 1983; 24. Primor *et al.*, 1978; 25. Kleerekoper, 1978; 26. Johnsen and Teeter, 1985; 27. Kleerekoper and Gruber, 1975, 28. Kajiura *et al.*, 2005; 29. Gardiner and Atema, 2010.

RECOMMENDED FURTHER READING

Clark, E. 1983. Shark repellent effect of the Red Sea Moses sole. Pp. 135–150 in Zahuranec, B. J. (Ed.), Shark Repellents from the Sea. Westview Press, Inc., Boulder.

Gardiner, J. M., and J. Atema. 2010. The function of bilateral odor arrival time differences in olfactory orientation of sharks. *Current Biology*, 20: 1187–1191.

Hodgson, E. S., and R. F. Mathewson. 1978. Electrophysiological studies of chemoreception in elasmobranchs. Pp. 227–267 *in* Hodgson, E. S. and R. F. Mathewson (Eds.), *Sensory Biology of Sharks, Skates, and Rays*. U.S. Government Printing Office, Washington, D.C.

Johnsen, P. B., and J. H. Teeter. 1985. Behavioral responses of bonnethead sharks (*Sphyrna tiburo*) to controlled olfactory stimulation. *Mar. Behav. Physiol.*, 11: 283–291.21.

Kajiura, S. M., J. B. Fornia, and A. P. Summers. 2005. Olfactory morphology of carcharhinid and sphyrnid sharks: does the cephalofoil confer a sensory advantage? *J. Morph.*, 264: 253–263.

Sisneros, J. A., and D. R. Nelson. 2001. Surfactants as chemical shark repellents: past, present, and future. *Env. Biol. Fish.*, 60: 117–129.

Zieske, E., B. Theisen, and S. H. Gruber. 1987. Functional morphology of the olfactory organ of two carcharhinid shark species. *Can. J. Zool.*, 65: 2406–2412.

CHAPTER 7

Sense of Hearing: Mechanoreception

The cartilaginous fishes are highly attuned to low-frequency underwater sounds, attracted by many of them and also frightened by some. The ocean is not at all a silent place, as it might appear when you dive with a mask and snorkle or SCUBA gear. This is because the oscillations of water molecules are small compared to the large back-and-forth movements of highly compressible air molecules, which easily resonate our membranous eardrums. The sharks, rays, and chimaeras live in an environment full of sounds. There is the constant crackling of snapping shrimp, living in burrows in the bottom but periodically emerging to snap their forward appendages outward to stun their prey. Crabs scuttling over the bottom make cracking sounds as they crush small clams with their claws. Male damselfish, gaily colored small fishes common on coral reefs, produce popping sounds by moving their jagged jaw bones against each other to advertise their readiness to mate. Impulsive booms are produced when groupers, snappers, tunas, densely spaced within schools, simultaneously beat their tails as they dart in all directions to avoid a predator charging within the school to seize a victim. Marine catfish produce drumming sounds by rapidly contracting and relaxing the muscles around their swim bladders. Then there is the monotonous and never-stopping hum produced

Bonnethead shark (*Sphyrna tiburo*) approaching an underwater speaker broadcasting low-frequency pulsed sounds in the shallow waters of the Florida Keys.

Spear fishermen were the first to notice that sharks were attracted by sounds. Donald Nelson augmented his modest salary as a graduate student at the marine laboratory of the University of Miami by spearing fish and selling them at a fish market in Miami. Upon spearing a fish, he observed that sharks arrived very rapidly and then circled the fish at the end of his spear.[1] Nelson considered the rapid appearance of the sharks inconsistent with blood attracting them because it slowly emanated from the fish and dispersed gradually in the slow-moving current. Furthermore, the often poor visibility eliminated the possibility that the sharks could see the fish struggling at the end of the spear. He reasoned that the sharks might be attracted by vibrations produced by the fish spasmodically beating its tail back and forth. He and a fellow graduate student, Samuel Gruber, recorded the pulsating sound of a grouper struggling at the end of a spear with a hydrophone, a microphone sensitive to minute water movements. The properties of the sound are apparent in a black trace displayed on two sound spectrograms (fig. 7.1b & fig. 7.1c). Each is a graphical representation in which the pitch, or frequency, of the sound is indicated on the left axis of the graph, rising from 20 vibrations per second, or Hz, on the bottom to 500 Hz at the top of the graph. The fifteen-second time period over which the sound was recorded is indicated on the bottom axis of the graph. The strength, or amplitude, of the sound is indicated by the shade of gray, whose shade varies from a light gray for a faint sound to a dark gray for a loud sound. The sounds generated by the struggling grouper consisted of intermit-

tent pulses with the majority of the energy in the sound from 20–150 Hz (fig. 7.1a). Based on the properties of the natural sound, the two graduate students created a similar artificial sound (fig. 7.1b). They played this artificial sound twenty-two times, each time for fifteen minutes, and during these periods eighteen sharks were attracted to the speaker. This number included nine bull sharks (*Carcharhinus leucas*), two hammerhead sharks (*Sphyrna lewini*), two lemon sharks (*Negaprion brevirostris*), and one tiger shark (*Galeocerdo cuvier*).

The next step was to ascertain from what distance sharks could be attracted to the struggling fish on the end of a spear. Warren Wisby, the professor with whom they studied, set out with one of them in an airplane to search for sharks in Biscayne Bay, Florida (fig. 7.1a). The sharks were easily visible from the sky over the white sandy bottom as dark silhouettes as they swam in shallow water. Gruber in the plane located a shark swimming in a relatively straight line and conveyed its position to Nelson, who was aboard a boat.[2] The latter then motored to one side of the shark and anchored the boat based on the former's directions. Nelson then lowered the speaker into the water and broadcast the pulsing sounds into the water. Gruber from the airplane observed nine different sharks abruptly turn to the side and accelerate toward the speaker. One turned at a distance as far as 800 m from the speaker. The sharks turned away when they came within 20 m from the speaker as if the sound appeared unnatural when they were close to its source.

by midshipmen, small fishes abundant in the shallow waters at temperate latitudes, vibrating their air bladders. Added to this low-frequency cacophony are the high-frequency clicks of dolphins as they echolocate a potential meal. Finally, every once in a while there might the loud shriek of a humpback whale or the monotonous rumble of a blue whale so low in frequency that it is hardly perceptible to humans. A bonnethead shark is shown here being attracted to the punctuated "bum . . . bum . . . bum" sound of a struggling fish emitted from a speaker while searching for source of the sound in shallow water.

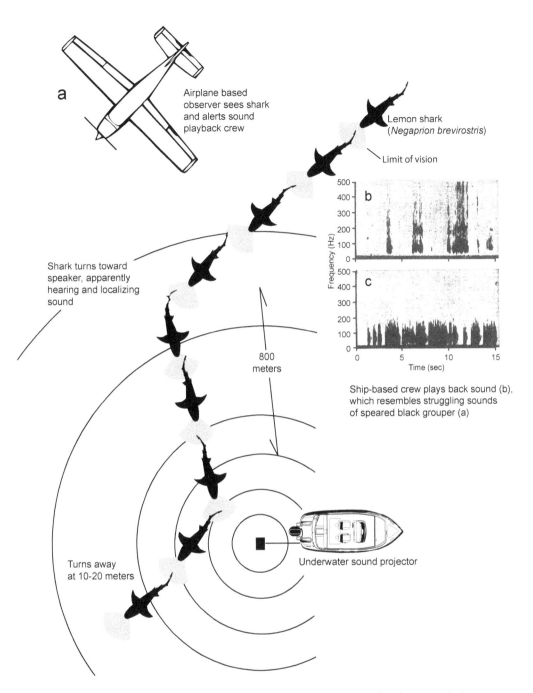

Fig. 7.1 (a) Experiment showing that sharks change course and swim toward a source of low-frequency pulsed sounds. The distance at which they responded to sounds was on the order of hundreds of meters. This exceeded the distance the fish would be visible in the ocean, which is on the order of tens of meters. Audiograms of (b) the sounds emitted by grouper struggling at end of spear and (c) artificially produced sound.

Another graduate student at the University of Miami, Arnold Banner, found that the smaller, juvenile sharks were attracted to their prey by the sounds they emit over a shorter distance. Lemon sharks not more than 70 cm long inhabit the shallowest waters of Biscayne Bay. They feed preferentially on the small flathead striped mullet (*Mugil cephalus*). He reasoned that the sound most likely to attract the sharks to this prey item would be the sound it produced when suddenly accelerating upon being frightened by a predator. He recorded two sounds, one from the mullet as it bolted forward after receiving an electric shock and another from a mullet moving atypically after the surgical removal of part of its tail.[3] Upon playback of both sounds, the small sharks abruptly changed their swimming course and swam directly toward the speaker at distances up to six meters and circled the speaker at a distance of one meter. Other sounds that attracted these newborn sharks were the drumming distress calls of the marine catfish (*Galeichthys felis*) and the crackling feeding sounds made by a school of sheepshead minnows (*Cyprinodon variegates*) as they crushed food particles in their jaws. The former sounds caused the sharks to veer from a straight swimming path and approach the speaker from a distance of 11 m and the latter from a distance of 16 m.

Unfortunately, little is known about auditory abilities of the rays and chimaeras. It is possible that they are less sensitive to distant sounds, as many have little need to detect prey at great distances. Many of the members of the order Rajiformes lie in wait on the bottom, unseen as their bodies match the color pattern of the substrate, and ambush their prey when they come too close. Other more mobile rays in the order Myliobatiformes locate their prey, often mollusks or crustaceans buried in the sand or mud, while foraging individually close to the bottom. It is doubtful that these prey items generate loud sounds while buried in the bottom substrate. The chimaeras of the order Chimaeriformes also forage on the bottom for prey that they crush with the plates of pavement-like teeth in their lower and upper jaws.

WITHDRAWAL FROM IMPULSIVE SOUNDS

The first sound found in nature to frighten sharks was the intense shriek-like sound emitted by killer whales (*Orcinus orca*) while hunting in groups. This sound begins with a continuous tone of 500 Hz, rises in quarter of a second to 2,000 Hz, and ends with the higher continuous tone (fig. 7.2a). You can make this loud sound yourself by shrieking "wheee . . . oooh . . . wheeee." Bill Cumming and Paul Thompson at the Naval Undersea Research and Development Center played back this scream to gray whales (*Eschrichtius robustus*) as they passed within 150–450 m of their boat near Point Loma off San Diego.[4] The reactions to the killer whale sounds were spectacular. Whales that were

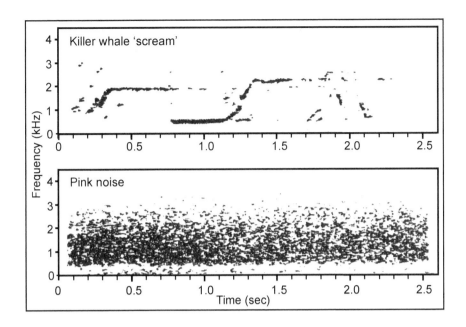

Fig. 7.2 (a) Audiogram of the killer whale scream and pink noise of the same frequencies that compose the scream. (b) Author observing lemon shark turn in response to the playback of the killer whale scream. Notice the cable leading the hydrophone, which measured the amplitude of the sound at a distance of 10 m from the underwater speaker.

at the surface suddenly changed direction and headed directly away from the sound projected from the boat. Many of the whales were swimming just outside the kelp. These whales fled into the heavy growth of the kelp forest and stayed there until the sounds stopped.

I played the killer scream to lemon sharks swimming around a circular channel together with two other control sounds (fig. 7.2).[5] The first control was composed of pink noise (fig. 7.2a), whose sound energy was equally distributed over the bandwidth of the scream but had none of the unique tonal modulation. The second control was a pulsed low-frequency sound, 150–300 Hz, similar to the sounds of a struggling fish that had attracted sharks in previous studies. Surprisingly, the pink noise frightened the sharks more often than the scream. Even the pulsed low-frequency sound frightened the sharks when played at high amplitudes despite being attractive at lower ones. Additional tests, in which the pink noise was increased to the same level at different rates, indicated that sounds with an abrupt, impulsive onset were most frightening to the sharks. It appears the lemon sharks were not frightened per se by the sound of the killer whale but exhibited a startle response to a sudden sound. This result might be anticipated because they rarely come in contact with each other—the lemon shark mainly inhabits tropical waters while the killer whale stays mainly in cold temperate and boreal waters. This reflexive withdrawal response from the sharks is similar to the reflexive forward movement of your body upon hearing a sudden, loud slam of a door behind you.

WATER PARTICLE DISPLACEMENTS AND PRESSURE OSCILLATIONS

How does the little lemon shark detect the explosive acceleration of mullet when frightened? Water particles are pushed in one direction for a distance equal to the sudden side movement of its tail. These particles come in contact with more distant water particles and push them farther in the same direction, and so forth with only a slight decrease in the distance of displacement of each group of water particles, each at a greater distance from the tail. The successive propagation of these one-way particle displacements away from the tail is due to the less compressible nature of water relative to highly compressible air. However, there are also pressure oscillations produced by the beat of the tail, due to the resulting compression and subsequent rarefaction of the water particles. The water particles oscillate back and forth about a fixed point without any net movement in one direction, yet they do transfer their energy to adjacent particles that oscillate around a fixed point at a greater distance from the tail. Thus, these oscillations are propagated far away from the tail with each successive group of particles moving back and forth slightly less.

The two stimuli, the one-way water particle displacements and two-way pressure oscillations, diminish, or attenuate, at different rates.[6] The one-way particle displacements in the proximity of the prey diminish by 1 divided by the cube of the distance from a vibrating source, such as the spasmodically bending tail of a black grouper struggling on the end of the spear or the explosive tail beats of a frightened mullet. Particle displacements near the prey diminish by 1 divided by the square of the distance from a pulsating source, such as the rapidly expanding and contracting swim bladder of the marine catfish that produces distress calls. On the other hand, the two-way particle displacements associated with the sound pressure oscillations near the prey decrease in intensity more slowly than the one-way displacements, by 1 divided by the distance from the source. The distance at which the amplitudes of one-way water particle displacements are greater than those of the back-and-forth displacements caused by pressure variations is the boundary between the acoustic near and far fields. The importance of this boundary is that the cartilaginous fishes have external receptors, free neuromasts and a lateral line, that sense these one-way movements when closer to the prey in the near field and an internal receptor, the inner ear, that senses pressure oscillations when farther from the prey in the far field. The boundary between the two fields is frequency dependent. It is equal to the wavelength of the sound. The near field of the lowest frequency of 10 Hz recorded from the struggling grouper extends to a distance of 150 m; whereas the near field of a highest frequency of 100 Hz extends only 15 m from the sound's source.

Humans produce sound by forcing air through the larynx and moving a membrane back and forth to produce the alternating compaction and expansion of air molecules. They perceive sound due to the back-and-forth movements of a membranous tympanum, or eardrum, coupled by bones to the cochlea of the inner ear. The cartilaginous fishes unlike humans do not possess an eardrum. Nor do they resemble some bony fishes in having a membrane-enclosed air cavity, the air bladder, that resonates in response to its alternating compaction and expansion of water and transfers these vibrations to the inner ear with a series of bones. The cartilaginous fishes, on the other hand, sense the movements of water by the back-and-forth movement of small cilia embedded within a gel-like layer, the cupula. These cilia move in response to mechanical disturbances of this medium, and hence their sense of hearing is more accurately described as mechanoreception.

External Mechanoreceptors Sensitive to Water Displacements

The mechanical movements of the medium are detected by a sensory epithelium, or macula, on which are located hair cells.[7] Each hair cell has an apical bundle of cilia, consisting of a single long kinocilium and up to sixty smaller

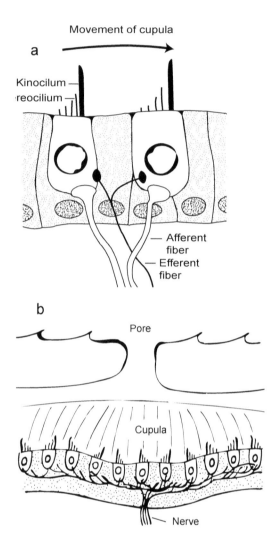

Movement of cupula

a

Kinocilum —
·reocilium —

— Afferent
fiber
— Efferent
fiber

b

Pore

Cupula

Nerve

Fig. 7.3 Two neuromasts with their afferent and efferent fibers shown in association with their supporting cells. The apex of the neuromast has a ciliary bundle with one kinocilium and many stereocilia. (a) Many neuromasts are embedded in a cupula, which moves back and forth in response to water particle displacements. (b) Pores enable water to enter the tubular canal, or lateral line, and vibrate the cupula just under the skin of the cartilaginous fishes.

stereocilia, which are graded in height with the largest stereocilia being next to the kinocilium (fig. 7.3a). The base of the hair cell forms a synapse, or contact, with an afferent nerve that transmits nerve impulses from the hair cell to the central nervous system and a junction with an efferent nerve that transmits nerve impulses from the central nervous system to the hair cell. The hair cell and its associated cells form a sensory organ called a neuromast. Many of these organs are embedded in a gelatinous mass, or cupula (fig. 7.3b). The hair cells move the less dense water to the more dense cupula, which in turn moves the ciliary bundles embedded within the macula in one or another direction. Movements of the cilia alter the electrical potential within the hair cell, and it then releases a neurotransmitter at the synapses to the afferent fiber leading to the cell. Displacement of the shorter stereocilia in the direction of the longer kinocilium produces an increase in the firing rate of the afferent nerve above its spontaneous rate, while displacement of the stereocilia in the opposite direction results in a decrease in the firing rate of the afferent nerve below its spontaneous rate.

Superficial Neuromasts, Lateral Line, and Vesicles of Savi

The cartilaginous fishes have a battery of receptors on their external surfaces that are sensitive to mechanical stimulation by water particle movements. These organs are classified by their anatomy and location.[8] They are thought to provide a sense of the closeness of individuals to objects in the environment as water they displace while swimming is reflected off these objects. They are also thought to sense general particle displacements generated by the tail movements of adjacent individuals in schools as well as their prey when approaching them closely. Superficial neuromasts, also called pit organs, are located on the skin of all cartilaginous fishes. These sensory receptors are located between the bases of adjacent dermal denticles. Free neuromasts on sharks are oval as shown for the sicklefin lemon shark (*Negaprion acutidens*) (fig. 7.4b), whereas those of rays are more elongated and elevated on the body as is evident on the cowtail ray (*Pastinachus sephen*) (fig. 7.4h).[9] The dorsal opening of each superficial neuromast permits direct contact of water with its cupula and the embedded kinocilia and stereocilia. This is apparent in a micrograph taken of a neuromast on the side of the body of the sicklefin lemon shark (fig. 7.4c). Some neuromasts are arranged in neat rows; others are scattered diffusely over broad surfaces of the body.[10] There are on the lemon shark two mandibular rows, lines of pit organs that cross the ventral surface of the body immediately posterior to the mouth and two umbilical rows that extend inward between the two pectoral fins (fig. 7.4a). Superficial hair cells are scattered over the bodies of sharks from the head to the posterior lobe of the tail. The number of neuromasts varies greatly among species, ranging from an average of 77 per side on a piked dogfish (*Squalus acanthias*) (fig. 7.4d) to an average of 600 per flank on a scalloped hammerhead shark (fig. 7.4f). The pit number depends on the shark's habitat and level of activity. For example, intermittent swimmers that are epibenthic and often swim near the bottom such as the spiny dogfish (fig. 7.4d) and the nurse shark (*Ginglymostoma cirratum*) (fig. 7.4e) have fewer superficial neuromasts, with most of them distributed above the lateral line. These hair cells are mainly sensitive to water displacements originating from above. Species that swim constantly in upper water column such as the scalloped hammerhead (see fig 7.4f) have many neuromasts located both dorsal and ventral to the lateral line; they are sensitive to stimuli originating from above as well as below.

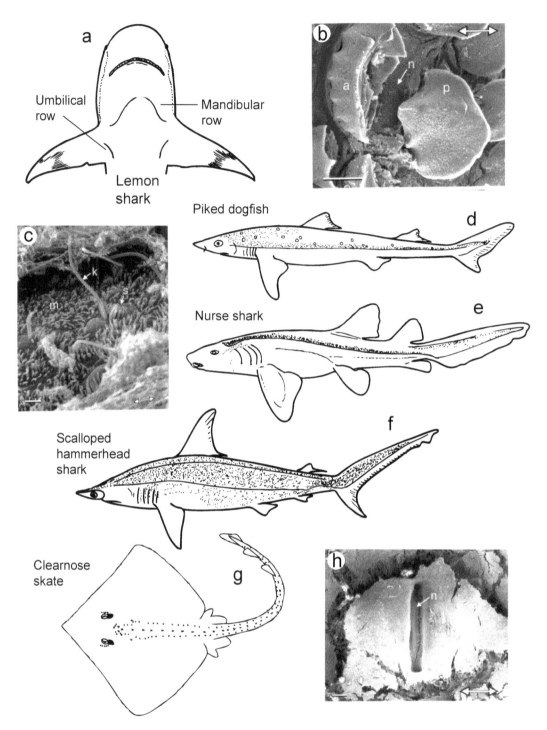

Fig. 7.4 Distribution of superficial neuromasts on sharks and rays. (a) Ventral surface showing the mandibular rows of hair cells posterior to jaw and umbilical rows of neuromasts between the pectoral fins of the lemon shark. Micrographs of (b) a free neuromast located on the mandible of the sicklefin lemon shark and (c) the kinocilium and adjacent stereocilium surrounded by microvilli of supporting cells. The distribution of neuromasts over the bodies of (d) the spiny dogfish, (e) the nurse shark, and (f) the scalloped hammerhead shark. (g) The distribution of neuromasts on the body of the clearnose skate along with (h) a micrograph of a neuromast of the side of the body of a cowtail ray. a = remnant of anterior denticle after trimming; k = kinocilium; m = microvilli; n = neuromast; p = posterior denticle.

In the rays, superficial neuromasts are often situated on small protuberances, or papillae, with the cupula with its cilia at the base of a central groove (see fig. 7.4h). Each protuberance rises half a millimeter above the surrounding skin and the cupula is exposed directly to water displacements. The typical distribution of neuromasts on the body of a ray is illustrated for the clearnose skate (*Raja eglanteria*) (fig. 7.4g).[10] Primary neuromasts, which are separated by a greater distance from each other, are distributed along the midline of a clearnose skate's body and extend from just behind the eyes all of the way to the caudal peduncle at the base of the caudal fin. Secondary neuromasts are located to either side of the primary ones and either even with or slightly posterior to the primary ones so that the grooves with their staircase arrangement of cilia range from being perpendicular to the head-to-tail axis to forming an oblique angle. This orientation furnishes insight into their function. The cilia are sensitive to water disturbances produced as the prey swims past the side of the skate as it remains motionless, half buried within the sand. The shoulders on either side of the groove prevent water movements from their own forward and backward movements within the sand from bending the hair cells. The clearnose skate also has rows of neuromasts on either side of its eyes. These grooves and their cilia, situated on that part of the ray most often uncovered by the sand, are oriented parallel to the snout-tail axis of the ray. They likely provide useful information about forward movements relative to the water column when the skate moves over the sand by performing walking movements of its pelvic fins.

The free neuromasts in chimaeras often appear as intermittent extensions of their continuous canals.[11] For example, there are lines of neuromasts that extend forward toward the snout and underneath it from the orbital canal that are apparent on a side view of the spotted ratfish (*Hydrolagus colliei*) (fig. 7.5a). Lines of neuromasts connect the oral and jugular canals on the underside of the head of this species (fig. 7.5d). These free neuromasts are neither wedged between dermal denticles as they are on sharks, nor situated within protuberances as they are on rays. The neuromast grooves are open to the water (fig. 7.5b). Their openness enables even the smallest water movements to stimulate the cilia in both the free and canal neuromasts. This type of neuromast provides a spatial sense to chimaeras, which move slowly, mainly in tight spaces, with their paddle-shaped fins; they rarely, if ever, swim quickly.

Both sharks and rays have neuromasts underneath the skin in long canals. These are either in contact with the external aqueous environment through tubules leading to surface pores or isolated from the external environment (fig. 7.3b).[7] Water within the canal moves to exert a frictional force displacing the gelatinous cupula with the embedded ciliary bundles relative to the stationary sensory epithelium. Each canal has many hair cells, each with stereocilia and a

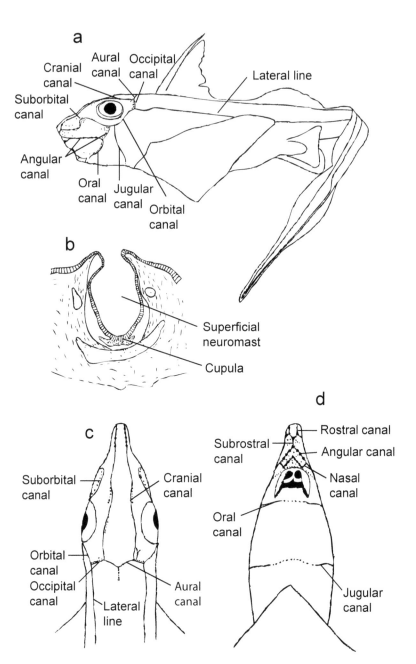

Fig. 7.5 (a) Side view of superficial neuromasts, cranial canals, and lateral canal system of the spotted ratfish with (b) enlargement of a neuromast within the intermittent anterior extension of the orbital canal. (c) Dorsal view and (d) ventral view of the cranial canals of the spotted ratfish.

single kinocilium, with numerous microvilli at the apical surface. The polarity of the hair cells is related to the direction of the staircase increase in the height of the successive stereocilia nearer to the kinocilium. This is usually parallel to the main canal axis, although a small number of hair cells may be oriented at an oblique angle to the axis. This is shown in a micrograph of the hair cells of the Atlantic stingray (*Dasyatis sabina*). The axis along which the main canal passes

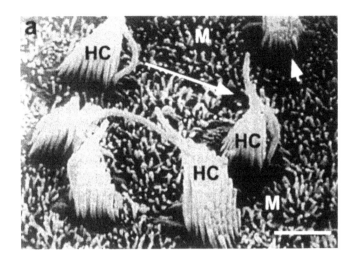

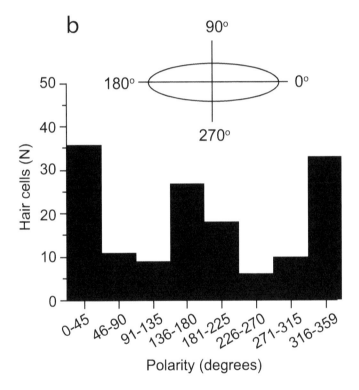

Fig. 7.6 (a) Scanning electron micrograph of the hyomandibular canal of the Atlantic stingray. The polarity of the majority of the hair cells is roughly parallel to the main axis of the canal (long arrow), but the polarity of a few hair cells is off the main axis (short arrow). HC = hair cells; M = microvilli. (b) Histogram of the frequency distribution of the polarities of hair cells in the neuromasts of the hyo-mandibular canal of the Atlantic stingray.

is indicated by a long arrow, an oblique orientation to this axis by a small arrow (fig. 7.6a). The main axis of the canal lies along the line between 0 and 180° (see oval in fig. 7.6b). Notice that the highest bars in the histogram, indicating the highest counts of hair cells, are on its sides and middle. The maximum numbers of hair cells are in two classes, 0–45° and 316–359°, nearest to 0° and in the two classes, 136–180° and 181–225°, closest to 180°. More than 75% of the hair cell polarities are oriented in a direction within 45° of the snout-to-tail axis of the ray.

Typically in sharks and rays neuromasts are organized in three systems of canals. They are present in the lateral canals that run from head to tail on either side of the body, the supraorbital canals forward of the eyes, and the infraorbital canals even with or posterior to the eyes. The supraorbital canal branches into the hyomandibular canal, which leads to the supratemporal canal, which unites with hyomandibular canal on the other side of the rostrum. The distribution of canals is similar underneath the rostrum except that a mandibular canal branches off from the hyomandibular canal and extends forward and parallel to the mouth.[10]

The lateral line of bonnethead sharks (*Sphyrna tiburo*), which swim along the bottom searching for prey buried in the substrate, extends from the back of the head to the tip of the tail with many lateral tubules extending dorsally and ventrally over the trunk (fig. 7.7a). The lateral tubules extend farther ventrally on the trunk near the caudal peduncle and on the caudal fin. This extensive system of pored canals on the shark's side may enable it to perceive the general water movements around it as it forages in the shallow bays of the southeastern Atlantic Ocean. All of the neuromast canals are connected to each other near the head. The superorbital canal has many lateral tubules that branch out from them and are concentrated along the anterior margin of the laterally expanded rostrum with the highest density of canals toward the center of the rostrum. The superorbital canal on the ventral side of the head branches into left and right canals that lead to the margin of the rostrum, where they turn perpendicularly and extend along the margin to the nares, with many lateral tubules leading to pores at the edge of the head. The infraorbital canal on the dorsal surface of the head extends inward toward the midline before extending outward again, with this section having many smaller canals branching out from it. Two infraorbital canals lead outward on the ventral surface of the head with many branched tubules. The hyomandibular canal on the dorsal surface consists of a small branch that extends from the infraorbital canal outward toward the eye and a second branch that leads posteriorly to the lateral line canal. The hyomandibular canal on the ventral surface is composed of a loop that extends from the eye to the lateral line with many smaller canals branching outward from it. The ample distribution of pored canals on both the dorsal and ventral surfaces of the rostrum enables the bonnethead shark to detect its prey by the water motions it generates. The canals are present across the entire expanded rostrum. This provides a larger surface for detecting particle displacements than the reef sharks possess on their conical rostrum. The bonnethead lives in shallow bays where it swims over the bottom searching at night for benthic crabs and shrimps moving over the sub-

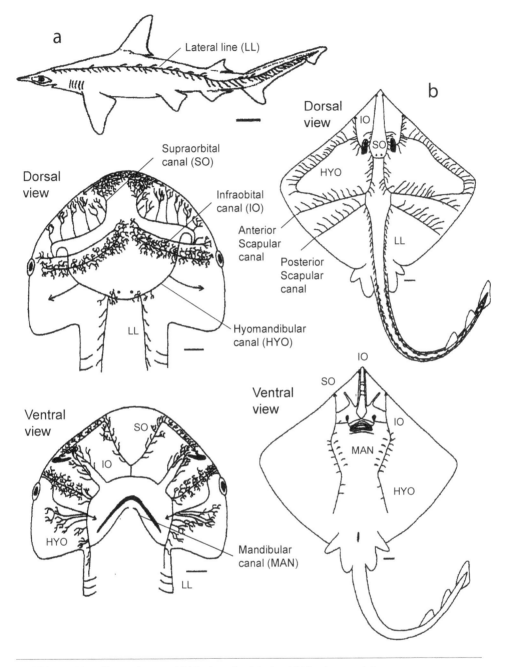

Fig. 7.7 Views of lateral line canal system of (a) the bonnethead shark and (b) the clearnose skate.

strate or clams buried within the substrate. The shrimp and crabs are detected by the water motions they generate as they accelerate suddenly upon being disturbed while the bivalves may be located in the jet of water emitted through their siphon while buried in the mud out of sight.

The clearnose skate (*Raja eglanteria*) lives in deeper water, where it feeds on clams, snails, and small flatfishes buried in the sand. The lateral line of the clearnose skate extends from the mid-disk to the upper lobe of the caudal fin (fig. 7.7b). The supraorbital, infraorbital, and scapular canals on the dorsal surface of the ray are straight and unbranching canals that extend outward toward the margin of the disk, where they terminate in single pores. The hyomandibular canal originates from the infraorbital canal and extends ventrally along the pectoral margin before it is directed inward to join the anterior branch of the scapular canal. This pore arrangement makes them sensitive to water motions above them while on the bottom lying in wait for above-swimming prey.[10] The same canals on the ventral surface have few pores, and it is likely that these receptors are more sensitive to touch than to water motions. The ventral hyomandibular canals do have a few single tubules extending outward on either side of the gills with single pore openings. The presence of mostly non-pored canals and absence of branching pored canals on the ventrum is consistent the skate's at times moving close to the bottom and detecting its prey by contact.

The canals of the chimaeras are all open to the water.[11] The lateral line canal extends midway on the side of the body from just posterior to the eye to the tip of the tail. The line drops rather suddenly on the tail and extends along its lower edge to its tip (fig. 7.5a). The lateral line bifurcates near the eye into the occipital canal, which branches upward, and orbital canal, which branches downward. The occipital canal separates into the aural canal on the top of the head and cranial canal that extends across the forehead to the tip of the snout (fig. 7.5a,c). The orbital canal bifurcates into the suborbital canal, which curves under the eye and extends to the tip of the rostrum, and the angular canal, which passes first downward and then forward to the tip of the rostrum forming a right angle. The oral canal extends from the angular canal downward to the ventral rostrum. The jugular canal branches downward from the orbital canal below the eye and continues to the ventrum of the rostrum (fig. 7.5a,d). The chimaera's many canals on the rostrum enable it to perceive directional movements of water near its head and likely aids in localizing its prey.

These neuromasts are enclosed in subepidermal pouches, which are present only on the ventral surface of some species of rays.[10] The vesicles are found on the Brazilian electric ray (*Narcine brasiliensis*). Each vesicle contains a large central neuromast and smaller peripheral neuromasts above and below it. All of the neuromasts have a common afferent nerve and are encased within a thin walled pouch, or pedicle (fig. 7.8a). The cupula of the central neuromast appears denser than the cupulae of smaller neuromasts. The walls of the pouch are thin, but pedicles of tough connective tissue on either side come together to form a roof for the vesicle.

The vesicles of Savi are on the dorsal surface of the electric ray in two bilateral rows leading from the eyes to the anterior margin of the rostrum (fig.

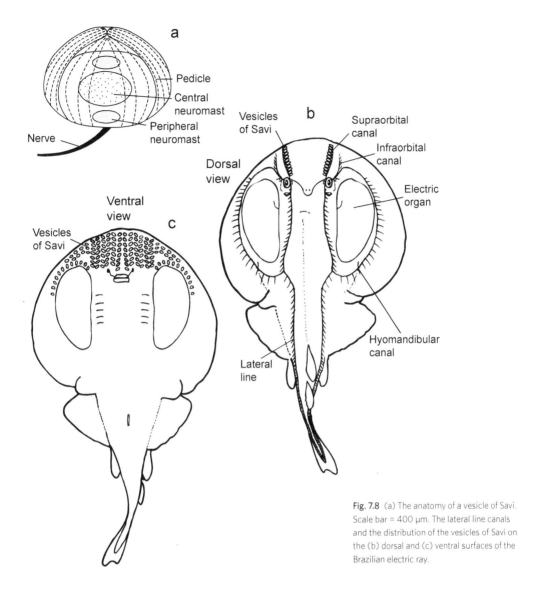

Fig. 7.8 (a) The anatomy of a vesicle of Savi. Scale bar = 400 μm. The lateral line canals and the distribution of the vesicles of Savi on the (b) dorsal and (c) ventral surfaces of the Brazilian electric ray.

7.8b). Each row is composed of 8 to10 vesicles that are oval with the long axis of each oriented at an angle roughly 30° to the ray's body axis. On the outer side and close to the vessels are non-pored supraorbital canals, and farther away are infraorbital canals with outward-extending tubules with terminal openings. There are vesicles on the ventral rostrum in front and to the side of the mouth. The hyomandibular canals encircle the electric organs with many tubules leading from them to the margin of the pectoral disk. There are no lateral line canals on the ventral surface of the electric ray.

The vesicles of Savi are thought to function as specialized tactile receptors that are sensitive to the displacement of the surrounding skin during contact with the substrate, a conspecific, or prey.[10] Their concentration on the rostrum in front of the mouth is consistent with their being used to locate prey near the mouth. It is unlikely that they are sensitive to water motion because they do not have a direct connection with the external environment, being enclosed within a pouch in the tissues of the ray.

Internal Mechanoreceptors Sensitive to Pressure Oscillations

As a sound wave is propagated outward by a catfish's oscillating swim bladder, energy is imparted to the water particles that move back and forth about a fixed point without any net movement in one direction. However, they transfer this energy to adjacent particles, which again oscillate around fixed points at a greater distance from the sound source. These oscillations that make up the distress call are propagated farther away from the swim bladder with each successive group of particles moving back and forth slightly less. At some distance, the relative amplitude of the back-and-forth movements of water exceeds the minute one-way displacements of water. This is the boundary between the near and far fields of sound propagation. The one-way displacements produced by the grouper's tail are dominant in the near field, where they are detected by the superficial and canal neuromasts. The rapidly expanding and contracting swim bladder of the marine catfish, on the other hand, produces a wave that propagates well out into the far field. The cartilaginous fishes have a labyrinth, or inner ear, on either side of their head that is sensitive to both the one-way and two-way water displacements in the near field. However, this organ is also sensitive to the pressure oscillations in the far field unlike the superficial and canal neuromasts. The reef sharks in Biscayne Bay turned toward the source of sounds of struggling fish at least 200 m from the underwater speaker that propagated them. This distance exceeded the 150 m boundary to near field of 10 Hz by at least 50 m.

Inner Ear, or Labyrinth

The inner ears are located in otic capsules surrounded by cartilage on either side of the braincase in the rear part of the chondrocranium. Each inner ear, also referred to as the labyrinth due to its elaborate internal network of passages, consists of three membranous semicircular canals with enlarged sacs at their bases (fig. 7.9a). There are also two large chambers, the utriculus and sacculus, with the latter continuing at one end and expanding to form the lagena. Each of these chambers contains a macula with many hair cells. These resemble the superficial and lateral line canal neuromasts in their staircase arrangement of many stereocilia, successively increasing in height toward a single kinocilium.

There is an additional sensory epithelium containing hair cells, the macula neglecta. There are two separate sensory epithelia in the macula neglecta in sharks but only one epithelium in rays. Each is located along a duct that extends from the base of the endolymphatic duct to the saccular chamber. There are many hair cells on each macula. The number of hair cells in this particular sensory epithelium ranges from 7,000 in the piked dogfish (*Squalus acanthias*) and bat eagle ray (*Myliobatis californica*) to 260,000 in the blacktail reef shark (*Carcharhinus melanopterus*).[12] Each macula is covered by a mucilaginous cupula, which contains numerous otoconia (calcium carbonate granules). The hair cells are moved back and forth by the cupula with its embedded otoconia relative to the stationary basal tissue due to their different density. This occurs in response to the particle displacements associated with pressure oscillations. Movements of the stereocilia toward the kinocilium on the hair cell produce an increase in the frequency of nerve discharge while movements in the opposite direction result in a decrease in the frequency of nerve discharge.[13]

Vibratory stimuli have been applied to the different macula within the labyrinth and their nerve output recorded by electrophysiological methods. The sensitivity to these vibrations is greatest when measured from cells within the macula neglecta of the thornback ray (*Raja clavata*), and hence, it has been implicated as the part of the labyrinth involved in sound detection by elasmobranchs.[14] Furthermore, the neural responses to auditory click stimuli recorded from the eighth nerve were only slightly reduced by cutting the nerves coming from the sacculus but were stopped completely by severing the nerves originating from the macula neglecta.[14] Finally, the largest neural responses in the specific area of the brain associated with processing of sound were evoked in the blacktip reef shark when a sound source was positioned over the macula neglecta. The neural output was less when the sound source was moved to either side of the region of head above the macula.[15] These studies indicate that the macula neglecta is the important portal for sound detection in cartilaginous fishes.

The macula neglecta is located on the sides the interior chamber, or lumen, of the posterior canal duct. Two of these ducts extend downward and outward from an ovoid area at the top of the head of the blacktip reef shark (*Carcharhinus melanopterus*). This organ can be seen from a top (fig. 7.9c) and a transverse (fig. 7.5d) view of the shark's head.[15] The loose connective tissue, indicated by stippling in the top view and cross-hatching in the transverse view, is between the skin and the depressed dorsal surface of the chondrocranium. The connective tissue is likely located here because it conducts the back-and-forth particle displacements associated with pressure oscillations to the labyrinth. The fenestra ovalis, an opening through the cartilaginous chondrocranium that connects this connective tissue with the posterior canal ducts, is apparent in a

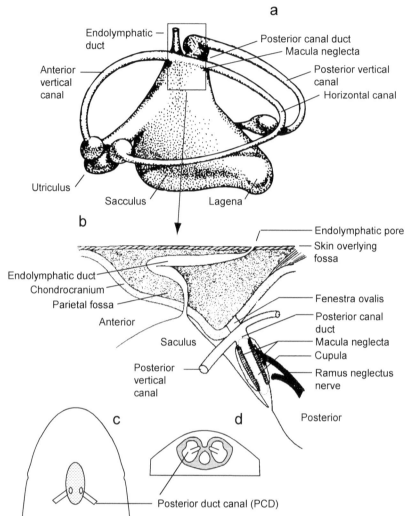

Fig. 7.9 (a) The inner ear of the carcharhinid shark. (b) Cross-section along long axis of a shark showing the two epithelia comprising the macula neglecta on either side of the posterior canal duct and the overlying parietal fossa. Two views of the posterior canal ducts: (c) dorsal view on the left and (d) transverse view on the right.

cross-section along the long axis of the shark's head showing the parietal fossa, the fenestra ovalis, and the epithelia making up the macula neglecta on the lumen, or expanded chamber, of the posterior canal duct (fig. 7.9b).[16] The particle displacements are transmitted through the fenestra ovalis into the posterior canal duct resulting in the vibration of the cupula, in which the hair cells are embedded. These minute movements increase the electrical discharge of the ramus neglectus nerve, which is a branch of the eighth cranial nerve.

The inner ears of a greater diversity of sharks and rays have now been examined many years after the pioneering studies by Jeffery Corwin and categorized into four distinct morphotypes, which also appear more related to the

Jeffery Corwin, while a graduate student in the Comparative Sensory Physiology Laboratory of Theodore Bullock at Scripps Institution of Oceanography, conducted pioneering studies leading to insight as to the functional significance of the macula neglecta. He related its anatomy to the different feeding tactics of sharks and rays.[12] The utriculus, sacculus, lagena, and semicircular canals do not differ grossly in size and shape between sharks and rays. This is evident if you examine the inner ears of the broadnose sevengill shark (*Notorynchus cepedianus*), nurse shark, blacktail reef shark, and bat eagle ray (fig 7.10). In contrast, the posterior canal duct, identified by an arrow on the labyrinth, differs greatly in size and shape among the species. The canal ducts of the sevengill and blacktip reef sharks are wider and their ducts situated closer to the parietal fossa and fenestra ovalis than the nurse shark and bat eagle ray. Sound is conducted through this path to the macula neglecta. Look on the labyrinths for the arrows that point to the ducts to appreciate their size in the broadnose sevengill and blacktail reef sharks compared to the nurse shark and bat eagle ray. Scanning micrographs illustrate that each sensory epithelium is usually composed of two maculae. A large epithelium forms the posterior wall of the lumen of the posterior canal duct while a smaller epithelium lines the anterior wall of the lumen. Only a single epithelium exists in the posterior area of the lumen of the bat ray's canal duct. The size of the macula and number of hair cells vary between the species. The broadnose sevengill shark and blacktail reef shark have expansive epithelia with many hair cells, 25,000 hair cells in a 2.5 kg specimen of the former and 260,000 in a 16 kg specimen of the latter. The nurse shark and bat ray have smaller epithelia with relatively few hair cells, 13,000 cells in a 34 kg specimen of the former species and 7,000 cells in a 7 kg specimen of the latter. Furthermore, there is a difference in the polarities of the hair cells within the epithelia. The polarities of the hair cells are indicated on the epithelia of the four elasmobranches by arrows pointing in the direction of the bundle's kinocilium. The polarities of the bundles in the sixgill and reef sharks are all aligned in one direction. The bundles on the anterior macula point one way and those on the posterior macula point in the opposite direction. On the other hand, the bundles on the maculae of the nurse shark and bat eagle ray are oriented in many different directions.

These differences in the morphology of the macula neglecta among elasmobranchs are not explicable by their ancestral relationships but make more sense based on their different feeding modes. One would not expect the location, size, and number of hairs in the macula of the more advanced blacktail reef shark to be more similar to those of the primitive broadnose sevengill shark than the more recently evolved nurse shark and bat eagle ray. A more plausible explanation relates to their different feeding tactics. The broadnose sevengill and blacktip reef sharks swim up in the water column, locate prey from a considerable distance, and upon nearing them chase them down and seize them. The nurse shark and bat eagle ray, on the other hand, spend most of their time swimming close to the bottom and feed either on gastropods on the bottom or excavate mollusks buried in the sand or mud. The maculae of the nurse shark and bat ray likely have fewer hair cells because they need not have great hearing sensitivity to detect their prey as they pass close to them. Furthermore, their hair cells are likely aligned in multiple directions to detect prey on either side of the body as they search for prey on the bottom. The many hair cells in a unidirectional alignment provide the blacktail reef and broadnose sevengill sharks with greater hearing sensitivity and the ability to orient directionally to their prey at a distance.

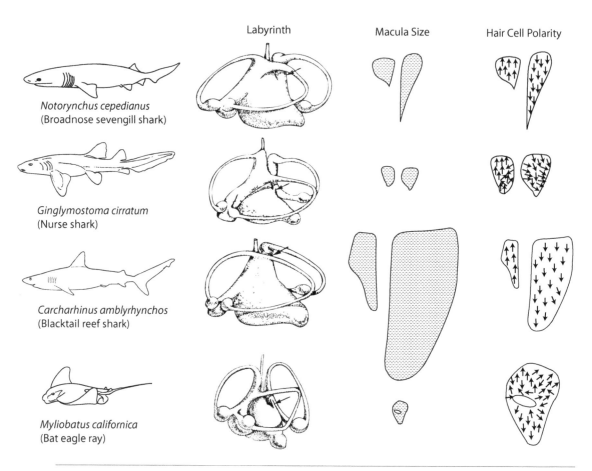

| Labyrinth | Macula Size | Hair Cell Polarity |

Notorynchus cepedianus
(Broadnose sevengill shark)

Ginglymostoma cirratum
(Nurse shark)

Carcharhinus amblyrhynchos
(Blacktail reef shark)

Myliobatus californica
(Bat eagle ray)

Fig. 7.10 The anatomy of the labyrinth, size and shape of the macula, and the polarities of the hair cells populating the maculae neglectae of the broadnose sevengill shark, nurse shark, blacktail reef shark, and the bat eagle ray.

feeding styles of the species than to their evolutionary histories.[17] The inter-specific variation in labyrinth morphology was described for twenty-nine species of sharks and rays. The constitution of the cartilage within the otic capsule protecting the inner ear of the sharks was found to differ from that of the rays—the former tissue was rigid and hard while the latter was soft and flexible. The ampullae at the end of the canals also differed between sharks and rays—the ampullae on the posterior canals of the former were more elongate and tubular than those of the latter species, which were spherical. The sharks and rays were separated into four groups based on quantitative measurements of the external morphology of their inner ears. The defining feature of the first group is the clearly visible canal ducts separating the otoconial organs, the utriculus, sacculus, and lagena, from the semicircular canals and the clearly defined otoconial organs. Fig. 7.11 presents a generalized diagram of this type of inner ear along with a micrograph of the inner ear of the blue-spotted sting-

Group	Typical Inner Ear	Species	Predicted Feeding Strategy	Micrograph of Inner Ear
1	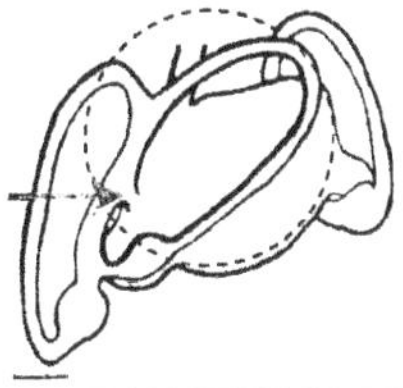	Bat eagle ray	Non raptorial	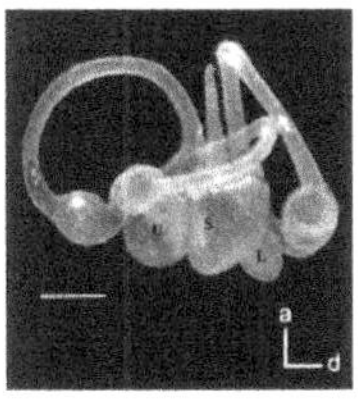
		Blue spotted stingray	Non raptorial	
		Estuary stingray	Non raptorial	
		Short-tail stingray	Non raptorial	
		Sparcely spotted stingaree	Non raptorial	
		Spotted eagle ray	Non raptorial	
		Stingaree sp.	Non raptorial	
2	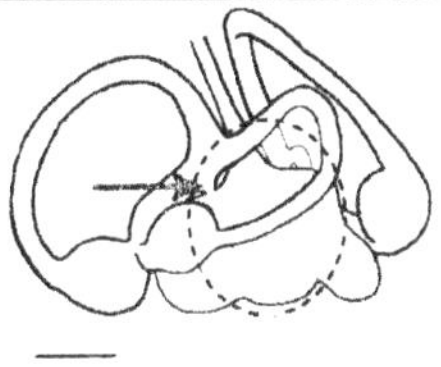	New Zealand torpedo	Sit and wait	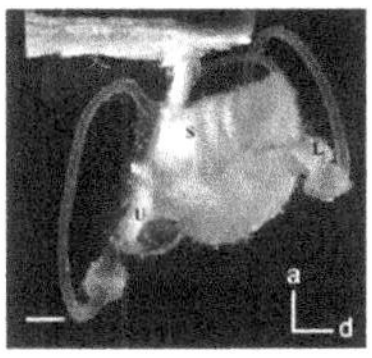
		Smooth butterfly ray	Raptorial	
		Tawny nurse shark	Non raptorial	
3	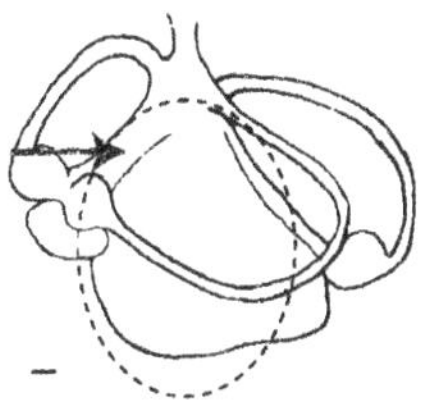	Australian sharpnose shark	Raptorial	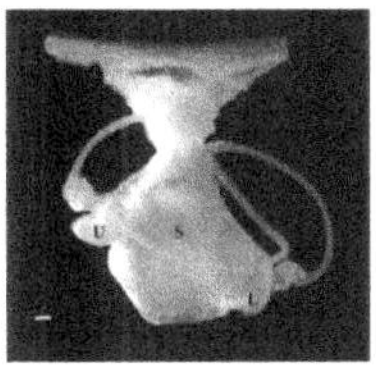
		Blue shark	Raptorial	
		Bramble shark	Raptorial	
		Broadnose sevengill shark	Raptorial	
		Dusky smooth-hound	Raptorial	
		Eastern shovelnose ray	Raptorial	
		Leafscale gulper shark	Raptorial	
		Longnose velvet dogfish	Raptorial	
		Sandbar shark	Raptorial	
		Southern lanternshark	Raptorial	
		Spotted wobbegong shark	Sit and Wait	
4	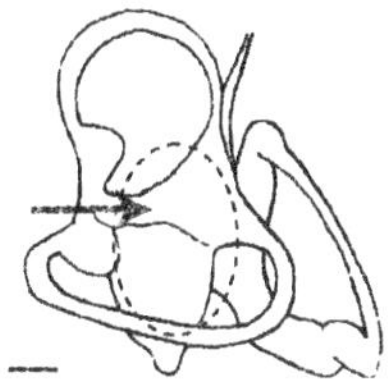	Birdbeak dogfish	Raptorial	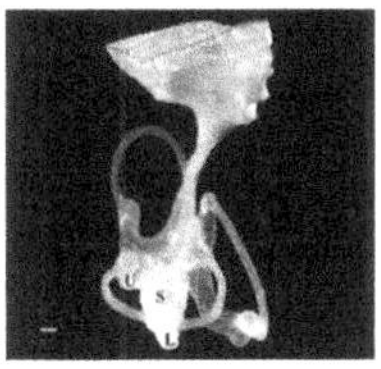
		Bluntnose sixgill shark	Raptorial	
		Brownbanded bamboo shark	Non raptorial	
		Giant shovelnose ray	Non raptorial	
		Piked dogfish	Raptorial	
		Porbeagle	Raptorial	
		Port Jackson shark	Non Raptorial	
		Thresher	Raptorial	

Fig. 7.11 Diagrams of four general types of inner ears, grouped by anatomical similarity, of sharks and rays with members of each group identified with a micrograph of the inner ear of a representative (bold type). An arrow indicates where the endolymphatic canal continues toward the parietal fossae, the natural path of pressure oscillations to the inner ear.

ray (*Neotrygon kuhlii*) (fig. 7.11). There were seven species of rays that had this morphotype, and they were all non-raptorial species, which did not chase and capture their prey but located either crustaceans or fishes walking or swimming close to the bottom or mollusks buried in the mud or sands. The inner ear of second group of sharks and rays is similar to that of the first group, but they have a larger sacculus, utriculus, and lagena. A generalized diagram of this ear type is shown together with a micrograph of the inner ear of the smooth butterfly ray (*Gymnura micrura*). Although this species is related in an evolutionary sense to the members of the prior group, it has a broader diet than the former, not foraging only on crustaceans and benthic prey, but actively chasing and capturing fishes. The third and fourth groups are separated from the former two groups because their semicircular canals are bound in part to the dorsal surface of the sacculus, giving these ears a triangular appearance. The members of the third group are characterized by a larger sacculus than that of the fourth group. A generalized diagram of this ear morphotype is given together with a micrograph of the inner ear of the sandbar shark (*Carcharhinus plumbeus*). Nine of ten members of this group were sharks, with only a single ray, the eastern shovelnose ray (*Aptychotrema rostrata*). The majority of this group such as the members of the family Carcharhinidae are raptorial foragers, feeding on mobile prey items such as cephalopods, fishes, and other sharks, whereas a few in the family Orectolobidae are ambushers, exhibiting a sit-and-wait predatory strategy. A generalized diagram is presented of the fourth ear morphotype, with a smaller sacculus, with a micrograph of the inner ear of the Port Jackson shark (*Heterodontus portusjacksoni*). The members of the fourth group are epibenthic, such as the bluntnose sevengill shark, brownbanded bambooshark (*Chiloscyllium punctatum*), piked dogfish, and Port Jackson shark, or more active, such as the porbeagle shark (*Lamna nasus*) and thresher shark (*Alopias vulpinus*).

Complicating any explanation of an affinity of lifestyle is the fact that some are more closely evolutionarily related than others and that some are ectotherms and others endotherms. A phylogenetic tree was created for the genera of sharks and rays included in the study showing the ear morphotype for each (fig. 7.12). The sharks and rays are separated by the first branching of the tree due to their evolutionary affinity. The sharks possess the third and fourth morphotypes of inner ear, while the rays have the first and second. However, there are exceptions to this rule. For example, sharks of the genus *Neburis* have inner ears of the second morphotype, while shovelnose rays of the genus *Aptychotrema* have inner ears with the third morphotype and those of *Glaucostegus* have labyrinths with the fourth morphotypes. In conclusion, there is no clear indication that the structure of inner ears of the sharks and rays is related solely to their behavioral and ecological lifestyles; there appear to be both phylogenetic and functional factors driving the inner ear morphological variation.

HEARING SENSITIVITY

The hearing sensitivity has been determined for some species of sharks, but not for the rays and chimaeras.[18] Thresholds, or the minimum intensity at which sounds can be detected, have been determined both for the one-way displacement of water particles associated with the tail beating of the mullet and the back-and-forth displacements associated with pressure oscillations emitted by the pulsating swim bladder of the marine catfish. The horn shark perceives one-way displacements ranging from 30 to 120 Hz with its maximum

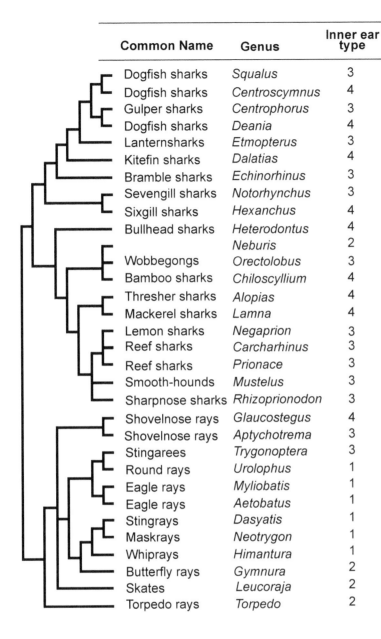

Common Name	Genus	Inner ear type
Dogfish sharks	*Squalus*	3
Dogfish sharks	*Centroscymnus*	4
Gulper sharks	*Centrophorus*	3
Dogfish sharks	*Deania*	4
Lanternsharks	*Etmopterus*	3
Kitefin sharks	*Dalatias*	4
Bramble sharks	*Echinorhinus*	3
Sevengill sharks	*Notorhynchus*	3
Sixgill sharks	*Hexanchus*	4
Bullhead sharks	*Heterodontus*	4
	Neburis	2
Wobbegongs	*Orectolobus*	3
Bamboo sharks	*Chiloscyllium*	4
Thresher sharks	*Alopias*	4
Mackerel sharks	*Lamna*	4
Lemon sharks	*Negaprion*	3
Reef sharks	*Carcharhinus*	3
Reef sharks	*Prionace*	3
Smooth-hounds	*Mustelus*	3
Sharpnose sharks	*Rhizoprionodon*	3
Shovelnose rays	*Glaucostegus*	4
Shovelnose rays	*Aptychotrema*	3
Stingarees	*Trygonoptera*	3
Round rays	*Urolophus*	1
Eagle rays	*Myliobatis*	1
Eagle rays	*Aetobatus*	1
Stingrays	*Dasyatis*	1
Maskrays	*Neotrygon*	1
Whiprays	*Himantura*	1
Butterfly rays	*Gymnura*	2
Skates	*Leucoraja*	2
Torpedo rays	*Torpedo*	2

Fig. 7.12 A phylogenetic tree, clustering the sharks and rays by evolutionary affinity, indicating the morphotype of the inner ear of each member of tree. Phylogenetic relatedness increases from left to right. Note that there are two branches farthest to the left separating the sharks from the rays. The inner ears of the sharks have a morphological structure typical of groups 3 and 4 and rays 1 and 2, but there are exceptions to this rule.

sensitivity at 80 Hz[19] (fig. 7.13a), whereas the lemon shark detects displacements of a frequency as low as 20 to 1,000 Hz with its sensitivity increasing to a maximum at 1,000 Hz.[20]

Thresholds of the detection of the back-and-forth particle movements associated with pressure oscillations have been determined for three species of sharks: the horn shark,[19] the lemon shark,[20] and the bull shark.[21] The horn shark is sensitive to pressure oscillations ranging from 20 to 120 Hz with its

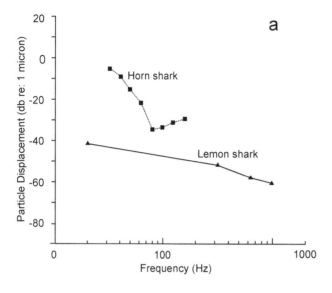

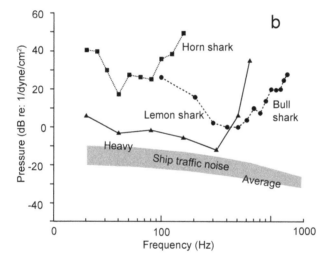

Fig. 7.13 Hearing sensitivities of sharks. (a) Particle displacement thresholds for the horn shark (solid squares) and lemon shark (solid triangles). (b) Sound-pressure thresholds for the horn shark (solid squares), lemon shark (solid triangles), and bull shark (solid circles) with an indication of average and high levels of the ambient noise produced by ship traffic.

greatest sensitivity at 40 Hz (fig. 7.13b); the lemon shark is sensitive from 10 to 900 Hz with its maximum sensitivity over a broad range of frequencies from the lower limit to 300 Hz, and the bull shark is sensitive from 100 to 1,200 Hz with its maximum sensitivity between 400 and 500 Hz. A possible explanation of the bull shark's absence of sensitivity at the lower frequencies was the inability of the sound transducer used in this early study to produce pressure oscillations at these low frequencies.

Based on these thresholds, it is apparent that sharks are most sensitive to low-frequency sounds. There is a diversity of such sounds underwater, where these low frequencies propagate the greatest distance. Hence, the ability to perceive them is critical to sharks' foraging success in the marine environment.

SOUND LOCALIZATION AND NOISE POLLUTION

There is evidence that sharks are capable of locating their prey from considerable distances, swimming toward them, and then pursuing and seizing them. Inshore sharks were attracted from considerable distances to a speaker that broadcast low-frequency pulsed sounds in Biscayne Bay. Art Myrberg and his colleagues were able to attract silky sharks (*Carcharhinus falciformis*) and oceanic whitetip sharks (*Carcharhinus longimanus*) circling a large buoy in the deepwaters off Andros Island in the Bahamas to a speaker from a distance of several hundred meters. The authors observed these sharks approach the speaker from the direction of the buoy while treading water within the protective confines of a shark cage. Terrestrial species locate their prey using hearing on the basis of the slight delay between when the sound wave impinges on the eardrum nearer to the source of the sound and when it reaches the eardrum farther from the source, and by minute differences in the loudness of the sound, which is slightly louder to the nearer ear. This mode of localization is more difficult for a predator in water, in which sound travels more rapidly and diminishes in intensity less rapidly with distance. Now the difference in the time of arrival of the sound wave is much less and the intensity gradient between the two closely separated inner ears is much smaller. Yet sharks appear capable of localizing their prey.

Two potential mechanisms have been proposed for how they accomplish this. According to one theory, the direction to the source of the sound is determined by a timing analysis between the direct and indirect sounds as they reach the shark's inner ears. A sound stimulus reaches the inner ear along two paths, a direct path between the source and the ear and an indirect path after reflecting off the ocean surface. The sound reflected off the soft sea surface will differ from the direct sound in the reversal of the phase of its particle displacement component. The two waves stimulate differently oriented hair cells in one or more maculae, and the shark then determines direction from this information.[22] Alternatively, it has been proposed that the parallel polarization of the hair cells in the macula neglecta enables a shark to resolve the direction to the prey. This is accomplished in conjunction with the reception of a sound by maculae in the sacculus in another part of the inner ear.[23]

More and more boats are taking to the oceans, and these boats are producing high levels of noise in the underwater oceanic environment. Furthermore, the United States Navy is periodically broadcasting broadband white noise and measuring the speed of its travel between widely separated locations to measure the temperature of the ocean in the intervening path. Sound travels faster in warmer water than in colder water. Hence, the time interval between the onset of the sound and its detection on the other side of the ocean is an indicator of the overall warmth of the ocean. These measurements can be used to determine whether the oceans are warming due to the buildup of carbon dioxide in the atmosphere. Finally, the Navy is now operating an extraordinarily high-intensity low-frequency sonar that is capable of reflecting off the hull of enemy submarines. All of these sources of sound are cumulatively making the oceans a noisy place. This noise may at some time limit the ability of the cartilaginous fishes to hear the sounds of their prey from great distances. Some level of masking, or loss of signal detection, is believed to occur if a signal-to-noise ratio of a biological sound to the ambient noise level is 20 decibels or less. In fig. 7.13b, the noise levels of ship traffic in the ocean are indicated by a gray band on the audiogram of the hearing sensitivities of the horn, lemon, and bull sharks to pressure oscillations. Note that the lower and upper edges of the band indicate the average and heavy levels.[24] The hearing sensitivity of the lemon shark varies from 5 to 10 decibels from the levels of noise at frequencies of less than 300 Hz. Hence, it is possible that the hearing ability of lemon sharks is reduced by the noise in the waters off Bimini, an island across the Gulf Stream from Miami. It is possible that the distance at which other sharks, rays, and chimaeras detect their prey may also be reduced by cacophony produced by the many ships that move between continents.

SUMMARY

Sharks, rays, and chimaeras can locate their prey from considerable distances. Adult reef sharks can detect a grouper struggling at the end of a spear from a distance of 200 m. Juvenile sharks can detect sudden accelerations of their prey or distress calls produced by vibrations of their swim bladder from shorter distances of up to 20 m. This sense is called mechanoreception because mechanical disturbances of the water are detected by two organ systems. The free neuromasts, canal neuromasts, and vessels of Savi on the external body are sensitive to one-way displacements of particles of water. The inner ear is sensitive not only to these displacements but also to two-way displacements of water particles associated with pressure oscillations. The former receptors are active in the near field, where the one-way oscillations exceed the two-way oscillations; the latter receptor is active in both the near and far field, where the two-way oscillations are dominant. The boundary between these fields for a very low frequency of 10 Hz is 150 m while for 100 Hz it is only 15 m. Used together, these two senses enable cartilaginous fishes to find their prey and avoid their predators from considerable distances. The anatomy of the free neuromasts, neuromast canals, and inner ear varies not only between the sharks and rays. For example, pelagic sharks that hunt prey in midwater have larger maculae in their inner ears with more polarized hair cells than do benthic sharks and rays. The increase in the amount of ambient noise in the ocean may reduce the ability of the chondrichthyan fishes to detect their prey from such great distances.

KEY TO COMMON AND SCIENTIFIC NAMES

Atlantic stingray = *Dasyatis sabina*; bat eagle ray = *Myliobatis californica*; birdbeak dogfish = *Deania calcea*; blacktail reef shark = *Carcharhinus amblyrhynchos*; blacktip reef shark = *Carcharhinus melanopterus*; blue shark = *Prionace glauca*; blue-spotted stingray = *Neotrygon kuhlii*; bluntnose sixgill shark = *Hexanchus griseus*; bonnethead shark = *Sphyrna tiburo*; bramble shark = *Echinorhinus brucus*; Brazilian electric ray = *Narcine brasiliensis*; broadnose sevengill shark = *Notorynchus cepedianus*; brownbanded bambooshark = *Chiloscyllium punctatum*; bull shark = *Carcharhinus leucas*; clearnose skate = *Raja eglanteria*; cowtail stingray = *Pastinachus sephen*; dusky smooth-hound shark = *Mustelus canis*; eastern shovelnose ray = *Aptychotrema rostrata*; electric ray = *Torpedo nobiliana*; estuary stingray = *Dasyatis fluviorum*; flathead gray mullet = *Mugil cephalus*; giant shovelnose ray = *Glaucostegus typus*; gray whale = *Eschrichtius robustus*; horn shark = *Heterodontus francisci*; killer whale = *Orcinus orca*; kitefin shark = *Dalatias licha*; leafscale gulper shark = *Centrophorus squamosus*; lemon shark = *Negaprion brevirostris*; marine catfish = *Galeichthys felis*; New Zealand torpedo = *Torpedo fairchildi*; nurse shark = *Ginglymostoma cirratum*; oceanic whitetip shark = *Carcharhinus longimanus*; piked dogfish = *Squalus acanthias*; pink whipray = *Himantura fai*; porbeagle shark = *Lamna nasus*; Port Jackson shark = *Heterodontus portusjacksoni*; sandbar shark = *Carcharhinus plumbeus*; scalloped hammerhead shark = *Sphyrna lewini*; sheepshead minnow = *Cyprinodon variegatus*; short-tail stingray = *Dasyatis brevicaudata*; sicklefin lemon shark = *Negaprion acutidens*; silky shark = *Carcharhinus falciformis*; smooth butterfly ray = *Gymnura micrura*; southern lantern shark = *Etmopterus granulosus*; sparcely-spotted stingaree = *Urolophus paucimaculatus*; spotted eagle ray = *Aetobatus narinari*; spotted ratfish = *Hydrolagus colliei*; spotted wobbegong shark = *Orectolobus maculatus*; tawny nurse shark = *Nebrius ferrugineus*; thresher shark = *Alopias vulpinus*; thornback ray = *Raja clavata*; tiger shark = *Galeocerdo cuvier*; whale shark = *Rhincodon typus*.

LITERATURE CITED

1. Nelson and Gruber, 1963; 2. Nelson, 1969; 3. Banner, 1972; 4. Cummings and Thompson, 1971; 5. Klimley and Myrberg, 1979; 6. Myrberg, 2001; 7. Montgomery, 1988; 8. Combs and Montgomery, 1999; 9. Peach and Marshall, 2009; 10. Maruska, 2001; 11. Reese, 1910; 12. Corwin, 1978; 13. Flock, 1967; 14. Lowenstein and Roberts, 1951; 15. Corwin, 1981; 16. Fay *et al.*, 1974; 17. Evangelista *et al.*, 2010; 18. Fay, 1988; 19. Kelly and Nelson, 1975; 20. Banner, 1967; 21. Kritzler and Wood, 1961; 22 Schuijf, 1975; 23. Corwin, 1977; 24. Myrberg, 1978.

Discussion Questions

1. How would you test whether a white shark might be frightened by the scream of the killer whale?

2. How would you determine what sounds are attractive to cartilaginous fishes? You should consult Banner (1972); he made a valiant attempt to accomplish this with juvenile lemon sharks but had only mixed success in his own opinion. One potential defect to the study was that he was forced to produce his prey sounds artificially such as by scraping the claws of crabs together to make a potentially attractive sound. It is very difficult to identify the origin of sounds recorded in the natural environment, hence he used few natural sounds in his study.

3. Pressure oscillations can be recorded by a hydrophone, or underwater microphone. How can you measure the one-way water particle movements to which the cartilaginous fishes are so sensitive?

RECOMMENDED FURTHER READING

Banner, A. 1972. Use of sound in predation by young lemon sharks, *Negaprion brevirostris* (Poey). *Bull. Mar. Sci.*, 22: 251–283.

Corwin, J. T. 1978. The relation of inner ear structure to the feeding behavior in sharks and rays. *Scanning Electron Microscopy*, 11: 1105–1112.

Evangelista, C., M. Mills, U. E. Siebeck, and S. P. Collin. 2010. A comparison of the external morphology of the membranous inner ear in elasmobranchs. *Jour. Morph.*, 271: 483–495.

Klimley, A. P., and A. A. Myrberg. 1979. Acoustic stimuli underlying withdrawal from a sound source by adult lemon sharks, *Negaprion brevirostris. Bull. Mar. Sci.*, 29: 447–458.

Maruska, K. P. 2001. Morphology of the mechanosensory lateral line system in elasmobranch fishes: ecological and behavioral considerations. *Environ. Biol. Fishes*, 60: 57–75.

Myrberg, A. A. 1978. Ocean noise and the behavior of marine animals: relationships and implications. Pp. 169–208 in Fletcher, J. L., and R. G. Busnel (Eds.), Effects of Noise on Wildlife. Academic Press, San Diego.

Nelson, D. R., and S. H. Gruber. 1963. Sharks: attraction by low-frequency sounds. *Science*, 142: 975–977.

Peach, M. B., and N. J. Marshall. 2009. The comparative morphology of pit organs in elasmobranchs. *Jour. Morph.*, 270: 688–701.

CHAPTER 8

Sense of Sight: Photoreception

The distance at which objects can be seen is much shorter in the ocean than on land. Taking off in an airliner at an altitude of 1,000–2,000 m, you can distinguish a flock of white egrets flying 30 m overground. Standing on a mountain top, you can see tens of kilometers in the distance. Humans, as well as most other terrestrial species, rely greatly on vision. In all but nutrient- and plankton-deficient waters, a human can see less than 10 m in the ocean. Light is quickly absorbed in water unlike in air. Much of the energy in the visible portion of the electromagnetic spectrum is absorbed at distances of less than 30 m from the surface in the clearest oceanic waters. The red wavelengths, which transmit great distances in air and to which human eyes are particularly sensitive, are absorbed first underwater as depth increases and the green wavelengths are absorbed last. This results in the ocean being dimly illuminated with a blue-green hue. I remember well, while a graduate student at Scripps Institution of Oceanography, searching for sharks so that I could affix ultrasonic transmitters to them. I would hold my breath, slowly kick my legs back and forth, and propel myself downward. Upon reaching the top of the Espiritu Santo Seamount, 25 m below the surface, I would rise to a horizontal position and

Scalloped hammerhead sharks swimming over the surface of a seamount. Notice that light is diminished at this depth and everything is blue. The energy in the red region of the visual spectrum is absorbed in the first 10 m of the water column, leaving only the blues and greens to penetrate to greater depths. Note that Snell's window is visible at the surface.

look around me at the ghostly hammerhead sharks slowly swimming across the surface of the seamount in this dimly lighted blue world.

The eyes of sharks are well adapted to perceive low-level blue light. The pigments in the photoreceptors of sharks, rays, and chimaeras are most sensitive to these wavelengths. Visibility in the ocean is usually limited to less than 10 m due to the absorption and reflection of light by suspended flocculent aggregations of organic particles known as marine snow and planktonic algae and animals. Energy in the blue region of the spectrum penetrates farthest in the open ocean, where there is less of this suspended material. However, the electromagnetic energy is quickly absorbed in coastal waters, where these materials are more concentrated. Here the green wavelengths penetrate farther down the water column than the blue wavelengths. It is not surprising that the eyes of juvenile lemon sharks are most sensitive to the green wavelengths. They grow up in shallow waters, often tinted green by the large stands of mangroves. The eyes of adults, which live in the clear waters passing over offshore coral reefs, are sensitive to the blue wavelengths. First to be discussed will be the nature of this physical stimulus in the oceanic environment, second will be the anatomy of the photoreceptor of the cartilaginous fishes, and third the relative sensitivity of their photoreceptors to different wavelengths.

UNDERWATER PHOTIC ENVIRONMENT

The light that we see is only a small portion of the electromagnetic spectrum.[1] Electromagnetic radiation is composed of small packets of energy (quanta) that travel in straight lines while vibrating at unique frequencies to generate specific wavelengths. These quanta range from gamma and x-rays with a wavelength of a less than 1 nanometer (nm) to the longest radio waves with a wavelength of 1 km. The human eye is sensitive to wavelengths varying from 380 to 700 nm with its greatest sensitivity at a wavelength of 550 nm. The electromagnetic radiation within this range of the spectrum has historically been referred to as light. The quanta in this visible region of the spectrum are usually called photons. We perceive the wavelength differences as colors, ranging from blues for the shortest wavelengths through the colors of rainbow to the reds for the longest wavelengths. In addition to color, we are sensitive to the amplitude of each wave. The amplitude of this perceived light is referred to as brightness. The levels of light vary over a broad range of magnitudes from the dim to the very bright, over which the iris, or portal to the eye, adjusts to produce the sensation of a constant gradation of brightness. "Irradiance" is a less specific term used for energy in those wavelengths within and outside the visible spectrum of wavelengths perceptible to humans. "Intensity" is the word used to describe the amplitude of the vibrations of quanta in this more inclusive range of wavelengths. We will use the terms "irradiance" and "intensity" when discussing chondrichthyan vision because the spectral sensitivity of these species differs from that of human beings.

Most of the electromagnetic radiation reaching the earth comes from the sun. Only a small fraction of it reaches earth from the moon and stars. Sunlight is composed of a wide range of wavelengths, from 250 to 10,000 nm, although 98% of the energy comes from wavelengths less than 3,000 nm (see top curve in fig. 8.1a). The energy radiated by the sun usually is defined as the density of quanta falling in a defined way on a specified surface. Biologists define irradiance as the quanta falling on a flat surface at an angle of 90° over a specified range of wavelengths. This range is usually broader than the narrow range of wavelengths perceptible to humans defined as light. The unit in which irradiance is expressed is milliwatts per square centimeter ($mW \cdot cm^{-2}$). The irradiance in the ultraviolet wavelengths reaching the surface of the ocean is reduced due to absorption by ozone in the atmosphere. The infrared wavelengths are absorbed by the carbon dioxide and water vapor in the atmosphere. The atmosphere also scatters much of the entering radiation so that only 85% of it reaches the earth's surface and only 15% as direct skylight. This radiation is composed mainly of blue wavelengths, the perceived color of the sky. The proportion of irradiance of various wavelengths changes as the sun descends from a position directly overhead at its zenith to a position near the horizon. This change in the spectrum of available irradiance is apparent from curves showing the distribution of irradiance. The uppermost curve indicates the spectrum of electromagnetic energy impinging upon the earth's atmosphere. The next highest curve shows the energy reaching the surface of the earth when the sun is positioned directly overhead at a solar altitude of 90°, the third highest when the sun is at an altitude of 30° with respect to the earth's surface, and the fourth highest curve when the sun is at a solar altitude of 6°. The amount of irradiance in the upper regions of the spectrum decreases through absorption as the radiation passes through successively greater sections of the earth's atmosphere as the sun lowers on the horizon. The result is that at dusk, it is mainly the blues that stand out in the environment visible to us.

Some irradiance will be reflected as a perpendicular beam of quanta from the sun even if it impinges directly upon a smooth sea surface. The fraction of the irradiance reflected increases with the decreasing solar altitude as the sun approaches the horizon. The solar irradiance not reflected passes from the air into the sea, but is refracted in the process. This bending of the wavelengths concentrates irradiance into a cone of a half angle of 48.5° that is perceptible at the surface to an observer below as a disk of bright light called Snell's window.

Irradiance is absorbed either by water molecules or suspended particulate matter once it passes through the sea surface and penetrates downward

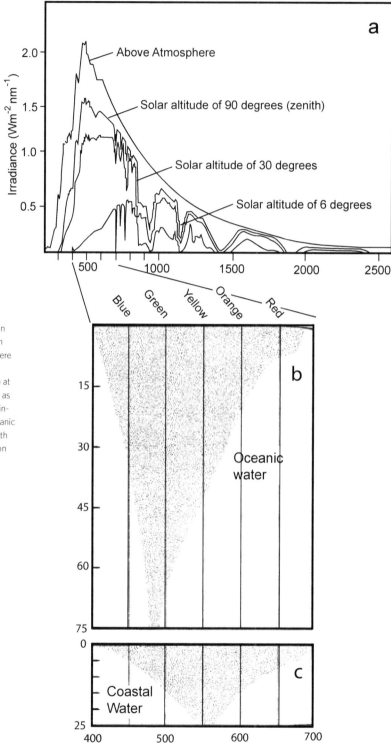

Fig. 8.1 (a) Spectral distribution of solar irradiance from the sun above and within the atmosphere and at different solar altitudes with the sun being at its zenith at 90°. The extinction of daylight as a function of wavelength with increasing depth in (b) clear oceanic water and (c) coastal water with increased turbidity and plankton biomass.

into the water column. The energy in every wavelength is absorbed at a different rate, resulting in a change in the relative amount of energy in different wavelengths at increasing depths in the ocean. Irradiance of a wavelength of 460 nm transmits farthest in clear oceanic water, whereas the energy in wavelengths either shorter or longer diminishes in intensity or attenuates at faster rates. This selective attenuation results in the bluish hue of oceanic waters. As irradiance passes downward into the ocean, water molecules absorb the energy in the longer wavelengths more readily than that in the shorter wavelengths.[2] Irradiance of the color red is absorbed almost completely by a depth of 10 m. This is apparent from the inflection on the right side of the stippled area in the shape of an upside-down triangle (fig. 8.1b) corresponding to a wavelength of 650 nm on the bottom axis of the graph showing the attenuation of irradiance with increasing depth. The irradiance of the color orange is no longer present at a depth of 20 m, blue at a depth of 30 m, yellow at a depth of nearly 40 m, and green at a depth of 65 m. Blue green irradiance of 470 nm penetrates farthest to depths exceeding 75 m. Irradiance of both short and long wavelengths is abundant in shallow clear water, but with increasing depth the irradiance present becomes more monochromatic in the blue region of the spectrum. Gelbstoff, or dissolved organic compounds, flocculent aggregations of marine snow, and phytoplankton, are more ubiquitous in coastal than offshore waters. These substances absorb irradiance of the blue wavelengths and scatter the longer wavelengths. Hence, coastal waters often have a green color, unlike the clear oceanic waters that are devoid of these absorptive substances. In the coastal waters there is also a tendency for irradiance left at increasing depths to be in the green-to-orange region of the spectrum.

ANATOMY OF EYE

The size of the eyes and their location on the head vary among the chondrichthyan fishes. For example, the eye of the bigeye thresher (*Alopias superciliosus*) is exceedingly large in order to maximize the amount of irradiance impinging upon it while spending most of its life in deepwater where little irradiance is present (fig. 8.2a). The eye of the blue shark (*Prionace glauca*) is smaller proportional to its body size since it spends much of the time near the surface, although it does make repeated dives to considerable depths to forage upon squid and fishes (fig. 8.2b). The eyes of these two species as well as other neritic and pelagic sharks such as the blacktip reef shark (*Carcharhinus melanopterus*) are located on either side of the head, which tapers inward toward the conical snout (fig. 8.2c). This orientation enables these sharks to see what is in front of them when swimming up in the water column. Their positions on the both sides of the head provide them with a visual field of 270° in front and

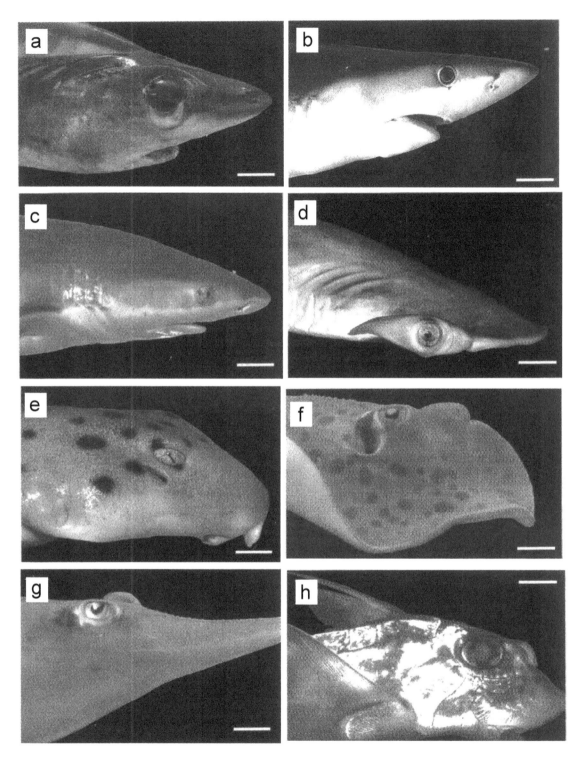

Fig. 8.2 Top view of five sharks, two rays, and one chimaera illustrating the variation in the size of the eye, its location on the body, and the shape of the body. The species are: (a) bigeye thresher, (b) blue shark, (c) blacktip reef shark, (d) scalloped hammerhead shark, (e) epaulette shark, (f) blue-spotted fantail ray, (g) giant shovelnose ray, and (h) spotted ratfish. Scale bars = (a) 80 mm, (b) 100 mm, (c) 15 mm, (d) 15 mm, (e) 10 mm, (f) 20 mm, (g) 15 mm, and (h) 30 mm.

to either side. This field of vision is increased when the head moves laterally when swimming. The blind area immediately in front of an individual extends less than a body length in front of it.[3]

The eyes of benthic species such as the epaulette shark (*Hemiscyllium ocellatum*) that spend much of the time lying on the bottom are located high on the head, the surface of which is more horizontal, and the eyes thus are directed upward to see what is swimming above (fig. 8.2e). The eyes of the rays such as the ribbontail stingray (*Taeniura lymma*) and the giant shovelnose ray (*Glaucostegus typus*), both of which are dorsal-ventrally flat and often remain partially buried in the sand or mud, have eyes on the top of their head that are often situated on protruberances in the chondrocranium, elevating them above the substrate in order to increase their vision above and around them (fig. 8.2f,g). The eyes of chimaeras such as the spotted ratfish (*Hydrolagus colliei*) are on the side toward the front of the head, adapting them for seeing objects ahead of them (fig. 8.2h).

The eyes of the cartilaginous fishes are highly adapted anatomically for sensing underwater irradiance, whose intensity varies by ten powers of ten from daytime to nighttime. The sclera, the external covering of most of the eye, is composed of thick cartilaginous tissue (fig. 8.3a).[4] The reef sharks of the family Carcharhinidae and hammerhead sharks of the family Sphyrnidae have a nictitating membrane, or eyelid, that extends upward from underneath the eye to cover the whole eye. This membrane protects the eye. It moves upward over the eye when the shark's head comes in contact with another object during feeding. The iris of cartilaginous fishes is able to change the shape of the pupil, or opening to the eye. This muscle retracts to increase the amount of irradiance entering through a widened pupil during dim conditions and expands to reduce the amount entering through a narrowed pupil in bright conditions. Most neritic or pelagic sharks have a pupil, which is contracted into a vertical slit with a small aperture at the bottom when adapted to high irradiance intensities and dilated into more round opening when adapted to low intensities. An example is the pupil of the blacktip shark (*Carcharhinus limbatus*). This pupil narrows to a vertical slit in high light intensities (fig. 8.4a) and widens into a football shape in low intensities (fig. 8.4b).[5] However, the shape of the dark-adapted pupil varies in these species of shark. The pupil of the nurse shark becomes dilated into an obliquely oriented, sausage-shaped pupil when adapted to dark conditions. The pupils of other sharks become small centrally located circular openings under many light levels. An advantage to this type of pupil is that it can be closed to a pinhole under high irradiance conditions. Hence, it is found in species that are active both at nighttime and during daytime, enabling them to cope with enormous fluctuations in light intensity. However, the asymmetry of this pupil results in the detail of an image being slightly

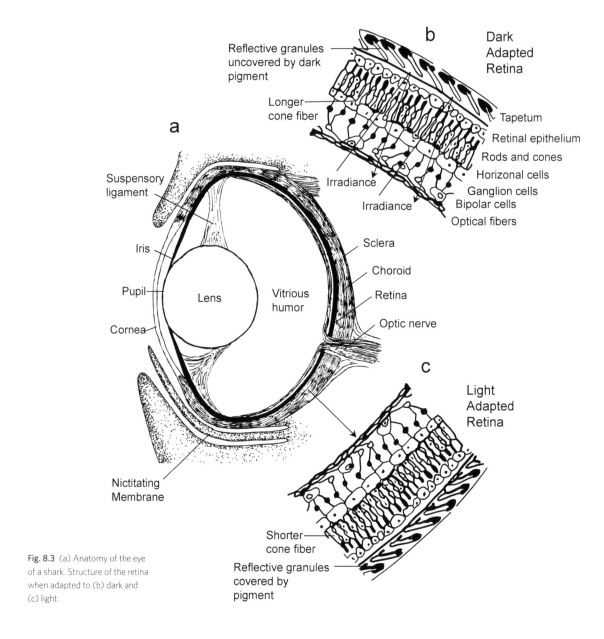

Fig. 8.3 (a) Anatomy of the eye of a shark. Structure of the retina when adapted to (b) dark and (c) light.

distorted along an axis perpendicular to the slit under high light levels. Pupil shape and size not only control the brightness of the image reaching the retina, but they also affect image quality. Rays of light passing through the periphery of the lens are brought into focus at a different level in the retina than rays of light passing through the center of the lens, resulting in a blurred image. A small central circular pupil improves image quality by impeding the passage of light through peripheral regions of the lens. Furthermore, the size of the aperture is inversely related to the depth of focus. Hence, there are fundamen-

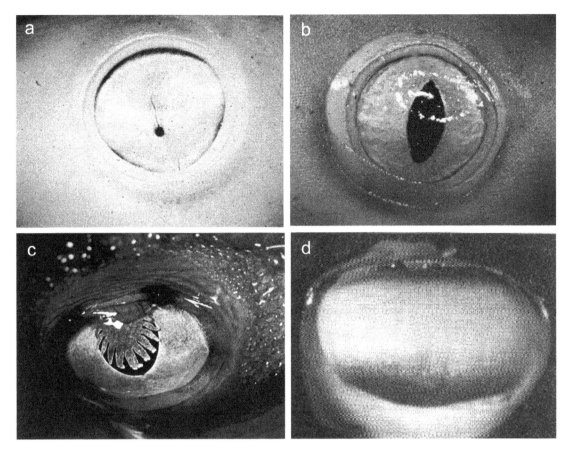

Fig. 8.4 (a) Pinhole iris of light adapted eye; (b) circular dilated iris of eye of dark-adapted blacktip shark; (c) multiple downward projecting fingers of operculum pupillare of thorny skate; (d) tapetum after the cornea and lens are removed from eye of giant shovelnose ray. Notice the wide horizontal band of reflective material lying beneath the retina.

tal tradeoffs between a particular pupil's ability to gather light and its ability produce a high-quality image on the retina.

Sharks that inhabit the deep sea have a large circular pupil with a less mobile iris.[5] The large pupil permits more electromagnetic energy to enter the eye under dim lighting, which is generated primarily from bioluminescent emissions from deep-sea prey. There is also little need to regulate the amount of irradiance entering the eye in the deep sea because the levels of irradiance are constant over time unlike during day and night in the surface waters. The rays have either crescent pupils or windowpane-like apertures separated by thin fingerlike flaps of skin. The thorny skate (*Raja radiata*) has the former shape to its pupil with the thin flaps of skin, between which irradiance enters the eye (fig. 8.3c). It has been suggested that the pupillary apertures may serve as focus indicators. If the eye is out of focus, these structures will produce many

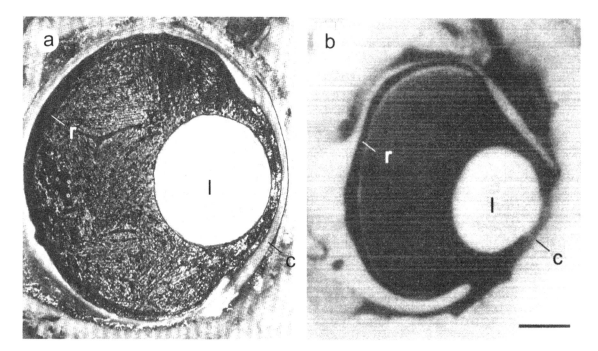

Fig. 8.5 The eyes of the sharks and rays. Micrographs of cross-sections of the eyes of (a) the lemon shark and (b) the bluntnose stingray. c = cornea; l = lens; r = retina.

laterally displaced images on the retina; if the eye is in focus, the images would merge into one on the retina.[6]

The eyes of the sharks are round, while those of rays are oval. This can be appreciated from micrographs of cross-sections of an eye of the lemon shark (fig. 8.5a) and bluntnose stingray (*Dasyatis sayi*) (8.5b). Light passes into the eye through the cornea, passes through the crystalline lens, and focuses on the light-sensitive retina, which contains two types of photoreceptors, rods and cones. The rods are the most common photoreceptor in the retina.[5, 6] They are distinguishable by their oval base and their long inner and outer segments, which are similar in diameter and rounded at their ends. The cones are less common; they have a conical base and a thick inner segment attached to a thinner outer segment that tapers to a point. Sections of the retina are shown for the lemon shark (fig. 8.6a) and giant shovelnose ray (fig. 8.6c). The section of retina in the micrograph of the lemon shark contains a single cone in the center with rods on either side. To the side of the micrograph is a diagram of the light-adapted cone in the micrograph as well a dark-adapted cone and a rod photoreceptor (fig. 8.6b). These two organelles have different functions. The rods are highly sensitive to quanta and enable cartilaginous fishes to distinguish light from dark objects in the presence of low irradiance during dusk and night. The cones are less sensitive to quanta, but may enable carti-

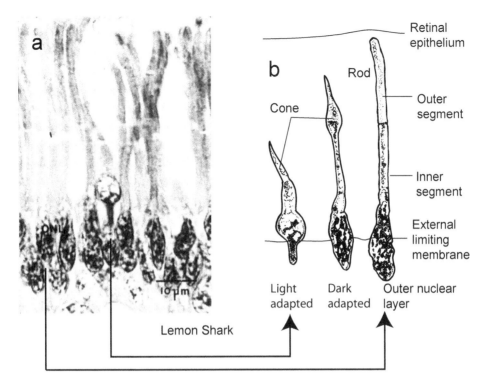

a

Lemon Shark

10 µm

b

Retinal epithelium

Cone

Rod

Outer segment

Inner segment

External limiting membrane

Light adapted

Dark adapted

Outer nuclear layer

Giant shovelnose ray

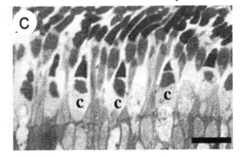

c

c   c   c

Fig. 8.6 (a) Micrograph of the retina of a lemon shark showing cones and rods. (b) Diagram of both light- and dark-adapted cone and rod photoreceptors (c) Micrograph of retina of giant shovelnose ray. c = cone receptor.

laginous fishes to distinguish colors during daytime as well as to distinguish an object from a background by its slightly different brightness. The cones are most abundant in shallow dwelling species that are active during the day. The lowest rod-to-cones ratios of 4:1 are found in the retina of the white shark (*Carcharodon carcharias*) and giant shovelnose ray (table 8.1). Note that you can pick out seven cones in the micrograph of the retina of giant shovelnose ray (see fig. 8.6c). The high number of cones indicates that members of both species forage much of the time in the presence of ample irradiance. Other species with slightly higher rod-to-cone ratios, but low enough to indicate that they are active during the day, are the lemon shark and the thresher shark (*Alopias vulpinus*), each with a ratio of 5:1, the South American freshwater stingray

TABLE 8.1. Ratios of rods to cones and peak spectral absorption of irradiance for sharks and rays with their habitats and depth ranges. Blank spaces indicate the absence of records.

Order Species	Common name	Ratio (rods:cones)	Rod l_{max} (nm)	Cone l_{max} (nm)	Habitatat[a]
Squaliformes					
Centrophorus squamosis[7,8]	Leafscale gulper shark		482–484		Continental slopes, 230–2,400 m
Centroscymnus coeliolepis[7,8]	Portuguese dogfish		472		Abyssal plain, 128–3,675 m
Deania calcea[7,8,9]	Birdbeak dogfish	>100: 1	484		Upper continental slope, 70–1,470 m
Squalus acanthias[10,11]	Piked dogfish	50:1	498–500		Continental, insular shelves, 0–600 m
Squatiniformes					
Squatina californica[12,13]	Pacific angelshark	P	488		Continental, insular shelves, 1–200 m
Carchariformes					
Carcharhinus falciformis[14]	Silky shark	11:1			Insular, offshore: shallow, 0–500 m
C. plumbeus[14]	Sandbar shark	13:1			Inshore, usually 20–55 m, occasionally 55–280 m
Mustelus henlei[15]	Brown smooth-hound	P	499		Continental and insular shelves, < 20 m
Negaprion brevirostris[16,17,18]	Lemon shark (juveniles)	12:1, 5:1	501		Bays on mudflats and mangroves, 0–5 m
Negaprion brevirostris[18,19]	Lemon shark (adults)		522		Inshore and coastal, 0–92 m
Prionace glauca[14]	Blue shark	8:1			Oceanic, pelagic, 0–350 m
Sphyrna lewini[20]	Scalloped hammerhead	P			Continental, insular shelves, 1–1000 ,
Galeus melastomus[21]	Blackmouth catshark		482		Upper continental slope, 200–500 m
Scyliorhinus canicula[7,12]	Small-spotted catshark	P	500–505		Continental shelves and upper slope, < 110 m

Order / Species	Common name	Ratio	λmax	Additional λmax	Habitat[a]
Orectolobiformes					
Ginglymostoma cirratum[22]	Nurse shark	7:1			Rocky and coralline reefs, 1–12 m
Lamniformes					
Alopias vulpinus[14]	Thresher shark	5:1			Nearshore to far offshore, 0–360
Carcharodon carcharias[14]	White shark	4:1			Nearshore to far offshore, 0–1300 m
Isurus oxyrinchus[14]	Shortfin mako shark	10:1			Nearshore to far offshore, 0–500 m
Rajiformes					
Dasyatis pastinaca[23]	Common stingray	P	510	476, 502, 540	Bays and flats, shallow water
Aptychotrema rostrata[24,25]	Eastern shovelnose ray	5:1	498	459, 492, 553	Bays and flats, shallow water
Glaucostegus typus[25,26]	Giant shovelnose ray	4:1	504	477, 502, 561	Bays and flats, shallow water
Myliobatiformes					
Myliobatis californica[27]	Bat eagle ray		500		Bays and sloughs, shallow water
Rhinoptera steindachneri[27]	Pacific cownose ray		500		Bays, shallow water
Potamotrygon motoro[28]	So. Am. freshwater stingray	6:1	499		Rivers and lakes, 0–2 m

[a]Habitats types and depth ranges taken from Compagno *et al.*, 2005.[29] P = rods and cones present.

(*Potamotrygon motoro*), with a ratio of 6:1, and the nurse shark (*Ginglymostoma cirratum*) and blue shark with ratios of 7:1 and 8:1, respectively. Cones are least abundant in deep dwelling or nocturnal species such as birdbeak dogfish (*Deania calcea*) and the piked dogfish (*Squalus acanthias*), which have rod-to-cone ratios of >100:1 and 50:1, respectively.

The lens is moved toward the retina in the back of the eye to focus on distant objects or away from the retina to focus on proximate objects (see fig. 8.3a). The suspensory ligament secures the lens from the top and it is moved back and forth by the expansion and contraction of the ligament underneath, and this movement results in the refraction of irradiance onto a particular region of the retina, where the rods and cones are concentrated.[5] The rods contact the retinal epithelium while both the rods and cones contact the internal layer of the horizontal cells, from which ganglion cells lead to bipolar cells (fig 8.3b,c). Optical fibers lead from these bipolar cells to the visual centers of the brain. The choroid of cartilaginous fishes contains a tapetum, a reflective layer composed of angled parallel cells containing guanine crystals. In dim conditions, these crystals reflect the quanta that have passed through the retina and have not yet been absorbed by the photoreceptors backward through the same layer so that the quanta have a second chance of being absorbed (fig. 8.3b). The tapetum can clearly be seen as a wide horizontal band of reflective material in a micrograph of the retina of an eye of the giant shovelnose ray from which the cornea and lens were removed (fig. 8.4d).[6] Furthermore, the cones become longer; this may enhance their sensitivity when fewer quanta come in contact with them during low light conditions.[5] The reflective layer can be occluded under bright light by the migration of dark pigment granules in front of the guanine crystals so that the granules absorb the quanta before they pass through the retina again (fig. 8.3c).

Visual Pigments

Located within the outer segments of the rods are irradiance-sensitive pigments. These organic dyes are composed of an opsin, a high–molecular weight protein bonded to a chromatophore, a short molecular chain related to either Vitamin A_1 or A_2.[5] The visual pigments are usually purple or pink unlike other retinal components, which are transparent; the color of a pigment fades as it absorbs photons upon exposure to quanta of a particular wavelength. The pigment associated with Vitamin A_1 is called rhodopsin. It is most sensitive to blue green irradiance ranging from 497–510—the wavelength that transmits farthest in clear surface waters. Again, this is apparent in the graph of irradiance transmission in oceanic water by the equal amounts of shading on either side just to the left of the line labeled green with a wavelength of 500 nm at depths ranging from 30 to 50 m (see fig. 8.1b). Rhodopsin is present in the rods of the elasmobranchs that live in the surface waters of the oceans. For example, piked dogfish (*Squalus acanthias*), possess this pigment in their retinas with a maximum sensitivity, or a λ_{max} of 498–500 nm (table 8.1). These sharks migrate over the continental shelf from the waters off South Carolina, where they live during winter, to the waters off Massachusetts, which they occupy during the summer. Adult lemon sharks, which inhabit the clear waters surrounding coral reefs in the Caribbean, also possess rhodopsin in their retina with a λ_{max} of 522 nm. Brown smooth-hound sharks (*Mustelus henlei*), which inhabit the coastal waters off California, have an almost identical λ_{max} of 499 nm. Skates of the order Rajiformes and rays of the order Myliobatiformes also have rhodopsin in their retinas. They occupy the same habitats as coastal sharks. Some are relatively mobile, searching prey buried in the sand or mud in shallow water, other are more sedentary, lying concealed on the bottom, where they ambush prey passing near them. The giant shovelnose ray, a member of the Rajiformes, lives in shallow waters less than 100 m deep in habitats ranging from the rocky intertidal to offshore continental and insular shelves in the central western Pacific Ocean. Adults are found in offshore waters while young individuals are found inshore on sand flats, around atolls, and in mangrove swamps. It possesses a rhodopsin pigment in its rod receptors with a λ_{max} of 504 nanometers (nm). The bat eagle ray, a member of the Myliobatiformes that inhabits sloughs, bays, and the nearshore waters off the western coast of North America, also has rhodopsin in its retina with a λ_{max} of 500 nm.

The deep-dwelling species have blue-shifted pigments in their rod photoreceptors. The Portuguese dogfish (*Centroscymnis coeliolepis*) has the most blue-shifted rhodopsin pigment in its rod photoreceptors with a maximum sensitivity of 472 nm. It inhabits waters as deep as 3675 m on the abyssal plain in the Atlantic, Indian, and Pacific Oceans. The birdbeak dogfish, blackmouth catshark (*Galeus melastomus*), and leafscale gulper shark (*Centrophorus squamosus*) have maximum sensitivities of 484 nm, 482 nm, and 482–484 nm, respectively. These three sharks inhabit similar deep habitats, the upper slopes of continental shelves to the edge of the abyssal plains at depths ranging from 500–2400 m.

The maximum spectral sensitivity of both rods and cones is determined using microspectrophotometry. This entails aligning an ultra-miniature measuring beam, only 1–2 μm wide, within an outer segment and recording the amount of light transmitted at each wavelength from 380 to 800 nm.[25] A baseline scan is made in similar fashion with a cell-free area adjacent to the measured cell. The baseline transmittance is then subtracted from the sample for each wavelength to generate a pre-bleach spectrum that is then converted to absorbance. The outer segment is then exposed to white light containing a

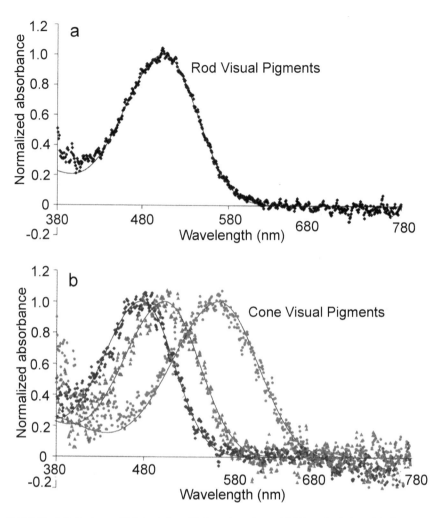

Fig. 8.7 Spectral absorbance of (a) visual pigment in the rod receptors and (b) three visual pigments in the cone photoreceptors of the giant shovelnose ray. The former possess a visual pigment with a λ_{max} of 504 nanometers (nm) (black diamonds); the cones possess visual pigments with λ_{max} value of 477 nm (blue diamonds), 502 nm (green triangles), and 561 (red squares).

broad band of wavelengths for a period of three minutes so that the pigment absorbs the electromagnetic energy and becomes bleached. A post-bleach spectrum is then obtained by scanning the same outer photoreceptor segment with the measuring beam, doing the same to the cell-free adjacent area, and subtracting the transmittance for each wavelength of the latter from the former. The post-bleach spectrum is deducted from the pre-bleach spectrum to form a bleaching difference spectrum for many outer segments to obtain an average value. The normalized absorbance of the rod pigment is shown as a function of wavelength for the rod photoreceptor of the giant shovelnose ray (fig. 8.7a). Notice that the peak of the rounded curve is around 500 nm. The rods in the

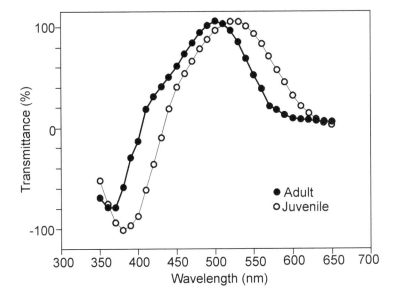

Fig. 8.8 Relative percentages of absorption of irradiance for different wavelengths of juvenile and adult lemon sharks.

retina of the adult shovelnose ray have a λ_{max} of 504 nm, as expected from a retina having a single type of rod with an A_1-based visual pigment, rhodopsin.[25]

Of particular interest is the spectral absorbance of the receptors in the cone-rich retina of the giant shovelnose ray. The outer segments have three pigments, a short-wavelength (blue-sensitive) pigment with a λ_{max} of 477 nm, a medium-wavelength (green-sensitive) pigment with a λ_{max} of 502 nm, and a long-wavelength (red-sensitive) pigment with a λ_{max} of 561 nm. This is evident from the different locations of the peaks of the three color-coded curves (fig. 8.7b). The presence of the three different types of cone photoreceptors, each with a distinct absorbance spectrum, raises the possibility that this species has trichromatic color vision. Two other species in the order Rajiformes, the common stingray (*Dasyatis pastinaca*) and the eastern shovelnose ray (*Aptychotrema rostrata*), also have three distinctly different peaks to the absorbance spectra from their cone photoreceptors, raising the possibility that color vision is ubiquitous among the skates.

The pigment associated with Vitamin A_2 is called porphyropsin. It is most suited for seeing in turbid, yellowish waters. This pigment has been isolated from rods in the retinas of juvenile lemon sharks, which inhabit the waters over the shallow flats off the coast of South Florida.[18] The pigments from both juveniles and adults were exposed to quanta of successively longer wavelengths to determine the percentage of absorption by this pigment of irradiance for a variety of wavelengths. The peak in the curve indicates the maximum absorption. For adults, the peak was at a wavelength of 500 nm (fig. 8.8). This was likely due to the prevalence of rhodopsin in their retinas. The peak in the spectral absorption curve of the juveniles was at a higher wavelength of 522 nm. This was because their retinas mainly had porphyropsin, which is better suited to absorbing quanta in the red-shifted spectral conditions of coastal waters (see fig. 8.1c). As juvenile lemon sharks grow larger, they leave the shallow flats close to shore and dwell in the clearer waters along the shelf. At this stage in their life, the pigments in their retinas are the blue-shifted rhodopsins, which are better suited for seeing the deep blue prevalent in the clear reef waters of the Gulf Stream.

Day and Night Vision

Juvenile lemon sharks appear to have both pigments that are ideal for daytime vision others suited for nighttime conditions.[5] The sensitivity of a shark to different wavelengths can be determined by threshold experiments, in which a shark is exposed to successively increasing intensities of irradiance of different wavelengths. A shark will respond to a mild shock by raising its nictitating membrane to cover and protect its eye as a human blinks an eye when startled. Irradiance of a particular wavelength is directed on the shark's eye accompanied by a mild shock under dim conditions. If the two stimuli, irradiance and electricity, are paired a sufficient number of times, the shark will eventually blink its nictitating membrane when only irradiance stimulus is presented. The individual can then be exposed to successively lower intensities of energy to determine the threshold of sensitivity to that particular wavelength. Three or more individuals are exposed to a wide range of wavelengths in order to record a spectral curve of the species relative sensitivities. The lemon shark's maximum spectral sensitivity is evident as a broad peak centered around 500 nm if it is kept in darkness prior to the experiment (see upper curve in fig. 8.9a). However, when the shark has been kept in a lit environment, the spectral sensitivity shifts 20 nm toward the red, and the curve is broadened to encompass a wider range of wavelengths (see lower curve in fig. 8.9b).[5] This shift makes the shark more sensitive to the wavelengths of 520 nm in the red-shifted conditions of shallow water. This effect is named the Purkinge shift. It is evidence for the presence of cone photoreceptors, which contain the red-shifted porphyropsin. The red-shifted sensitivity in high irradiance conditions is termed photopic vision, and the blue-shifted sensitivity in low-irradiance conditions is called scotopic vision. Cone photoreceptors enable elasmobranchs to see under bright light conditions when the pigments in their rod photoreceptors are bleached and incapable of image detection. The primary benefit of a duplex retina, possessing both rods and cones, is to extend the range of intensities over which the visual system can function effectively. Since irradiance intensity varies over 10 logarithmic units, or powers of ten,

Samuel Gruber's finding a spectral shift to red during daytime provided evidence of the importance of cones within the retina of sharks. His first anatomical studies, performed on juvenile lemon sharks, indicated a cone-to-rod ratio of 1:12 (see table 8.1).[16] His graduate student, Joel Cohen, later found an even lower ratio of cones to rods of 1:4 in the retinae of these sharks.[17] These results were consistent with a dual function of rods and cones. The former, which dominate in the retina, enable the lemon shark to see vague shapes at low levels of irradiance. The cones in the retina would enable the shark to better discriminate the shape of objects and perhaps even perceive their color in bright conditions. Another experiment conducted by Gruber provided further evidence for the distinctly different functioning of rods and cones. He trained juvenile lemon sharks to blink in response to their perception of a change from an on-and-off pattern of flashes changed to a continuously fused pattern.[5] Both the rates of the flicker and the intensity were increased, while recording that point at which the stimulus appeared to be constant—this threshold was termed the "flicker fusion rate." If the retina had only one photoreceptor, Gruber expected to find a smooth curve when he plotted the flicker fusion rates for a wide range of intensities. If two distinct photoreceptors are in the retina, a discontinuity would be apparent in the smooth curve, signaling the crossover between rod-based vision at low intensities to cone-based vision at higher intensities. The curve of the critical fusion frequency for varying intensities of irradiance of the nocturnally active gecko is smooth (fig. 8.9b). There are only rods in the retina of this species, and these enable geckos to see only the vague shape of their prey over a narrow range of low levels of illumination. The curve for the daytime-active iguana is also smooth. There are only cones in its retina. Hence, the iguana can make out the shape of its prey better during a wider range of higher levels of illumination. On the other hand, an inflection exists in the curve of the flicker fusion rate of lemon sharks at an illumination intensity of four log units. The intensity at which this change in the flicker rate occurred was nearly that same as that observed for humans. This was strong evidence for the lemon shark possessing a duplex retina, in which the rods and cones performed different functions.

Gruber then attempted to find out whether the cones in the lemon shark's retina enabled it to perceive color. He and his colleagues built a single-choice maze, in which two guillotine-type doors were equipped with optical fibers that could be illuminated with irradiance of the same intensity but a different composition of wavelengths (fig. 8.10). The sharks were first rewarded with food when they chose the door emitting quanta of a slightly higher intensity. The subjects of the experiments could discriminate between minute differences in illumination. The next objective was to demonstrate color perception by the juvenile lemon sharks. This entailed training them to swim through a door emitting quanta of a different wavelength yet of the same intensity as quanta emitting from another door. Gruber was eventually able to condition one juvenile lemon shark to pick the door on the basis of irradiance of a different wavelength emitted while keeping intensity constant in both doors. However, this preliminary result was never published in a scientific journal because only a single subject was tested before the loss of funding.

You might expect sharks to see colors. Their eyes are well developed and function over a wide range of light intensities. As you learned earlier, three species of rays have three different pigments in their cones, evidence of color vision—a trait likely to provide distinct evolutionary advantages. Nathan Hart and other Australian researchers performed single-cell microspectroptometry on the photoreceptors of seventeen species of sharks.[30] Although the wavelengths of maximum absorbance of both the rod and cone photoreceptors varied among species, there was only one long-wavelength-sensitive cone receptor in the retina of each species. For example, only a single curve denotes the spectral absorbance of the rod photoreceptors in the retina of the blacktip shark with a λ_{max} of 506 and a single curve for the absorbance of its cone photoreceptors with a λ_{max} of 532 nm (fig. 8.11). It is thus likely that the blacktip and other sixteen species have monochromatic vision. This type of vision is rare in terrestrial vertebrates due to their living in an atmosphere of air containing light of all the colors in the rainbow. Color vision thus would confer a considerable evolutionary advantage to these species. However, the band of light available in the marine environment narrows quickly with depth until only blue green irradiance is present at depths greater than 60 m in oceanic waters (see fig. 8.1b). For this reason, there might not have been a strong evolutionary impetus for the sharks, many of which dive to considerable depths, to evolve color vision—unlike the rays that inhabit shallower water with a broader spectrum of light present. The sharks may have arrived through convergent evolution at the same visual design as the aquatic mammals such as the whales, dolphins, and seals, which have monochromatic vision with a single green-sensitive cone type.

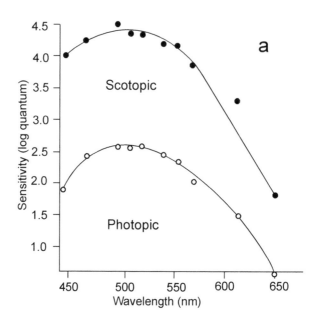

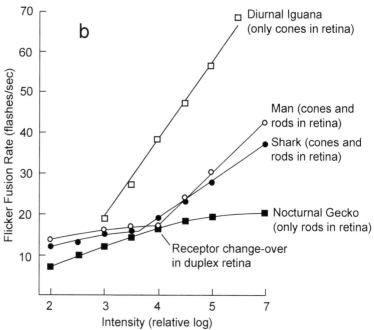

Fig. 8.9 (a) Spectral sensitivity of adult lemon sharks under low irradiance intensity, the scotopic condition, and high intensity, the photopic condition. (b) The effect of irradiance intensity on critical flicker fusion rate for the iguana, gecko, lemon shark, and human.

from bright light of daytime to dim conditions on a moonless night, it is not unexpected that most vertebrates possess a duplex retina. However, a secondary function of the cones of many vertebrates is in color discrimination. If two or more cone types are present, each containing a different pigment, the cartilaginous fish may be able to compare the outputs from these specialized cell types and extract chromatic information from the retinal images in their eyes.

Fig. 8.10 Juvenile lemon shark attempting to choose between two doors, each with optical fibers producing irradiance of the same intensity but composed of different wavelengths.

There is a long history of research by scientists interested in knowing whether sharks perceive color. Perhaps, this was likely encouraged at first by admirals in the United States Navy, who were concerned that the yellow on parachutes, referred to as yum-yum yellow in jest, might attract sharks to a pilot who bailed out of his fighter.

VISUAL CAPABILITIES

The photoreceptors are more concentrated in certain regions of the retina to meet the visual needs of the cartilaginous fishes.[6] An increased density of photoreceptors is accompanied by an increase in both bipolar and ganglion cells in that region of the retina. These regions, or *areae* of increased receptor density, enable that part of the retina to increase the resolution of its spatial sampling of the image. Their extent and location in the retina is closely correlated with

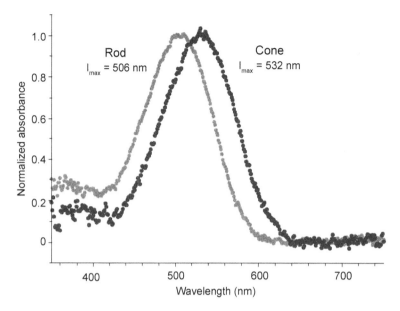

Rod
I_{max} = 506 nm

Cone
I_{max} = 532 nm

Fig. 8.11 Absorbance spectra of the rod and cone receptors of the blacktip shark. Note that there is only one pigment in the rod photoreceptors with a λ_{max} of 506 nm and one pigment in the cone photoreceptors with a λ_{max} of 532 nm. The single pigment only permits the shark to distinguish images on the basis of brightness.

their foraging ecology. There are two types of *areae* in the chondrichthyan fishes. The first, the *area centralis*, is a circular region, where the density of receptors is greatest in the middle and decreases in a radial direction. The second, the *area horizontalis* or visual streak, consists of a band across the retina with an increased density of photoreceptors. *Area centrales* are present in the retinae in animals that inhabit open environments with unobstructed views. *Area horizontales* are characteristic of animals that occupy more structurally complex environments with more limited views.

Most of the elasmobranch species have horizontal visual streaks of increased ganglion cell density.[6] Most benthic sharks and rays have a long visual streak located across the dorsal surface of the retina. The increased density of receptors in the visual streak of the Bigelow's ray (*Raja bigelowi*) is apparent as central bands of denser stippling across the retina (fig. 8.12d).[31] Many of these benthic species, especially the skates and rays, remain camouflaged and partially buried in the substrate with their eyes protruding to provide an unobstructed view of their surroundings. The dorsal location of the streak provides them with wide field of vision that would reduce their need to move their eyeballs from side to side—an action that alerts prey or predators of their presence. The streak also provides better visual acuity along the horizon so they can see crustaceans and fishes moving just above the bottom. Neritic species that swim up in the water column, such as the blackmouth catshark (*Galeus melastomus*) have visual streaks more centrally located on the retina (fig. 8.12b).[31] The central location of the streak provides an enhanced ability to locate objects approaching from in front rather than from below near the bot-

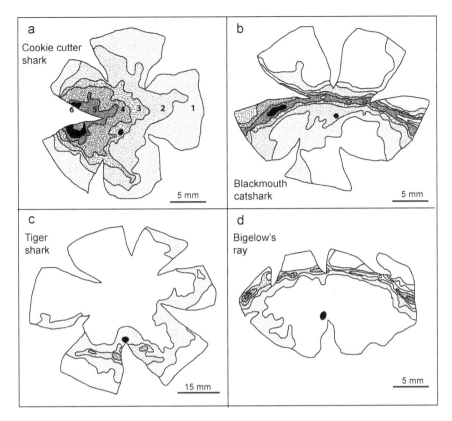

a
Cookie cutter shark

6 5 4 3 2 1

5 mm

b

Blackmouth catshark

5 mm

c
Tiger shark

15 mm

d
Bigelow's ray

5 mm

Fig. 8.12 The dorsal visual streaks of the (a) cookie cutter shark, (b) blackmouth catshark (c), tiger shark, and (d) Bigelow's ray. Note that the concentrations of ganglion cells increase on a scale from 1 to 6 and proportional to the density of stippling in each contour of diagram.

tom. The visual streak of the tiger shark (*Galeocerdo cuvier*) is located on the ventral surface of the retina (fig. 8.12c).[31] This would provide the better spatial resolution in the upper visual field when looking upward toward the surface. This likely enables the species, while swimming just over the bottom, to distinguish prey swimming at the surface. Large adults of this species feed on sea turtles, birds, seals, and sea lions, and at times humans, all of which are usually swimming at the surface of the ocean.[32] The white shark has an *area centralis* located slightly below the center of the retina, and this affords an upper frontal view with the highest spatial resolution.[33] The white shark, with an *area centralis* confined to a small region of its retina, must depend more on the movement of its eyes to carefully examine objects. Hence, it has large extraocular eye muscles[34] and moves its eyes to fixate on an object.[35]

Sharks certainly use vision in their final dash to seize their prey. The angelshark dashes upward from the bottom and seizes prey swimming above it. Similarly, the white shark swims either near the bottom or at a depth at which it is indistinguishable from the surface, and accelerates upward toward the surface to seize its prey. The latter shark is capable of seeing from a depth of 20 m in clear water conditions an object at the surface 15 cm in diameter.[35] Although the white shark detects a silhouette of its prey at the surface against downwelling light when feeding, the majority of cartilaginous fishes view their prey along a horizontal line of sight.[36] The horizontal line of sight in the clear ocean waters contains mostly blue quanta, whereas the vertical line of sight in coastal waters contains mostly blue green quanta. The pigments in the eye of the majority of elasmobranchs are adapted to vision along these two axes. Vision appears most useful to the cartilaginous fishes in relatively clear coastal or oceanic waters. However, this sense is in no way as important as it is to terrestrial species, which rely greatly on vision to find mates and prey in the clear atmosphere of air that exists on land.

SUMMARY

The size of the eyes and their location on the head vary among the chondrichthyan fishes. The eyes of the cartilaginous fishes are highly adapted for sensing underwater irradiance, whose intensity varies over ten powers of ten from daytime to nighttime. The iris is capable of changing the shape of the pupil, or opening to the eye. This muscle retracts to increase the amount of irradiance entering through a widened pupil during dim conditions or expands to reduce the amount entering through a narrowed pupil in the presence of bright irradiance. Most neritic or pelagic sharks have a pupil that contracts into a vertical slit with a small aperture at the bottom when adapted to high-irradiance intensities and dilated into a more round opening when adapted to low intensities. Sharks that inhabit the deep sea have a large circular pupil with a less mobile iris. The large pupil permits more electromagnetic energy to enter the eye under dim lighting, which is generated primarily from bioluminescent emissions from deep-sea prey. The rays have either crescent pupils or windowpane-like apertures separated by thin finger-like flaps of skin. It has been suggested that the pupillary apertures may serve as focus indicators. Light passes into the eye through the cornea, travels through the crystalline lens, and focuses on the light-sensitive retina, which contains two types of photoreceptors, rods and cones. The choroid of cartilaginous fishes contains a tapetum, a reflective layer composed of angled, parallel cells containing guanine crystals. In dim conditions, these crystals reflect the quanta that have passed through the retina and not yet been absorbed by the photoreceptors backward through the same layer so that the quanta have a second chance of being absorbed.

Located within the outer segments of the rods are irradiance-sensitive pigments composed of an opsin, a high–molecular weight protein bonded to a chromatophore, a short molecular chain related to either Vitamin A_1 or A_2.

The pigment associated with Vitamin A_1 is called rhodopsin. It is most sensitive to blue green irradiance ranging from 497–510 nm—the wavelength that transmits farthest in clear surface waters. It is in the rods of the elasmobranchs that live in the surface waters of the oceans. The maximum spectral sensitivity of both rods and cones is determined by microspectrophotometry. Of particular interest is the spectral absorbance of the cones in the cone-rich retina of the giant shovelnose ray. The outer segments appear to have three pigments: a short-wavelength (blue-sensitive) pigment with a λ_{max} of 477 nm, a medium-wavelength (green-sensitive) pigment with λ_{max} of 502 nm, and a long-wavelength (red-sensitive) pigment with a λ_{max} of 561 nm. The presence of the three different types of cone, each with a distinct absorbance spectrum, raises the possibility that this species has trichromatic color vision. Although the wavelength of maximum absorbance of both the rods and cones varies among species of sharks, there is only one long-wavelength-sensitive cone receptor in the retina of each species. For example, the blacktip shark has one pigment with a λ_{max} of 506 nm in its rods and another pigment with a λ_{max} of 532 nm in its cones. It is thus likely that the blacktip and other sharks have monochromatic vision. The band of light available in the marine environment narrows quickly with depth until only blue green irradiance is present at depths greater than 60 m in oceanic waters. For this reason, there might not have been a great evolutionary impetus for the sharks, many of which dive to considerable depths, to evolve color vision—unlike the rays that inhabit shallower water with a broader spectrum of light. The sharks may have arrived through convergent evolution at the same visual design as the aquatic mammals such as the whales, dolphins, and seals, which have monochromatic vision with a single green-sensitive cone type.

* * *

KEY TO COMMON AND SCIENTIFIC NAMES

Bat eagle ray = *Myliobatis californica*; Bigelow's ray = *Raja bigelowi*; bigeye thresher shark = *Alopias superciliosus*; birdbeak dogfish = *Deania calcea*; blackmouth catshark = *Galeus melastomus*; blacktip reef shark = *Carcharhinus melanopterus*; blacktip shark = *Carcharhinus limbatus*; blue shark = *Prionace glauca*; bluntnose stingray = *Dasyatis sayi*; brown smoothhound shark = *Mustelus henlei*; common stingray = *Dasyatis pastinaca*; cookie cutter shark = *Isistius brasiliensis*; eastern shovelnose ray = *Aptychotrema rostrata*; epaulette shark = *Hemiscyllium ocellatum*; false-eyed skate = *Raja oscellata*; giant shovelnose ray = *Glaucostegus typus*; horn shark = *Heterodontus francisci*; leafscale gulper shark = *Centrophorus squamo-*

Discussion Questions

1. What is the comparative usefulness of vision to a shark or ray versus a terrestrial predator in detecting their prey?

2. What additional experiments could be performed to demonstrate color vision in chondrichthyan fishes?

3. What new research might be proposed to learn more about the anatomy and functioning of the chimaerid eye, about which little is known?

4. What types of studies should be performed on chondrichthyan fishes to elucidate the value of vision to their lifestyle?

sus; lemon shark = *Negaprion brevirostris*; nurse shark = *Ginglymostoma cirratum*; oceanic whitetip shark = *Carcharhinus longimanus*; Pacific angelshark = *Squatina californica*; Pacific cownose ray = *Rhinoptera steindachneri*; piked dogfish = *Squalus acanthias*; Portuguese dogfish = *Centroscymnis coeliolepis*; ribbontail stingray = *Taeniura lymma*; sandbar shark = *Carcharhinus plumbeus*; scalloped hammerhead shark = *Sphyrna lewini*; shortfin mako shark = *Isurus oxyrinchus*; silky shark = *Carcharhinus falciformis*; small-spotted catshark = *Scyliorhinus canicula*; South American freshwater stingray = *Potamotrygon motoro*; spotted ratfish = *Hydrolagus colliei*; thornback ray = *Raja clavata*; thorny skate = *Raja radiata*; thresher shark = *Alopias vulpinus*; tiger shark = *Galeocerdo cuvier*; white shark = *Carcharodon carcharias*.

LITERATURE CITED

1. Drew, 1983; 2. McFarland, 1986; 3. Hueter *et al.*, 2004; 4. Johnson, 1978; 5. Gruber and Cohen, 1976; 6. Hart *et al.*, 2006; 7. Denton and Nicol, 1964; 8. Denton and Shaw, 1963; 9. Kohbara *et al.*, 1987; 10. Stell, 1972; 11. Beatty, 1969; 12. Ali and Anctil, 1976; 13. Crescitelli *et al.*, 1985; 14. Gruber *et al.*, 1975; 15. Sillman *et al.*, 1996; 16. Gruber *et al.*, 1963; 17. Cohen, 1980; 18. Cohen *et al.*, 1990; 19. Bridges, 1965; 20. Anctil and Ali, 1974; 21. Bozzano *et al.*, 2001; 22. Hamasaki and Gruber, 1965; 23. Govardovski and Lychakov, 1977; 24. Litherland, 2001; 25. Hart *et al.*, 2004; 26. Collin, 1988; 27. Munz, 1965; 28. Munz and McFarland, 1973; 29. Compagno *et al.*, 2005; 30. Hart et al., 2011; 31. Bozzano and Collin, 2000; 32. Lowe *et al.*, 1996; 33. Gruber and Cohen, 1985; 34. Demski and Northcutt, 1996; 35. Strong, Jr., 1996; 36. Hueter, 1991.

RECOMMENDED FURTHER READING

Cohen, J. L., R. E. Hueter, and D. T. Organisciak. 1990. The presence of porphyropsin-based visual pigments in the juvenile lemon shark (*Negaprion brevirostris*). *Vision Res.*, 30: 1949–1953.

Gruber, S. H., and J. L. Cohen. 1976. Visual system of the elasmobranchs: state of the art 1960–1975. Pp. 11–105 *in* Hodgson, E. S. and R. F. Mathewson (Eds.), Sensory Biology of Sharks, Skates, and Rays. U.S. Government Printing Office, Washington D.C.

Gruber, S. H., and J. L. Cohen. 1985. Visual system of the white shark, *Carcharodon carcharias*, with emphasis on retinal structure. *S. Calif. Acad. Sci., Mem.*, 9: 61–72.

Hart, N. S., T. J. Lisney, and S. P. Collin. 2006. Visual communication in elasmobranchs. *Pp.* 337–392 in Ladich, F., S. P. Collin, P. Moller, and B. G. Kapoor (Eds.), Communication in Fishes. Science Publishers, Enfield.

Hart, N. S., T. J. Lisney, N. J. Marshall, and S. P. Collin. 2004. Multiple cone visual pigments and the potential for trichromatic colour vision in two species of elasmobranch. *Jour. Exp. Biol.*, 207: 4587–4594.

Hart, N. S., S. M. Theiss, B. K. Harahush, and S. P. Collin. 2011. Microspectrophotometric evidence for cone monochromacy in sharks. *Naturwissenschaften*, 98: 193–201.

Hueter, R. E. 1991. Adaptations for spatial vision in sharks. *J. Exp. Zool. Suppl.*, 5: 130–141.

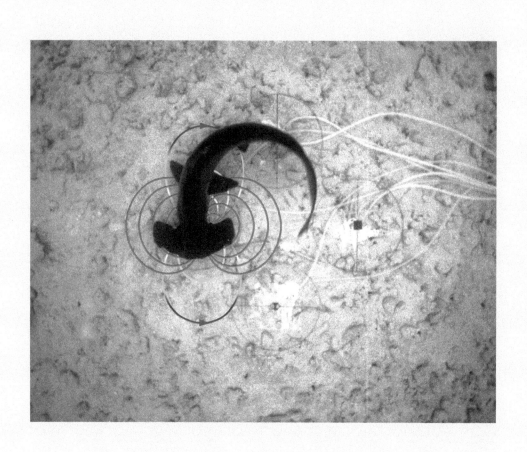

CHAPTER 9

Sense of Electromagnetic Fields: Electroreception/Magnetoreception

Dissolved within the waters of the oceans are large amounts of sodium and chloride ions. The ionic composition of water permits electrons to pass between materials that differ only slightly in their tendency to shed electrons. Air, on the other hand, is practically nonconductive. It conducts electricity only in the form of lightning when the electron differential between two points is on the order of many thousands of volts. You may have witnessed the flow of electrons or current as your teacher passed a copper wire held in one hand past a strong magnet held in the other hand in an elementary physics course. The electron flow is indicated by the movement of a needle on a voltmeter with leads attached to either end of the wire. This flow of electrons also creates a magnetic field, which is revealed if your teacher spread iron filings over a smooth surface with the wire on top of it—the current flow causes filings to align in one direction along lines of force with fewer filings farther away from the wire where the magnetic field is weaker. Hence, physicists consider these fields to be electromagnetic.

Cartilaginous fishes have sensory organs, the ampullae of Lorenzini, that are capable of detecting minute electromagnetic fields. These are produced by the electrical discharge as prey flex their muscles and draw saltwater, a conductor,

Scalloped hammerhead shark biting two electrodes on circular translucent plates that produce an electric dipole. Shown are equipotential lines that indicate the gradient in the tendency of electrons to flow (red) and lines indicating the density of electron flow (blue).

through their gills within the magnetic field of the earth. The sharks' ampullary organs are concentrated near the mouth yet extend to the edges of the underside of the snout. These jelly-filled pores with canals leading to sensory cells within the body enable the cartilaginous fishes to detect prey buried half a meter deep in the sand. The pores are present all the way to the edges of the laterally expanded rostrum of the hammerhead sharks and the wide pectoral fins of rays. This increases the area of the bottom substrate, where they can detect buried prey while swimming over the bottom.

The cartilaginous fishes employ their sense of the electrical component of these fields when quite close to their prey. Here it is likely used instead of chemoreception although a scent may be present in the immediate vicinity of the prey. This shift between senses as the predator approaches the prey is apparent from experiments carried out in shallow water on the dusky smooth-hound (*Mustelus canis*), a small coastal shark, and on the blue shark (*Prionace glauca*), which is more oceanic in its distribution.[1] Liquefied herring was dispensed at nighttime under dim lighting from a tube next to a pair of electrodes that were both mounted on a polyvinyl plate set on the bottom of a polyvinyl tank and camouflaged by a covering of sand. The smooth-hound approached the experimental apparatus singly or in small groups and invariably attacked only those electrodes between which passed an electrical current. The smaller smooth-hound accelerated toward the electrodes when 15 cm away, the larger ones when 30 cm distant. A similar experiment was carried out on blue sharks in the deeper waters off of Martha's Vineyard. An odor source and a pair of electrodes were attached at either end of a horizontal polyvinyl bar attached to a line suspended from a boat. Again, the sharks appeared to be attracted initially by the chemical scent, but changed direction and accelerated toward the electrodes through which current was passed when they approached the apparatus to within 100 cm.

Finally, ion-rich tissues of a cartilaginous fish generate an electrical field as the conductive body of the shark moves forward in the earth's magnetic field. The shark can use its ampullary organs to sense the resulting electron flow. Swimming forward in one direction results in the perception of a constant voltage, and the shark can use this sensory rule to maintain a single heading and swim for a period of time in a straight course as an airplane pilot flies an airplane directly between two cities. Sharks may also perceive local patterns in the magnetic field, which exist on the seafloor, and follow these patterns as a helicopter pilot follows a highway in a sinuous manner as it winds through the hills.

SALTWATER, A HIGHLY CONDUCTIVE MEDIUM

The charged nature of sodium and chloride ions in ocean water enables it to conduct electrons, and water flowing through the earth's magnetic field induces a flow of these electrons. The electron density varies in different substances, yet electrons are able to freely flow between these substances in this conductive medium. In this respect, water is very different than air, which is nonconductive and requires large electron differentials for electrons to flow between two points. It is this ionic property of the water in the oceans that permitted the cartilaginous fishes to evolve a sense, electroreception, absent in terrestrial species.

Understanding the electric sense of the cartilaginous fishes requires a rudimentary understanding of electricity and the terms "voltage," "resistance," and "current." Electricity is all about the flow of electrons, and this process can be compared to the flow of water, another more tangible property of the environment. Electrons are often concentrated in one medium, and their flow into and through another medium depends on the resistance of that medium, just as water within a reservoir and released through a gate will flow through the riverbed at a rate based on its width and rockiness. The larger the concentration of electrons, the greater the tendency of them to flow, just as more water in the reservoir would result in a greater flow of water out of the reservoir and downstream through the river. This tendency for electrons to flow, which increases with an increasing concentration of electrons, is termed voltage. One source of a high concentration of electrons is a battery. A current of electrons flows once a switch is opened to let the electrons pass through a continuous circuit including the electron source. Only so many electrons can pass through the pathway depending on its opposition to electron flow, termed the resistance of the circuit. Ohm's law expresses the relationship that the current that flows is equal to the voltage divided by the resistance to that flow ($I = E/R$). Electrons flow through a space between an anode (negative pole), which has a higher concentration of electrons, to a cathode (positive pole), with a lower concentration of electrons. The ocean is an unbounded three-dimensional conductor, so most electrons flow directly between the poles, fewer in a path slightly to either side of the direct path, and fewer even farther from that path. These paths of differing current flow are indicated by successively thinner lines of current drawn between the two poles at increasing distances from the direct path. Another way of describing a dipole is based on the tendency of electrons to flow toward the other dipole. These lines of equipotential are denoted by oval shapes with lines that become thinner at increasing distances from the pole.

The cartilaginous fishes perceive electric fields with the ampullae of Lorenzini, which are apparent on the dorsal and ventral surfaces of the rostrum as clusters of pores, each with a tubule leading to a common junction. There are five networks of these canals, which are symmetrically distributed on both sides of the laterally elongated head of the scalloped hammerhead shark (*Sphyrna lewini*) (fig. 9.1a). The superficial ophthalmic anterior tubules originate near the tip of the rostrum and branch forward to each side of the medial notch on its forward margin. The superficial ophthalmic posterior tubules start forward of the mouth and extend outward to either side of the centerline of the rostrum. The buccal tubules originate on either side of the mouth and extend inward to surround the anterior margin of the mouth as well as branch outward to pores opening along the side and back margins of the rostrum. The hyoid tubules branch outward from the side of mouth with pores surrounding the lateral and posterior margins of the rostrum. The mandibular tubules are located slightly behind the mouth. The concentration of pores around the mouth indicates that the ampullae of Lorenzini function in the seizure and ingestion of prey.

The ampullae vary in their distribution over the bodies the sharks and rays. The organs exist on both the dorsal and ventral surfaces of all species. The canals extend farther outward to either side of the body on the laterally expanded rostrum of the hammerhead sharks and expanded pectoral disk of the rays than the majority of sharks, which have a narrow, conical snout. For example, the superficial ophthalmic anterior tubules (green lines) of the scalloped hammerhead shark extend outward farther along expanded anterior margin of its head both on the dorsal and ventral surfaces than the tubules on the conical head of the sandbar shark (*Carcharhinus plumbeus*) (figs. 9.2a & 9.2b).[2] The buccal canals (blue lines) mainly extend outward on the rostrum of the hammerhead, whereas they extend inward toward the mouth on both the dorsal and ventral surfaces of the head of the sandbar shark and extend farther posterior on the latter surface. In the hammerhead shark, the ophthalmic posterior tubules (red lines) on the ventral surface extend not only inward toward the mouth but outward toward the middle of the anterior margin of the expanded rostrum. Many hyoid tubules (pink lines) extend outward to the anterior and posterior margins of the pectoral disk in the brown stingray (*Dasyatis lata*) (fig. 9.2c) Again, the main difference between these three species is that the canals of the sandbar shark extend farther back on the body while those of the hammerhead and brown stingray extend farther toward the sides of the body.

Each ampulla of Lorenzini consists of a canal leading to an expanded sac

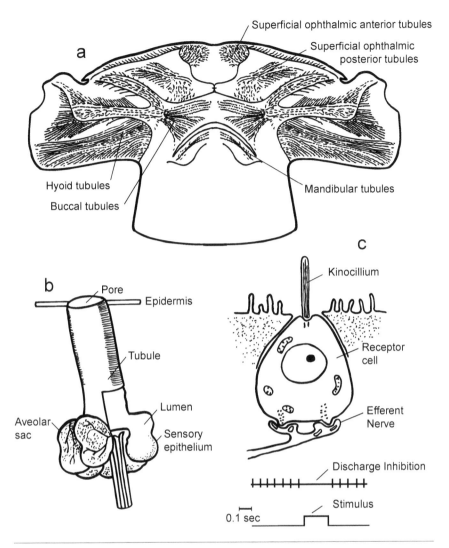

a

Superficial ophthalmic anterior tubules

Superficial ophthalmic posterior tubules

Hyoid tubules

Buccal tubules

Mandibular tubules

b

Pore

Epidermis

Tubule

Lumen

Aveolar sac

Sensory epithelium

c

Kinocillium

Receptor cell

Efferent Nerve

Discharge Inhibition

Stimulus

0.1 sec

Fig. 9.1 (a) The many tubules of the ampulla of Lorenzini in the rostrum of the scalloped hammerhead. (b) The alveolar sac of an Ampulla of Lorenzini. (c) The receptor cell with its kinocilium and efferent nerve. Shown underneath is the inhibitory effect of a stimulus consisting of an outside-positive voltage on rate of discharge of efferent nerve.

containing a receptor cell. At the base of each canal, roughly 1 mm in diameter, is a vase-shaped alveolar sac (fig. 9.1b).[3] A sensory epithelium covers the walls of the alveolar sacs at the base of the canal that consists of many receptor cells interspersed with supporting cells. A single kinocilium projects into the lumen from each receptor cell and an efferent neuron leads from each cell (fig. 9.1c).[4] The inner space within the canal is filled with a gel, which has little resistance and permits electrons to pass from the saltwater into the lumen of the alveolar sacs in the same density as they are present in the surrounding oceanic water.[4] The receptor cells are packed so closely together in the sensory epithelium

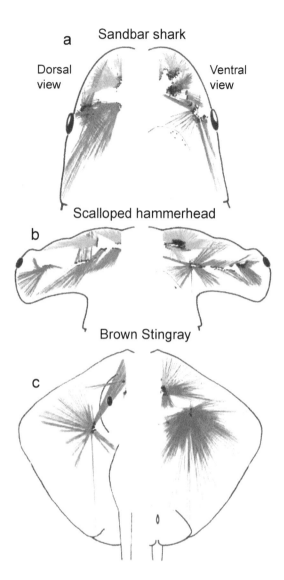

a Sandbar shark

Dorsal view

Ventral view

b Scalloped hammerhead

Brown Stingray

c

Fig. 9.2 The distribution of ampullary tubules on (a) the sandbar shark, (b) the scalloped hammerhead shark, and (c) the brown stingray. Canals with pores on the dorsal surface shown on the right side; canals with pores on the ventral surface shown on the left. The superficial ophthalmic anterior canals are represented by green, superficial ophthalmic posterior canals by red, buccal canals by blue, hyoid canals by pink, and mandibular canals by light blue.

that few electrons can pass this impermeable barrier into the tissues of the cartilaginous fish, where the concentration of electrons is less. Hence, a voltage differential is created between the lumen of the ampulla and the surrounding body tissue. As prey such as a flatfish use their muscles to draw saltwater through their mouth, past their gills, and out their gill slits, the activity of the muscles as well as the movement of the saltwater create an electric field. As the shark swims over the flatfish, buried within the sand, these electrons pass down the tubule into the lumen, and a greater voltage differential is created across this epithelium. This decreases the rate of transmitter release between the receptor cell and the efferent nerve, resulting in the inhibition of the rate of nerve discharge.[5]

The sensitivity of these groups of ampullary organs to the surrounding electrical fields is dependent upon their distribution on the dorsal and ventral surfaces of the body. This is apparent from spherical projections of the sensitivity of the ampullae in the sandbar shark, scalloped hammerhead shark, and brown stingray.[2] The elevation on the ordinate of the graph indicates the electrical sensitivity of the ampullary canals in a 90° arc extending above and below the body of the shark or ray. The azimuth on the abscissa refers to a 360° arc around the entire body of the animal, starting directly on the anterior to the body, a 90° rotation clockwise around the body on its right side, another 90° rotation toward the posterior, and a final rotation toward the left side. Green circles are clustered on the graph for the sandbar and scalloped hammerhead shark above and below an elevation of 0° on either side of an azimuth of 0° (figs. 9.3a & 9.3b). This indicates that the highest sensitivity of the pores from the superficial ophthalmic anterior tubules to an object is in front of the body. There are clusters of red triangles from 30° to –20° to either side of the medial point toward the posterior of the head, indicated by 180°. These points of sensitivity are located on either side of the mouth, where the pores from the ophthalmic posterior tubules are located. Pink diamonds are clustered from 10° to –10° from either side of the 0° azimuth outward to each side of the pectoral disk of the brown stingray, from 270° on the left side to 90° on the right anterior margin (fig. 9.3c). This indicates that the pores from hyoid canals are located along the entire anterior edge of the flattened body, where they can detect the electric field of prey and mates. Pink triangles and red triangles are also concentrated in a flattened X shape extending outward from an azimuth of 180°. These pores to the tubules from the hyoid and superficial ophthalmic posterior groups are on the underside of the pectoral disk on either side of the mouth. The tubules are distributed across the body to maximize the chance of detecting a mate or prey item by its electromagnetic fields as the chondrichthyan fish moves forward in the water column. Both a benefit and a cost are associated with an expanded sensory surface. On the one hand, through the lateral expansion of the rostrum and the lateral expansion of pectoral fins, the scalloped hammerhead and brown stingray have increased the width of their sensory field and are more likely to detect mates and prey by their electromagnetic fields. On the other hand, any deviation from the fusiform body plan of the majority of sharks likely results in poorer energetic efficiency during sustained swimming or a reduction in the speeds attainable during burst swimming.

One can better understand how the organs are adapted to their environment by contrasting the morphology of the ampullae on the body of a ray of the genus *Dasyatis*, which lives in the ocean, with those of a ray of the genus *Potamotrygon*, which lives in a river or lake (fig. 9.4).[6] The former is immersed in highly conductive saltwater, the latter in less conductive freshwater. Solid

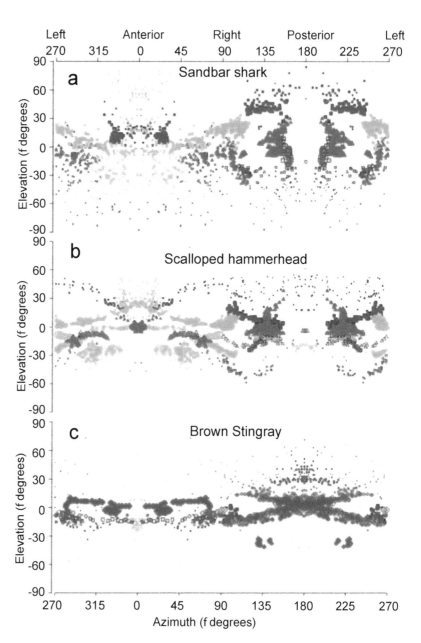

Fig. 9.3 Spherical projections for the ampullary groups in (a) the sandbar shark, (b) the scalloped hammerhead shark, and (c) the brown stingray. The elevation indicates the sensory ability in a 90° arc above and below the horizontal plane; the azimuth refers to the 360° angle from the anterior directly in front of the elasmobranch to its right, posterior, and left sides. Note that the ampullary groups are designated by different colors and symbols with area sensed by the ophthalmic anterior canals indicated by green circles, ophthalmic posterior canals by red triangles, buccal canals by blue squares, and hyoid canals by pink diamonds.

lines of current indicate the movement of electrons from the anode on the left to the cathode on the right. Dashed lines of equipotential denote varying concentrations of electrons, varying from a voltage of +7 near the cathode on the left to 0 in the center to –7 near the anode on the right in an imposed electric field. The body of each ray is shown as an oval with a single alveolar sac passing through its epidermis. The saltwater ray has a high internal resistance of 200 ohms/cm, while the surrounding saltwater has a low resistance of only 20 ohms/cm. On the other hand, the freshwater ray has the same internal resis-

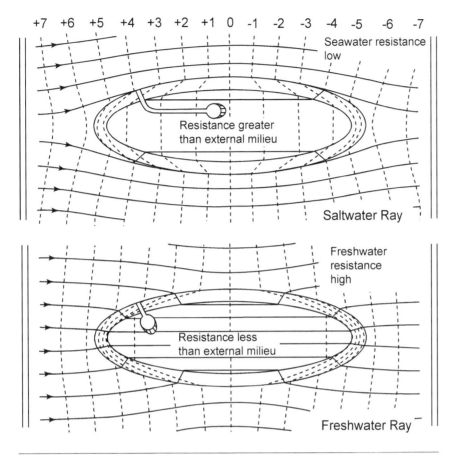

| +7 | +6 | +5 | +4 | +3 | +2 | +1 | 0 | -1 | -2 | -3 | -4 | -5 | -6 | -7 |

Seawater resistance low

Resistance greater than external milieu

Saltwater Ray

Freshwater resistance high

Resistance less than external milieu

Freshwater Ray

Fig. 9.4 The electrical properties of the ampullae of Lorenzini on the body of a ray of the genus *Dasyatis* that inhabits saltwater (above) and ray of the genus *Potamotrygon* that lives freshwater (below).

tance but is immersed in freshwater with a high resistance of 6,000 ohms/cm. For the saltwater ray, the lines of current pass around the body—the electrons pass more easily through the saltwater with less resistance. However, the lines of voltage equipotential bend toward the body of the saltwater ray and then diverge within it due its higher internal resistance. A few equipotentials cross the ampullary canal in the dermis of the saltwater ray. However, the canals of saltwater rays extend far into their internal tissues to create a greater voltage gradient to be sensed by the receptor cells in the sensory epithelium (see additional equipotential lines across the canal). Both the voltage gradient across the skin and internal body contribute to the effective stimulus. This explains the long ampullary canals of the sandbar shark, scalloped hammerhead shark, and brown stingray (figs. 9.1 & 9.2). On the other hand, the lines of current converge within the body of the freshwater ray, which has a lower internal resistance than freshwater, and the electrons thus pass more freely through the

Early in his career, Adrianus Kalmijn conducted a series of experiments that proved that sharks detect prey by perceiving the electrical fields emanating from their bodies.[6] The first step in these experiments was for Kalmijn to build an agar chamber to enclose the plaice, consisting of a flat cavity surrounded by thick sidewalls and a thin roof reinforced with foam plastic (fig. 9.5a). The walls were fitted with an inlet and outlet tube permitting seawater to circulate through the cavity. The purpose of the agar chamber was to permit the electrical field from the plaice to pass outward unimpeded while preventing other attractive stimuli from leaving the chamber. He then put a plaice in the chamber and buried it under the sand. The spotted catsharks, which he kept in captivity in his laboratory, were motivated to search for food by his adding a mixture of water and macerated fish to the tank. The sharks began to actively move around the tank actively searching for food. When they passed within 15 cm of the agar chamber, they turned sharply toward it and approached closely—even trying to dig up the flatfish as sand was observed squirting out of their gills (fig. 9.5c). After vigorous attempts at getting to the fish, they moved away and left an open patch of sand immediately over the chamber. This cleared spot always was just above the head region of the plaice, and it would be here that the prey generated an electric field either from muscle activity or from saltwater passing through its mouth and out its gills. It was obvious from this experiment that the shark did not need visual contact with the plaice to detect it buried under the sand.

Kalmijn wondered whether the subjects of his experiment could be using smell to locate their prey. Sharks and rays do have an acute sense of smell, as described in the earlier chapter on chemoreception. This seemed unlikely to Kalmijn because, to start with, they responded immediately upon introduction of the plaice into the tank, while an odor would have taken considerable time to diffuse through the roof of the chamber. Furthermore, the sharks showed little interest in the outlet tube, through which the odor might be expected to emanate. In order to eliminate this possibility, he placed a small nylon bag containing pieces of whiting in the chamber instead of the plaice and buried the chamber under the sand. The sharks now swam close to the outlet tube seeking food but ignored the roof of the agar chamber (fig. 9.5d). Might the sharks locate the plaice by perceiving water displacements or pressure oscillations emitting from the chamber? The spotted catsharks had both a lateral line and inner ear sensitive to these two stimuli. The plaice would create such disturbances by the movement of the mouth and gills during respiration. The agar chamber should attenuate these stimuli somewhat, but to eliminate them altogether, a thin polyethelene film was spread over the chamber and then buried under the sand. The sharks, stimulated with a liquefied fish extract, now searched all over the bottom of the tank without noticing the plaice although they passed it closely (fig. 9.5e).

Could the catsharks be locating the plaice by detecting an electrical field associated with it? First of all, it was necessary to know whether an electrical field could permeate throughout the agar chamber, in particular the plastic-strengthened roof. He placed two electrodes by themselves under the sand and next put the electrodes into the chamber 5 cm apart. He then turned on the power supply and generated electrical sine waves with a frequency of one cycle per second under both conditions. The intensity of fields passing through the agar chamber differed little in magnitude than when uncovered in the sand—the agar chamber appeared not to block an electrical current passing through it. Kalmijn also measured the electrical field when the chamber was covered with the polyethelene sheet and found that it prevented an electrical field from emanating from the chamber. This would explain why the sharks ignored the plaice when the chamber was covered with the plastic sheet. But would the sharks react to a small electrical field of one cycle per second, the field characteristic of the species, when produced by electrodes buried alone under the sand (fig. 9.5f)? Yes, the sharks reacted to the electrodes with the same vigorous feeding responses that they had exhibited when the plaice was in the chamber uncovered by the polyethelene sheet! The sharks could locate the precise location of the electrodes, which were exposed once the sharks inhaled the sand near the electrodes and discharged it through their gill slits in their vain attempt to get at the imaginary prey.

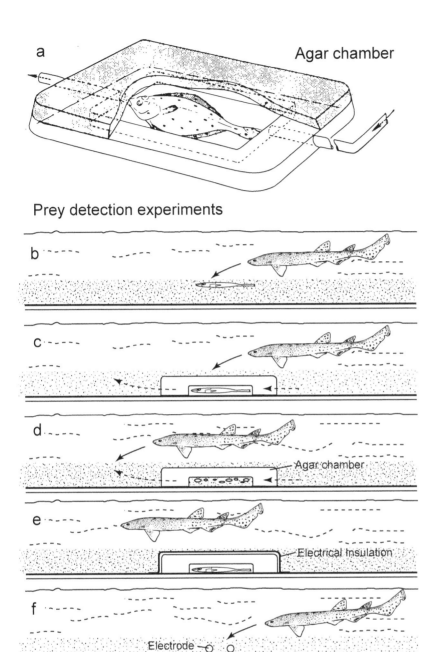

a

Agar chamber

Prey detection experiments

b

c

d

Agar chamber

e

Electrical insulation

f

Electrode

Fig. 9.5 (a) An agar chamber, in which a plaice was enclosed, was utilized to eliminate visual and mechanical cues used by small-spotted catsharks in prey detection. Five experiments (b-f) performed with catsharks to demonstrate their responsiveness to the electrical fields emanating from a plaice buried in the sand.

body of the ray. Furthermore, the lines of equipotential bend away from the internal body because its resistance is lower than that of the external environment. The lines converge within the epidermis due to its very high resistance compared to that of the rest of the body. This results in the effective stimulus being the voltage drop across the skin. Hence, the ampullary canals of the freshwater ray are very short compared with those of marine species of rays.

The sense of electroreception was discovered by Adrianus Kalmijn in an elegant series of experiments performed with the small-spotted catshark (*Scyliorhinus canicula*) and the European plaice (*Pleuronectes platessa*), a common item in this predator's diet.[7] He had observed spotted catsharks easily locate flatfish buried in the sand and feed upon them (fig. 9.5b).

Additional experiments have been carried out to describe the behavior of sharks as they detect and approach an electric field simulating that emitted by their prey. There are four swimming patterns by which juvenile scalloped hammerhead and sandbar sharks approach an artificial dipole field (figs. 9.6a–9.6d).[8] This field is characterized by lines of current flow, shown by concentric circles above and below the two electrodes, and equipotential lines, denoted by concentric circles on either side of the electrodes. The juveniles either approach the dipole in a straight line, make a single turn toward the source of the dipole, overshoot the dipole and turn back to approach it from the reverse direction, or circle the dipole eventually to approach it from a direction different from their first approach. Individuals of both species most often localize the dipole after making a single turn, 51% of the time for the hammerheads versus 55% for the sandbar sharks. The next most common approach is along a straight line with the former displaying this response 34% of the time and the latter 38% of the time. The hammerheads overshoot and/or exhibit a spiral track, 11% and 4% of the time, while the sandbar sharks overshoot the dipole 8% and did not make a spiral track. Hence, the juvenile scalloped hammerheads exhibit a greater diversity in their orientation to the prey than the sandbar shark. Furthermore, the scalloped hammerhead sharks exhibit a uniform decrease in the number of field detections with increasing distance, with a significant proportion of these distances at greater than 10 cm from the dipole source (fig. 9.6f). The sandbar sharks exhibit the greatest number of orientations nearer to the source, from 5 to 10 cm away (fig. 9.6e). This may be one reason behind the evolution of the lateral expansion of the rostrum in hammerheads. It enables them to detect the weak electric fields emitted from their prey at a greater distance than those sharks without a laterally enlarged rostrum.

MATE DETECTION

Not only do cartilaginous fishes detect prey by the electrical fields that emanate from their bodies, but also they can find their mates during courtship from the fields emanating from their bodies. Male Haller's round rays (*Urolophus halleri*) detect females from the bioelectric fields that they emit during the mating season from January to March in the shallow waters in Kino Bay of the Gulf of

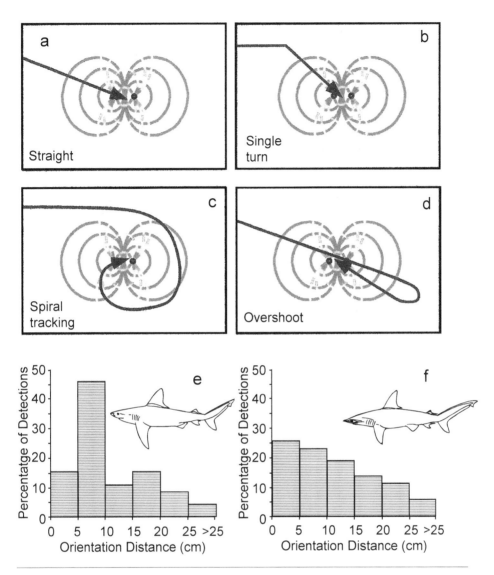

Fig. 9.6 (a-d) Four behavioral patterns used by the scalloped hammerhead and sandbar sharks to approach a dipole electrical field simulating that emanating from their prey. Distances at which (e) scalloped hammerhead and (f) sandbar sharks begin to orient toward the electrical field.

California.[9] Males swim near the bottom. Upon approaching within a distance of a meter from a buried female, they change their direction and swim directly toward that female. Each male then inspects a female, which is out of his view, by moving its rostrum over the margin of her disk until resting near the pelvic fins, where the male may sense a pheromone released through her vent (see 1–3 in fig. 9.7a). The male ray then uncovers the buried female and initiates courtship by seizing her pectoral disk in his mouth. Bioelectric potentials, or voltage changes, measured from near the female were compared to a reference

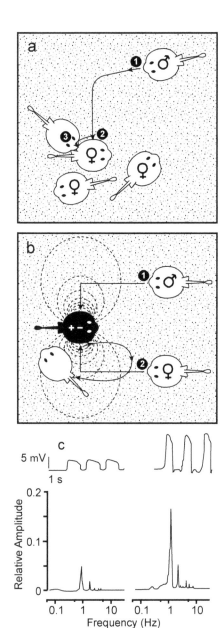

Fig. 9.7 Male round stingrays detect buried females by the bioelectric fields that they emit. (a) Male shown changing direction (1), swimming toward female (2), and nudging her near the pelvic fins (3). (b) The responses of male (1) and female (2) round stingrays to artificial electrical field simulating that of the female ray. (c) Voltages recorded at a spiracle (upper left) and gill (upper right) over time and the relative amplitudes of the different frequencies at both respiratory sites (bottom left and right). The peak frequency is one per second, which is identical to the rate at which water is drawn through the mouth, over the gills, and out the gill slits in one respiratory cycle.

voltage measured a large distance away. The positive voltage oscillations were of a greater magnitude near the gill slits (upper right) than the spiracle (upper left in fig. 9.7c). This is because there is more muscle activity associated with the gills and perhaps more water is drawn through the gill slits than the spiracle. The movement of more conductive material in the earth's main dipole field induces a greater flow of electrons. The most common frequency of the oscillating voltages is one cycle per second. This is apparent from the location of the two largest peaks on lower left and right graphs. This frequency is identical to the respiratory rate at which the mouth is opened to inhale water, pass it over the gills, and flush it out through the gill slits.

The electrical waveform recorded from a female was simulated with two electrodes in the presence of males to demonstrate the attractiveness of these bioelectrical fields (fig. 9.7b). The current flow is shown by lines passing between the two electrodes with the more widely separated lines representing successively less current flow at greater distances from the electrodes. The male round stingrays exhibited the same behavior in response to the field produced by the electrodes as when approaching a female hidden under the sand. The males abruptly turned toward the dipole, approached it closely, and inspected the model of the round stingray (see upper male stingray). Females also turned toward the dipole and swam toward it, but then moved to within a 10 to 100 cm radius of the electrodes and settled on the bottom (see lower female stingray).

ORIENTATION AND NAVIGATION

As we learned earlier, the tissues of a cartilaginous fish contain large amounts of ions, and thus the body is a conductor just as saltwater. For example, as a scalloped hammerhead shark swims forward within the horizontal component of the earth's dipolar magnetic field, it induces electron flow or current through it and in the water around it (fig. 9.8b).[6] The motionally induced electric field is confined mainly to the body of the shark due to the highly conductive water surrounding the shark. The jelly within the ampullary canals has a low resistance and thus transmits the electrons to the vicinity of the sensory epithelium in the alveolar sac at the base of the ampullary canals increasing the voltage across this membrane. The voltage is sensed by the receptor cells. A shark such as a scalloped hammerhead can swim in a straight line by avoiding a change in voltage sensed by the ampullae on its laterally elongated rostrum.

THE EARTH'S MAGNETIC FIELD

Understanding how the hammerhead accomplishes this feat of orientations requires a rudimentary understanding of geomagnetism. The earth's magnetic field is the sum of contributions from two magnetic sources, the earth's inner core and the outer crust. Circulation of electrically conductive liquid in the earth's core creates a dipolar moment in the earth's field (fig. 9.9b).[11] The direction of flow is indicated by ovals with arrows within the concentric region of the outer core, which is 2,900 km from the surface of the earth (fig. 9.9a).[12] The large arrow in the diagram indicates the strong field induced by this circulation. The largest gradient in the strength of the main magnetic field usually

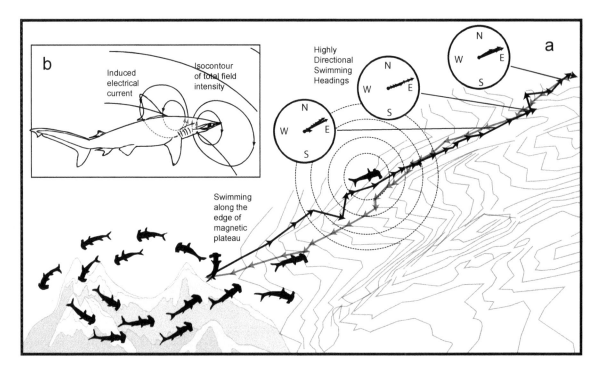

Fig. 9.8 (a) A scalloped hammerhead was tracked 20 km away from the Espiritu Santo Seamount, while a miniature sensor on an ultrasonic transmitter recorded its headings. The track of the shark, composed of positions determined at 15 min intervals, is superimposed upon the local pattern of magnetization. The highly directional nature of its swimming is evident from the similarity of the directions of ten successive headings given for three successively distant locations from the seamount (arrows in circles). (b) Electrical current shown that is induced as the hammerhead swims within the earth's magnetic field.

is in a north-south direction. It is denoted by lines of force that form circles around the earth with increasing latitude (fig. 9.9b). The earth's magnetic field is weakest at the equator and strongest at the poles. The shark could swim in a straight line by maintaining a constant voltage in its ampullary organs as its swimming path continues to pass through these lines of force at the same angle.

The second source of the earth's magnetic field is magnetic minerals, oxides of iron and titanium, in the earth's crust, no more than 4 km from the sea surface. These materials, which are deposited simultaneously on either side of the spreading axis of two crustal plates to produce magnetic lineations, create anomalies, or distortions in the dipolar main field.[13] These strong and weak bands of magnetization, with steep intensity gradients between them, alternate in a north-south direction and are ubiquitous to all ocean basins. They alternately contain magnetic particles with dipole moments either parallel or antiparallel to the current orientation to the earth's magnetic dipole axis. The boundaries between the adjacent steep and weaker gradients could serve as geographic reference roads along which marine animals might migrate

Adrianus Kalmijn was the first to hypothesize that sharks have a compass sense utilizing the earth's geomagnetic field as a reference for their movements.[6] Swimming at 100 cm/sec, the vertical component of the voltage gradient induced within the ampullae of Lorenzini would be as high as 0.4 µV/cm, which is well within their electrical sensitivity. I have demonstrated that sharks have the ability to swim in a highly directional manner.[10] A pole spear was used to insert a small barb, connected by a tether to an ultrasonic beacon, into the dorsal musculature of a hammerhead shark. This transmitter had a miniature heading sensor to detect whether the shark swam in a straight line, consisting of a pin with two supermagnets attached near one end and a circular plastic disk near the other end with a transparency gradient, permitting successively more light to pass through along its circumference as it was rotated 360 degrees in one direction. Two light emitters below the disk produced light, which passed through this disk, and was recorded by two light sensors above the disk. The supermagnets kept the pivot always aligned with the earth's magnetic field, yet any change in the direction the shark swam allowed more or less light to pass through the disk because, unlike the disk, the light emitters and sensors rotated with the shark. This miniature device recorded whether sharks swam in a straight line without the help of seeing the bottom. Ten successive headings of one migrating hammerhead are shown in each of the circular diagrams at three distances, each successively farther from the seamount (see arrows in circular compasses in fig. 9.9). The shortest arrow denotes the first compass heading, the next longest the second heading, and so forth. This shark swam in an extraordinarily constant heading even at a distance of nearly 20 km from

the seamount before reversing its direction and returning. One can only appreciate the challenge of accomplishing this remarkable feat without the guidance of an automobile's headlights illuminating the highway with white lines on either side of the driving lane.

What is also significant was that the hammerheads appeared to follow the same path to and from the seamount. Three hammerhead sharks were tracked, one for eleven days.[10] The three sharks repeatedly followed six paths when leaving and returning to the seamount. A magnetic survey was later conducted at the seamount to reveal the patterns in the local magnetization so that the paths of the sharks could be compared to them. The repeatedly traveled paths of the sharks coincided with ridges and valleys in the geomagnetic topography. The most prominent feature in this topography near the seamount was the edge between the western slope and plateau. The track of the shark with the highly directional swimming movements followed this feature. The outward and return paths of this shark coincided for two-thirds of the track, which was in a north-south direction. This coincided with edge of a magnetic lineation, characterized by an abrupt change from increasing magnetic intensity in an easterly direction over the slope to little further change over the plateau. The hammerheads also moved eastward along other valleys and ridges of the geomagnetic plateau. These paths can be likened to the network of roads and highways leading to and from a city. Hence, the same hammerhead shark may be able to repeatedly swim in a directed manner to a favored feeding ground, as someone in the suburbs might go to work in the city during the day to work and return home at night for dinner.

between temperate and tropical environments. In fact, the seamount is located on the edge of a band of strong magnetization.

These same magnetic minerals also form magnetic maxima and minima that lead away from seamounts produced by lava flows that extend outward for distances in the order of tens of kilometers. These ridges and valleys are caused by basalt that flows away from the cone of the volcano laden with mag-

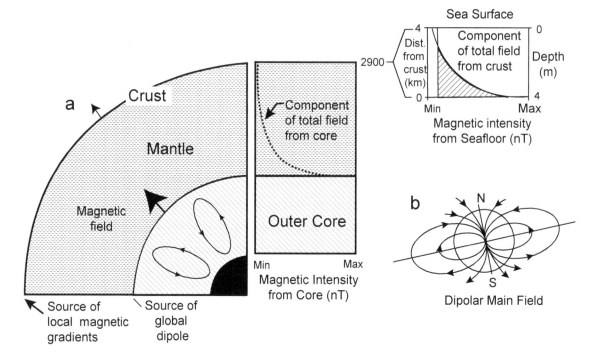

Fig. 9.9 (a) Cross-section of the earth showing the movement of magma in the outer core, which generates the earth's main field, and the relative strength of the field at the boundaries to the outer core and crust (see arrows). Insert illustrates the greater proportional increase of the field strength due to local magnetization at increasing depths in the ocean. (b) Diagram of magnetic lines of force and poles (N = north; S = south) of the earth's dipolar main field.

netite, minute magnetic particles with a dipolar field, during eruptions occurring over the long geological history of the earth. The magnetite particles are either parallel or antiparallel to the earth's current main field, depending upon the orientation of the earth's magnetic pole at the time of the eruption. The ridges and valleys leading away from a seamount or volcanic island can provide a reference that enables the sharks to return to their sites of aggregation after traveling to distant locations where they have found abundant prey.

The ability of a cartilaginous fish to guide itself using local maxima and minima in geomagnetic field intensity is called "geomagnetic topotaxis." *Topo* refers to the relationship of an animal's movement to topography, defined in the dictionary as "the configuration of a surface including its relief and the position of its natural . . . features."[14] *Taxis* refers to an individual's attraction to a physical property in the environment. In this case, the elasmobranch actively tracks magnetic ridges and valleys: features of relief in the surface of the geomagnetic field. It is essential to distinguish topotaxis from the compass sense, which is the ability to maintain a heading using directional references such as the sun, the moon, stars, or the earth's dipolar main field.

These two orientation mechanisms may best be distinguished from each other with an analogy to the quite different methods by which humans operate large airplanes and helicopters.[10] The airplane pilot navigates between two widely separated geographical points by steering in a direction relative to the northward orientation of the compass magnet, utilizes knowledge of the difference between the former direction and that of the destination, and returns by flying in a complementary direction. If the wind speed is strong and perpendicular to the plane's course, the airplane is deflected from the direction of the destination point, and the pilot changes course based upon knowledge of wind velocity to compensate for this deflection. The resulting flight path is often slightly curved. Similar paths are not expected from an animal with a compass sense. A helicopter, on the other hand, is often flown in relation to local features and therefore is navigated differently. The helicopter pilot visually follows a road, valley, or ridge, resulting in a path that can be sinuous. The winding path of the helicopter depends upon that of the reference feature. Straight roads give straight flight paths; winding roads give winding paths.

Both the dipolar nature of the earth's main field and the local distortions would be of great value to sharks in guiding their movements. The main field would be an ideal directional reference, while the local distortions would be an ideal spatial reference. At the sea surface, the local distortions relative to the overall dipole moment of the earth are small. The shark migrating over great distances could use the dipolar nature of the earth's main field as a reference. It could swim in a single direction either by keeping the induced field perceived by its receptors constant or maintain a constant differential between the intensity of the fields detected by bilaterally separated receptors. This may explain why many sharks such as the blue, mako, and white sharks regularly swim at the surface and move great distances in a straight line.[12]

However, with increasing depth these small field anomalies increase because the contribution to the total field from the crust, only 4 km from the surface, increases disproportionately to that of the core, 2,900 km from the surface. This is evident from the comparison of the two curves of magnetic intensity, one of which is derived from the crust and the other from the core (see upper right insert, fig. 9.9a). The curve depicting the magnetic contribution from the crust passes across the curve for the core as depth increases (see the intersection between the two lines and the cross-hatching in inset). Yo-yo swimming of sharks (also known as bounce dives) may be a behavioral tactic to distinguish the local gradients from the main field. The shark would be better able to detect guiding magnetic topography by descending in the water column when migrating to and from the seamount. Individuals could distinguish the local gradients from the main field by gliding downward until the field rotates and increases so that the magnetic ridge or valley is perceptible over the main field. The individual would have to periodically rise in order to reestablish its field of reference.

There are now experimental studies that indicate sharks perceive properties of the ambient electromagnetic field of the earth. Thornback rays (*Raja clavata*), were trained to distinguish the polarity of a uniform electric field produced by electrodes at either side of the tank.[15] This was done by feeding an individual when it moved into a corral to one side of the tank under one polarity. Once the ray learned to choose the same corral most of the time, the uniform field was reversed to the opposite polarity. The thornback ray now moved into a corral on the other side of the tank as predicted, if the behavior were performed with knowledge of the direction of current flow in the tank. This experiment will be described in more detail in the next chapter, which focuses in part on the learning capabilities of cartilaginous fishes. Sandbar sharks also have been shown to detect an increase in the earth's magnetic field.[16] They were kept in a tank encircled by a coil of wires through which current was passed when they were fed in a particular area of the tank. The increase in the strength of the earth's main field by the addition of an artificial magnetic field was eventually sufficient to induce heightened activity associated with feeding in the absence of the food reinforcement. This could serve as a behavioral assay to determine the threshold of magnetic field detection for different species. However, future experiments should take into account the empirical observations of those features in the geomagnetic field to which sharks have been shown to orient. For example, such an experiment could entail training sharks to move along a miniature version of a magnetic ridge produced within a tank, similar to but smaller than one associated with the magnetization of the basalts within a lava flow leading outward from a seamount.

HIERARCHY OF SENSES

The sharks, rays, and chimaeras use all of their senses in a hierarchical manner to find their prey and mates and avoid their predators. One or more of their senses, chemoreception, mechanoreception, photoreception, and/or electromagnetoreception, is capable of perceiving stimuli at distances up to several kilometers away (fig. 9.10). You learned in the sixth chapter, that a white shark could use its sense of smell to locate a dead whale using a form of chemical taxis to an odor corridor originating at a distance of six kilometers. In the seventh chapter, you learned that adult reef sharks detect with their inner ear pressure oscillations, emitted by a struggling fish, at distances as far as 800 m away, abruptly turn toward the source of the sound, and swim directly toward it. As they approach within 150 m of the struggling fish, they can perceive

The distance hierarchy of sensory stimul that attract a shark to its prey, mate, or predator

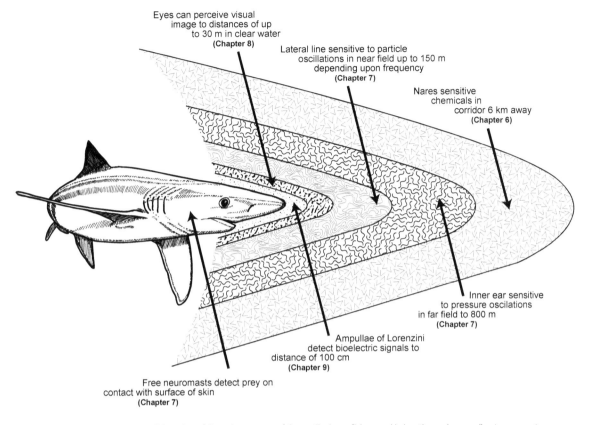

Eyes can perceive visual image to distances of up to 30 m in clear water
(Chapter 8)

Lateral line sensitive to particle oscillations in near field up to 150 m depending upon frequency
(Chapter 7)

Nares sensitive chemicals in corridor 6 km away
(Chapter 6)

Inner ear sensitive to pressure oscilations in far field to 800 m
(Chapter 7)

Ampullae of Lorenzini detect bioelectric signals to distance of 100 cm
(Chapter 9)

Free neuromasts detect prey on contact with surface of skin
(Chapter 7)

Fig. 9.10 The hierarchy to the range of detection of the various senses of the cartilaginous fishes used in locating and responding to prey, mates, or predators.

one-way particle displacements with their lateral line and use these stimuli to approach even closer to the prey item. In the ninth chapter, you learned that they would be able to see their prey at most at a distance of 30 meters in the clearest oceanic waters, but more likely at a distance less than 10 meters in coastal waters because the marine snow and abundant plankton reduces visibility greatly. The ampullae of Lorenzini detect prey, based on the bioelectric fields that they emit, only when the predator is very close to its prey, generally less than 100 cm in the case of juvenile sandbar and scalloped hammerhead sharks. The free neuromasts and canal neuromasts, which are present on the body of all cartilaginous fishes, may also be used in prey handling. They are very sensitive to minute, one way water particle movements. Hence, the chondrichthyan fishes have a diversity of senses, which can be used in conjunction with each other, to detect and to move either toward their prey or mates as well as to avoid their predators.

SUMMARY

The cartilaginous fishes perceive electric fields with the ampullae of Lorenzini, which are apparent on the dorsal and ventral surfaces of the rostrum as clusters of pores, each with a tubule leading to a common junction. There are five networks of these canals on the bodies of the sharks and rays. At the base of each canal, roughly 1 mm in diameter, is a vase-shaped alveolar sac. A sensory epithelium, which consists of many receptor cells interspersed with supporting cells, covers the walls of the alveolar sacs at the base of the canal. A single kinocilium projects into the lumen from each receptor cell and an efferent neuron leads from each cell. The inner space within the canal is filled with a gel that has little resistance and permits electrons to pass from the saltwater into the lumen of the alveolar sacs in the same density as they are present in the surrounding oceanic water. The receptor cells are packed so closely together in the sensory epithelium that few electrons can pass through into the tissues of the cartilaginous fish, where the concentration of electrons is lower. Hence, a voltage differential is created between the lumen of the ampulla and the surrounding body tissue. As prey such as a flatfish use their muscles to draw saltwater through its mouth, past its gills, and out its gill slits, the activity of the muscles as well as the movement of the saltwater create an electric field. As the shark swims over the flatfish, buried within the sand, these electrons pass down the tubule into the lumen and a greater voltage differential is created across this epithelium. This decreases the rate of transmitter release between the receptor cell and the efferent nerve, resulting in the inhibition of the rate of nerve discharge.

The electric sense enables cartilaginous fishes to detect prey that are buried under the substrate during daytime, but also to detect species hiding in crevices in coral and rocky reefs during nighttime. Male rays can detect females based on the electrical fields that they emit. This sense also enables sharks to detect the patterns of the earth's magnetization and use them to provide guidance not only during long-distance migrations but also during daily feeding migrations. Sharks migrating over great distances can use as a reference the dipolar nature of the earth's main field. They could swim with a constant heading by keeping constant the induced field perceived by their bilaterally separated receptors. Magnetic minerals embedded in basalt produce magnetic maxima (ridges in a topographic sense) and minima (valleys) associated with lava flows that extend outward from seamounts for distances in the order of tens of kilometers. These may provide guidance for sharks to return to their sites of aggregation after traveling to distant locations where they have found abundant prey. The electric sense was certainly critical to the long-term evolutionary success of the cartilaginous fishes. It may even explain the evolution of the lat-

1. Why is electroreception not common in terrestrial species?

2. How would you design an experiment to test whether cartilaginous fishes orient to magnetic minima and maxima in the geomagnetic field?

3. The cartilaginous fishes have electroreceptors, and hence we attribute changes in their behavior in response to alterations in the magnetic field to this sense. However, members of other taxa without electroreceptors have been shown to be sensitive to magnetic fields. The homing ability of birds was affected by placing tiny magnets on them with reversed polarities. Whales have been reported to strand where magnetic lineations lead to shore. How would one demonstrate unequivocally that the cartilaginous fields use their ampullae of Lorenzini to detect magnetic fields, not some unknown receptor shared with members of these other taxa?

4. There is some evidence that electroreception was more widespread in the past geological ages. Why might this be so? Hint: this might have something to do with plate tectonics and the extent of the oceans.

erally elongated rostrum in the hammerhead shark or wide pectoral disk of the rays. It enables them to sense a larger magnetic gradient across their bodies.

* * *
** ** **

KEY TO COMMON AND SCIENTIFIC NAMES

Blue shark = *Prionace glauca*; brown stingray = *Dasyatis lata*; dusky smooth-hound shark = *Mustelus canis*; Haller's round ray = *Urolophus halleri*; plaice = *Pleuronectes platessa*; sandbar shark = *Carcharhinus plumbeus*; scalloped hammerhead shark = *Sphyrna lewini*; shortfin mako shark = *Isurus oxyrinchus*; small-spotted catshark = *Scyliorhinus canicula*; smooth hammerhead shark = *Sphyrna zygaena*; thornback ray = *Raja clavata*; white shark = *Carcharodon carcharias*.

LITERATURE CITED

1. Kalmijn, 1982; 2. Rivera-Vicente *et al.*, 2011; 3. Tricas and Sisneros, 2004; 4. Heiligenberg, 1993; 5. Waltman, 1966; 6. Kalmijn, 1974; 7. Kalmijn, 1971; 8. Kajiura and Holland, 2002; 9. Tricas *et al.*, 1995. 10. Klimley, 1993; 11. Elasser, 1946; 12. Klimley *et al.*, 2002; 13. Skiles, 1985; 14. Gove, 1966; 15. Kalmijn, 1982; 16. Meyer *et al.*, 2005.

RECOMMENDED FURTHER READING

Kajiura, S. M. 2001. Head morphology and electrosensory pore distribution of carcharhinid and sphyrnid sharks. *Environ. Biol. Fishes*, 61: 124–133.

Kajiura, S. M., and K. N. Holland. 2002. Electroreception in juvenile scalloped hammerhead and sandbar sharks. *Jour. Exp. Biol.*, 205: 3609–3621.

Kalmijn, A. J. 1982. Electric and magnetic field detection in elasmobranch fishes. *Science*, 218: 916–918.

Klimley, A. P. 1993. Highly directional swimming by scalloped hammerhead sharks, *Sphyrna lewini*, and subsurface irradiance, temperature, bathymetry, and geomagnetic field. *Mar. Biol.*, 117: 1–22.

Meyer, C. G., K. N. Holland, and Y. P. Papastamatiou. 2005. Sharks can detect changes in the geomagnetic field. *Jour. Royal Soc. Interface*, 2:129–130.

Rivera-Vicente, A. C., J. Sewell, and T. C. Tricas. 2011. Electrosensitive spatial vectors in elasmobranch fishes: implications for source localization. *PLoS ONE*, 6 (1): e16009; doi:10.1371, *pone.* 0016008.

Skiles, D. D. 1985. The geomagnetic field; its nature, history, and biological relevance. Pp. 43–102 *in* Kirschvink, J. L., D. S. Jones, and B. J. MacFadden (Eds.), Magnetite Biomineralization and Magnetoreception in Organisms. Plenum Press, New York.

Tricas, T. C., S. W. Michael, and J. A. Sisneros. 1995. Electrosensory optimization to conspecific phasic signals for mating. *Neurosci. Letters*, 2002: 129–132.

CHAPTER 10

Brain Organization and Intelligence

The sharks have until recently had an undeserved reputation among the public for being dumb feeding machines. This impression as an indiscriminate predator has been largely created by cinematographers, whose films often depict white sharks swimming within a pool of mammalian blood and macerated fish up to a protective cage and opening their jaws wide in attempts to seize the frightened occupants within the cage. Olfactory stimulants such as macerated fish and mammalian blood are liberally dispensed in the water to attract members of this large and dangerous shark species close enough to be filmed. Hence, it is not unexpected that they might consider those in the cage to be the source of the chemical attractant. The sharks are usually fed often during filming to positively reinforce this feeding behavior. Furthermore, an attempt is made to attract many sharks to feed at once, resulting in social facilitation or the increased frenetic activity associated with competition for food as evident when a group of dogs feed together. The sharks were even encouraged to bite the cage in the classic film *Blue Water, White Death* by attaching pieces of meat to the bars of the cage. The sharks were really not attempting to bite the humans within the cage but the horse meat attached to it.

A small male nurse shark named Huey taught me that

Huey, a nurse shark, was taught to bump different colored targets to be fed a piece of fish.

cartilaginous fishes are not stupid while I was a graduate student at the Rosen-stiel School of Marine and Atmospheric Sciences.[1] I caught Huey with my hands while swimming with mask and snorkel in the shallow waters around Key Biscayne, brought him back to the laboratory, and put him into an aquar-ium. Did this little nurse shark have a capacity to learn? Would he learn if fed a tasty piece of fish after performing a particular behavior? I lowered a wooden handle with a white Plexiglas square into the aquarium, gently pushed it against his snout, and then dropped a piece of fish into the tank. Huey vigor-ously swam up to it and swallowed it whole. After pressing the paddle against his snout and feeding him a few times, he learned to swim forward to bump the paddle himself when it was held in front of him in order to be positively reinforced with a piece of fish. He was then presented with either a black or a white target but fed only after bumping the black one. Sure enough, he quickly stopped paying any attention to the white target and only pushed his snout against the black one in order to be fed. Huey was really a smart creature and learning rapidly but growing fast at the same time. He was soon too big for the aquarium so I returned him to the shallow waters near the laboratory, yet wor-rying a bit at the time that he might find it hard to forage properly after being pampered in my aquarium. As you will learn in this chapter, sharks, rays, and chimaeras can learn as quickly as birds and mammals and have brains of a com-parable size and elaboration.

BRAIN ORGANIZATION

Not only did many in the general public believe that the chondrichthyan fishes were primitive, but so did some of the first scientists who studied their anat-omy.[2] These anatomists were not aware that similar structures might evolve at the same time and falsely believed that the evolution of the cartilaginous skel-eton predated that of the bony skeleton. Hence they believed the cartilaginous fishes were a predecessor of the bony fishes. Furthermore, fishes and sharks were assumed to represent earlier and simpler stages in the evolution of mam-mals. This myth was easily perpetrated because students mainly dissected the piked dogfish (*Squalus acanthias*) or the small-spotted catshark (*Scyliorhinus canicula*) in their high school or university biology classes. The piked dogfish is a member of the Squalomorphi, the primitive superorder of sharks. The spotted dogfish is one of the most primitive members of the Galeomorphi, the advanced superorder of sharks. Examination of the brains of either of these two relatively primitive species provides little appreciation for the complexity of the brains and sensory organs of more advanced reef sharks of the genus *Carcharhinus*, hammerhead sharks of the genus *Sphyrna*, and eagle rays of the genus *Myliobatis*.

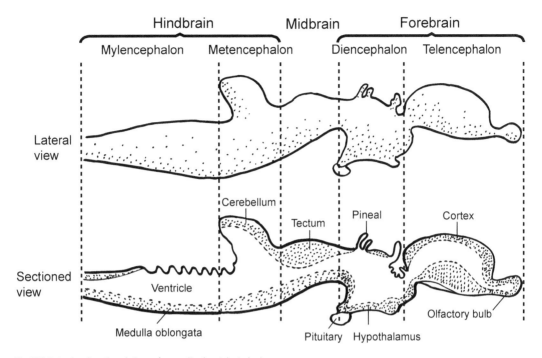

Fig. 10.1 Lateral and sectioned views of generalized vertebrate brain.

The size of the brain has always been considered an indicator of intelligence in vertebrates. Yet this is a simplistic view because the brain consists of various distinct regions, each serving different neurological functions. The complexity of each region is dependent upon the species' mode of locomotion such as swimming, flight, or walking as well as the relative importance of its senses such as smell, hearing, vision, or electroreception. The usefulness of each sense varies depending on whether the species lives in water such as amphibians, bony fishes, and cartilaginous fishes, or in air such as birds and many mammals. The vertebrate brain consists of three bulbous regions, the hindbrain, midbrain, and forebrain (fig. 10.1). In the most basic condition, the hindbrain, or rhombocephalon, concerns itself with the sensory functions of balance, mechanoreception, and electroreception.[3] The rear portion, the mylencephalon or medulla oblongata, controls respiration and serves as a relay station for the efferent fibers leading from the ear. The forward portion, the metencephalon, consists of an important outgrowth, the cerebellum, that coordinates and regulates motor activities, particularly those that are reflexive such the maintenance of posture or escape responses. The nerve from the lateral line and ampullae of Lorenzini leads to this section of the brain where dorsal nucleus processes information from the electroreceptive system. The midbrain, or mesencephalon, receives input from the optic nerve and processes this information in an enlarged dorsal region, the optic tectum. The forebrain, or prosencephalon, has two regions, the diencephalon and telencephalon. The diencephalon serves as a major relay station between the sensory areas in the hindbrain and midbrain and the higher brain centers in the anterior region of the forebrain. The pituitary gland, an important endocrine organ, is a ventral protuberance from the diencephalon. The floor of the diencephalon, the hypo-thalamus, and the pituitary gland are involved in hormonal control of parts of the brain. Another endocrine gland, the pineal organ, is a dorsal outgrowth of the diencephalon. It is light sensitive. A clear region in the chondrocranium of the cartilaginous fishes covers the pineal organ and permits light to reach it. It may have a role in the setting of the biological clock. The most anterior region of the brain is the telencephalon. Here information from the sensory systems is integrated, including that from olfactory bulbs, which are connected directly to the forebrain. The telencephalon is enlarged and its surface convo-luted, or foliated, in some vertebrates. It is then referred to as the cerebrum. An increase in the size of this region is associated with more complex behav-ioral repertoires and sociality among species.

ANATOMY OF BRAINS OF SHARKS, RAYS, AND CHIMAERAS

The anatomical structures of the brains of sharks, rays, and chimaeras differ from each other.[2] Despite having a common ancestor, the brain of the chi-maeras differs in two main ways from the brains of sharks and rays. First, chi-maeras have small olfactory lobes that arise directly from the end of the fore-brain, whereas the olfactory lobes from sharks and rays are larger and at the end of peduncles extending from the sides of the forebrain. For example, the forebrain of the spotted ratfish (*Hydrolagus colliei*) is elongate and divides into two hemispherical telencephalic processes leading to the two round olfactory lobes consisting of olfactory epithelium (fig. 10.2a). In contrast, the bat eagle ray (*Myliobatis californicus*) and the dusky smooth-hound (*Mustelus canis*) have two thin olfactory peduncles extending from the lateral pallium at the outer edge of the central nucleus of the telencephalon to large oval olfactory lobes (fig. 10.2b,c). This difference may reflect the chimaeras' lesser reliance on the sense of smell than that of sharks and rays. Second, the forebrain of the chi-meras is elongate and lacks the bulbous expansion of the forebrain, the central nucleus, and the lateral pallial formations, which bridge the two distinct hemi-spheres of the telencephalon. This area of the brain is thought to be devoted to visual discrimination and social activity.

The sharks have two major patterns of brain organization, one exhibited by the primitive squaloform and the other by the advanced galeoform sharks.[2] The former consist of the sixgill and sevengill sharks (Hexanchiformes),

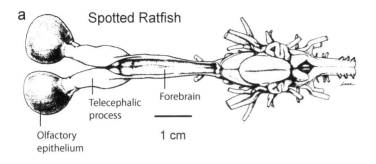

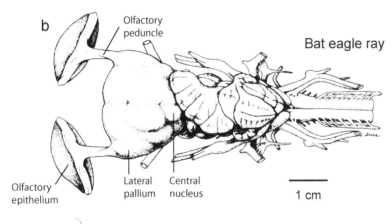

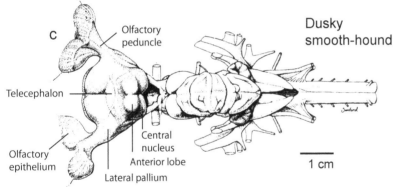

Fig. 10.2 Dorsal view of the brains of the (a) white-spotted ratfish, (b) bat eagle ray, and (c) dusky smooth-hound.

saw sharks (Pristiophoriformes), angelsharks (Squatiniformes), and dogfishes (Squaliformes). The latter group is composed of the bullhead sharks (Heterodontiformes), carpet sharks (Orectolobiformes), and requiem sharks (Carcharhiniformes). The squaloform sharks are more similar to each other in their genetic makeup and anatomy than are the galeoform sharks. The squaloform sharks evolved along a divergent line starting with the sixgill sharks, which were first present in the seas 200 million years ago, at the beginning of the lower Jurassic. The galeoform sharks evolved along a separate line starting with the horn sharks, which also appeared first at the same geological time period.

The cerebellum of the squaloform sharks has a smooth surface, while that of the galeoform sharks is convoluted and highly foliated.[2] This difference is apparent when comparing the cerebellum of the broadnose sevengill shark (*Notorynchus cepedianus*), a squaloform (fig. 10.3a) to those of galeoforms such as the white shark (*Carcharodon carcharias*) (fig. 10.3b) and bonnethead shark (*Sphyrna tiburo*) (fig. 10.3c). The optic tectum of the squaloform sharks is well-developed and is well exposed on the dorsal side of the brain, whereas that of the galeoform sharks is far less developed and is overlapped by the cerebellum. Conversely, the telencephalon of the squaloform sharks is poorly developed, whereas that of the galeoform sharks is highly developed and complex in its organization. Note the large telencephalon of the dusky smooth-hound with its bilateral central nuclei (fig. 10.2c) and the even larger bilateral central nuclei of the bonnethead shark (fig. 10.3c). There are a few exceptions to this rule such as the horn shark, a galeoform, whose brain resembles that of the squaloform sharks. This may be due to its being the first to evolve on a divergent line from a common ancestor along with the sixgill sharks. The greatest brain complexity in the sharks is in the reef and hammerhead sharks. Their larger cerebellum may be needed to process information from their highly developed sense of electroreception used both to detect prey and mates as well as to navigate using the earth's magnetic field as a reference. Their expanded telencephalon can be attributed to their complex behavioral repertoires and social systems.

The rays resemble more the galeoform sharks in the more complex organization of their brain. The skates of the order Rajiformes and torpedo rays of the Torpediniformes have a relatively simple or slightly convoluted cerebellum and a medium-size telencephalon. Conversely, the stingrays of the Myliobatiformes have a highly convoluted cerebellum and a large telencephalon, whose size rivals that of the reef and hammerhead sharks. At present, all measures of neural complexity indicate that the reef sharks, hammerhead sharks, and eagle rays have the most advanced brain development among the cartilaginous fishes. This is an example of convergent evolution, where members of the two groups independently developed highly complex brains. With this has come an elaboration of the electroreceptive sensory system, which provides them with the ability to detect prey and navigate better in the limited visibility of the oceans. Furthermore, the elaboration of the forebrain is consistent with their possessing complex behavioral repertoires and social behavior.

The white shark's brain is smaller and less specialized than that of the reef and hammerhead sharks (fig. 10.3b).[4] Its forebrain has a very small central nucleus and overall size, indicating that it is less capable of visual discrimination and exhibits less social activity. Although convoluted, the cerebellum is relatively small for a galeoform shark. Thus, the white shark is likely to rely less on

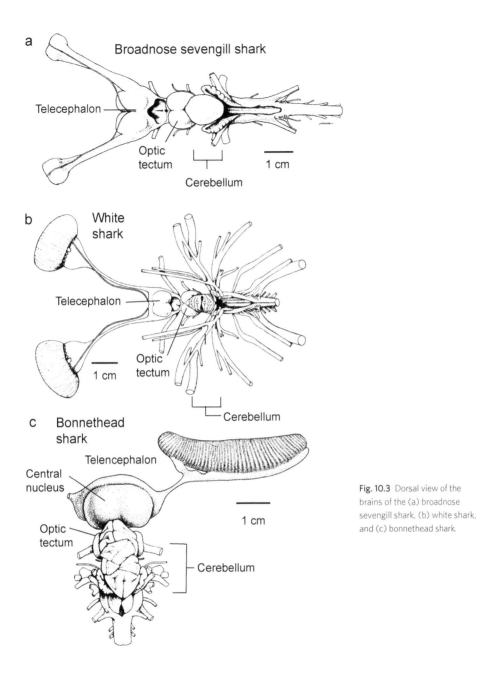

a

Broadnose sevengill shark

Telecephalon

Optic
tectum

Cerebellum

1 cm

b

White
shark

Telecephalon

Optic
tectum

1 cm

Cerebellum

c   Bonnethead
shark

Telencephalon

Central
nucleus

Optic
tectum

Cerebellum

1 cm

Fig. 10.3 Dorsal view of the brains of the (a) broadnose sevengill shark, (b) white shark, and (c) bonnethead shark.

electroreception and mechanoreception in finding prey than other galeoform sharks. Yet the white shark does have the largest olfactory bulb mass relative to total brain mass of any shark. This olfactory elaboration is consistent with a reliance on locating pinniped and whale carcasses by following their chemical trails. Consistent with the species' use of vision when foraging is its relatively large eyes, the well-developed muscles surrounding them, and the presence of a rete, or network of blood vessels for warming them. Finally, the eye of the white shark has an area centralis of high rod concentration that adapts the species for daytime predation.

DIFFERENCES IN DEVELOPMENT OF REGIONS OF BRAIN

The relative sizes of the different regions of the brain vary greatly across taxonomic groups. The differences among the chimaeras, sharks, and rays are greatest in the forebrain, which coordinates the information received from the different sensory systems and enables individuals to respond to stimuli in the environment as well as supports the development of complex behavior. The relative percentage of the brain mass composed by the telencephalon of the spotted ratfish, 31%, is comparable to that of the percentage of the brain mass in squaloform sharks such as the piked dogfish, 24%, Caribbean lantern shark (*Etmopterus hillianus*), 29%, and chain catshark (*Scyliorhinus rotifer*), 35% (see bars with clear fill in fig. 10.4). However, these relative sizes are low in comparison to the 51% for the sandbar shark (*Carcharhinus plumbeus*) and 52% for the scalloped hammerhead shark (*Sphyrna lewini*). The telencephalons of the rays are also very large, comprising 39% of the brain mass for the bullnose eagle ray and 47% for the roughtail stingray (*Dasyatis centroura*).

Although variation in the relative sizes of the divisions of the brain is related in part to evolutionary divergence, there are similarities in the degree of development of the parts of the brain related to their common lifestyles, particularly among the more advanced galeoform sharks and the rays.[5] These relationships are based less on the similarity in the sizes of various brain areas than on the degree of complexity in the surface of the cerebellum. This required the development of an index quantifying the degree of complexity based on the length, numbers, and depth of the folds in the cerebellar corpus, or surface. There were five scores indicating more or less development of the cerebellum (fig. 10.5). A score of 1 was assigned to a cerebellum with a smooth surface, bilateral symmetry, and little foliation with small branching grooves leading away from the larger grooves on the surface. A cerebellum that had shallow grooves running parallel to each other with little foliation received a score of 2. Scores 3 and 4 were assigned to brains with deeper grooves, more branching of these deep grooves, and more foliation leading from the groves. Finally, a maximum score of 5 was given to a cerebellum with deep, branched grooves, extreme foliation, and with distinctive asymmetrical sections. The gradient in the degree of complexity to the surface of the cerebellum is very apparent on the brains of the five sharks and rays shown in the illustration.

The relation between brain foliation and lifestyle is most apparent in the galeoform sharks of the orders Lamniformes and Carcharhiniformes. Members of the families in these two orders are grouped either by their order or by

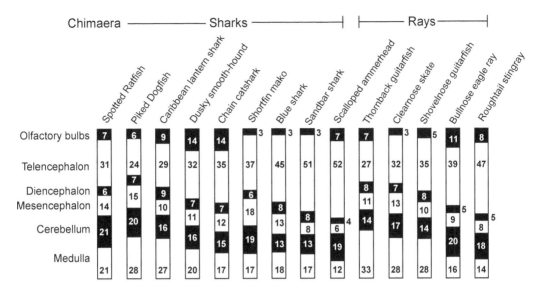

Fig. 10.4 Relative size of major brain divisions in a diversity of cartilaginous fishes.

three different lifestyles. The benthic sharks remain on the bottom or swim very close to it, the neritic sharks swim higher in the water column but also live in the shallow continental waters, and the pelagic sharks inhabit the deep oceanic waters. Each species, grouped either by order or by lifestyle, is denoted by a bar in a graph (fig. 10.6) with its fill indicating the family it comes from and the height of the bar the degree of foliation to its cerebrum. There are differences in the degree of foliation among the species within the lamniform and carcharhiniform sharks (fig. 10.6a). Among the carcharhiniforms, the catsharks (Scyliorhinidae), with scores of 1, have the lowest foliation, the houndsharks (Triakidae) and requiem sharks (Carcharhinidae) have intermediate foliation scores of 2 to 4, and the hammerhead sharks (Sphyrnidae) have scores of 5, the highest foliation of all of the families.

However, the foliation indexes are similar among sharks with each of the three lifestyles (fig. 10.6b). All of the benthic sharks have foliation development at level 1; the majority of neritic sharks an intermediate degree of foliation of 3 with a few species at either 2 or 4; and the pelagic species have foliation scores of 4 and 5. The eleven species of pelagic sharks are from six families, and all but one of them have foliation scores of 4 and 5, indicating that high foliation adapted them to life in the pelagic environment.

Both the sharks and the holocephalans can be grouped according to brain complexity to determine whether it might adapt them for a particular lifestyle. The degree of brain development of each species is apparent from scores that are given in six columns (fig. 10.7), one for the relative mass of each of the

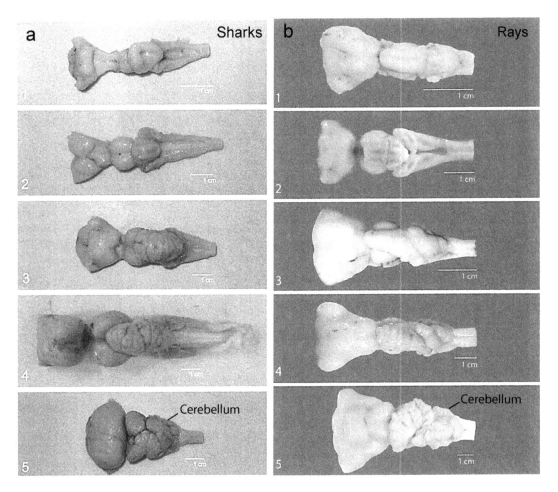

Fig. 10.5 Top views of the brains of five species of (a) sharks and (b) rays, illustrating the foliation index with values ranging from 1 to 5. The grades of cerebellar foliation are: (1) no foliation, smooth surface, symmetry; (2) minimal foliation, shallow and parallel grooves without branching; (3) moderate foliation, shallow to moderately deep grooves with slight branching; (4) very foliated, moderate to deep, branched grooves with symmetry; and (5) extremely foliated, deep and branching grooves, distinct sections, and asymmetry.

five major brain divisions and one for the foliation index, to the left of a dendrogram clustering the species into groups according to the values in the columns. A "+" indicates that particular area of the brain is of greater structural complexity than the average for all species, whereas "A" denotes an average mass. The foliation scores again range from 1 to 5 based on the development of the cerebellum. The algorithm sorted the sharks and holocephalans into six groups, which differed greatly from groupings based on molecular and anatomical evidence (see fig. 2.13). This exercise demonstrates that evolutionary divergence alone does not explain the variation in the organization of the shark and holocephalan brains. However, you must remember that there is a tendency for more evolutionarily related sharks to occur together in a group.

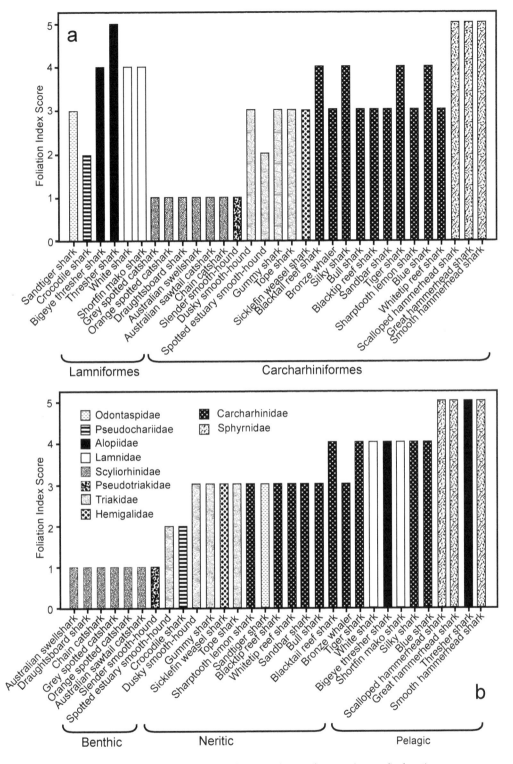

Fig. 10.6 Foliation index scores of six species of sharks from the order Lamniformes and twenty-five from the order Carcharhiniformes. The sharks are grouped by (a) evolutionary affinity and (b) lifestyle and habitat.

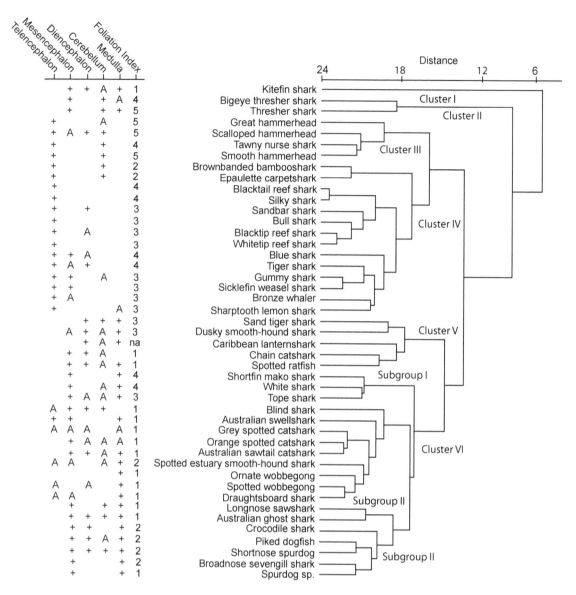

Fig. 10.7 Clustering dendrogram that separates 43 species of sharks and 2 species of holocephalans into groups by the development of each of the five major areas of the brain relative to that of the whole brain and the degree of cerebellar foliation. A+ indicates that particular area of the brain is of greater than the average for all species; an A denotes it is of average size; a blank space indicates that it is below average. The na for the Caribbean lanternshark indicates that no measurements were available. The dark lines in the cluster diagram indicate that the grouping is a statistically significant one.

The bulk of the species, forty-five of them, were grouped into four clusters, III, IV, V, and VI. It is interesting that the two species of thresher sharks had a unique pattern of brain development, characterized by a relatively large mesencephalon and a highly foliated cerebellum (see '+' and 4 and 5 scores in fig. 10.7). This could be related to their evolution of a unique form of feeding by stunning prey with a blow delivered with the elongated upper lobe of their caudal fin. The remaining four groupings are separated into two distinct lineages, based on the relative size of their telencephalon. The species in clusters III and IV have larger than average telencephalons with higher degrees of cerebellar development with foliation scores of 3 or higher. These are all advanced galeomorphs, predominantly in the order Carcharhiniformes although some are in the Orectolobiformes, and include the species with the largest brains such as members of the families Carcharhinidae and Sphyrnidae. Many of these species live in the complex sensory environments of coral reefs such as the blacktail reef shark (*Carcharhinus amblyrhynchos*), blacktip reef shark (*Carcharhinus melanopterus*), whitetip reef shark (*Triaenodon obesus*), tiger shark (*Galeocerdo cuvier*), and three hammerhead species from the genus *Sphyrna*. Most of these species exhibit the carangiform mode of swimming, enabling them to actively chase and capture their prey (see swimming styles, fig. 3.12). They forage mainly on fishes, including smaller sharks, and cephalopods in a three-dimensional environment. Some of them also exhibit complex social behaviors. The species in clusters V and VI mainly have smaller than average telencephalons, an average cerebellum size, low to average foliation scores of 1 to 3, and an enlarged mesencephalon and/or medulla. All of these species are sluggish species that are benthic, resting much of the time on the bottom, or benthopelagic, cruising just above the ground searching for benthic prey. Species such as the Australian swellshark (*Cephaloscyllium laticeps*), draughtsboard shark (*Cephaloscyllium isabellum*), ornate wobbegong (*Orectolobus ornatus*), and spotted wobbegong (*Orectolobus maculatus*) swim with an anguilliform mode of swimming or a modified, slow-moving carangiform style as exhibited by the Australian sawtail catshark (*Galeus boardmani*), and chain catshark (*Scyliorhinus rotifer*). All of these species live in a more two-dimensional environment close to the bottom, feeding mainly on mollusks, crustaceans, and fishes associated with the substrate. In conclusion, there seems to be a relationship between the level of cerebellum foliation and both mode of locomotion and swimming speeds in the sharks. The slow-moving sharks, which undulate their entire bodies in an anguilliform motion, have low levels of foliation; the fast-moving sharks, which undulate the posterior half or third of the bodies in carangiform or thunniform motion, have higher levels of foliation. Unlike the sharks, the holocephalans swim by oscillatory movements of their pectoral fins. Although not highly foliated, their cerebellum is relatively large, which may be due to the dexterity and enhanced locomotory abilities derived from the increased use of the pectoral fins necessary to feed on their diet of bony fishes, crustaceans, and polychaetes.

The rays also can be grouped according to brain complexity to see whether they might be suited to a particular lifestyle.[6] The degree of brain development of each species is apparent from scores that are given in six columns (fig.

10.8), one for the relative mass of each of the five major brain divisions and one for the foliation index, to the left of a dendrogram clustering the species into groups based on the values in the columns. The twenty-one species were grouped into two clusters. Again, the relative size of particular divisions of the brain is closely correlated with their evolutionary origin. However, there also appears to be a relationship with lifestyle. The first cluster contains eleven species of stingrays from the order Myliobatiformes. Nine of these species are benthopelagic, often swimming up in the water column despite foraging on the bottom. Only two stingarees are benthic. All of these species, excepting the stingarees, possess a relatively large telencephalon as well as an average to large cerebellum. It is also highly foliated with index scores ranging from 4 to 5, as might be expected from their occupying a complex habitat. The second cluster is composed of nine skates of the order Rajiformes, an electric ray of the Torpediniformes, and two stingrays in the order Myliobatiformes. All of these species are benthic in their lifestyle and live in noncomplex environments. Their brains have a smaller than average telencephalon, a relatively large medulla, and little foliation on their cerebellum with index scores ranging between 1 and 3, as might be expected from their living in a noncomplex habitat.

It is possible to make specific inferences about particular shark species based on the relative size of different regions of their brain.[7] The blacktail reef shark, which chases and captures prey up in the water column, has a brain with different large regions than does the chain catshark, which forages close to the bottom feeding on mollusks and crustaceans (figs. 10.9a & 10.9b). Four parts of the brain process very different sensory information. These regions are identified with different fill patterns in a diagram of the brain of the reef shark (10.9c) and the catshark (10.9d). The olfactory bulbs are used in chemoreception, the posterior lateral line lobes in mechanoreception, the optic tectum in photoreception, and the anterior lateral line lobes in electroreception. The brains of the two species can be compared by the percentage of the total mass contributed by each region. The two species have about the same proportion of brain mass devoted to processing sensory information using their auditory and electroreceptive senses. However, in the blackfin reef shark the optic tectum makes up 47% of the mass of the brain processing sensory information (see 10.9e), while in the chain catshark, the optic tectum contributes only 21% (see 10.9f). This conspicuous difference is probably because the reef shark relies mainly on vision to locate its prey and mates up in the water column, whereas the catshark uses the same sense less when foraging near the bottom. The olfactory bulbs of the catshark make up a larger proportion of the brain mass devoted to sensory processing than those of the reef shark, 62% in the former

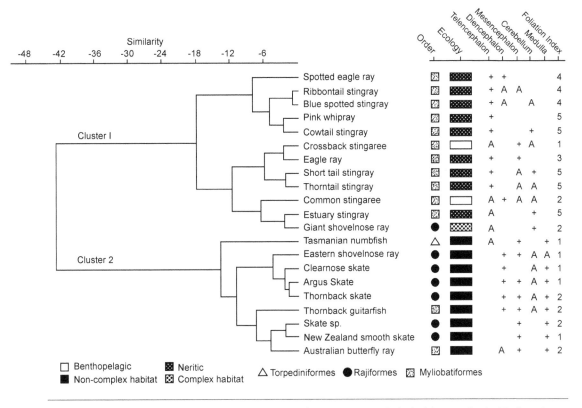

Fig. 10.8 Clustering dendrogram that separates twenty-one species of rays into groups on the basis of the mass of each of the five major areas of the brain relative to that of the whole brain and the degree of foliation on their cerebellum. Different fill patterns indicate the order to which they belong and type of habitat they occupy.

versus 36% in the latter. The chain catshark likely relies more on its sense of smell to locate prey as it swims along the bottom. This is an illustration of how the functional anatomists make inferences on the behavior of sharks and rays based on the structure and size of particular parts of their brains.

The overall brain sizes of members of different species of sharks can be plotted as a function of the body size. A comparison of brain to body size is most valid when the species have a comparable body size. For example, the brain weight of a squaloform species such as the piked dogfish can be compared on the graph with the brain weight of a galeoform species of similar size such as the dusky smooth-hound (*Mustelus canis*) (see small separation between them on ordinate in fig. 10.10a). The dogfish has a brain mass of roughly 4 g and the smooth-hound a brain mass of 9 g, whereas the former's body weight is 7 kg and the latter's weight is 9 kg. Comparisons such as this one reveal that the primitive squalomorph sharks generally possess low brain-to-body ratios and the more advanced geoleomorph sharks possess high ratios. The evolu-

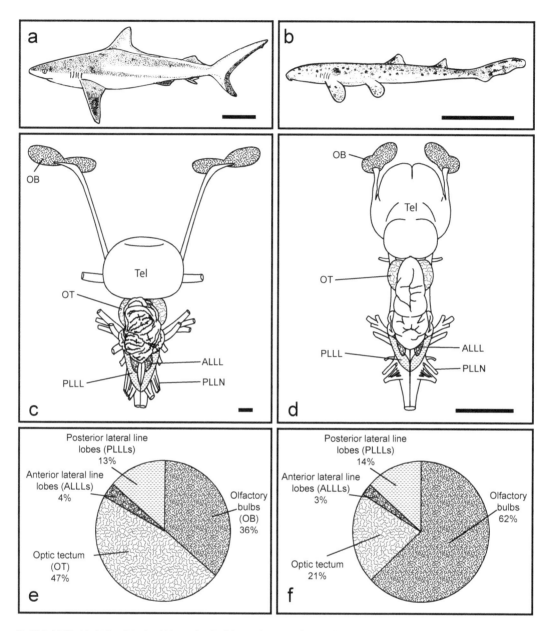

Fig. 10.9 (a) The blacktail reef shark, which searches for fishes and captures them up the water column, and (b) the chain catshark, which forages close to the bottom feeding on mollusks and crustaceans, have very different brains, whose regions differ due to their different foraging strategies. Four regions of the brain are identified by different fill patterns (c = blacktail reef shark, d = chain catshark) as well as their relative masses (e = blacktail reef shark, f = chain catshark).

tion of the modern sharks has resulted in a twofold to sixfold increase in brain size.[2] The points for the rays are scattered throughout the minimum convex polygon on the plot of brain-to-body mass. Members of the order Rajiformes such as the clearnose skate have a low brain-to-body ratio, but members of the order Myliobatiformes such as the American stingray have among the high-

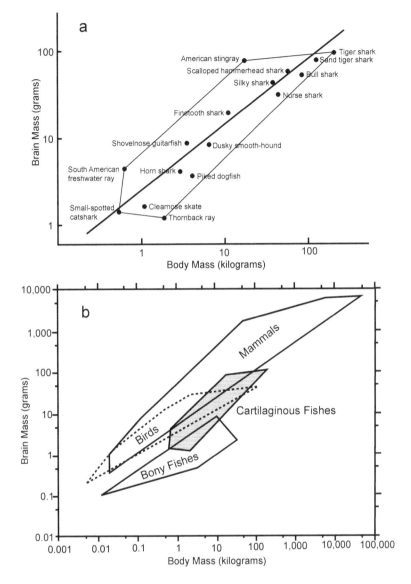

Fig. 10.10 (a) Minimum convex polygon drawn around brain masses plotted against body mass. This gives you an idea of the relative sensory and intellectual capacity of the sharks and rays. The coefficient of allometry, 0.8 (indicated by the solid line and its slope), indicates a rough proportional relationship of brain with body size with growth (1.0 equals a perfect correlation). (b) Brain versus body mass indicated by minimum convex polygons drawn between values for members of four vertebrate classes: bony fishes, cartilaginous fishes, birds, and mammals.

est of the ratios for the cartilaginous fishes. The ground sharks of the order Carcharhiniformes such as the silky and scalloped hammerhead sharks have very high brain-to-body ratios. This is more evident when one compares the encephalization quotients for the different species. This indirect measure of intelligence is the ratio of actual brain size to an imaginary brain size using an equation relating the two brain masses while excluding size in the comparison. The piked dogfish, a squaloform, has an encephalization quotient of 0.62 while the scalloped hammerhead, a galeoform, has a value of 1.38 (table 10.1). The clearnose skate has value of 0.57, whereas the American stingray a value of 1.47. The freshwater ray has the highest value of 2.77.

Order Species	Common name	Brain mass (g)	Body mass (kg)	Encephalization quotient
Squaliformes				
Squalus acanthias	Piked dogfish	3.87	4.20	0.62
Squatiniformes				
Squatina californica	Pacific angelshark	2.06	6.00	0.24
Heterodontiformes				
Heterodontus francisci	Horn shark	4.30	2.93	0.90
Orectolobiformes				
Ginglymostoma cirratum	Nurse shark	31.65	45.30	0.85
Lamniformes				
Carcharias taurus	Sandtiger	82.55	123.00	1.05
Carchariformes				
Carcharhinus falciformis	Silky shark	43.32	36.24	1.37
Carcharhinus isodon	Finetooth shark	18.75	10.87	1.47
Carcharhinus leucas	Bull shark	54.36	83.80	0.92
Carcharhinus obscurus	Dusky shark	20.76	12.00	1.51
Galeocerdo cuvier	Tiger shark	107.50	200.00	0.95
		20.74	23.50	0.91
Mustelus canis	Dusky smooth-hound	8.31	6.50	0.96
Scyliorhinus canicula	Small-spotted catshark	1.38	0.57	0.99
Sphyrna lewini	Scalloped hammerhead	59.88	55.71	1.38
Rajiformes				
Raja eglanteria	Clearnose skate	1.66	1.10	0.57
Rhinobatos productus	Shovelnose guitarfish	9.11	3.62	0.91
Torpediformes				
Torpedo marmorata	California electric ray	1.55	1.87	0.31

TABLE 10.1. *continued*

| Order | | Brain mass | Body mass | Encephalization |
Species	Common name	(g)	(kg)	quotient
Myliobatiformes				
Dasyatis centroura	Roughtail stingray	19.86	5.66	1.24
Dasyatis pastinaca	Common stingray	24.87	6.80	1.28
Dasyatis sabina	Atlantic stingray	19.86	5.66	1.24
Raja clavata	Thornback ray	24.87	6.80	1.24
Potamotrygon motoro	So. Am. freshwater ray	4.51	0.63	2.77

CAPACITY TO LEARN

One way to assess intelligence is by assessing the animal's ability to learn, or to change its behavior on the basis of experience. Lemon sharks protect their eyes by rotating their nictitating membrane upward when exposed to a mild electrical shock. If a light is flashed at the same time as the shock is applied, after fewer than one hundred trials, the lemon shark learns to blink only in response to the flash of light. Hence, the shark moves its nictitating eyelid upward not in an involuntary manner but in a voluntary response to a stimulus to which it would not ordinarily respond in this way. This form of learning is termed respondent conditioning and has been used to ascertain whether the shark can perceive irradiance of different wavelengths or sounds of different frequencies.

Another form of learning is instrumental conditioning, which consists of increasing the frequency of a particular behavioral pattern by positively reinforcing it, usually by feeding the individual after it exhibits a desired behavior. Small-spotted catsharks, *Scyliorhinus canicula*, have been conditioned to discriminate the direction of the flow of electrons in saltwater in the presence of a constant magnetic field.[8] The purpose of this experiment was to determine whether sharks are able to discriminate the direction of the flow of saltwater, a moving conductor, in the presence of the earth's main field and use this information to guide their movements in the ocean. A uniform electrical current was generated through a body of saltwater by permitting electrical current to enter and leave the system through two electrodes enclosed within plastic bottles. Two sets of eighteen salt bridges, each consisting of a latex tube filled with an electrolytic gel with the same resistance as saltwater, led from each bottle to the opposite sides of a circular plastic fence in the tank so that cur-

The relative size of the brain regions differs among the vertebrate classes. Vertebrates living in air, such as the birds and terrestrial mammals, have a large optic tectum because they rely mainly on vision. The cartilaginous fishes have a smaller optic tectum because vision is less important to them underwater in the limited visibility. However, their cerebellum is large because the efferent nerve impulses from their electroreceptors are processed within this region of the brain. This sensory system depends upon the conductivity of ocean salts for its effectiveness. Another reason for the large size of the cerebellum is that the

efferent output from the mechanoreceptors is processed there. Aquatic species are capable of detecting vibrating objects in water at a greater distance than terrestrial species in air. The sum of the sizes of the different regions of the brain can be used as an indicator of mental development. The brain mass of the cartilaginous fishes exceeds that of most bony fishes of similar body size[2] (see fig. 10.10b). The brain-to-body mass ratios of cartilaginous fishes are greater than a third of the bird species and greater than those for some mammalian species.

rent flowed between them in a straight line (see lines with arrows indicating the direction of the current flow in fig. 10.11a).

Haller's round rays, *Urolophus halleri*, were then taught to seek a reward, a small piece of cut herring, by choosing to enter one of two corrals from the left side. Each corral opened toward the outside of the tank. The current flow between the two electrodes was 8 μA; the voltage difference between the two electrodes was 5 nV/cm. The stingrays earned a food reward by swimming into the enclosure with the field set at a chosen field polarity, positive or negative, and they were gently prodded with a plastic tube when choosing the wrong one. Between each trial, the field polarity and consequent direction of the flow of electrons were switched randomly within the tank. The electrical field was produced under three different magnetic conditions. The first was a null field, in which the earth's main field was cancelled by passing a current through a large electrical coil surrounding the apparatus. The second field consisted of only the vertical component of the earth's field, which has three vectoral components— two on the horizontal plane and one on the vertical plane. The third condition was the normal magnetic field present where the rays had been collected.

Three rays were tested under each condition. Fig. 10.11b shows the number of correct versus incorrect choices made by one of the three rays. A line indicating the score of the ray extends upward flanked by two oblique lines. The line on the right indicates that 50% of the choices were correct and the line on the left indicating that 75% of choices were correct. The passage of the line across the left line indicates that the ray had learned to discriminate

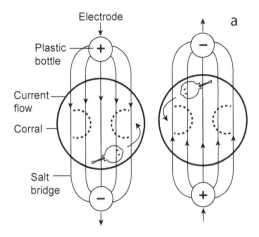

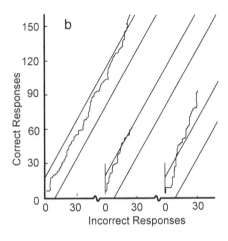

Fig. 10.11 (a) Experimental apparatus used to determine whether round stingrays can distinguish a flow of electrons, or voltage gradient, within a body of saltwater. (b) Plot of correct versus incorrect choices for one stingray relative to one direction of current flow or voltage gradient within the body of saltwater in the presence of a null field (left), the vertical component of the earth's field (middle), and the normal earth's main field (right).

the direction of current flow. The first ray learned to detect the electrical field under the null magnetic field after 218 trials, of which 140 were correct and 78 were incorrect responses (see first curve). This is apparent from the respective numbers on the ordinate and abscissa where line crossed the left oblique line. The ray took fewer trials, 59, to detect the electrical field when the vertical component of the earth's field was present (see second curve to right) and even fewer, 47, to discriminate the electrical field in the presence of the normal earth's field (see third curve).

In another example of instrumental learning, nurse sharks (*Ginglymostoma cirratum*) were kept in a tank separated into two areas by a gate, which led into the part of the tank with a black rectangular target suspended within it. The gate was opened and a nurse shark was lured toward the target with a filet of fish on the end of a pole until it bumped the target. The shark was then given food. After three days of trials, the nurse shark would press the target regularly without having to be led to it.[9] The gate was then opened 15–20 times during a

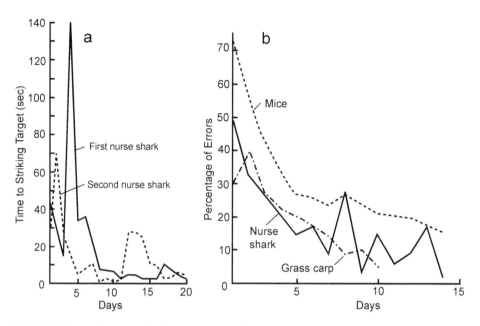

Fig. 10.12 (a) The mean latency, or time between the raising of gate and two nurse sharks hitting a target, is shown as a function of days of training. (b) Learning curve shown for nurse sharks for discriminating a lighted target from a nonlighted target, *Tilapia* taught to discriminate horizontal stripes from vertical bars, and the common mouse showing the average performance by eight mice performing a similar discrimination.

day, and the time it took the shark to move toward and press the target was recorded in seconds before it was fed. It took five days for one shark (see dashed line) and eight days for another shark (see solid line) for the time taken to press the target to decrease from 70 and 140 seconds, respectively, to less than ten seconds (fig. 10.12a).

In a related learning task, the nurse sharks were taught to bump an illuminated target and avoid an adjacent nonilluminated target when the orientation of the two with respect to each other was randomized. The percentage of mistakes made by the nurse sharks per day decreased to a level of 15% after five days and 1% after fifteen days (see solid line in fig. 10.12b). This rate of learning was similar to the average performance of five grass carp (*Tilapia heudeloti*) in a related problem involving visual discrimination as well as in a similar experiment performed with eight mice. Note the similarity of the three learning curves. The solid line indicates the ability of the shark to complete the task, the dashed-dotted line the performance of the grass carp, and the dashed line the performance of the mice. The similarity in the success of cartilaginous fishes in this learning experiment to that of a mammal is consistent with the similarity of their brain-to-body mass ratios. Combined, these observations are at odds with the long-held impression that the cognitive abilities of sharks are less than those of other vertebrates.

SUMMARY

The sharks have two major patterns of brain organization, one exhibited by the primitive squaloform and the other by the advanced galeoform sharks. The former consist of the six- and sevengill sharks, saw sharks, angelsharks, and dogfishes. The latter group is composed of the bullhead sharks, carpet sharks, and ground sharks. The squaloform sharks have a smooth surface of the cerebellum, whereas that of the galeoform sharks is convoluted and highly foliated. The brains of the rays resemble those of the galeoform sharks in their more complex organization. The skates and torpedo rays have a relatively simple or slightly convoluted cerebellum and a medium-size telencephalon. Conversely, the stingrays have a highly convoluted cerebellum and a large telencephalon, whose size rivals that of the reef and hammerhead sharks.

Although variation in the relative sizes of the divisions of the brain is related in part to evolutionary divergence, there are similarities in the degree of development of the parts of the brain related to their common lifestyles—particularly among the more advanced galeoform sharks and the rays. These relationships are based less on the similarity in the sizes of various brain areas than on the degree of complexity in the surface of the cerebellum. The members of the genera of *Carcharhinus* and *Sphyrna* have larger telencephalons than average with higher degrees of cerebellar development. Many of these species live in the complex sensory environments of coral reefs. Most exhibit the carangiform mode of swimming, adapting them to actively chase and capture their prey. They forage mainly on fishes, including smaller sharks, and cephalopods in a three-dimensional environment. The other ground sharks have smaller telencephalons than average, an average cerebellum size, low to average foliation, and a large mesencephalon and/or medulla. These are sluggish benthic species, resting much of the time on the bottom, or benthopelagic, cruising just above the ground searching for benthic prey. Some of these species swim with an anguilliform mode while other use a modified slow-moving carangiform style. These species live in a more two-dimensional environment close to the bottom, feeding mainly on mollusks, crustaceans, and fishes. In conclusion, there seems to be a relationship between the level of cerebellum foliation and both mode of locomotion and swimming speeds in the sharks.

The rays also can be grouped according to brain complexity to see whether it might adapt them for a particular lifestyle. The stingrays, excepting the stingarees, are benthopelagic, often swimming up in the water column despite foraging on the bottom. These species possess a relatively large telencephalon as well as an average to relatively large cerebellum. They also generally have a highly foliated cerebellum, as might be expected due to their occupying a complex habitat. The skates, electric rays, and a few stingrays are benthic in their lifestyle and live in noncomplex environments and have a smaller than average telencephalon, a relatively large medulla, and little foliation, as might be expected from their noncomplex habitat.

It is therefore not a surprise that the ratios of brain mass to body mass of the more advanced sharks such as the scalloped hammerhead and bat eagle ray are comparable to those of birds and even mammals. Consistent with this is the ability of sharks to learn to discriminate between targets that are illuminated nor not illuminated at a rate comparable to a bony fish (such as the grass carp) or a mouse. Although the cartilaginous fishes may owe much of their evolutionary success to their keen underwater senses of mechanoreception and electroreception, the more advanced species such as the members of the genus *Carcharhinus* and *Sphyrna* have a well-developed brain comparable to those of birds and mammals that enables them to have a diverse behavioral repertoire and complex social system.

✳✳ ✳✳ ✳✳

KEY TO COMMON AND SCIENTIFIC NAMES

Argus skate = *Dipturus polyommata*; Atlantic stingray = *Dasyatis sabina*; Australian butterfly ray = *Gymnura australis*; Australian sawtail catshark = *Galeus boardmani*; Australian swellshark = *Cephaloscyllium laticeps*; bat eagle ray = *Myliobatis californicus*; bigeye thresher shark = *Alopias superciliosus*; blacktail reef shark = *Carcharhinus amblyrhynchos*; blacktip reef shark = *Carcharhinus melanopterus*; blind shark = *Brachaelurus waddi*; blue shark = *Prionace glauca*; blue-spotted stingray = *Dasyatis kuhlii*; broadnose sevengill shark = *Notorynchus cepedianus*; bronze whaler shark = *Carcharhinus brachyurus*; brownbanded bambooshark = *Chiloscyllium punctatum*; bullnose eagle ray = *Myliobatis freminvillei*; bull shark = *Carcharhinus leucas*; Caribbean lantern shark = *Etmopterus hillianus*; chain catshark = *Scyliorhinus rotifer*; clearnose skate = *Raja eglanteria*; common stingaree = *Trygonoptera testacea*; cowtail stingray = *Pastinachus sephen*; crocodile shark = *Pseudocarcharias kamoharai*; crossback stingaree = *Urolophus cruciatus*; draughtsboard shark = *Cephaloscyllium isabellum*; dusky smooth-hound shark = *Mustelus canis*; eagle ray = *Myliobatis tenuicaudatus*; eastern shovelnose ray = *Aptychotrema rostrata*; epaulette shark = *Hemiscyllium ocellatum*; estuary stingray = *Dasyatis fluviorum*; finetooth shark = *Aprionodon isodon*; ghost shark = *Callorhinchus milii*; giant shovelnose ray = *Glaucostegus typus*; grass carp = *Tilapia heudeloti*; gray spotted catshark = *Asymbolus analis*; great hammerhead shark = *Sphyrna mokarran*; gummy shark = *Mustelus antarcticus*; Haller's round ray = *Urolophus halleri*; horn shark = *Heterodontus francisci*; kitefin shark = *Dalatias licha*; longnose sawshark = *Pristiophorus cirratus*; New Zealand smooth skate = *Dipturus innominatus*; nurse shark = *Ginglymostoma cirratum*; orange spotted catshark = *Asymbolus rubiginosus*; ornate wobbegong shark = *Orectolobus ornatus*; piked dogfish = *Squalus acanthias*; pink whipray = *Himantura fai*; ribbontail stingray = *Taeniura lymma*; sandbar shark = *Carcharhinus plumbeus*; sandtiger shark = *Carcharias taurus*; scalloped hammerhead shark = *Sphyrna lewini*; sicklefin lemon

Discussion Questions

1. Why do you think that the cartilaginous fishes, in particular the sharks, have gotten the reputation of being simplistic and robotic in their behavior?

2. Can you make a cogent argument that cartilaginous fishes have considerably developed brains and learning abilities?

3. Do you think it fair to infer the behavioral capabilities of the cartilaginous fishes based on their brain anatomy? This method of inference is usually referred to as functional neuroanatomy.

shark = *Negaprion acutidens*; shortfin mako shark = *Isurus oxyrinchus*; shortnose spurdog = *Squalus megalops*; short tail stingray = *Dasyatis brevicaudata*; shovelnose guitarfish = *Rhinobatos productus*; sicklefin weasel shark = *Hemigaleus microstoma*; silky shark = *Carcharhinus falciformis*; skate. = *Dipturus* sp.; slender smooth-hound shark = *Gollum attenuatus*; small-spotted catshark = *Scyliorhinus canicula*; South American freshwater stingray = *Potomotrygon motoro*; spotted eagle ray = *Aetobatus narinari*; spotted estuary smooth-hound shark = *Mustelus lenticulatus*; spotted ratfish = *Hydrolagus colliei*; spotted wobbegong shark = *Orectolobus maculatus*; spurdog = *Squalus* sp.; Tasmanian numbfish = *Narcine tasmaniensis*; tawny nurse shark = *Nebrius ferrugineus*; thornback guitarfish = *Platyrhinoidis triseriata*; thornback skate = *Okamejei lemprieri*; thorntail stingray = *Dasyatis thetidis*; thresher shark = *Alopias vulpinus*; tiger shark = *Galeocerdo cuvier*; tope shark = *Galeorhinus galeus*; white shark = *Carcharodon carcharias*; whitetip reef shark = *Triaenodon obesus*.

LITERATURE CITED

1. Klimley, 1978; 2. Northcutt, 1978; 3. Pough *et al.*, 2002; 4. Demski and Northcutt, 1996; 5. Yopak, *et al.*, 2007; 6. Lisney et al., 2008; 7. Hart *et al.*, 2006; 8. Kalmijn, 1982; 9. Aronson *et al.*, 1967.

RECOMMENDED FURTHER READING

Aronson, L. R., F. R. Aronson, and E. Clark. 1967. Instrumental conditioning and light-dark discrimination in young nurse sharks. *Bulletin of Marine Science*, 17: 249–256.

Demski, L. S. and R. G. Northcutt. 1996. The brain and cranial nerves of the white shark: an evolutionary perspective. Pp. 121–130 *in* Klimley, A. P., and D. G. Ainley (Eds.), Great White Sharks: The Biology of *Carcharodon carcharias*. Academic Press, San Diego.

Kalmijn, A. J. 1982. Electric and magnetic field detection in elasmobranch fishes. *Science*, 218: 916–918.

Lisney, T. J., K. E. Yopak, J. C. Montgomery, and S. P. Collin. 2008. Variation in brain organization and cerebellar foliation in Chondrichthyans, *Brain, Behav., Evol.*, 72: 262–282.

Northcutt, R. G. 1978. Brain organization in the cartilaginous fishes. Pp. 117–193 *in* Hodgson, E. S., and R. F. Mathewson (Eds.), Sensory Biology of Sharks, Skates, and Rays. U.S. Government Printing Office, Washington.

Yopak, K. E., T. J. Lisney, S. P. Collin, and J. C. Montgomery. 2007. Variation in brain organization and cerebellar foliation in Chondrichthyans: sharks and holocephalans. *Brain, Behav., Evol.*, 69: 280–300.

CHAPTER 11

Courtship and Reproduction

I remember well the male nurse sharks constantly pursuing females in the Shark Channel of the Miami Seaquarium. One day while standing by my sound system that broadcast sounds into the water to attract or repel sharks, I witnessed a male shark closely following a large female lurch forward with his mouth wide open and seize her by the back margin of her pectoral fin. Her pectoral fins were ragged with many notches along its posterior edge, likely due to the repeated bites of males. The female reacted by suddenly turning in front of the male and rolling over on to her back so that she lay crosswise in front of him. I ran to where the two sharks were, which was just under a bridge that crossed the channel. I jumped over the side of the bridge, clung to the hand railing with one hand, and leaned outward to see what would happen next. I observed that the male was now on top of the female. She was still lying on her back. He spasmodically thrust his ventral midsection forward and then rolled over on to his back. It was now possible to see one of his claspers inserted into the female's cloaca. I had been lucky enough to witness courtship and copulation—one of the first to ever witness this rarely witnessed event among a pair of sharks. I now knew the answer to a long-standing debate over whether

Male nurse shark, identifiable by darker coloration, chasing female and then seizing her pectoral fin to induce her to roll on to her back during courtship in the Dry Tortugas.

the male shark inserts one or two claspers into the female—one clasper was inserted into the cloaca of the female nurse shark.

The success of the cartilaginous fishes over evolutionary time may in large part be due to their internal mode of reproduction. The male inserts sperm into the female's reproductive tract to fertilize her eggs, which develop in some species into eggs enclosed within cases or in other species into full-term embryos. The eggs are deposited by females in safe locations, and the embryos protected within the tough egg cases are able to develop to advanced state, capable of swimming and feeding, before hatching. Those embryonic sharks that are directly released into the external environment usually resemble adults and are immediately able to feed. Internal fertilization ensures that the eggs, toward which the female has allocated much of her energy upon becoming sexually mature, are not wasted by being broadcast haphazardly into the water column but will likely develop into full-term embryos within the uterus of the female.[1] In addition, the male can pick the female with whom he will sire young and can deliver his sperm with little wastage. The majority of the fertilized eggs develop into embryos within the highly protective environment of the female uterus. The embryos when full-term may constitute up to a third of a pregnant female's body mass. Thus, there is great biological investment in producing so few young. Contrast this with the mating practices of the majority of bony fishes within the marine environment. A female usually swims upward into the water column, exposing herself to the risk of predation, to release thousands of eggs in a diffuse cloud that are fertilized indiscriminately by sperm released by many males. The males follow her upward and release their sperm as close as possible to her and her cloud of eggs. The fertilized eggs then develop into embryos. Upon hatching, the larvae must find food quickly after absorbing their sac of energy-rich stores or perish by hunger. They may also die in vast numbers due to predation by a diverse group of planktivores including jellyfish, bony fish, whale sharks (*Rhincodon typus*), basking sharks (*Cetorhinus maximus*), and baleen whales.

REPRODUCTIVE ANATOMY AND PHYSIOLOGY

The reproductive anatomy of the sharks and rays will be described in detail but not that of the chimaeras because their reproductive system closely resembles that of the elasmobranchs.[2] The description of the reproductive system of the sharks will be based on the anatomy of a galeomorph, the blue shark (*Prionace glauca*), and a squalomorph, the piked dogfish, *qualus (acanthias)*. Our description of the reproductive system of the rays will be based on the anatomy of the Atlantic stingray (*Dasyatis sabina*).

Male Sharks and Rays

The male shark's reproductive system consists of testes, where spermatozoa develop, and genital ducts, through which the sperm pass to the urogenital papilla and then into the clasper that is inserted into the female cloaca. The male's reproductive system also includes the siphon sac, a water-filled bottle-shaped chamber surrounded by muscles on the ventrum of the male. Water is forced out of the sac by the contraction of the surrounding muscles into an internal chamber within the clasper. Here water mixes with and breaks up spermatophores, small clusters of spermatozoa in oval capsules, and a mixture of spermatozoa and water travels the length of the clasper before being released through the hypopyle, an opening on its distal end, into the female's cloaca.

Each male blue shark has two smooth and rounded testes (fig. 11.1a),[3] which are embedded in the anterior section of the long and irregularly shaped epigonal organ. The testes protrude from the top of the epigonal organ in the requiem and hammerhead sharks of the genera *Carcharhinus* and *Sphyrna*. The epigonal organ provides support for the testis. Each testis is cylindrical with rounded ends and contains within it the seminiferous ampullae that produce spermatozoa. Extending inward from the dorsal surface of each testis is a flat narrow partition of connective tissue called the mesorchium. This extends from the testis across to the epididymis and contains a series of narrow sperm-transporting tubules, the ductus efferens, embedded within it. These tubules are close to the forward edge of the mesorchium at the head of the epididymis. Just forward and adjacent to the testis, its spherical head narrows into a neck, and it then expands towards the rear to form a long strap-like organ. The epididymis consists of small tubules that eventually enlarge to become the ampulla ductus deferens. These tubes follow a convoluted path so the organ has a cerebriform, or brainlike, appearance. The epididymis functions in the storage and transfer of spermatozoa. The multiple convoluted tubules in the epididymis converge into the single straight tubule of the ductus deferens, adjacent to the ureter. Its diameter increases where it passes under the kidney. Also, its outer wall becomes thicker and more convoluted before leading to the ampulla ductus deferens. It is the site of the formation and storage of spermatophores. The ureter becomes entwined with the ductus deferens and follows parallel in its convoluted course as they both approach the urogenital sinus. This is the external opening from the male reproductive system. Both the ductus deferens and ureter enlarge into finger-like papillae, which project into the anterior wall of the urogenital sinus (fig. 11.1b). The entire reproductive system is paired until this point. The urogenital sinus projects into the common cloaca through a single large finger-like sac, the urogenital papilla. The sperm sac, which like

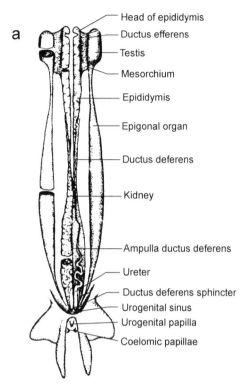

a

- Head of epididymis
- Ductus efferens
- Testis
- Mesorchium
- Epididymis
- Epigonal organ
- Ductus deferens
- Kidney
- Ampulla ductus deferens
- Ureter
- Ductus deferens sphincter
- Urogenital sinus
- Urogenital papilla
- Coelomic papillae

Fig. 11.1 Reproductive system of the male blue shark. (a) Overall view of the reproductive organs; (b) expanded view of the urogenital organs and kidney (b).

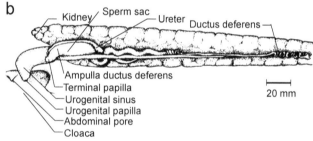

b

Kidney / Sperm sac — Ureter — Ductus deferens —

- Ampulla ductus deferens
- Terminal papilla
- Urogenital sinus
- Urogenital papilla
- Abdominal pore
- Cloaca

20 mm

the ductus deferens stores spermatozoa, leads into the anterior end of the urogenital sinus through an opening between the terminal papillae of the ductus deferens and ureter. It lies along the midline of the body and extends forward into the kidney, where it becomes narrower and threadlike before ending.

The claspers are densely calcified scroll-shaped appendages that are derived from the pelvic fins of the sharks, rays, and chimaeras. Spermatophores, stored in the ampulla ductus deferens, pass into a tubular channel at the base of the clasper and into the cloaca of the female during copulation. The sperm is forced through the tubular channel in the clasper by the contraction of a muscular sac under the skin of the male. In the mature male blue shark, this sac is just posterior to its pectoral girdle and extends posteriorly until it ends on the surface of the pectoral fin. The clasper of the blue shark does not have a backward projecting spur or hook but secures itself within the cloaca by spreading open its rhipidion, an umbrella-like terminal organ, composed of radiating elements with sharpened edges. These embed themselves within the wall of the cloaca like an umbrella opening in a soft tube.

The clasper is described in great detail for the piked dogfish.[4] It is highly ornamented not only with a spur and a spine but also with a rhipidion to ensure that the clasper stays in the uterus until the sperm is delivered. The basal support for the pelvic fin is provided by the pelvic girdle, which is attached to the propterygium and the metapterygium (fig. 11.2a). The pelvic fins extend outward from these two elements laterally and are composed of radial elements that articulate with fin rays, or ceratotrichia. The claspers are attached to the posterior of the metapterygium by two intermediate elements, the joint and beta cartilages. The principal skeletal piece in the clasper is the stem cartilage. It is round in cross-section at its base and becomes progressively more oval or flattened toward its distal end. Attached to it here are two small elements, the dorsal and ventral marginal cartilages, that are apparent on dorsal and ventral views of the clasper and pectoral fin (figs. 11.2a & 11.2b). Fused to these elements are four terminal elements, the claw and rhipidion on the dorsal side of the clasper and the spur and the ventral terminal cartilage on the ventral side. The rhipidion is a thin ear-shaped cartilaginous plate. The claw is a partially calcified cartilaginous element that is hard and sharp on its inner edge, whose distal end curves outward like a hook. It looks like an upside-down question mark. The ventral terminal cartilage is a long, broad, flat cartilaginous element with upwardly curved edges that form a shallow trough that forms the end of the clasper grove. It is through here that the spermatozoa pass on their way to the cloaca. The spur is a long, slender, and partially calcified element that curves outward to end in a sharp point.

When the male piked dogfish copulates with a female, it bends one clasper medially across its body and slightly forward before inserting it into the cloaca of the female. After insertion, the clasper's terminal elements are rotated at right angles to the stem cartilage to anchor the clasper firmly within the cloaca (fig. 11.2c). The bending of the whole clasper and the elements at its end are independent of each other. The three flexor muscles, the dorsal, middle, and ventral (fig. 11.2d), contract in concert to rotate the clasper across the midline of the body and slightly forward over an arc of 90–100° without turning over (fig. 11.2c). Thus, the upper side of the clasper with its open grove remains oriented upward. The terminal cartilages are projected outward by the contraction of the dilator muscle, while the same elements move back into their resting positions as the muscle relaxes. The ventral terminal cartilage and rhipidium both bend perpendicularly to the stem cartilage. Simultaneously, the claw rotates 180° on its long axis to move in the same direction as the ventral

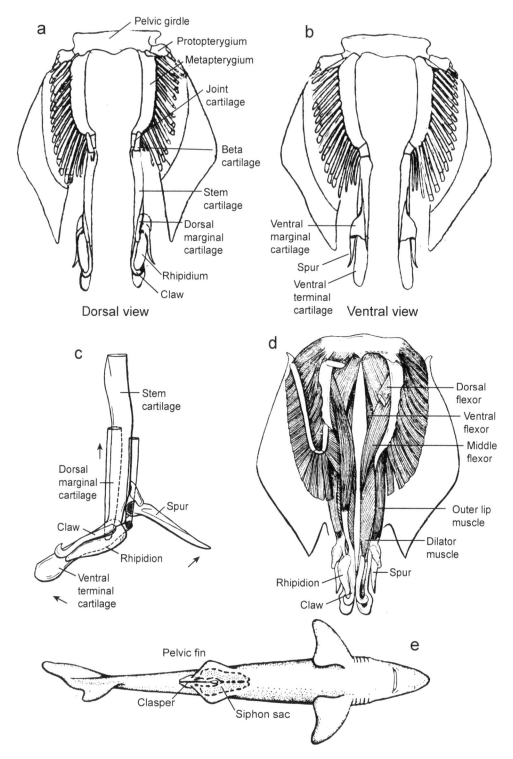

Fig. 11.2 The pelvic girdle, claspers, and siphon sac of the piked dogfish. (a) Dorsal and (b) ventral views of the cartilaginous skeleton associated with the pectoral girdle and claspers. (c) An expanded view of clasper, illustrating the rotation of the rhipidion, ventral terminal cartilage, hook, and spur. (d) Musculature associated with the girdle and claspers. (e) Ventral view of the male piked dogfish showing the location of the siphon sacs.

terminal cartilage and rhipidion. Also at the same time, the spur moves outward in the opposite direction. The claw and spur act together to anchor the clasper in the female's cloaca. They thus ensure the intromittent organ stays inside the female's cloaca while a mixture of spermatozoa and water is forced through the clasper's dorsal grove.

The male shark has two siphon sacs, which are subcutaneous bladders in the pelvic region on either side of the midline (fig. 11.2e). They are located between the outer skin and belly musculature. Each sac opens into the grove at the base of a clasper. The siphon sac of the spiny dogfish extends from slightly forward of the base of the pelvic fin to its posterior margin and holds within it saltwater with spermatozoa. The sac can contract 85% of its original length, which is 12% of the shark's body length. The male shark pumps water into his clasper by swimming forward while bending it across his body. Once this organ is inserted into the female's cloaca, the male contracts his ventral compressor muscle to force the stored water out of the siphon sac through a small opening, the apopyle, into the dorsal grove at the base of the clasper. Here it mixes with sperm injected from the urogenital papillae and moves along the dorsal groove in the clasper to a distal opening, the hypopyle. Although there is no direct connection between the urogenital papilla and the apopyle, the male rotates its clasper in such a way so that the papilla and apopyle are in close contact. In this way, the spermatophores are efficiently transferred from one organ to the other.[1]

The reproductive organs of the rays are arranged in a wider but shorter space in the body cavity than in the sharks because their bodies are wider and shorter than those of the sharks due to the lateral elongation of their pectoral fins into the pectoral disk. The reproductive system the Atlantic stingray has been described in detail.[5] The testes of the males of this species are dorsoventrally compressed and suspended from the dorsal wall of the body cavity by the mesorchium (fig. 11.3). Although both testes are functional, the right testis is smaller in most rays due to its sharing space in the body cavity with the stomach and intestine. The epigonal organ of the rays is not straight like that of the sharks but curves around many irregularly shaped testicular lobes that project inward from the flat testicular surface. The sperm migrate from the seminal follicles at the center of each lobe outward and downward to the periphery of each lobe, where they collect in ducts and pass to a single duct, the vas efferens. The spermatozoa then pass through the epididymis and Leydig's gland on the way to the seminal vesicle, from which they are passed into the male clasper during copulation.

The claspers of rays are similar to those of sharks. On the other hand, the claspers of the chimaeras are quite different from the claspers of the elasmobranchs. For example, the clasper of the spotted ratfish (*Hydrolagus colliei*),

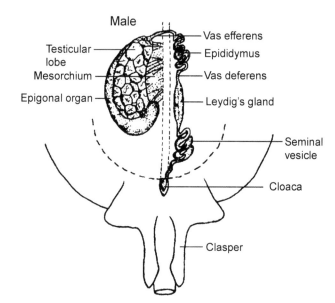

Male

Testicular lobe
Mesorchium
Epigonal organ

Vas efferens
Epididymus
Vas deferens
Leydig's gland

Seminal vesicle

Cloaca

Clasper

Fig. 11.3 Reproductive system of the male Atlantic stingray.

bifurcates into two appendages, the anterior and external claspers (see fig. 3.3b). There is a hook at the end of the internal radius that attaches to the wall of the female's cloaca securing the clasper and enabling the male to inseminate the female during copulation. The spotted ratfish also has an anterior clasper and a posterior clasper. The anterior either protrudes outward with its hooklets directed forward to grasp the female during copulation or is withdrawn into a slit-like pouch.

Spermatogenesis

The spermatozoa, or sperm cells, develop in masses of germinal tissue in seminiferous follicles within the testis. In the blue shark, the development of spermatocytes proceeds in the seminiferous fossicles from the surface on one side of the testis to the surface on the other side. Here the spermatozoa are collected in the ductus efferens (see arrows in fig. 11.4a). This form of spermatogenesis, or sperm development, is termed diametric and is present in the reef and hammerhead sharks of the order Carcharhiniformes.[3, 6] Aggregations of sixty to seventy spermatozoa exist within each spermatocyte and their nuclei are oriented outward near the circumference of the mesochordium (fig. 11.4b). The individual spermatozoa disperse from these aggregations once they are fully mature within the testis of the blue shark (fig. 11.4d). They pass across the mesochordium in the many small tubules comprising the ductus efferens that lead between the testis and head of the epididymis (see fig. 11.1a). In immature males, the whole epididymis is usually straight and thin walled.[1] In

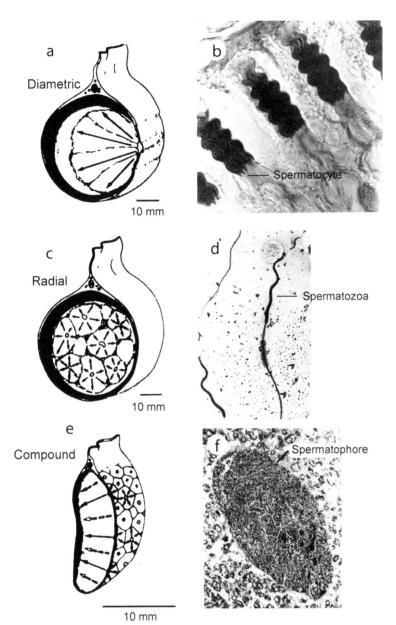

a
Diametric

10 mm

b
— Spermatocyte

c
Radial

10 mm

d
— Spermatozoa

e
Compound

10 mm

f
Spermatophore

Fig. 11.4 Spermatogenesis, or sperm formation, in the blue shark. The three modes of sperm development within the testis are (a) diametric, (c) radial, and (e) compound. (b) Aggregations of spermatozoa in the seminiferous follicles in the testis. (d) Spermatozoa are liberated from the aggregations and cross the ductus efferens into the upper epididymis, where (f) they later coalesce into spermatophores within the ductus deferens.

mature males, the anterior of the organ becomes highly convoluted and thick walled and enlarges into an ampulla at its terminus. The spermatozoa are stored within the epididymis, which is usually very swollen in mature males.[7] Once the spermatozoa reach the posterior epididymis, they aggregate again in the ductus deferens with their heads together and their longitudinal axes parallel in groups of sixty to seventy. Many thousands of these groups coalesce into spermatophores, white ovoid capsules 0.5–2.0 mm across at their largest diameter, in the ductus deferens (fig. 11.4f). In the blue shark, the spermatophores eventually break down and individual spermatozoa are released from the urogenital papilla into the apopyle of the clasper.

There is a second mode of testicular development in the sharks and rays.[6] These seminiferous fossicles are comprised of separate lobules, each separated by connective tissue (fig. 11.4c). This mode of spermatogenesis is termed radial because the germination zone is at the center of each lobule and the spermatozoa originate in an outward direction from the center to the circumference of each lobule. This form of testicular development is present in the basking shark (*Cetorhinus maximus*), the shortfin mako shark (*Isurus oxyrinchus*), the sandtiger shark (*Carcharias taurus*), and the white shark (*Carcharodon carcharias*), which are all members of the order Lamniformes. In the rays, the spermatozoa develop in a compound manner that exhibits elements of the diametric and radial modes of development common to elasmobranchs (fig. 11.4e). The spermatozoa originate in a radial manner from lobules but migrate to one side of the testis to the other.

Sexual Maturation

The sexual maturation of male sharks requires the complete development of the male organs that deliver sperm into the cloaca of the female.[8] First, the distal end, or head, of the clasper must be fully formed with the rhipidion, hook, and spur able to spread outward to secure the clasper in place within the female cloaca. Second, the stem cartilage, the long cartilaginous element that supports the head of the clasper, must be sufficiently rigid through calcification to push the head of the organ into the female's uterus. Third, the base of the clasper must rotate so that its head can be directed forward. Fourth, the siphon sacs must be sufficiently enlarged to hold enough water to flush the spermatozoa through the clasper and into the female's uterus.

Puberty is the developmental process by which individuals acquire the ability to breed and propagate. Among the sharks, this process is perhaps best described for the bonnethead shark (*Sphyrna tiburo*) a small hammerhead species abundant in shallow waters close to shore along the eastern and western coasts of North and South America in subtropical and tropical latitudes. The anatomical and endocrine changes associated with puberty have been closely monitored both for captive males and wild males during the second year of their life.[9] Every month these males were captured to measure their body and clasper lengths and to take blood to measure the concentrations of reproductive hormones. Growth curves are presented for four individuals for which these measurements were taken every month over a period of eight months from February to November (fig. 11.5a). The males initially grew at a steady rate until their growth slowed in September once they reached a length

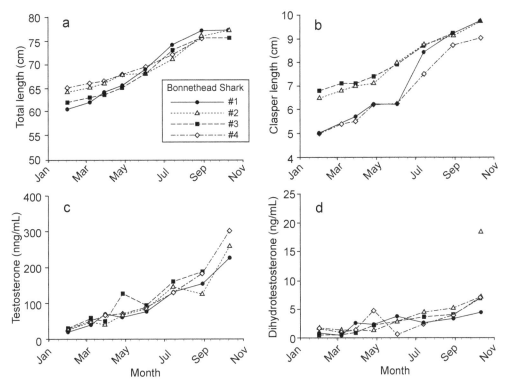

Fig. 11.5 Changes in the (a) total length, (b) clasper length, (c) serum concentration of testosterone, and (d) dihydrotestosterone of four captive male bonnethead sharks while reaching puberty during the second year of their lives.

of 75–80 cm. The pattern to the growth of the claspers, used in copulation, differed from that of overall body length. This was particularly apparent for the claspers of Sharks 1 and 4, which grew slowly from February to June but more rapidly from June through November. Notice the change in the slopes of the curves connecting the solid circles and clear circles in the growth curves for these two sharks (fig. 11.5b). These sharks reached maturity in September, when their claspers were rigid from calcification and semen was present within them. This acceleration in clasper growth is simultaneous with a steady increase in the concentration of testosterone from June to September for the same two sharks (fig. 11.5c). The increase in the concentration of dihydrotestosterone in the serum of the sharks over this same period was not as great but the concentrations were still significantly greater than the early concentrations in a statistical sense (fig. 11.5d). There were also similar increases in two other reproductive hormones, 17β-estradiol and progesterone.

Measurements of body and clasper length and the concentrations of reproductive hormones were also performed on bonnethead sharks captured in the coastal waters using gill nets. In addition, histological samples were taken of the tissues in their testis, epididymis, and seminal vesicle. The histologi-

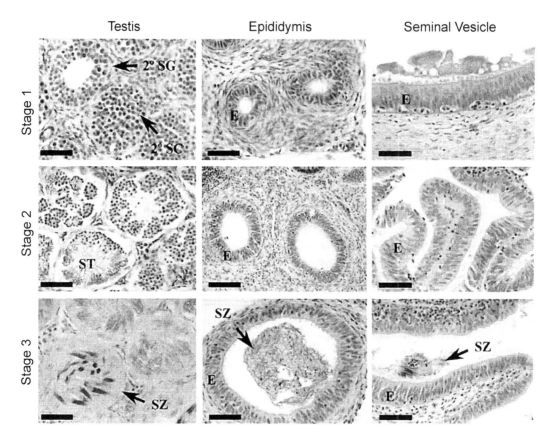

Fig. 11.6 Three progressive stages of development in the tissues in the male testis, epididymis, and seminal vesicle of male bonnethead sharks while reaching puberty. 2° SG = secondary spermatagonia; 2° SC = secondary spermatocytes, ST = spermatids, SZ = spermatozoa, E = epithelium. The scale bar = 50 μm.

cal samples were classified in three stages. The first classification comprised sharks that were in the earliest stage of gonadal development with secondary spermatogonia or spermatocytes in the testis but only epithelium in the epididymis and seminal vesicle farther down the reproductive tract (fig. 11.6). The second category included sharks undergoing spermatogenesis or the active development of spermatozoa, that was evident by the presence of spermatids in the testis but still only epithelium in the epididymis and seminal vesicle. The third stage was indicative of the individual having undergone puberty and reached maturity by the presence of tightly aggregated spermatozoa in the testis, epididymis, and seminal vesicle. The temporal changes in reproductive development are evident upon viewing the stage of histological development of the sharks captured at different times of the year (fig. 11.7). The different stages are assigned different symbols on these graphs, a solid circle and a clear inverted triangle for Stages 1 and 2 for immature sharks and a solid square for

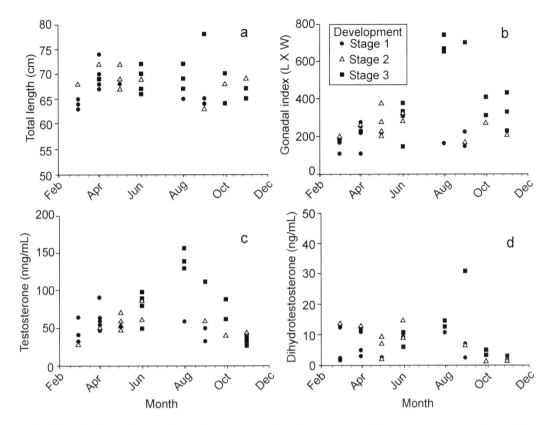

Fig. 11.7 (a) Total length, (b) gonadal index, (c) serum concentrations of testosterone, and (d) dihydrotestosterone associated with the three stages of development in three reproductive organs of wild, male bonnethead sharks while reaching puberty.

Stage 3 for mature sharks. Note the preponderance of circles and triangles from March to May on the plots of total length, gonadal index, and concentrations of testosterone and dishydrotestosterone—over 90% of the males were categorized as having incomplete maturation of the testes and gonadal ducts. However, full maturity appeared to be reached by early summer as evident by the appearance of solid squares in June on the four graphs. The size of the testes, based on measurements of their length, was particularly high in August and September (see solid squares for the gonadal index in fig. 11.7b). Again, there were very high concentrations of testosterone and dihydrotestosterone during the same two months (see solid squares in fig. 11.7c & 11.7d).

Male bonnetheads from the Tampa Bay region regularly undergo puberty between the first and second year following their birth. These studies indicate that hormones may promote the initiation of puberty of both captive and wild individuals of this species, yet it is unknown yet just what factors may trigger the increases in these reproductive hormones. These factors could be environmental such as temperature or length of day or the availability of food in their environment.

Female Sharks and Rays

The ovary is a large, tear-shaped organ attached to an expansion of the epigonal organ near its anterior terminus (fig. 11.8). The ovary lies in the anterior of the abdominal cavity not far from the shark's liver and gall bladder. In the blue shark, only the right ovary is present and functional.[3] In a mature female, the ovary is large. There are within it usually 100–130 large ova that are more prominent than the remaining hundreds of smaller immature follicles present in the organ. The epigonal organ is a bilateral strap-like organ, which extends the length of the peritoneal cavity. It is attached to the body wall by a mesentery or thin web-like tissue. In the blue shark, it is round in the central body cavity and flat at the posterior end. The ostium is the funnel-shaped opening to the oviduct. It is located at the anterior end of the peritoneum cavity, where it splits into the right and left oviducts. The oviduct of the blue shark expands from the ostium to become the oviducal gland. This heart-shaped organ has two short horns on either side of the anterior part of its white surface.

The oviducal gland has functions both in shell formation and in sperm storage. This gland has also been termed the nidamental, or shell gland, because within it develop the leathery shells that encase the embryos of some sharks, many rays, and all of the chimaeras.[10] The anterior region of the gland is composed of cells specialized in the secretion of albumen, a constituent of the shell of the egg case. This zone of the gland is large in egg-laying species but is small in species such as the blue shark that produce fully developed young. The posterior of the oviduct is composed of mucus secreting tubules. They originate at the margins of the gland and extend inward parallel in their orientation until they meet a central canal. These vessels pass down the length of the gland. As the blue shark matures, the gland grows larger and changes in shape from oval to heart-shaped. Spermatozoa were found in the oviducal glands of 79 of 160 female blue sharks collected over a three-year period in the northeastern Atlantic Ocean.[3] The oviducal glands of females from 160 to 180 cm long had tubules full of spermatozoa. These had likely recently mated with males. The walls of the oviduct become furrowed below the oviducal gland in a region called the isthmus. Here the oviduct returns to its original diameter before becoming wider in females that are reproductively active. The oviduct leads to the uterus, which is oval in cross-section and thick walled. The two uteri extend caudally until they join to become the cloaca. The entrance to this organ in immature female blue sharks has a thin, circular membrane, the

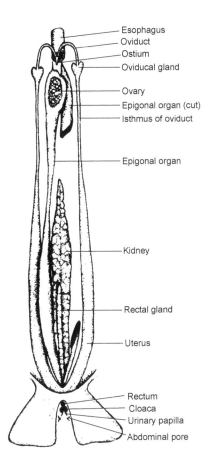

Esophagus
Oviduct
Ostium
Oviducal gland

Ovary
Epigonal organ (cut)
Isthmus of oviduct

Epigonal organ

Kidney

Rectal gland

Uterus

Rectum
Cloaca
Urinary papilla
Abdominal pore

Fig. 11.8 Reproductive system of female blue shark.

hymen. The rectum, through which feces pass, opens into the forward wall of the cloaca.

The ovaries of the female American stingray are elongate organs that are attached to the body cavity by mesorchium and are anterior and central to the surrounding epigonal organ (fig. 11.9).[5] Only the left ovary is active and produces ova, unlike the two functional male testes. The right ovary is small relative to the surrounding epigonal organ. The ovaries of mature stingrays contain numerous small oocytes embedded within connective tissue throughout the year, but they begin to grow larger in October. Vitellogenesis, or yolk production, begins at this time; the ova increase in size and become yellow and orange in color while the epigonal organ diminishes in size. The ova reach their maximum diameter in March, at which time the largest ova are released from the ovary into the ostium and oviduct. Thus, there is a brief two-month period between when the ova become mature and copulatory activity ceases toward the end of May, when the ova of a female can be fertilized by sperm from a male.

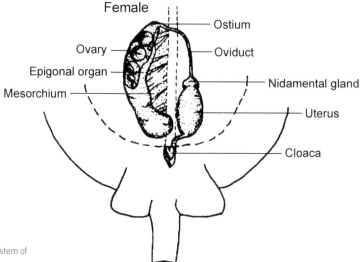

Female

Ostium

Ovary

Oviduct

Epigonal organ

Nidamental gland

Mesorchium

Uterus

Cloaca

Fig. 11.9 Reproductive system of female Atlantic stingray.

Sexual Maturation and Embryonic Development

The hymen is penetrated during the first attempt at copulation. However, that copulation may not result in fertilization unless the female is sexually mature. The presence of fully matured ova in the ovary is the most widely used indicator of reproductive maturity. Mature ova with a diameter of 18 mm are found in the female blue sharks of 180 cm or longer in fork length—the distance between the tip of the snout and the notch midway between the upper and lower lobes of the tail fin.[3]

The fertilized egg develops into an embryo, which either becomes encased within an egg shell and is deposited in a safe place in the external environment or is nourished further within the uterus and released into the environment as a fully developed miniature replicate of the adult. The former mode of embryonic development is termed oviparity, while the latter is referred to as viviparity. All of the primitive sharks of superorder Squalomorphii—the Hexanchiformes, Pristiphoriformes, Squatiniformes, Echinorhiniformes, and Squaliformes—exhibit different modes of viviparity (fig. 11.10).[12] With respect to the advanced sharks of the superorder Galeomorphii, about half of the carpet shark species of the Orectolobiformes and three-quarters of the reef shark species of the order Carcharhiniformes are viviparous. The bullhead sharks in the order Heterdontiformes are oviparous egg layers as are many of the catsharks of the family Scyliorhinidae in the order Carcharhiniformes. The mackerel sharks or the order Lamniformes have a unique mode of embryo nourishment, oophagy, in which embryos feed on developing eggs within the uterus. The majority of the rays are viviparous with the only exception being

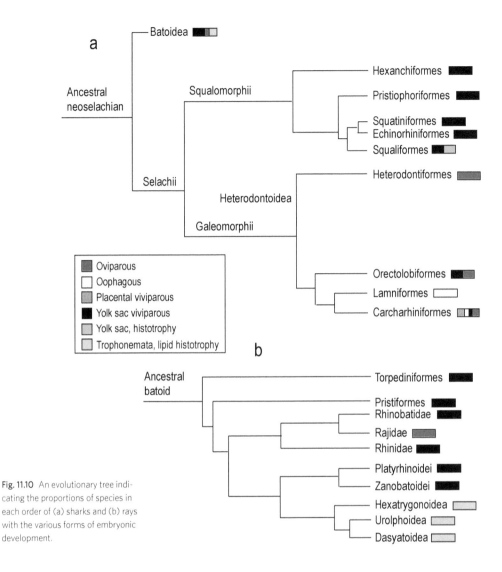

Fig. 11.10 An evolutionary tree indicating the proportions of species in each order of (a) sharks and (b) rays with the various forms of embryonic development.

the skates of the order Rajiformes, which are oviparous. Finally, all of the species in the order Chimaeriformes are oviparous.

Oviparity is characteristic of those orders of cartilaginous fishes containing species less than a meter long that would have few young if born as fully developed miniature sharks.[12] The fecundity of small oviparous species exceeds that of viviparous species of a similar size by a factor of ten or more in both the sharks and rays. The chain catshark (*Scyliorhinus rotifer*) and small-spotted catshark (*Scyliorhinus canicula*), whose maximum sizes are 48 and 100 cm, produce 29–190 and 44–55 eggs per year (table 11.1). In contrast, the hook-tooth dogfish (*Aculeola nigra*) and little gulper shark (*Centrophorus uyato*), whose maximum sizes are 60 and 100 cm, give birth to three or more neonates each

TABLE 11.1. Mating periodicity, type of embryonic development, gestation period, and range of lengths of neonates at birth of selected sharks, rays, and chimaeras.[a]

Order Species	Common name	Max. size (cm)	Development	Embryonic nourishment	Mating period (year)	Gestation (month)	Fecundity (pups, eggs)	Size at birth (cm)
Hexanchiformes								
Hexanchus griseus	Bluntnose sixgill	482	Viviparous	Yolk sac			22–108	60–75
Squaliformes								
Aculeola nigra	Hook-tooth dogfish	60	Viviparous	Yolk sac, histotrophy			>3	13–14
Centrophorus uyato	Little gulper shark	100	Viviparous	Yolk sac, histotrophy			1	40–50
Centroscymnus crepidater	Longnose velvet dogfish	90	Viviparous	Yolk sac, histotrophy			4–8	28–35
Deania profundorum	Arrowhead dogfish	70	Viviparous	Yolk sac, histotrophy			5–7	
Etmopterus hillianus	Caribbean lanternshark	50	Viviparous	Yolk sac, histotrophy			4–5	9
Euprotomicrus bispinatus	Pigmy shark	27	Viviparous	Yolk sac, histotrophy			8	6–10
Isistius brasiliensis	Cookie cutter shark	50	Viviparous	Yolk sac, histotrophy			6–12	
Somniosus microcephalus	Greenland shark	400	Viviparous	Yolk sac, histotrophy			10	
Squalus megalops	Shortnose spurdog	71	Viviparous	Yolk sac, histotrophy	2	24	2–4	20–24
Squatiniformes								
Squatina dumeril	Sand devil	152	Viviparous	Yolk sac			>25	
Heterodontiformes								
Heterodontus francisci	Horn shark	122	Oviparous			7–9	1/duct/2 wks	

Orectolobiformes								
Ginglymostoma cirratum	Nurse shark	430	Viviparous	Yolk sac, histotrophy	2		21–28	20–30
Rhincodon typus	Whale shark	2,000	Viviparous	Yolk sac, histotrophy			300	58–64
Lamniformes								
Alopias vulpinus	Thresher shark	760	Viviparous	Oophagy/cannibalism	2	9	2–4	114–150
Carcharias taurus	Sandtiger shark	320	Viviparous	Oophagy/cannibalism	2	8–9	1–2	100
Carcharodon carcharias	White shark	792	Viviparous	Oophagy/cannibalism	2	>12	2–10	120–150
Cetorhinus maximus	Basking shark	900	Viviparous	Oophagy/cannibalism	2	>12	5–6	150–200
Isurus oxyrinchus	Shortfin mako	400	Viviparous	Oophagy/cannibalism	3	15–18	10–18	60–70
Carcharhiniformes								
Carcharhinus falciformis	Silky shark	350	Viviparous	Placenta	2	12	2–15	57–87
Carcharhinus melanopterus	Blacktip reef shark	200	Viviparous	Placenta		8–16	2–4	33–50
Galeorhinus galeus	Tope shark	193	Viviparous		3	12	6–52	30–36
Mustelus canis	Dusky smooth-hound	150	Viviparous	Placenta	1	10	4–20	
Negaprion brevirostris	Lemon shark	340	Viviparous	Placenta		10–12	5–17	
Scyliorhinus canicula	Small-spotted catshark	100	Oviparous				29–190/yr	
Scyliorhinus rotifer	Chain catshark	48	Oviparous			7	44–53/yr	
Scyliorhinus torazame	Cloudy catshark	50	Oviparous				1/duct/2 wks	8
Prionace glauca	Blue shark	400	Viviparous	Placenta	2	9–12	4–135	35–44

TABLE 11.1. *continued*

Order Species	Common name	Max. size (cm)	Development	Embryonic nourishment	Mating period (year)	Gestation (month)	Fecundity (pups, eggs)	Size at birth (cm)
Rhizoprionodon terraenovae	Atlantic sharpnose shark	110	Viviparous	Placenta	1	10–11	1–7	29–37
Sphyrna lewini	Scalloped hammerhead	430	Viviparous	Placenta		9–12	12–41	39–57
Sphyrna tiburo	Bonnethead	150	Viviparous	Placenta	1	4.5–5.0, 9	6–9	35–40
Triaenodon obesus	Whitetip reef shark	213	Viviparous	Placenta		>5	1–5	52–60
Torpediformes								
Narcine brasiliensis	Brazilian electric ray	54	Viviparous	Trophonemata			4–15	
Torpedo nobiliana	Electric ray	180	Viviparous	Trophonemata		12	<60	23
Pristiformes								
Pristis pectinata	Smalltooth sawfish	760	Viviparous	Yolk sac, histotrophy			15–20	61
Rajiformes								
Amblyraja radiata	Starry ray	150	Oviparous				2–88/year	3.4–8.9
Dipturus batis	Blue skate	285	Oviparous				40/year	10.6–24.5
Leucoraja erinacea	Little skate	54	Oviparous				30/year	4.6–6.3
Raja clavata	Thornback ray	105	Oviparous				60–140/year	5.0–9.0
Raja eglanteria	Clearnose skate	65	Oviparous				60/year	5.5–9.9
Myliobatiformes								
Aetobatus narinari	Spotted eagle ray	330	Viviparous	Trophonemata			4	17–35
Dasyatis americana	Southern stingray	200	Viviparous	Trophonemata			2–10	

Dasyatis pastinaca	Common stingray	57	Viviparous	Trophonemata			6	
Dasyatis sabina	Atlantic stingray	61	Viviparous	Trophonemata			1–4	
Dasyatis sayi	Bluntnose stingray	100	Viviparous	Trophonemata				
Manta birostris	Giant manta	910	Viviparous	Trophonemata			2	110–130
Myliobatis californica	Bat eagle ray	180	Viviparous	Trophonemata	1	12	2–5	
Platyrhinoides triseriata	Thornback guitarfish	91	Viviparous	Trophonemata			1–15	
Potamotrygon motoro	S.Am. freshwater stingray	50	Viviparous	Trophonemata			16	
Pteroplatytrygon violacea	Pelagic stingray	160	Viviparous	Trophonemata		4	2–9	15–25
Rhinoptera bonasus	Cownose ray	213	Viviparous	Trophonemata				
Rhinobatos lentiginosus	Atlantic guitarfish	75	Viviparous	Trophonemata			6	
Rhinobatos productus	Shovelnose guitarfish	119	Viviparous	Trophonemata		12	6	
Rhinoptera javanica	Flapnose ray	150	Viviparous	Trophonemata			1	~60
Urolophus halleri	Haller's round ray	58	Viviparous	Trophonemata				
Zapteryx exasperata	Banded guitarfish	83	Viviparous	Trophonemata		3–4	4–11	
Chimaeriformes								
Chimaera monstrosa	Rabbit fish	150	Oviparous			9–12		10
Harriotta raleighana	Pacific longnose chimaera	120	Oviparous					10–13
Hydrolagus colliei	Spotted ratfish	97	Oviparous				2 on stalk	

[a]Type of embryonic development based on fig. 3.1 & fig. 3.2 in Musick and Ellis (2005)[11] and table 8.1 in Lutton *et al.* (2005)[13]; data on mating periodicity, gestation period, and range of lengths of neonates from "Reproduction" in FishBase (see www.fishbase.org/search.php).

year. The clearnose skate (*Raja eglanteria*) and thornback ray (*Raja clavata*), whose maximum lengths are 65 and 105 cm, produce 60–140 and 60 eggs each year. On the other hand, the Atlantic guitarfish (*Rhinobatos lentiginosus*) and shovelnose guitarfish (*Rhinobatos productus*), whose maximum sizes are 75 and 119 cm, give birth to only six neonates each year. Even the oviparous mode of reproduction has advantages over broadcast spawning of eggs and sperm in the open ocean. The young of these fishes avoid being larvae, which are very vulnerable to predation while drifting in the currents and suffer a high rate of starvation because they must feed soon after they absorb their yolk sac.

The embryos of oviparous sharks and rays are encased in tough leather-like cases. There are two forms of this reproductive mode: single and multiple oviparity.[11] One egg case from each of the oviducts is deposited into the environment, usually in pairs, in the former reproductive mode. However, over the mating season the total number of egg cases delivered into the environment can exceed 100. For example, small-spotted catsharks produce as many as 190 egg cases each year. The starry ray (*Amblyraja radiate*) deposits as many as 88 cases into the environment per year and the thornback ray 140 egg cases per year. This is the only type of reproduction for the bullhead sharks and the skates. The carpet sharks exhibit single oviparity along with viviparity, as do the majority of catshark species. Multiple oviparity entails retaining up to ten egg cases within the oviducts at a time and depositing them all at once into the environment. This mode of reproduction exists only in a few species of catsharks and even fewer carpet sharks.

The cartilaginous fishes often wedge their egg cases between rocks or affix them to bottom structures with the long tendrils that are often present on the cases. Adult horn sharks move their egg cases to different locations every week.[14] The motivation for this behavior may be to protect them from predation. The egg cases produced by species such as the Mexican horn shark (*Heterodontus mexicanus*) are ideally adapted for securing in a rocky crevice. They have a ridge that coils around a cylindrical center, in which the embryo is enclosed (fig. 11.11a).[15] Extending outward from one end are numerous coiled tendrils. The coiled ridge of the egg case acts like the threads of a screw, permitting the adult to secure the egg in a location by rotating it. The coiled tendrils also extend around a structure and hold the egg case fast to it. The egg case of the swellshark (*Cephaloscyllium ventrum*) is flat and rectangular with horns extending outward and inward from each corner; the horns enable its ends to become fastened to structures in the environment. The egg case of the deepsea ray (*Bathyraja interrupta*) has inward-facing horns at its edges (fig. 11.11b).[15] The egg case of the spotted ratfish has a knife-like dorsal and ventral ridge along its entire length with sharp projections on either end both for protection and for ease of securing between rocks (fig. 11.11c).[16] These egg cases

Fig. 11.11 Egg cases of the (a) Mexican horn shark, (b) deep sea ray, and (c) white-spotted ratfish.

are extruded from the oviduct over a period of eighteen to thirty hours during parturition.[17] The egg cases then remain attached to the female for a period of four to five days. The cases are suspended in the water below her by slender extension of the egg case called an elastic capsular filament. The egg cases are eventually deposited either in the mud or on a gravel bottom.

The embryos develop within the egg case for a period ranging from three to twelve months in most species of cartilaginous fishes.[1] The embryos find their nourishment while within the case from the yolk sac. In less than a month, small openings, usually slits, appear at either end of the case. Fully oxygenated water is drawn through these openings into the egg case. An embryo in an advanced developmental state increases the rate of water exchange and oxygenation by beating its tail back and forth. The case is tough and composed of a thick, thorny material. However, this does not afford certain protection from predators such as large shell-bearing gastropods, rays, or sharks. Furthermore, the young upon hatching are smaller than those of live bearing sharks. The oviparous sharks are mostly bottom dwellers and are smaller than other sharks and rays. Ovipary is likely the ancestral mode of reproduction for the cartilaginous fishes. Viviparity likely evolved to nourish the embryo over a longer period of time in the uterus of the female so that it could develop into a more advanced state prior to parturition.[1]

The viviparous sharks retain embryos within the uterus throughout their entire period of development until they resemble adults. These sharks reproduce either placentally, with a connection through which nourishment can pass from the female to the developing embryo, or aplacentally. The two modes of development are really the opposite poles of a continuum with many intermediate forms.[18] The embryos of only few species of cow and frilled sharks in the order Hexanchiformes and the angelsharks of the order Squatiniformes gain nourishment solely from their yolk sac. The yolk sac becomes smaller as the embryo grows larger and digests the yolk (fig. 11.12a). Many more species of sharks and rays gain their nourishment not only from the yolk but also from mucus secreted from cells on the surface of uterus. This mode of nourishment is termed limited histotrophy.[19] There is a steady progression in the degree of histotrophy from limited vascularity with nonsecreting villi, or small projections, in the uteri of primitive species to the proliferation of secretory cells that produce the nutritive mucus in more advanced sharks. This mode of reproduction is exhibited by the species of dogfish sharks in the order Squaliformes. It is also found in a more advanced state in few more advanced galeoform sharks of the orders Orectolobiformes and Carcharhiniformes (see species labeled "yolk sac, histotrophy" in the embryonic nourishment column in table 11.1). In the stingrays of the order Myliobatiformes, many large projections with many villi, called trophonemata, on them extend outward from the walls of the uterus toward the embryo (see table 11.1).[20] This increases the surface area on the uterus for respiratory exchange and secretion of mucus mainly composed of lipids, which are high in energy content.

The most advanced form of embryonic development is the placental mode (fig. 11.12b).[1, 19] The yolk sac elongates, its surface becomes covered with a network of blood vessels. The sac eventually narrows into a tube with a highly

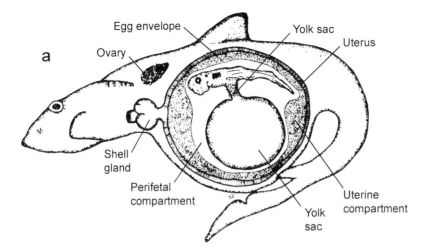

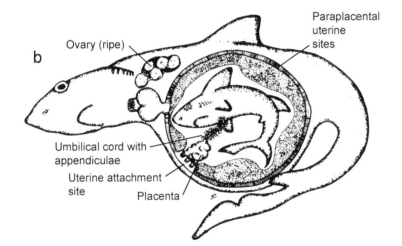

Fig. 11.12 Placental development in sharks. (a) At an early stage in gestation, the embryo is nourished through a yolk stalk from the yolk sac. (b) At an advanced stage of gestation, the embryo is nourished through an umbilical chord. This placental form of development is the most advanced form of embryonic development in sharks.

vascularized surface that contacts the surface of the uterine wall and forms a placental conduit through which nutrients are carried within the bloodstream of the female into the body of the embryo. This results in a continuous supply of energy to the embryos, which is dependent only on the health of the female and her success in finding prey. This mode of reproduction is found in many members of the reef and hammerhead sharks of the order Carcharhiniformes.

REPRODUCTIVE CYCLE

The most successful reproductive strategy ensures the production of the greatest number of young and the most efficient utilization of the maternal resources.[21] The female must occupy a unique ecological niche. She must possess a lifestyle that ensures success as a predator, growth to adulthood, the

The embryos of the mackerel sharks in the order Lamni-formes first derive nourishment from their yolk sac but then feed either on other eggs or embryos in a less advanced state of development. This reproductive strategy favors having fewer but larger young. The female shortfin mako shark gives birth to only ten to eighteen pups, while the white shark gives birth to ten or fewer neonates. The eggs in these oophagous, or egg-eating, species are usually small, 5-7 mm in diameter, with little yolk so they provide nour-ishment for an embryo for less than a month.[1] The embryo, which is no larger than 5 cm long after its yolk sac has been absorbed, commences then to feed on eggs that were not fertilized during copulation. The sandtiger shark (*Carcharias taurus*) exhibits the extreme of this developmental mode—being embryophagous[21]; i.e., it feeds not only on eggs but also on embryos. By the time of parturition, there re-mains only one large full-term embryo in each uterus. The sandtiger embryo consumes all of its companion embryos within the first 100 days after conception. During the next 100 days, it feeds upon newly developing ova. A hungry embryo, growing 100-335 mm in length over the last 100 days of its development, feeds on eleven to thirteen egg capsules for every 6 mm increment of growth. The embryo increases its mass by a maximum of 15% and its length by 15% per day during the ovophagous phase and 6% in mass and 1% in length per day during the embryophagous phase of its development. The young of the sandtiger are larger at birth than those of other sharks relative to the size of the adult. The two juveniles, one released from each uterus, may exceed 1 m in length and their mass may constitute one third of the weight of the mother at birth.[1]

nourishment of her embryos, and the deposition of her eggs or young in a unique environment. Here the young must be relatively safe and food plen-tiful. The duration of the reproductive cycle is known only for a relatively few species of cartilaginous fishes, and for most of those it is two years (see table 11.1). However, a few small shark species such as the Atlantic sharpnose (*Rhizoprionodon terraenovae*), the dusky smooth-hound (*Mustelus canis*), the bonnethead, and the bat eagle ray (*Myliobatis californica*), can complete their reproductive cycle every year.[22] The tope (*Galeorhinus galeus*) has a longer re-productive cycle. It appears to reproduce every third year because its egg fol-licles when examined exist in three different size ranges.[23]

Blue sharks that are four or five years of age arrive at their breeding and feeding grounds on the continental shelf off southern New England during May and June.[7] Here the males begin courtship activities with the females, as is evident from the many scars observed on females at this time of year. Court-ship continues through November. Mating behavior may continue throughout the year in the southernmost range of the blue shark off the Bahamas. The largest of the females, who carry spermatozoa stored within their oviducal glands, during their sixth year remain offshore during spring and wait until the following spring to permit their eggs to be fertilized. Their fertilized eggs develop into full term in nine to twelve months. Hence, the females, now in their seventh year, give birth to pups from April to July at two-year intervals.

The reproductive cycle, and the endocrinological changes associated it, is best known for the bonnethead shark.[24] The species exhibits placental viviparity and has a gestation period of only 4.5 to 5.0 months, one of the shortest known for sharks. The reproductive cycle of this species can be divided into nine distinct stages. These stages were highly synchronized for bonnethead sharks despite living at two different locations, the shallow waters of the Gulf of Mexico in Pine Island Sound and Tampa Bay off the southwestern coast of Florida. The reproductive events in the latter location occurred seven to ten days later than in the former location (note displacement to the right of the events in the lower time scale in fig. 11.13a). Courtship and copulation, determined from the presence of external mating wounds on the females, is carried out during November at both locations. Mating thus occurs four to five months before preovulation and ovulation, the third and fourth stages of reproduction. The former stage is characterized by the initiation in the growth of follicles while the uterus is still small; in the latter stage many ova have become fully developed to partially expand the uterus. The diameters of the ova in the uterus increase during preovulation and reach their maximum size during the ovulation (fig. 11.13e). Sperm from the males is stored in the oviducal gland of the females over this prolonged period, as is evident in histological samples taken on a monthly basis over this time period. The amount of sperm in the gland first begins to decrease during postovulation, the fourth stage, while the uterus of the female becomes full, and this amount continues to decrease during early pregnancy, the fifth stage, when embryos first appear during May within the uterus. The exact time of fertilization in the reproductive tract of the female is not known for certain but is inferred to be just after ovulation, when the amount of sperm in the oviducal gland begins to decrease. During implantation, the sixth stage, a placenta starts to develop. The placenta is fully developed during late pregnancy, the seventh stage of reproduction. The females give birth to neonates during August, and the uterus then becomes flaccid. Postpartum, the ninth stage, occurs during September and October after parturition. The uterus is now small with small follicles beginning to grow within it. What is curious about the reproductive cycle of the bonnethead is that the females store sperm within their oviducal gland for a prolonged period before permitting fertilization of the eggs within their uterus. A reason proposed for this is that mature bonnetheads in southwestern Florida leave the estuaries during the winter months, and possibly the males and females make separate migrations to different feeding grounds. This storage of sperm would permit ovulation and fertilization to be synchronized among females without the necessity of a reunion with the males. In conclusion, the time between mating and parturition is nine months as in many other viviparous sharks. However, the storage of sperm in the oviducal glands accounts for half that period. Hence, gestation, defined as the period between fertilization and parturition, takes only about 4.5 months in the species.

The levels of reproductive hormones were determined for females throughout the reproductive cycle. The levels of 17β-estradiol were significantly elevated during mating and immediately prior to ovulation with the highest concentrations recorded during preovulation; the lowest levels occurred during early pregnancy (fig. 11.13b). The concentration of progesterone steadily increased to its highest level during postovulation and decreased to its lowest level during late pregnancy (fig. 11.13c). The level of testosterone increased significantly during mating and preovulation and decreased to its lowest level during parturition. The progesterone appears to reach peak concentrations during postovulation after peaks in 17β-estradiol and testosterone during mating and preovulation. The high levels of progesterone may enable the uterus to compartmentalize, an important precondition for implantation. Another possible role is to inhibit vitellogenesis, the development of follicles in the ovary.

Female southern stingrays mate with males continuously over a period of seven months in spite of a brief two-month period of ovulation.[5] Spermatozoa are present in the cloaca, uterus, and nidamental gland of the females from November to May while sperm are absent in these sexual organs from June to September. Mating scars are found on female rays over this seven-month period. The females' eggs, however, appear to be fertilized only during March and April despite the prolonged mating period. It is not yet understood why mating activity occurs prior to when eggs are fully developed and fertilized. One possible explanation for early copulation is that the sperm of the males are stored in the oviducal gland of the female until they are later released in spring, when the ova are ready to be fertilized. However, sperm are found in the gland only immediately before ovulation in March and April. Sperm storage is usually attributed to species that are nomadic and segregate by sex, such as the blue shark, to adapt to the infrequent contact between the sexes.[6] The southern stingray does not segregate by sex in the Gulf of Mexico. This reproductive riddle has yet to be answered; its resolution awaits further study. Gestation usually takes three to four months with the females giving birth to young within egg cases from late June to early August.

The reproductive cycle, and the endocrinological changes associated with it, is also well known for the clearnose skate.[26] The reproductive period for these skates in the eastern Gulf of Mexico begins during October and November, when mature rays enter the coastal waters. This appears to be associated with the development and maturing of the ovarian follicles for the ensuing reproductive season. Courtship and copulation occur from the middle of De-

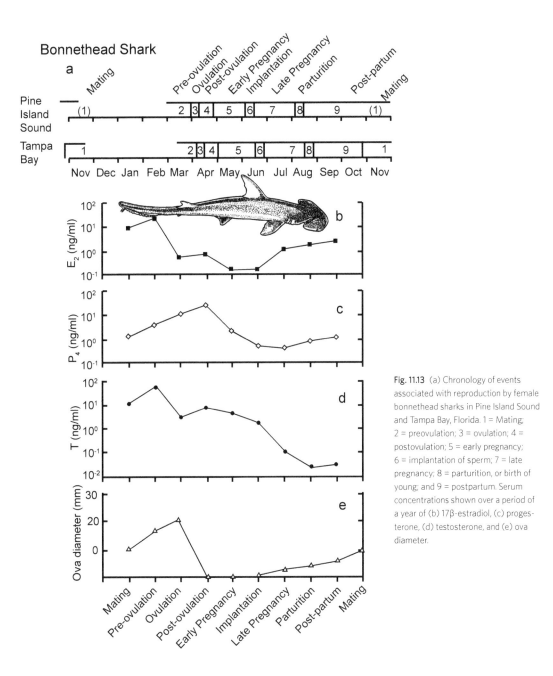

Bonnethead Shark

Fig. 11.13 (a) Chronology of events associated with reproduction by female bonnethead sharks in Pine Island Sound and Tampa Bay, Florida. 1 = Mating; 2 = preovulation; 3 = ovulation; 4 = postovulation; 5 = early pregnancy; 6 = implantation of sperm; 7 = late pregnancy; 8 = parturition, or birth of young; and 9 = postpartum. Serum concentrations shown over a period of a year of (b) 17β-estradiol, (c) progesterone, (d) testosterone, and (e) ova diameter.

cember until early April with a peak in activity during January and February (fig. 11.14a). The adults begin to migrate offshore into the deeper and colder waters of the gulf from late February to early March, presumably due to the warming of the inshore waters at this time. They are absent from the coastal waters until the following fall. Hence, their annual reproductive cycle of the clearnose skate could be observed only with captive individuals. Following copulation, spermatozoa move through the reproductive track to the paired

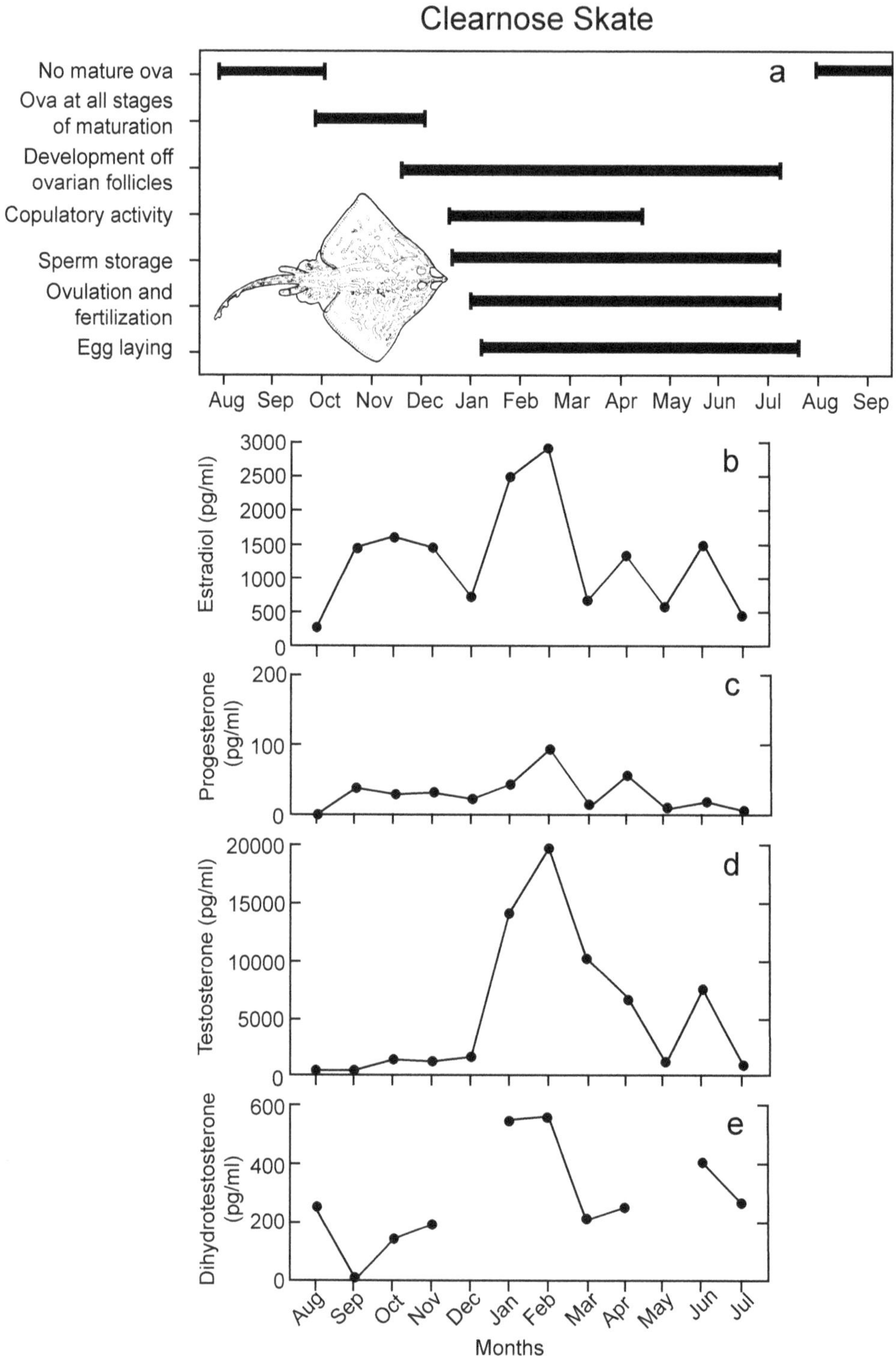

Fig. 11.14 (a) Chronology of the events associated with reproduction by clearnose skates in the nearshore waters off Sarasota, Florida. Serum concentrations shown over a period of a year for (b) 17β-estradiol, (c) progersterone, (d) testosterone, and (e) dihydrotestosterone.

oviducal glands, where they are stored from late December until early July. Ova mature and move through the oviducts to the oviducal glands, where fertilization occurs as well as encapsulation within an egg case. Egg laying commences during January and continues into early to mid-July, if the females are maintained in captivity at a constant temperature of 20° C. This ray, like all other members of the order Rajiformes, displays single oviparity. They lay eggs in pairs, one from each side of the symmetrical reproductive tract, at intervals of four to five days between successive egg-laying events.[27]

The hormonal concentrations within the blood of female skates were determined by taking monthly samples of serum. The levels of 17β-estradiol were significantly elevated during two periods, September through November and January and February (fig. 11.14b). These increases were likely associated with the development of ovarian follicles during October and November and the start of mating activity during January and February. In contrast, during the summer, when the females are reproductively inactive, the levels of 17β-estradiol were the lowest of the year. The concentration of progesterone, testosterone, and dihydrotestosterone were also high during January and February, coinciding with the peak in mating activity. It is apparent that progesterone is the maturation-inducing steroid in the clearnose skate. Of particular interest is the elevation of the concentration of progesterone after each oviposition of a pair of eggs by a female of this species.

Sexual Segregation

A general characteristic of the populations of live-bearing sharks is segregation of the sexes.[28] This has been inferred from the unequal numbers of individuals of each sex in catches of many species such as the white, sandbar, school, lemon, blue, and scalloped hammerhead sharks. Shark populations may be divided into three social units: subadults of both sexes, mature males, and mature females. Certain aspects of the biology of the scalloped hammerhead provide some insight into the origin of this sexual asymmetry. Female scalloped hammerhead sharks separate from the males at an intermediate size when they migrate offshore while the males remain inshore. This results in their achieving greater feeding success than the inshore-based males. The consequence is that they grow more rapidly to maturity than males. Sexual segregation may have evolved because females need to be of a larger size to nourish their embryos. These fully developed neonates may comprise almost a third of the female's body mass. Conversely, males need only to produce small numbers of spermatozoa to successfully fertilize a single female's ovum. Female scalloped hammerheads captured in the Gulf of California were mature at larger sizes than males. It is possible that females grow more rapidly than males because they feed with greater success in the offshore waters of the gulf. The result is that males and females reach maturity at a similar age, permitting reproductive activities to occur over the same period of their lifetime.

COURTSHIP AND COPULATION

The sequence of courtship behaviors leading to copulation in sharks was first described for nurse sharks in captivity[29] and later in the shallow waters of the Bahamas and Dry Tortugas.[30, 7] The latter is an archipelago extending away from the tip of the Florida peninsula. Courtship is initiated when one or more male nurse sharks swim alongside a female with their tails sweeping synchronously back and forth (fig. 11.15a). Their bodies remain on the same plane as the female less than two pectoral widths distant with the male abreast of the female or just behind the posterior margin of her pectoral fin. Occasionally their caudal or pectoral fins touch as they swim together. The duration of this behavior varies from a few to as many as 90 minutes. One of the males seizes the margin of either pectoral fins in his mouth (fig. 11.15b-11.15c). He bites down on her fin, and in response she pivots her anterior torso over an arc of ~90° in front of the male. At that time, she straightens out her body while coming to rest at a right angle to the body of the male (fig. 11.15d). He moves forward slightly with her pectoral fin in his mouth in order to rotate her onto her back. He then releases her fin from his jaws and pushes her anterior torso with his snout around until she is lying parallel beside him on her back. The female remains motionless and rigid with her pectoral fins outstretched all of this time (fig. 11.15e). The male then swims on top of the female and inserts his clasper (fig. 11.15f). In some instances, the male rolls on to his back revealing his clasper inserted into the female's cloaca (fig. 11.15g). The male rarely relinquishes his grip on the female's pectoral fin once he initiates courtship. The male often curls his caudal fin around her torso and uses her body for leverage while inserting his reproductive organ into her vent. The female remains quiescent and immobile during clasper insertion. The male's torso undulates as he compresses the cavity between his outer and inner belly. Forced out of this sac is water that mixes with sperm as it passes through a canal in the center of the clasper into the cloaca of the female. Males compete among each other to bite the female's pectoral fin. Four to six males have been observed to approach a female from one side and lunge forward in attempts to seize her pectoral fin (fig. 11.16a). In this group encounter, one male situates himself between the female and group of males in an act of altruism with his body at a right angle while a second male inserts his clasper from the other side (fig. 11.16b).[31]

The female shark advertises her reproductive readiness and attracts males by releasing a pheromone, a chemical attractant, from her reproductive system.

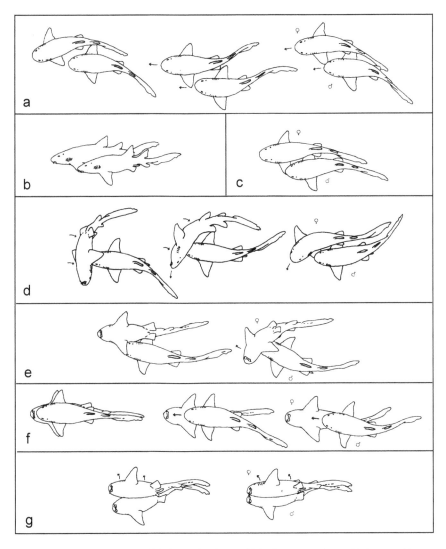

Fig. 11.15 Ethogram, or species-specific behavioral catalogue, of behavior patterns observed during courtship and copulation in the nurse shark in the Miami Seaquarium. (a) Parallel swimming, (b) pectoral biting, side view, (c) pectoral biting, top view, (d) pivot and roll, (e) nudging, (f) male on top of female, (g) male and female lying on back.

Two or three males have been observed to continuously follow the same female nurse shark in captivity.[29] The males always swam just behind the female's cloaca, the opening or vent to her reproductive system. These males may be attracted to her by a chemical secreted in the uterus and released through her cloaca. An adult male whitetip reef shark was observed on a coral reef to follow a female slowly along a winding path while he moved his tail back and forth slowly keeping his nares close to her vent as if perceiving a scent.[33] The female held her tail upwards while slowly moving her tail back and forth slowly. This unnatural

Fig. 11.16 (a) Five male nurse sharks surround a single female with two individuals trying to bite either of her pectorals. (a) A male exhibiting altruistic behavior by positioning his body at a perpendicular angle to provide support for a copulating male and female.

posture presumably permitted the male to place his snout closer to her cloaca in order to smell the odorant, which further enhanced his motivation to mate. Further evidence of pheromone use was the sudden abrupt turn and movement of a single male blacktip shark around a rock to follow the path of a female despite his view of the female being obstructed by the rock. He later joined the female and they swam in a coordinated manner with their tails sweeping back and forth in unison. The male kept his head just posterior of her vent, through which a pheromone would likely be secreted into the environment.

Both male and female nurse sharks are promiscuous and have been observed to mate with more than one individual of the other sex over the mating season.[7] Tissue samples were taken from thirty-two pups within the uterus of a single female.[32] The DNA within each sample was amplified first using the polymerase chain reaction, and then the sequences of genes were determined for the amplified segments of DNA using restriction fragment length polymorphism analysis. A mating between this heterozygous female and only one heterozygous male would result in a maximum of four distinct genotypes among the offspring. Fourteen distinct genotypes were found among the thirty-two pups. This indicates that there were more than three fathers for this brood of nurse sharks.

Male sharks and rays commonly seize the female with their jaws during courtship. This behavior serves two functions that ensure a successful copulation.[7] First, it serves as a behavioral releaser. This stimulus induces the female to remain motionless while the male inserts his clasper into her cloaca and fertilizes her egg with his sperm. Second, it provides the support needed for him to twist his body around and insert his forward-oriented intromittent organ into her vent. Members of other shark species seize parts of the female's anatomy other than the pectoral fin. Male blue sharks mainly seize the main torso of the female, inflicting semicircular jaw impressions, tooth slashes, and individual tooth nicks.[8] The first and most common jaw imprint consists of a semicircular pattern of distinct tooth marks. Two such mating scars, one behind and another to the right of the dorsal fin, are shown on an adult blue shark (fig. 11.17a).[7] The tooth slashes consist of straight or curved cuts caused by the male dragging one or more teeth across the body of the female. When healed, these slashes become pigmented black and can often be seen through the shark's dermal denticles. Triangular tooth nicks are often present in clusters in straight or curved lines on a female's body. The percentages of each of the three types of bites inflicted on various parts of the body have been recorded from adult females caught in the Bay of Biscay off England.[34] The semicircular tooth impressions are present on 60% of the females, tooth slashes on 56% of them, and tooth nicks on 80% of them. The highest percentages of the semicircular bites are present on the flank and dorsum near the pectoral with few on the underside. The tooth slashes are inflicted slightly farther back on the side and top of the body. Individual tooth nicks are found over the entire body. The adult female blue sharks develop a thicker dermis than males during mating season (fig. 11.17b).[7] The dermis consists of the skin and underlying

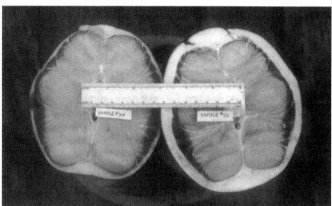

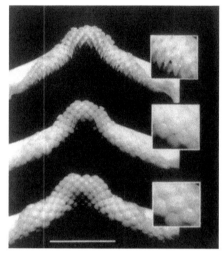

Fig. 11.17 (a) Male-inflicted bites on torso and fins of a female blue shark; (b) cross-section of the torso of a male (left) and female (right) blue shark showing the greater thickness of the muscle on the dorsum of the female than the male. (c) The cuspate teeth in upper jaw of male Atlantic stingray that develop during the mating season (upper), the molariform teeth present in the upper jaw of males outside the mating season (middle), and the similar teeth in the upper jaw of females present during the entire year (lower).

connective tissue. This thicker dermis enables females to heal more quickly from the multiple lacerations inflicted by males during courtship. The shape and size of the teeth of male blue sharks do not differ from the shape and size of the teeth of females, but male catsharks of the genera *Halaelurus* and *Apristurus* have teeth that are longer and have fewer cusps than the teeth of females of these species. These teeth are likely used to hold the female firmly during copulation.[7]

The mating behavior of the rays has best been documented for the eagle rays from observations of individuals kept in captivity.[35] One or more males swim alongside a female for long periods of time while staying immediately behind her pectoral fins. She sometimes accelerates in order to outdistance a group when it is chasing her. The successful male positions himself over her

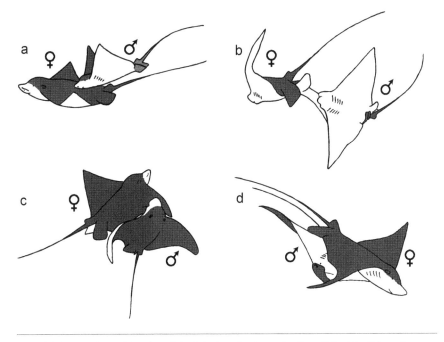

Fig. 11.18 Ethogram of courtship and copulation behaviors of eagle rays kept in captivity at the Okinawa Expo Aquarium. (a) Nibbling, (b) pectoral biting, (c) male torso rotation, and (d) abdomen to abdomen.

midtorso (fig. 11.18a). He then lowers his head so that his jaws come in contact with the dorsum and they open and close repeatedly to give the impression that he is gently biting the dorsum of the female. The female now swims slower. The male then seizes either the anterior or posterior margin of either pectoral fin and holds it in his jaws (fig. 11.18b). He then rotates his body so that the ventral surface of his torso is facing upward close to the underside of her torso (fig. 11.18c). The male then swims upside down with his underside close to the female's belly, rotates his clasper forward, and inserts it into the female's cloaca (fig. 11.18d). Either the right or left clasper is inserted into the female's vent. The duration of insertion varies from thirty to ninety seconds. Four different males mated with a single female during a period of an hour in the Okinawa Expo Aquarium. Eventually, she gave birth to two babies.

The difference between male and female teeth is more pronounced and widespread among the rays. The teeth in the center of the jaws of males are usually sharper and more pointed than those of females. In the first three to four tooth rows in the jaws of male Atlantic stingray, the teeth develop multiple prominent cusps during the spring and summer mating season (upper jaw in fig. 11.17c).[36] These teeth likely provide the male with additional support to enable him to bend his flattened body around the female and hold on tightly so that his clasper does not prematurely release from her during copulation.

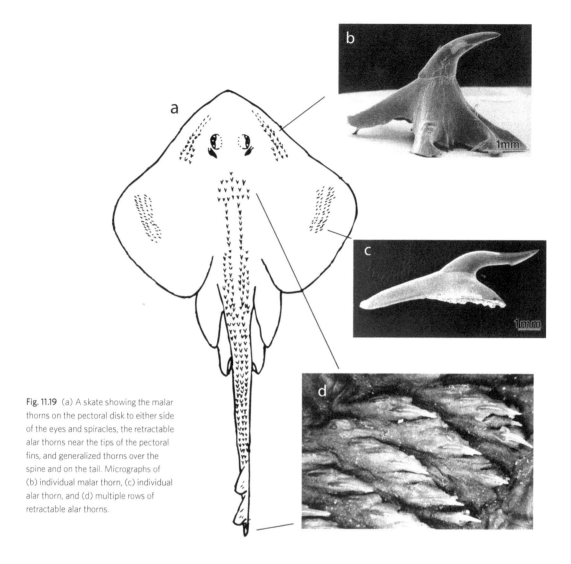

Fig. 11.19 (a) A skate showing the malar thorns on the pectoral disk to either side of the eyes and spiracles, the retractable alar thorns near the tips of the pectoral fins, and generalized thorns over the spine and on the tail. Micrographs of (b) individual malar thorn, (c) individual alar thorn, and (d) multiple rows of retractable alar thorns.

During the rest of the year the male teeth are molariform or blunt in shape (middle jaw). The teeth of female Atlantic stingrays remain molariform year round (lower jaw). There is little impetus to compromise their important function in crushing their prey to facilitate reproduction.

Male skates of the order Rajiformes have two distinct tooth-like hooks on their pectoral disk, the alar and malar thorns (fig. 11.19a).[37] They are likely used to stimulate the female during courtship. These modified dermal denticles develop on the pectoral fin of a male upon his becoming sexually mature. Patches of rows of uniformly sized alar thorns cover the far end of his pectoral fin. Each alar thorn consists a crown oriented in a parallel alignment in the same direction as its long base (fig. 11.19c). They form a row and are all oriented in a posterior direction. They can retract into a groove in the ray's

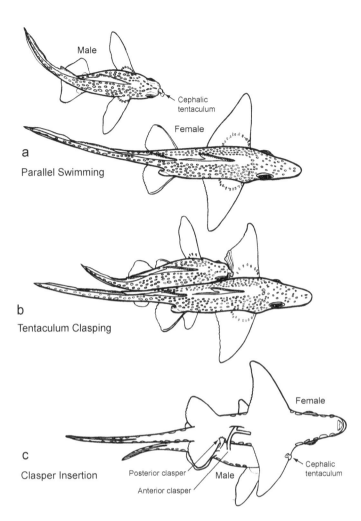

Fig. 11.20 Ethogram of courtship and copulation of the spotted ratfish. (a) Parallel swimming, (b) tentaculum affixation, (c) clasper insertion.

dermis while swimming to reduce hydrodynamic drag (fig. 11.19d). The malar thorn, unlike the alar thorn, is oriented obliquely upward rather than parallel to the dorsal surface of the body (fig. 11.19b). Patches of them are present near the eyes or spiracles. The malar thorn is rigidly erect and does not retract into a groove in the dermis like the alar thorn.

The courtship of the chimaeras has only recently been observed for the spotted ratfish,[38] and a description has yet to be published in the literature. Male chimaeras have flat teeth for crushing bivalves with hard shells or crustaceans with hard external skeletons. These teeth would be ineffective at seizing and holding a female during copulation. Yet males of the spotted ratfish have been observed to bite the pectoral fin of the female during courtship. On the other hand, they have a cephalic tentaculum, a stalked appendage with a bulbous end covered with small hooks (see fig 3.3b). This is usually withdrawn into a depression on the forehead but can be protruded to direct its hooks outward and moved to either side. The male spotted ratfish follows the female while remaining just behind her pectoral fin (fig. 11.20a). He then accelerates momentarily so that he is next to her, swimming in a synchronous manner, with their bodies touching while his cephalic tentaculum extends outward and grasps on to her pectoral fin (fig. 11.20b). The male ratfish then coils his body around the female, secures himself to her flank by extending his posterior, or prepelvic, clasper and affixing it to the side of her body, and only then inserts the calcified internal radius with its hook at the end of the bifurcate clasper into her cloaca. This behavior, in which the tentaculum is attached to the pectoral fin and clasper inserted within the cloaca, can be accomplished only because the mature male is much smaller than the mature female. The two then swim together with the clasper inserted in the female for up to two hours. There is evidence that the tentaculum is affixed at times to the surface of the dorsum near the base of the dorsal fin. Scars have been observed on the backs of female spotted ratfish consisting of five to fifteen lacerations.[39] The anterior scars are punctures while posterior ones are scratches. The number of penetrations corresponds to the number cusps along the frontal edge of the hook-bearing club. This relationship was demonstrated by placing a tentaculum next to the region of the body of the female bearing the lacerations. The presence of more than one scar on a female indicates that a male or several males likely attempt to copulate with a particular female multiple times. The social systems and courtship behaviors are described in table 11.2 for a diversity of elasmobranch species.

SUMMARY

The success of the cartilaginous fishes over evolutionary time may in large part be due to their internal mode of reproduction. The male inserts sperm into the female's reproductive tract to fertilize her eggs, which develop in some species to eggs enclosed within cases or in other species to full-term embryos. The eggs are deposited by females in safe locations, and the embryos protected within the tough egg cases are able to develop to an advanced state, capable of swimming and feeding, before hatching. Those embryonic sharks that are directly released into the external environment usually resemble adults and are immediately able to feed. Internal fertilization ensures that the eggs, toward which the female has allocated much of her energy upon becoming sexually mature, are not wasted by being broadcast haphazardly into the water column but will likely develop into full-term embryos within the uterus of the female. In addition, the male can pick the female with whom he will sire young and can deliver his sperm with little wastage. The majority of the fertilized eggs develop into embryos within the highly protective environment of the female

TABLE 11.2. The social system, courtship, and copulatory behavior of the sharks, rays, and chimaeras.

Order / Species	Common name	Social system	Male follows female	Male bites female	Male moves female; female acceptance	Male copulates with female
Heterodontiformes						
Heterodontus francisci	Horn shark			Bites pectoral fin[40]	Copulation 30–40 min[40]	Wraps body around female[40]
Orectolobiformes						
Ginglymostoma cirratum	Nurse shark	Polyandry, polygyny[7]	Moving side by side[29]	Bites pectoral fin[29,30]	Pivots and rolls on dorsum when bitten[29,30]	Lying on back,[29] heads to substrate, tails up[30]
Lamniformes						
Carcharias taurus	Sandtiger shark		Circles female, rotates clasper, noses vent[41]	Bites flank of body and pectoral fins[41]	Cupping and flaring of female's pectoral fins[41]	Swam side by side, clasper inserted for 2 minutes[41]
Carcharhiniformes						
Carcharhinus melanopterus	Blacktip reef shark		Head stays near vent[42]			
Negaprion brevirostris	Lemon shark		Keep body axes parallel[43]			Coordinated swimming with clasper insertion[43]
Scyliorhinus retifer	Chain catshark			Bites pectoral fin[44]	Copulation 15 sec to 4 min[44]	Wraps body around female[44]
Scyliorhinus torazame	Cloudy catshark			Bites pectoral fin[35]		Wraps body around female[35]
Sphyrna lewini	Scalloped hammerhead		Thrusts midsection with clasper rotation[45]			
Triaenodon obesus	Whitetip reef shark			Bites pectoral fin[35,46]	Copulation 15 sec to 4 min[46]	Undulate, tails held up from bottom[46]

continued

TABLE 11.2. *continued.*

Order Species	Common name	Social system	Male follows female	Male bites female	Male moves female; female acceptance	Male copulates with female
Rajiformes						
Raja eglanteria	Clearnose skate				Fin undulations[27]	Swings tail beneath female[27]
Myliobatiformes						
Aetobatus narinari	Spotted eagle ray	Polyandry[35]	Rapid chase near tail[35]	Gouging bite to dorsum[35]	Multiple matings[35]	Abdomen to abdomen[35]
Dasyatis americana	Southern stingray	Polyandry[47]	Multiple male follow[47]	Grasps pectoral tips[47]	Rapid succession[47]	
Manta birostris	Giant manta		Rapid chase near tail[48]	Grasps pectoral tips[48]	Attachment is brief[48]	Abdomen to abdomen[48]
Myliobatis californica	Bat eagle ray					Abdomen to abdomen, Wing beat synchrony[49]
Rhinoptera bonasus	Cownose ray			Gouging bite to dorsum[49]	Multiple matings[49]	Abdomen to abdomen[49]
Rhinoptera javanica	Flapnose ray	Polyandry[35]		Gouging bite to dorsum[35]	Multiple matings[35]	Abdomen to abdomen[35]
Urolophus halleri	Haller's round ray			Females insert spine in males[50]		
Chimaeriformes						
Hydrolagus colliei	Spotted ratfish		Swims beside body[38]	Tentaculum grasps pectoral[38]	Copulation lasts 1–2 hours[38]	Body rotated underneath torso of female[38]

uterus. The embryos when full-term may constitute up to a third of a pregnant female's body mass. Thus, there is great biological investment in producing so few young.

All of the primitive sharks of superorder Squalomorphii—the Hexanchiformes, Pristiphoriformes, Squatiniformes, Echinorhiniformes, and Squaliformes—exhibit different modes of viviparity. With respect to the advanced sharks of the superorder Galeomorphii, about half of the carpet shark species of the Orectolobiformes and three-quarters of the reef shark species of the order Carcharhiniformes are viviparous. A general characteristic of the populations of live-bearing sharks is segregation of the sexes. This has been inferred from the unequal numbers of individuals of each sex in catches of many species such as the white, sandbar, school, lemon, blue, and scalloped hammerhead sharks. The duration of the reproductive cycle is known only for a relatively few species of cartilaginous fishes, and for most of those it is two years. However, a few small shark species and some rays can complete their reproductive cycle every year. The tope has a longer reproductive cycle. It appears to reproduce every third year.

The sequence of courtship behaviors leading to copulation in sharks was first described for nurse sharks in captivity and later in the wild. Courtship is initiated when one or more male nurse sharks swim alongside a female with their tails sweeping synchronously back and forth just behind the posterior margin of her pectoral fin. One of the males seizes the margin of either pectoral fin in his mouth. He bites down on her fin, and in response she pivots her anterior torso over an arc of ~90° in front of the male. At that time, she straightens out her body while coming to rest at a right angle to the body of the male. He moves forward slightly with her pectoral fin in his mouth in order to rotate her onto her back. He then releases her fin from his jaws and pushes her anterior torso around with his snout until she is lying parallel beside him on her back. The female remains motionless and rigid with her pectoral fins outstretched all of this time. The male then swims on top of the female and inserts his clasper. Mating in rays is similar, yet differs for the chimaeras. Male chimaeras have flat teeth for crushing bivalves with hard shells or crustaceans with hard external skeletons. These teeth would be ineffective at seizing and holding a female during copulation. On the other hand, they have a cephalic tentaculum, a stalked appendage with a bulbous end covered with small hooks. This is usually withdrawn into a depression on the forehead but can be protruded to direct its hooks outward and moved to either side. The male spotted ratfish follows the female while remaining just behind her pectoral fin. He then accelerates momentarily so that he is next to her, swimming in a synchronous manner, with their bodies touching while his cephalic tentaculum extends outward and grasps on to her pectoral fin. The male ratfish then coils his body around the female, secures himself to her flank by extending his posterior, or prepelvic, clasper and affixing it to the side of her body, and only then inserts the calcified internal radius with its hook at the end of the bifurcate clasper into her cloaca.

* * *

KEY TO COMMON AND SCIENTIFIC NAMES

Arrowhead dogfish = *Deania profundorum*; Atlantic guitarfish = *Rhinobatos lentiginosus*; Atlantic sharpnose shark = *Rhizoprionodon terraenovae*; Atlantic stingray = *Dasyatis sabina*; banded guitarfish = *Zapteryx exasperata*; basking shark = *Cetorhinus maximus*; bat eagle ray = *Myliobatis californica*; blacktip reef shark = *Carcharhinus melanopterus*; blue shark = *Prionace glauca*; blue skate = *Dipturus batis*; bluntnose sixgill shark = *Hexanchus griseus*; bluntnose stingray = *Dasyatis sayi*; bonnethead shark = *Sphyrna tiburo*; Brazilian electric ray = *Narcine brasiliensis*; Caribbean lantern shark = *Etmopterus hillianus*; chain catshark = *Scyliorhinus rotifer*; clearnose skate = *Raja eglanteria*; cloudy catshark = *Scyliorhinus torazame*; common stingray = *Dasyatis pastinaca*; cookie cutter shark = *Isistius brasiliensis*; cownose ray = *Rhinoptera bonasus*; deepsea ray = *Bathyraja interrupta*; dusky smooth-hound shark = *Mustelus canis*; eagle ray = *Aetobatus narinari*; electric ray = *Torpedo nobiliana*; flapnose ray = *Rhinoptera javanica*; giant manta ray = *Manta birostris*; Greenland shark = *Somniosus microcephalus*; Haller's round ray = *Urolophus halleri*; hook-tooth dogfish = *Aculeola nigra*; horn shark = *Heterodontus francisci*; kitefin shark = *Dalatias licha*; lemon shark = *Negaprion brevirostris*; little gulper shark = *Centrophorus uyato*; little skate = *Leucoraja erinacea*; longnose chimaera = *Rhinochimaera atlantica*; longnose velvet dogfish = *Centroscymnus crepidater*; Mexican horn shark = *Heterodontus mexicanus*; nurse shark = *Ginglymostoma cirratum*; Pacific longnose chimaera = *Harriotta raleighana*; pelagic stingray = *Pteroplatytrygon violacea*; pigmy shark = *Euprotomicrus bispinatus*; piked dogfish = *Squalus acanthias*; rabbit fish = *Chimaera monstrosa*; sandbar shark = *Carcharhinus plumbeus*; sand devil = *Squatina dumeril*; sandtiger shark = *Carcharias taurus*; scalloped hammerhead shark = *Sphyrna lewini*; shortfin mako shark = *Isurus oxyrinchus*; shortnose spurdog = *Squalus megalops*; shovelnose guitarfish = *Rhinobatos productus*; silky shark = *Carcharhinus falciformis*; small-spotted catshark = *Scyliorhinus canicula*; smalltooth sawfish = *Pristis pectinata*; South American freshwater stingray = *Potamotrygon motoro*; southern stingray = *Dasyatis americana*; spotted eagle ray = *Aetobatus narinari*; spotted ratfish = *Hydrolagus colliei*; starry ray = *Amblyraja radiata*; swellshark = *Cephaloscyllium ventrum*; thresher shark = *Alopias vulpinus*; thornback guitarfish = *Platyrhinoides triseriata*; thornback ray = *Raja clavata*; tope shark = *Galeorhinus galeus*; whale shark = *Rhincodon typus*; white shark = *Carcharodon carcharias*; whitetip reef shark = *Triaenodon obesus*.

LITERATURE CITED

1. Carrier *et al.*, 2004; 2. Wourms, 1977; 3. Pratt, 1979; 4. Gilbert and Heath, 1972; 5. Maruska *et al.*, 1996; 6. Pratt, 1988; 7. Pratt and Carrier, 2001; 8. Clark and Von

Discussion Questions

1. Contrast the live-bearing reproductive strategy of the cartilaginous fishes with that of the egg-laying bony fishes, weighing costs and benefits toward reproductive success.

2. Why do many species of sharks segregate by sex, while few species of bony fishes do the same?

3. The reproductive biology of the chimaeras is not well known. Much of what is known is from the classical monographs of Bashford Dean, written at the beginning of the twentieth century. What studies can you think of that are needed to provide a more complete understanding of the reproductive biology of this group of species?

Schmidt, 1965; 9. Gelsleichter *et al.*, 2002; 10. Metten, 1941; 11. Musick and Ellis, 2005; 12. Callard *et al.*, 1995; 13. Lutton *et al.*, 2005; 14. McConnaghey 2000; 15. Hamlett and Koob, 1999; 16. Didier, 2004; 17. Ebert, 2003; 18. Wourms, 1977; 19. Hamlett, 1993; 20. Hamlett and Hysell, 1998; 21. Gilmore *et al.*, 2005; 22. Conrath and Musick, 2002; 23. Walker, 2005; 24. Manire *et al.*, 1995; 25. Hueter and Manire, 1994; 26. Rasmussen *et al.*, 1999; 27. Luer and Gilbert, 1985; 28. Klimley, 1987; 29. Klimley, 1980; 30. Carrier, *et al.*, 1994; 31. Pratt and Carrier, 2005; 32. Saville *et al.*, 2002; 33. Johnson and Nelson, 1978; 34. Stevens, 1974; 35. Uchida, *et al.*, 1990; 36 Kajiura and Tricas, 1996; 37. McEachran and Konstantinou, 1996; 38. VanDykhuizen, 2011; 39. Dean, 1906; 40. Dempster and Herald, 1961; 41. Gordon, 1993; 42. Johnson and Nelson, 1978; 43. Clark, 1963; 44. Castro *et al.*, 1988; 45. Klimley, 1985a; 46. Tricas and Lefeuvre, 1985; 47. DeLoach, 1999; 48. Yano *et al.*, 1999; 49. Tricas, 1980; 50. Michael, 1993.

RECOMMENDED FURTHER READING

Carrier, J. C., H. L. Pratt, Jr., and J. I. Castro. 2004. Reproductive biology of elasmobranchs. Pp. 269-285 *in* Carrier, J. C., J. A. Musick, and M. R. Heithaus (Eds.), Biology of Sharks and Their Relatives. CRC Press, Boca Raton.

Carrier, J. C., H. L. Pratt, Jr., and L. K. Martin. 1994. Group reproductive behaviors in free-living nurse sharks, *Ginglymostoma cirratum. Copeia*, 1994: 646-656.

Hamlett, W. C., and T. J. Koob. 1999. Female reproductive system. Pp. 398-443 *in* Hamlett, W. C. (Ed.), Sharks, Skates, and Rays. Johns Hopkins University Press, Baltimore.

Kajiura, S. M., and T. C. Tricas. 1996. Seasonal dynamics of dental sexual dimorphism in the Atlantic Stingray, *Dasyatis sabina. Jour. Exp. Biol.*, 199: 2297-2306.

Klimley, A. P. 1987. The determinants of sexual segregation in the scalloped hammerhead shark, *Sphyrna lewini. Environ. Biol. Fish.*, 18: 27-40.

Klimley, A. P. 1980. Observations of courtship and copulation in the nurse shark, *Ginglymostoma cirratum. Copeia*, 1980: 878-882.

McEachran, J. D., and H. Konstantinou. 1996. Survey of variation of alar and malar thorns in skates (Chondrichtheyes: Rajoidea). *Jour. Morphol.,* 228: 165-178.

Musick, J. A., and J. K. Ellis. 2005. Reproductive evolution of chondrichthyans. Pp.45-79 *in* Hamlett, W. C. (ed.), Reproductive Biology and Phylogeny of Chondrichthyes: Sharks, Batoids, and Chimaeras. Science Publishers, Inc., Enfield.

Pratt, H. L. 1979. Reproduction in the blue shark, *Prionace glauca. Fish. Bull.*, 77: 445-470.

CHAPTER 12

Feeding Behavior and Biomechanics

I still vividly remember observing sharks feeding on northern elephant seals from the vantage of Lighthouse Hill on Southeast Farallon Island. I would rise well before dawn, put on my long underwear, jeans, heavy shirt, and jacket that kept me warm during the cool days of late autumn, and slowly ascend the hill from switchback to switchback until I reached the concrete platform at its peak before the sun rose above the horizon. Here was a small building below a large light that was slowly rotating, casting its beam in all directions to alert mariners of the presence of this insular hazard to navigation. I opened the door to the lighthouse, grasped the tripod with its mounted theodolite, an instrument used to determine the direction and distance to the predatory white shark and its prey, and placed it where it had nearly a complete view of the waters off the rocky coastline of this island roughly a mile across. I then placed next to it another tripod with a video camera equipped with a high-power zoom lens. Then came the long, almost interminable wait until a shark fed on a seal. Each hour I would count the number of seals in a series of twenty-four zones stretching around the entire island to determine whether the sharks fed more frequently along one part of the coast where more seals were more abundant on the beaches.

(a) White shark feeding on a northern elephant seal at the south Farallon Islands; (b) whale shark feeding off Cabo Catoche in the Yucatan.

Then abruptly my attention would be drawn to the cries of seagulls as they began to take off from the island. They would first circle the island and then move off in one direction toward something. Looking in that direction, I would see a large crimson pool of water not far from shore. The birds would eventually descend and hold their positions, wings fluttering, while they picked up small pieces of the seal that was being eaten. My automatic response was to quickly seize the video camera, rotate it while looking in the telescopic gun sight for the pool of blood, and once I found it to start recording. It was now essential to describe in words the events seen so that this narration would aid in interpreting the recorded images. Often the round pool of blood would become elongated in one direction, evidence that the shark was carrying its prey somewhere rather than taking a quick bite and promptly letting go of it. Then after a minute or two, the seal's carcass appeared at the surface with a large bite removed from its body. The seal was first beheaded in close to 40% of the predations observed during a four-year study at Southeast Farallon Island. This observation perplexed me until I reasoned that the seal had surfaced to breathe close to shore and afforded an unseen white shark enough time to swim underneath and accelerate upward to seize before it could submerge to make its final approach to the sanctuary of the island's beaches. What also struck me as odd was that little blood flowed out of the seal's large wound while it floated at the surface. This led me to hypothesize the white sharks might carry seals in their jaws until they had become exsanguinated (drained of their blood supply). This tactic of the predator would ensure that the prey died and would remain immobile on the surface and easy to locate and consume. Soon the shark appeared at the surface and slowly swam with the tip of its tail flopping back and forth up to the seal and consumed the rest of its body in a large bite.

Upon witnessing my first seal predation my knees grew weak and a cold sweat came to my forehead. I couldn't but help to emotionally identify with the seal when it was bitten by the shark—was it because the seal and I were similar in size and both of us were mammals? Later in the laboratory I analyzed the tapes of over a hundred such predations on seals and sea lions to develop an understanding of the shark's predatory tactics. At this time, I forced myself to look at the screen with the objectivity one would have looking a spider feeding on a moth. My time viewing these precious videos of white shark feeding behavior was an exciting opportunity for an ethologist such as myself to learn more about the behavior of this large marine predator that had heretofore rarely been witnessed feeding in its natural habitat.

The majority of cartilaginous fishes are not apex predators such as the white shark, which feeds on dolphins, seals, and sea lions near the top of the food chain. Most are tertiary consumers, occupying the same trophic level as the prey of the white shark. These species feed upon small fishes, cephalopods, and mollusks that in turn feed on zooplankton, the primary consumers of phytoplankton in the oceans. There are few species of planktivorous sharks and rays: only three sharks, the basking (*Cetorhinus maximus*), whale (*Rhincodon typus*), and megamouth (*Megachasma pelagios*) sharks; the giant manta (*Manta birostris*); and a few smaller rays of the family Mobulidae in the order Myliobatiformes. These planktivores swim with their mouths wide to draw plankton drifting in the current into their mouth cavities so that they can swallow them. Tiny animals are caught in baleen-like projections on the gill arches and then swallowed. Only one cartilaginous species, the cookie cutter shark (*Isistius brasiliensis*) is a parasite. These small sharks remove small plugs of fat from elephant seals, dolphins, and whales while they forage in the deep sea.

FILTER FEEDING ON PLANKTON

The anatomy of the feeding apparatus, diet, and mechanism of planktonic feeding is well described for the whale shark.[1] They often swim with their mouths open and their heads slightly above the water surface so that either part or all of the dorsum between the rostrum and first dorsal fin are exposed (see photograph at beginning of chapter). They swim at an average of 1.1 m/second, equivalent to the speed of a steady walk. Their mouth opening is elliptical with the width greater than the height. The flow around the lateral edges of the mouth creates a bow wave, and this forces more water to enter the mouth. Sometimes the sharks cough while feeding on the surface, expelling material forcefully out of their mouths. Whale sharks feeding off Cabo Catoche in the Yucatan fed at the surface mostly early and late during the day, rarely at night. They feed in this region mainly on sergestid shrimps, calanoid copepods, chaetognaths, and fish larvae.

APPROACH, SEIZURE, AND HANDLING PREY

The majority of cartilaginous fishes are active predators, either tertiary or quaternary consumers. There are three steps in active predation: approach, seizure, and handling of the prey.[3] A predator must first approach closely without its prey escaping. It then must then seize the prey. Either the prey is swallowed whole and passed downward into the esophagus or pieces of its flesh are removed by a predator using its teeth in a sawing action and swallowed separately.

There are three modes of approaching prey utilized by predators: speculation, ambush, and stalking, which may be used alone or in tandem.[4] The most

Phil Motta, a professor in the Department of Integrative Biology of the University of South Florida, and his colleagues described in detail the anatomy whale shark's feeding apparatus and the manner in which they feed, and determined just how much plankton they extract from the water column. Water is drawn into the mouth as they swim slowly, passed through filtering pads supported by the branchial arches in their pharyngeal cavity, and is expelled through their gills (fig. 12.1a). There are twenty large mesh-like filtering pads that completely occlude, or cover, the cavity, and retain most of the planktonic organisms that pass by them. The filtering pads in front of the mouth are elongated while the most posterior pad is triangular (fig. 12.1b). The mesh extends to either side of a series of parallel vanes on each pad (fig. 12.1c). The water passes through the reticulated mesh on either side of the primary and secondary vanes and is then directed toward the gill tissue for oxygen extraction (see rm, pv, sv, and gt in figs. 12.1d & 12.1e). The sharks spent nearly a third of each day feeding on the dense concentrations of plankton. Motta and his colleagues estimated, based on measurements of the gape of the mouth and concentrations of plankton in the water, that a whale shark 622 cm long would pass 614 m^3 of water through its filtering apparatus each hour.

This would enable this large shark to ingest 2763 g of plankton each day. This would result in a daily energetic ration of 28,121 kilojouls, comparable to a large meal including a couple of Big Macs and a large milkshake at a McDonald's restaurant.

Basking sharks are unlike the whale and megamouth sharks, which use suction in addition to forward motion to force plankton into the mouth. The former rely solely on forward movement to ram filter their prey. Due to energy loss from the high drag caused by opening their mouths wide and expanding their pharyngeal cavities (see photograph at beginning of chapter 1), there is a threshold prey density of 0.62 g of wet plankton per m^3, below which they should cease feeding because a net energy gain cannot be achieved.[2] This plankton density is much lower than the average 4.5g·m^{-3} recorded at the highly productive coast of the Yucatan, where offshore winds cause upwelling of nutrients that result in dense local plankton blooms. The planktivorous sharks and rays probably avoid feeding at an energetic loss by migrating to areas of high productivity in the ocean basins. This tendency to aggregate at biological hotspots likely makes it more possible to protect these magnificent animals on a global scale.

common form of hunting can be termed speculation. It consists of a predator searching in an area, where prey items are expected to be abundant. Tiger sharks (*Galeocerdo cuvier*) congregate in Shark Bay along the western shore of Australia during the summer months when their prey, sea snakes and dugongs, are common in the local waters.[5] The foraging efficiency of a predator increases if it minimizes the distance traveled in search of its prey—thus the predator will usually forage by speculation in a small area characterized by a high density of prey. Tiger sharks arrive during June and July at French Frigate Shoals in the northwestern Hawaiian Islands at the same time that Laysan albatross form a large colony and give birth to their young. When juvenile albatross first learn to fly, they occasionally fall in the water close to the shoreline. Tiger sharks aggregate in the waters near shore and rise to the surface to swallow the fledglings.[6]

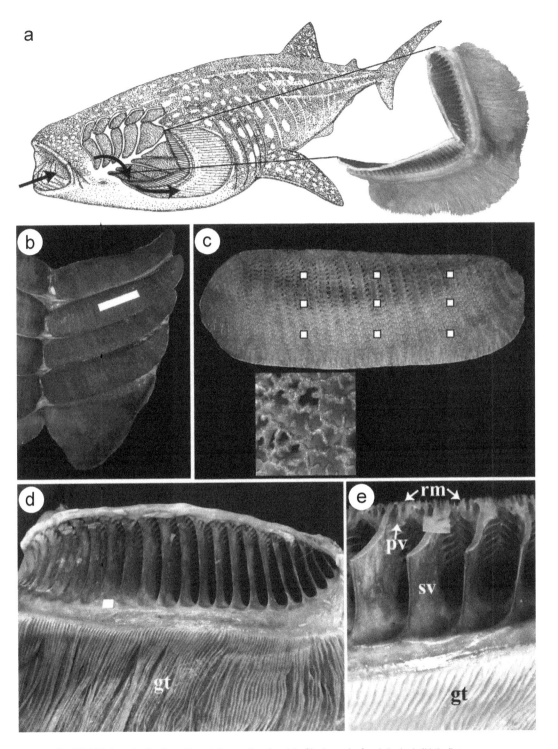

Fig. 12.1 (a) Schematic of pathway through the mouth and past the filtering pads of a whale shark; (b) the five filtering pads; (c) a single pad with an enlargement of the surface; (d) external view of the first upper left pad; (e) section of the pad showing the reticulated mesh. gt = gill tissue; pv = primary vane; rm = reticulated mesh; sv = secondary vane.

Speculative foraging is not confined to predators, but is also the mode of feeding of planktivores such as the whale shark. They aggregate each year during March and April off the central western coast of Australia to feed on the multitude of eggs expelled from coral polyps during their season of reproduction.[7] Similar aggregations of whale sharks form at coral reefs off the coast of Belize. Snappers of the family Lutjanidae and groupers of the family Serranidae converge there to spawn collectively, producing dense concentrations of highly nutritious eggs.[8]

An ambush predator conceals its presence while lying in wait for the prey to swim close.[4] This mode of predation is common among the dorsoventrally compressed angelsharks of the Squatiniformes and skates of the Rajiformes. Members of these orders remain motionless on the bottom with a color pattern matching the surrounding background. Pacific angelsharks (*Squatina californica*) lie partially buried in mud or sand with their terminal mouth oriented upward ready to draw in an above-swimming fish or squid. They do this by rapidly opening their mouths while expanding their branchial, or gill, chamber to create a negative pressure and draw their prey into their mouths.[9] These sharks prefer to lie on the bottom near reefs populated by small fish, where they can easily ambush unsuspecting prey. Individuals migrate between different hunting grounds at nighttime.[10] Ambushing is also the predatory tactic used by the Pacific electric ray (*Torpedo californica*) to stun and capture its prey. Individuals of this species lie motionless on the bottom during the day indistinguishable from background, from which they propel themselves upward by a few rapid beats of the tail and over a crustacean or small fish. They then bend their pectoral fins around it and discharge their electric organs.[11] The torpedo rays then swim over to an immobilized prey item and then rotate it using their pectoral fins so that it can be swallowed headfirst. At nighttime, the rays slowly drift above the bottom and lunge toward small fish swimming below. The rays then orient the underside of their electric organ–bearing pectoral fins toward the fish and discharge their organs to stun it with a debilitating electrical field (fig. 12.2).[12]

The third hunting strategy, stalking, is an elaboration of the ambushing strategy because the predator utilizes camouflage to approach the prey undetected and then rushes forward to make a sudden assault.[4] White sharks congregate close to shore at Southeast Farallon Island during October and November[13] after making long-distance migrations in the Pacific Ocean from as far away as the Hawaiian Islands.[14] Here they capture juvenile northern seals, one and two years old, as they converge upon a few beaches, which they colonize in great numbers. Observers stationed at the top of a hill on the island watched continuously for shark predation events during daylight over a period of five years during the fall season, when the juvenile seals are most abundant

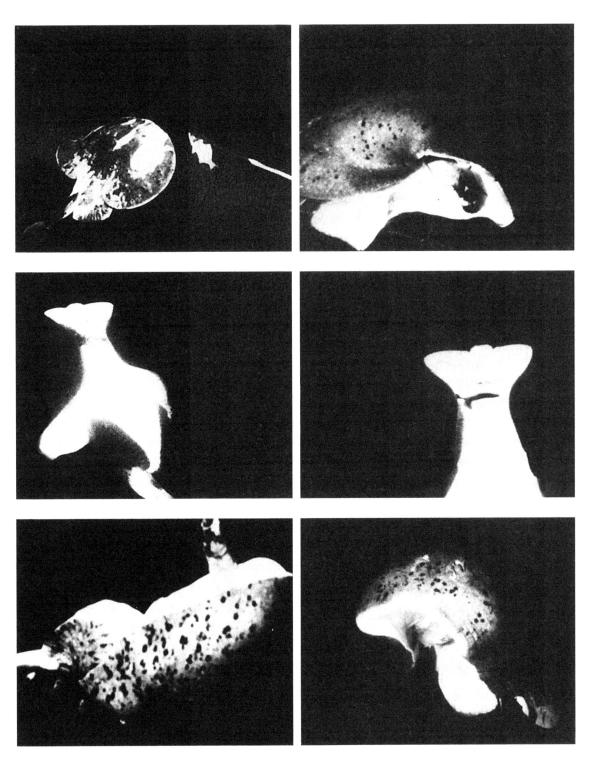

Fig. 12.2 Pacific electric ray shocking a fish during nighttime. (a) A female ray being presented a reef fish at the end of a spear; (b) posture of an electric ray after lunging for a prey fish; (c, d) a ray near the completion of a forward somersault having positioned the prey near its mouth with peristalsis-like foldings of the disk; (e) a jack mackerel partially enveloped in the disk of an upside-down ray; (f) the same ray at the completion of the somersault with the fish now fully enveloped and near the mouth.

on the beaches. More than 80% of the predatory attacks seen were in a zone from 25 m to 450 m from shore and were concentrated around points where elephant seals entered or left the water from their colonies (fig. 12.3a).[13]

The white sharks stalk seals and sea lions as they pass to and from their shore-based colonies. The common name "white shark," referring to the color of its underside, is misleading with regard to its mode of prey capture. The color of the shark's dorsum varies from solid to mottled gray, and this coloration matches the hues of the rocky bottom or the murky water below its prey. This camouflage enables white sharks to approach their prey close enough from below to see its silhouette against the surface. The seal is particularly vulnerable when making a dive after breathing at the surface. The shark, in a position underneath the seal, accelerates upward to seize it by the head as the seal begins its descent on its return to the island. Over 40% of seal carcasses observed after initial seizure by a white shark at the South Farallon Islands had bites on the head and anterior trunk of the body.[15]

Analysis of video records have provided evidence of how these predators immobilize and consume their prey (figs. 12.3b & 12.3c).[15, 16] A large, elongating bloodstain is often observed at the surface, and this indicates that the white sharks often carry their prey underwater over a great distance while they are bleeding profusely. The sharks appear to bite down when the seal is no longer bleeding in order to ensure that it is immobile and can not escape once released from their grip (fig. 12.3b). This mode of immobilization is termed exsanguination or blood deprivation. The shark is forced to let go of the seal after taking a bite. It is at this time that other sharks converge upon the carcass. They compete with the primary attacker for the prey by participating in a ritualized combat that involves splashing water on each other with their tails.[17] The sharks rarely immobilize California sea lions on the first strike because they manage to wrest themselves free from the shark's grip by biting and scratching (fig. 12.3c). The white sharks almost always need to seize them a second time to immobile them.

The shark seizes the prey with a sequence of behaviors that occur rapidly in comparison to the approach. The sequence typically takes less than a second and can be divided into four phases although they are continuous actions.[18] This process is illustrated with a time series of photographs of prey seizure by the Caribbean reef shark (*Carcharhinus perezi*) (fig. 12.4a). First, the mouth is opened slightly in a preparatory phase to pass water through its gills to extract oxygen. Second, the lower jaw is lowered while the head is raised in an expansive phase. The mouth is opened as the labial cartilages attached to the lower and upper jaws move apart and the branchial chamber is expanded to create an inward suction. Third, a compressive phase begins when the mouth is opened its widest. The lower jaw is now elevated and the upper jaw

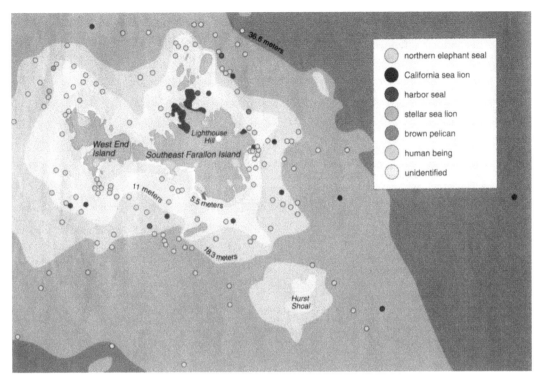

Fig. 12.3 (a) Spatial distribution of predatory attacks of white sharks on northern elephant seals and California sea lions in the coastal waters surrounding the south Farallon Islands; (b) predatory attack on a northern elephant seal that is immobilized after the first bite; (c) predatory attack on a California sea lion, which is not killed by the initial bite and must seized a second time to immobilize it.

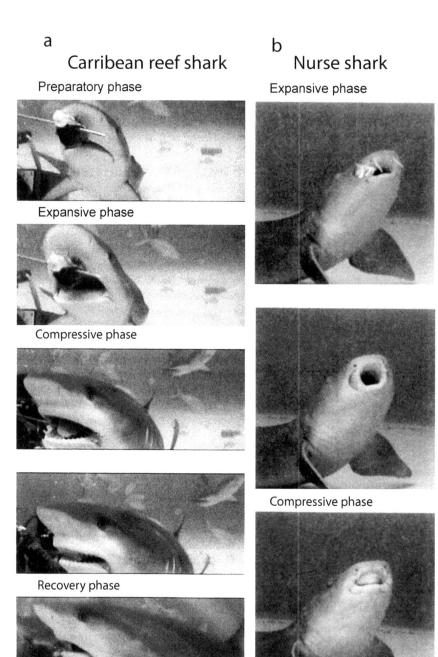

a
Carribean reef shark

Preparatory phase

Expansive phase

Compressive phase

Recovery phase

b
Nurse shark

Expansive phase

Compressive phase

Fig. 12.4 (a) Ram feeding by the Caribbean reef shark and (b) suction feeding by the nurse shark.

protruded forward and downward toward the lower jaw. At this time, the head is lowered. The prey is either seized between the teeth or drawn within the buccal, or mouth, cavity by the end of the compressive phase. Fourth, the upper jaw is retracted while the other elements in the jaw return to the original resting positions during the recovery phase. This ram feeding is the most common mode of feeding by the reef and hammerhead sharks in the order Carcharhiniformes and mackerel sharks in the Lamniformes.

Suction feeding is a less common mode of ingesting prey. It is utilized mainly by the carpet sharks of the order Orectolobiformes and the skates and stingrays of the Rajiformes and Myliobatiformes. This process is illustrated with a time series of photographs of prey seizure by the nurse shark (*Ginglymostoma cirratum*) (fig. 12.4b). A negative pressure or suction is generated in the branchial cavity to draw prey into the mouth. These sharks lower their jaws and lift their heads at the beginning of an expansive phase to draw prey into their mouth with a strong suction. The upper jaw is then protruded and moved downward toward the lower jaw during a compression phase. The suction feeders have a small mouth, little teeth, and large labial cartilages. They also have highly developed muscles for opening their mouths rapidly and expanding their buccal and branchial cavities.[18]

EVOLUTION OF JAWS

The extent of attachment of the jaws to the cranium has lessened over evolutionary time in the cartilaginous fishes. The ancient sharks had long upper and lower jaws that were attached to the braincase at two places.[19] These attachment points were above the nose in the ethmoidal region of the skull and behind the eye socket. Further support for the jaws was provided by the hyomandibula, which evolved from the first gill arch. It fit closely around the posterior margin of the palatoquadrate, or upper jaw, but was not physically attached to it by ligaments. This jaw structure is described as amphistylic due to the close association of the palatoquadrate and hyomandibula. Hence, the use of "amphi" from the Latin word *amplector* meaning "embrace" (see figs. 2.5a & 2.5b). The body and fins of these primitive sharks, whose mouth was at the terminus of the head, were not that different than those of modern sharks, and hence it is likely that they were fast and active swimmers capable of chasing and capturing their prey. Yet the three cusped teeth in the jaws of these sharks were useful only for seizing and swallowing small prey. They needed to shake their heads with the item held firmly in the jaws to tear off a piece of tissue from a larger animal.

The modern sharks, or neoselachians, have subterminal mouths located slightly underneath the snout. They possess shorter jaws that are attached less rigidly to the braincase and upper jaw. The upper jaw of the primitive squalomorph sharks has an ethmoid process, a protrusion that fits closely below the eye into a cavity in the wall of the orbital shelf of the neurocranium. There are two points of attachment between the upper jaw and the braincase, the ethmoidal and orbital articulations (fig. 12.5a). The lower jaw is attached to the

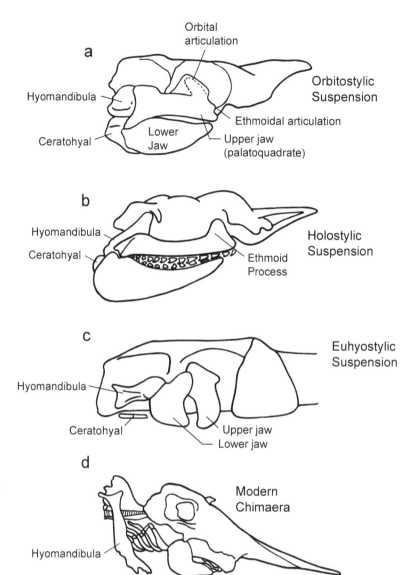

Fig. 12.5 Jaw types of sharks, rays, and chimaeras. (a) The primitive squalomorph sharks have an orbitostylic jaw suspension, (b) the advanced galeomorph sharks a hyostylic suspension, (c) the rays an euhyostylic suspension. (d) The chimaeras have jaws closely articulated with the braincase as in primitive sharks.

hyomandibula by a ligament, which permits the lower jaw to move back and forth relative to the upper jaw to saw off pieces of tissue from the prey. This type of jaw is called orbitostylic, referring to the attachment of the upper jaw to the orbit of the cranium.[18]

The ethmoid process on the upper surface of the upper jaw of an advanced galeomorph shark is attached to its braincase only by a ligament (fig. 12.5b). This permits the shark to protrude its upper jaw outward further during the compressive phase of prey seizure. The lower jaw is then moved back and forth while the upper jaw protrudes outward. This jaw mechanism, termed holostylic from the latin word for complete, enables the shark to saw off a large piece of tissue from its prey. The rays have the most flexible jaw structure. Their upper jaw is supported only by the lower jaw, which in turn is attached by a ligament to the hyomandibula (fig. 12.5c). This modified gill arch is able to move even more freely because it is not connected to the other gill arches. This is called euhyostylic, "eu" indicating that it is the "true" and "complete" form of jaw suspension. The chimaeras have jaws well adapted to seizing sessile invertebrates and crushing their exoskeletons to feed on them. The mouth of the chimaeras is subterminal. The upper jaw is articulated closely with the braincase along its entire length; the lower jaw is attached at the base of the braincase (fig. 12.5d). This high degree of articulation provides them with the ability to direct considerable force downward to crush hard objects such as crustaceans and mollusks.

BIOMECHANICS OF FEEDING

The biomechanics of feeding are better understood for the sharks than for the rays. The actions of the jaw during prey seizure are best described for ram-feeding carcharhinid sharks such as the lemon shark, *Negaprion brevirostris*.[20] The orbital process on the anterior end of the upper jaw of the lemon shark is attached loosely to the front of the neurocranium by the elastic ethmopalatine ligament (fig. 12.6a). The upper and lower jaws in the mouth of the lemon shark are attached to the hyomandibula, which is oriented at an angle directed backward and downward in relation to the neurocranium. The posterior end of the upper jaw is attached by the internal and external hyomandibular-palatoquadrate ligaments to the hyomandibula below its articulation with the neurocranium. The upper surface of the base of the lower jaw is attached by medial hyoid-mandibular ligament to the hyomandibula; the lower base of the lower jaw by the external hyoid-mandibular ligament. The mandibular knob on the lower jaw is braced against the distal end of the hyomandibula. Additional ligaments from the mandible to the ceratohyal and from the ceratohyal to the hyomandibula provide additional support.

The complex relationships between the skeletal elements and their attached muscles during jaw movement are best understood from a time series of diagrams in which each skeletal element is represented by a simple shape (fig. 12.7).[21] The skeletal elements and muscles, shown in fig. 12.6, are now indicated by polygons and solid lines. The muscles that are actively involved in each stage of jaw movement are identified in red and their contraction or relaxation indicated with small arrows. The directions to the movements of the skeletal elements are shown with large arrows. The shark initially opens its mouth by contracting its epiaxialis muscle (fig. 12.7a). At the same time the coracomandibularis, coracoarcualis, and coracohyoideus muscles contract,

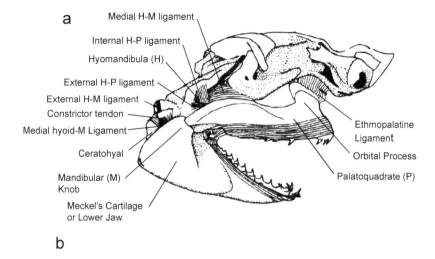

a

Medial H-M ligament

Internal H-P ligament

Hyomandibula (H)

External H-P ligament

External H-M ligament

Constrictor tendon

Medial hyoid-M Ligament

Ceratohyal

Mandibular (M) Knob

Meckel's Cartilage or Lower Jaw

Ethmopalatine Ligament

Orbital Process

Palatoquadrate (P)

b

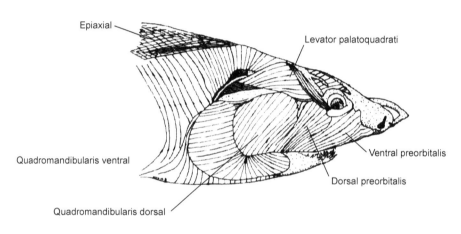

Epiaxial

Levator palatoquadrati

Quadromandibularis ventral

Ventral preorbitalis

Dorsal preorbitalis

Quadromandibularis dorsal

Fig. 12.6 (a) Side view of neurocranium and jaws of the lemon shark. (b) Side view of the head and muscles of the lemon shark. Note that the direction of the muscle fibers is indicated on the muscles, with contraction in a direction parallel to the parallel lines.

forcing the ceratohyal backwards and Meckel's cartilage downward to depress the lower jaw. The expansive stage is begun by the contraction of ventral preorbitalis muscle that extends from the anterior of the chondrocranium to the base of the palatoquadrate and the contraction of the levator palatoquadrati, which extends from slightly posterior on the braincase to slightly anterior on the palatoquadrate (fig. 12.7b). The contraction of these two muscles forces both jaws forward. At the same time, the contraction of the dorsal preorbitalis, which passes from a medial position in the upper jaw to the base of both jaws, forces the upper jaw outward and away from the braincase. The dorsal and ventral quadratomandibularis both contract to keep the ventral base of the two jaws together and in place. The orbital processes of the palatoquadrate are forced

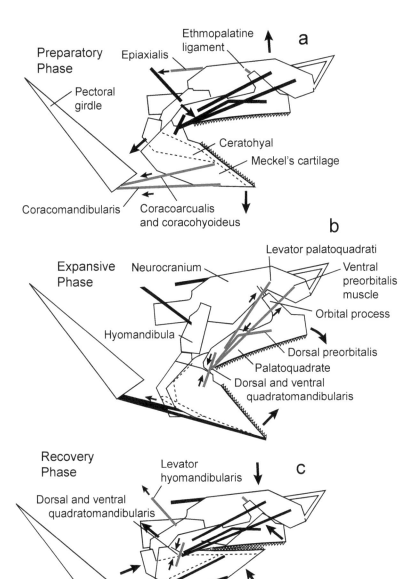

Fig. 12.7 A model of the movements of the jaws during (a) expansive, (b) compressive, and (c) recovery stages of prey seizure by the ram-feeding lemon shark. The thick dark lines indicate the muscles and the small arrows the direction of their contraction. Large arrows indicate the direction of movement of the elements of the jaw. The muscles whose movements are described in the text are identified by their red coloration.

forward, sliding on the ethmopalatine groove. The resulting force drives the upper jaw forward and the lower jaw downward. The cord-like ethmopalatine ligament becomes taut, and this halts any further jaw protrusion. Finally, the recovery phase begins as the palatoquadrate is retracted into its cranial seat (fig. 12.7c). This is caused by the contraction of the levator hyomandibularis, which draws the upper jaw inward and contraction of the ethomopalatine ligament, which draws the palatoquadrate and its orbital processes back into the ethmoid groove, while at the same time relaxing the other jaw muscles.

The feeding biomechanics of rays are best described for the suction-feeding Atlantic guitarfish (*Rhinobatus lentiginosus*).[22] Again the movements of the elements of the jaw in response to the contraction of muscles are indicated in a simple block diagram. The upper jaw is retracted by the contraction of the levator palatoquadrati muscle immediately prior to prey seizure during the preparatory phase (fig. 12.8a). The coracomandibularis contracts during the expansive stage and opens the ray's mouth with the upper jaw being raised and lower jaw being lowered (fig. 12.8b). Well into the expansive stage, the hyomandibula is drawn downward by the contraction of the coracohyomandibularis and depressor hyomandibula. Their combined action expands the buccal and branchial cavities, and the vacuum draws in water together with the prey item into the mouth. The lower jaw begins to elevate and the upper jaw protrudes the most during the compressive stage (fig. 12.8c). The protrusion of the jaws occurs in part in response to the contraction of the medial preorbitalis muscle, which is attached toward the anterior of the braincase and at the base of the lower jaw. At the same time, the quadratomandibularis, attached above and below the base of the two jaws, contracts to force the base of the lower jaw upward and protrude the upper jaw outward. Both outer jaws are then protruded downward from the braincase. As the medial preorbitalis muscle continues to draw the jaws forward, the upper jaw protrudes and the lower jaw rises until the two jaws come together. In the recovery phase, the palatoquadrate is pulled backward toward the cranium by the levator palatoquadrati and the hyomandibula is retracted by the levator hyomandibularis (fig. 12.8d). The chondrocranium is finally returned to its resting position by the contraction of the epaxialis levator rostri muscles.

ANATOMY AND FUNCTION OF TEETH

The composition of a shark's tooth is similar to that of the dermal denticle (see fig. 3.7b). The tooth consists of the mineralized tissue, dentine, interspersed with blood vessels and surrounded by highly mineralized enamel on the crown of the tooth.[23] This layer extends above the connective tissue in which the tooth is embedded in the jaw. Multiple teeth are aligned next to each other in rows on either side of the symphysis, the adjoining point between the left and right jaws. This is indicated by a dashed vertical line superimposed on the lower jaw of the piked dogfish (*Squalus acanthias*) (fig. 12.9a). There are fourteen teeth, indicated by numbers, distributed symmetrically on either side of the symphysis in the lower jaw.[24] There are eight rows of teeth, denoted by Roman numerals, in the same jaw. Each tooth begins its development as a dome-shaped dental papilla that is a projection in the dental lamina, a fold of epithelium lying along the inside edge of the lower and upper jaws.[23] The

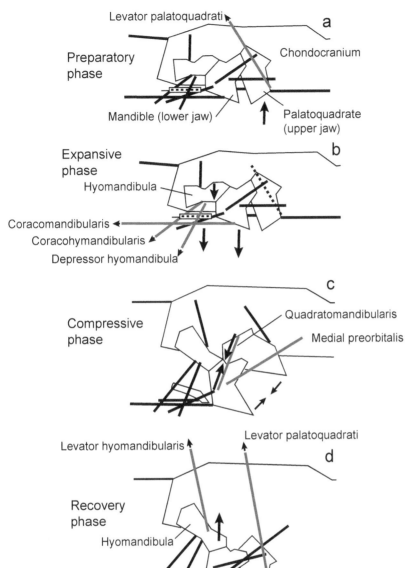

Fig. 12.8 A model of the movements of the jaws during the (a) preparatory, (b) expansive, (c) compressive, and (d) recovery phases of suction feeding by the Atlantic guitarfish.

developing tooth moves forward to become part of the next outermost row of teeth while becoming covered by a thin layer of dentine. A dome-shaped dental papilla then grows inside of the first row of teeth along the inside edge of the jaws that will become part of the next row of teeth. As each tooth moves toward the front of the jaw, its dentine layer increases in thickness and becomes surrounded by an enameloid layer. This outer layer increases in thickness as the tooth moves forward in the jaw. A cross-section of the lower jaw of the blacktail reef shark (*Carcharhinus amblyrhynchos*) reveals the procession of

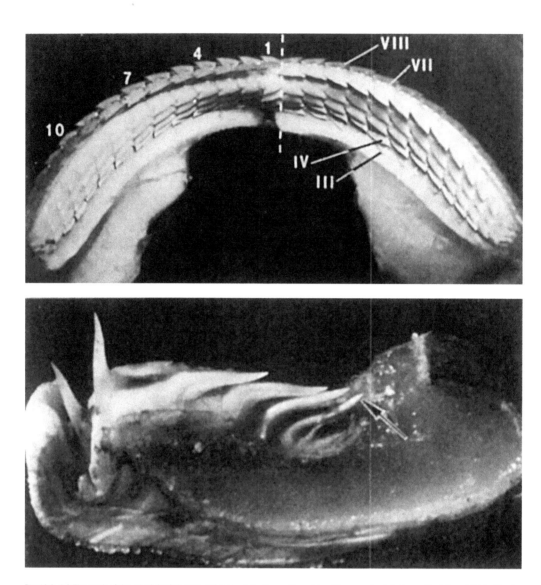

Fig. 12.9 (a) The teeth of the piked dogfish arranged symmetrically on either side of the symphysis between the left and right side of jaw. The number of teeth in each side of the jaw indicated with Arabic numbers; the successive rows denoted by Roman numerals. (b) A cross-section of a single file of teeth from successive rows from the lower jaw of the blacktail reef shark. Note that the most anterior tooth has dropped into the trough between the jaws and the lips of the shark, and would likely have been shed before long.

teeth of advancing stages in growth toward the front of the jaw (fig. 12.9b).[25] An arrow on the jaw draws attention to the mineralized crown of a developing tooth that lies in a flattened position pointing backward in the jaw. The currently functioning tooth is oriented upward. The outermost tooth often falls forward and recedes into the trough between the jaw and lip of the shark. The number of tooth rows in the right and left jaw varies from 1 per jaw in some rays to as many as 300 in the planktivorous whale shark. In most sharks there are 20–30 rows of teeth.[18]

The base of each tooth in a shark's mouth is embedded within the connective tissue covering the cartilaginous jaw. It is continually pulled forward as it matures and the connective tissue migrates anteriorly.[23] The functional teeth eventually become detached from the jaw, either falling to the bottom of the sea or becoming embedded within the tissues of a prey item. The teeth are replaced by the next generation of teeth behind them. The rate of tooth replacement, or the time taken to replace one row by the next row, varies among species. The shedding rate range is 8–10 days per row in the lemon shark, 9–12 days in the leopard shark (*Triakis semifasciata*), 9–28 days in the summer and 51–70 days in the winter for the nurse shark, 28 days for the horn shark (*Heterodontus francisci*).[18] Tooth loss occurs mainly during feeding. A fragment of a tooth may break when it fails to penetrate the tough flesh or hard bone of the prey. Alternatively, a whole tooth may be stripped from the shark's jaws upon retraction because it has become lodged within either tissue. Variation in rate of tooth loss depends on how frequently the shark or ray feeds and the softness or hardness of the tissues of its prey. For example, the tissues of cephalopods are softer and less sinewy than the muscles of bony fishes. Hence, one might expect less tooth loss in members of species feeding mainly on the former and more loss for species that prey largely on the latter. In truth, the rate of tooth loss varies not only with species but also with age and sex. This is because the diets of males and females or individuals of different age cohorts vary due to their occupying separate habits. Even the diets of these individuals can change with time of year as the abundance of a particular prey may be available only during one season with certain water temperatures. Whole teeth and tooth fragments of white sharks are often embedded in the muscular tissues of sea otters that drift ashore in Northern California.[26] A white shark may release the otter because of its low energy value,[16] yet its teeth still break off embedded in the tough muscular tissue of rejected prey item.

Many of the rays have teeth adapted for crushing prey either with a hard exoskeleton such as crustaceans or hard shell such as bivalves. The bat eagle ray (*Myliobatis californica*) has a central row of large, thick, flattened, hexagonal teeth flanked on either side by three lateral rows of smaller crushing teeth. The spotted eagle ray (*Aetobatus narinari*) has a single file of pavement-like fused teeth in both its lower and upper jaws (fig. 12.10).[18] The teeth fuse together as they move forward in the jaws. The bat eagle ray has three to ten rows of unworn teeth behind three rows of functional teeth toward the anterior of the jaws. The posterior teeth move forward as the anterior teeth become detached from the jaw. The anterior lower teeth in the spotted eagle ray move out of the crushing zone yet remain attached to the plate. They protrude from the jaw to form a spade-like appendage used to excavate prey from the substrate.

The teeth of Paleozoic sharks were adapted for piercing the skin, grasping

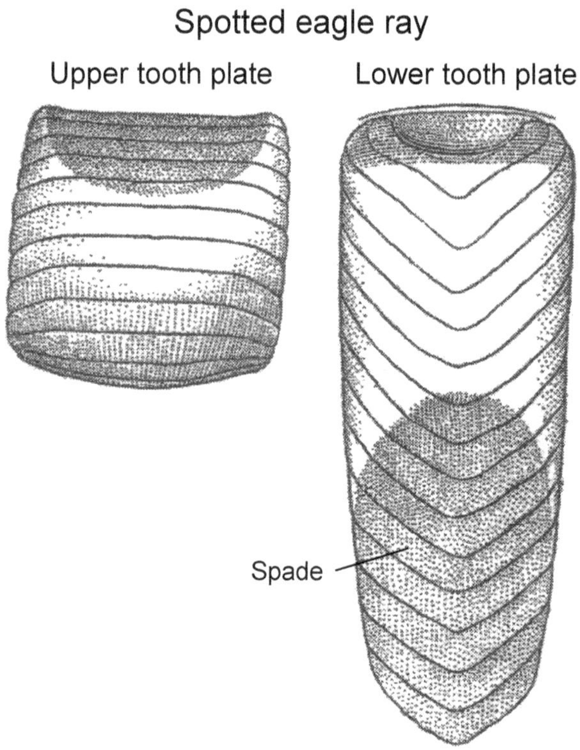

Spotted eagle ray

Upper tooth plate   Lower tooth plate

Spade

Fig. 12.10 Upper and lower plates of fused teeth in the spotted eagle ray. Notice the extension of the lower plate that protrudes from the mouth and can be used to dislodge prey buried in the substrate.

the prey tightly, and removing chunks of tissue by shaking their head back and forth vigorously. The teeth of the oceanic sharks of the genus *Cladoselache* had a broad base with a large central cusp flanked by smaller cusps on either side. The diplodus teeth of the freshwater sharks of the genus *Xenacanthus* had a central cusp reduced in size and flanked by two enlarged outer cusps. The majority of the hybodonts and ctenacanths had two or more large, pointed cusps for seizing prey with their teeth. However, some hybodonts also possessed flattened teeth with many smaller cusps in the posterior of their jaw. These pavement-like teeth were presumably used for crushing prey. This heterodonty, or duality in tooth type, is expressed in modern species of sharks, in particular in horn sharks of the order Heterodontiformes. The jaws of the horn sharks contain pointed, maple leaf–shaped teeth in the anterior of their jaws that are used to grip their prey and flat and molar-like teeth in the posterior of their jaws to crush it (fig. 12.11a).

There is great diversity to the teeth of modern sharks. The teeth that are used to grasp prey prior to swallowing it are usually small and multicusped and arranged in many rows. These are found in the carpet sharks of the order Orectolobiformes such as the nurse shark (fig. 12.11b)[18] and many rays that feed on crustaceans, gastropods, and bivalves that are protected with a hard external

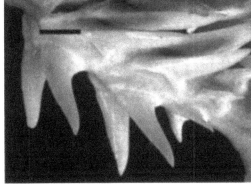

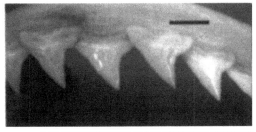

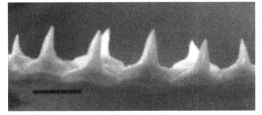

Fig. 12.11 Shark teeth adapted for different modes of feeding. (a) Seizing anterior and crushing posterior teeth of the horn shark; (b) many small rows of triangular teeth used in suction feeding by nurse shark; (c) long and slender teeth with cusps oriented backward for seizing and swallowing prey intact of shortfin mako; (d) upper teeth for cutting and (e) lower teeth for holding prey of sandbar shark.

shell.[27] Teeth used for seizing prey and forcing them quickly into the mouth are usually long and slender with sharp cusps that are oriented backward such as in the shortfin mako shark (*Isurus oxyrinchus*) (fig. 12.11c), and the sandtiger sharks of the genus *Odontaspis*. Two distinct types of teeth are needed to hold prey firmly in the jaws while sawing off a chunk of tissue. There are usually teeth with a pointed cusps and smooth sides for impaling and gripping the prey in either the upper or lower jaws. On the opposite jaw are triangular teeth with serrated sides for sawing off chunks of tissue. The former jaw is held stationary while the latter is moved back and forth as the head is shaken. Many of the primitive squalomorph sharks have the gripping teeth on the upper jaw and

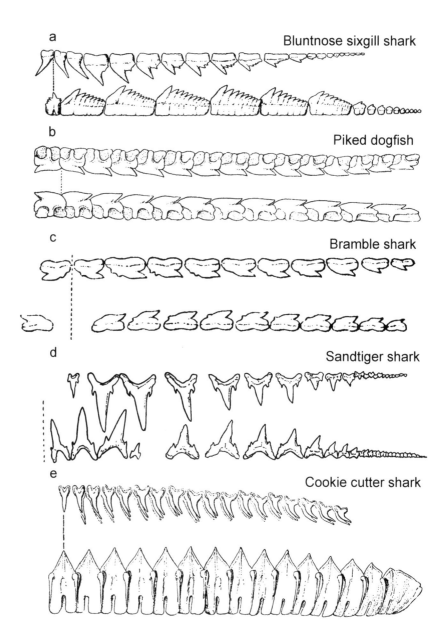

a Bluntnose sixgill shark

b Piked dogfish

c Bramble shark

d Sandtiger shark

e Cookie cutter shark

Fig. 12.12 Arrangement of teeth from the side of the jaw to the center of five species of sharks.

sawing teeth on the lower jaw. Most of the more advanced galeomorph sharks such as the sandbar shark (*Carcharhinus plumbeus*) have gripping teeth on the lower jaw and the sawing teeth on the upper jaw (figs. 12.11d & 12.11e).

However, there is an even greater diversity to the shapes of sharks' teeth.[28] The bluntnose sixgill shark (*Hexanchus griseus*), a squalomorph, has large pointed cusps on the teeth in the upper jaw for holding the prey in place and many cusp-like serrations on the teeth in the lower jaw for sawing off pieces of prey (fig. 12.12a). The piked dogfish has closely interlocking teeth with a single

Most sharks cut pieces of flesh from their prey with serrated triangular teeth. The tooth is drawn across an object on the horizontal plane (see large arrow, fig. 12.13a).[18] The force (F) exerted against the object by the tooth can be separated into a force in the x plane (F_x) and y plane (F_y). These forces are opposed by a direct force exerted by the object against the tooth surface (σ) and a shear force (τ) [see arrows] toward the distal tip of the tooth. The combined action of these two forces results in the object being deflected toward the tip of the tooth. As serration a comes into contact with the object, the stress is concentrated on the tip of the serration (fig. 12.13c). This results in penetration of the tooth margin into the object as it does for serrations b and c. The object is then severed as it moves across the edge of the tooth by the combined direct and shear force.

The teeth of the tiger shark are unique among the cartilaginous fishes. This species has serrations on the anterior and posterior margins of the teeth in both the upper and lower jaws with an acute-angled notch on the distal end of each crown (fig. 12.13b). The serrated surface of the more curved side of the teeth likely slices through soft tissue as prey moves across the serrations (fig. 12.13e). A reverse force is exerted by the notched side of the tooth, which serves to tear through harder substances such as collagen, cartilage, and bone.[18] An object encountered on the shoulder or tip of the tooth (locations 1 and 2) is driven toward the notch, which is narrow and thin, producing a cutting action. This is similar to that produced by scissors with a notch at the base of the two blades. The same tooth applies an additional slicing force as the jaw is moved in the across the curved and serrated surface on the other side of the tooth (location 3). The teeth on the right jaw are a mirror image of those on the left jaw. Hence, both cutting and slicing forces can be exerted simultaneously as the upper and lower jaws are moved back and forth relative to each other. Tiger sharks cut through the fish fed to them in captivity with fewer movements of the head than lemon and sandbar sharks. The Hawaiians were aware of the effective cutting action of the shark's tooth and made knives by fastening large teeth such as those of the tiger and white shark to a wooden handle (fig. 12.13d).[29]

cusp on both the upper and lower jaws (fig. 12.12b). The teeth of the bramble shark (*Echinorhinus brucus*) are similar, but spaced farther apart in each jaw (fig. 12.12c). The teeth of the sandtiger shark (*Carcharias taurus*) have long pointed cusps like those of the shortfin mako shark, and they are likely used to seize and hold prey after a rapid chase (fig. 12.12d). Finally, the cookie cutter shark, also a squalomorph, has pointed upper teeth for gaining purchase while it moves its sharp triangular teeth in its lower jaw back and forth to dislodge of plug of flesh from a seal or whale (fig. 12.12e).

SUMMARY

The majority of cartilaginous fishes are not apex predators such as the white shark, which feeds on dolphins, seals, and sea lions near the top of the food chain. Most sharks, rays, and chimaeras are tertiary consumers, occupying the same trophic level as the prey of the white shark. These species feed upon

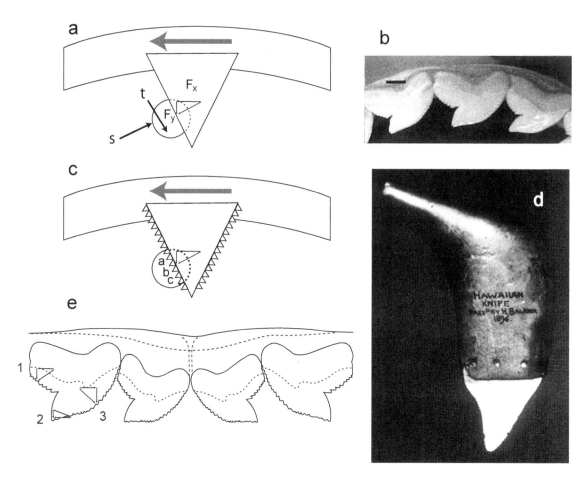

Fig. 12.13 Proposed mechanism for the cutting action of serrated and notched shark teeth. Model of cutting for (a, c) triangular and (e) notched teeth. (b) Photograph of teeth of tiger shark; and (d) knife with white shark tooth serving as a cutting surface used by Hawaiian warriors.

small fishes, cephalopods, and mollusks that in turn feed on zooplankton, the primary consumers of phytoplankton in the oceans. There are only a few species of planktivorous sharks and rays: three sharks, the basking, whale, and megamouth sharks; the giant manta; and a few smaller rays of the family Mobulidae. These species swim with their mouths wide to draw plankton drifting in the current into their mouth cavities so that they can swallow them. Tiny animals are caught in baleen-like projections on the gill arches and then swallowed by these planktivores. Only one cartilaginous species, the cookie cutter shark, is a parasite; it removes small plugs of fat from elephant seals, dolphins, and whales while they forage in the deep sea.

There are three steps in active predation: approach, seizure, and handling of the prey. A predator must first approach closely enough to seize its prey without its escaping. There are three modes of approaching prey: speculation, ambush, and stalking, which may be used alone or in tandem. Prey seizure occurs more rapidly than approach. The sequence typically takes less than a second and can be divided into four phases, although they are continuous action in the ram feeders. Suction feeding, used mainly by the carpet sharks, skates, and stingrays, is a less common mode used to ingest prey.

The extent of attachment of the jaws to the cranium has lessened over evolutionary time in the cartilaginous fishes. The ancient sharks needed to swallow their prey whole due to the rigid attachment of the jaws to the cranium. The jaws of the modern sharks trend toward a less rigid attachment to the cranium. The upper jaws of the less advanced modern sharks have two attachment points between the upper jaw and the cranium, but the lower jaw is attached to the hyomandibula by a ligament. This permits both the upper and lower jaws to separate more than the jaws of ancient sharks. The upper and lower jaws of the more advanced modern sharks and rays are attached to the cranium and the hyomandibula, respectively, only by a ligament. This loose support permits the upper jaws to be distended outward and opened wide in order to remove chunks of flesh from the prey. The combination of this mechanism with two types of teeth, one set for gripping and the other for sawing, enables sharks to cut through flesh with relative ease. The teeth of many rays, on the other hand, are adapted for crushing prey either with a hard exoskeleton such as crustaceans or hard shell such as bivalves. The behavioral tactics, biomechanics of jaw suspension, and dentition of the sharks, rays, and chimaeras have all contributed to their success in the modern oceans.

* * *
** ** **

KEY TO COMMON AND SCIENTIFIC NAMES

Atlantic guitarfish = *Rhinobatus lentiginosus*; basking shark = *Cetorhinus maximus*; bat eagle ray = *Myliobatis californica*; blacktail reef shark = *Carcharhinus amblyrhynchos*; bluntnose sixgill shark = *Hexanchus griseus*; bramble shark = *Echinorhinus brucus*; Caribbean reef shark = *Carcharhinus perezi*; cookie cutter shark = *Isistius brasiliensis*; dusky smooth-hound shark = *Mustelus canis*; horn shark = *Heterodontus francisci*; lemon shark = *Negaprion brevirostris*; leopard shark = *Triakis semifasciata*; megamouth shark = *Megachasma pelagios*; northern elephant seal = *Mirounga angustirostris*; nurse shark = *Ginglymostoma cirratum*; Pacific angelshark = *Squatina californica*; Pacific electric ray = *Torpedo californica*; piked dogfish = *Squalus acanthias*; sandbar shark = *Carcharhinus plumbeus*; sandtiger shark =

1. Getting close enough to seize the prey is the classic problem that must be solved by the predator. Review the diversity of the feeding behavior for the sharks, rays, and chimaeras.

2. Look through Bigelow and Schroeder (1948),[28] which illustrates tooth series for a diversity of sharks. Make inferences from the shape of the teeth about the feeding tactics of those species with very unusual tooth shapes based on knowledge of their feeding habits found in Compagno *et al.* (2005).[30]

3. Suppose that you are going market a set of knives using teeth of sharks and rays for their cutting surface. What species' teeth would be best for cutting meat? What species' teeth would be most useful for cutting bread? What teeth could be used in a knife used to cut a filet of fish? What teeth might be used in a knife used to make a puree?

Carcharias taurus; sea otter = *Enhydra lutris*; shortfin mako shark = *Isurus oxyrinchus*; spotted eagle ray = *Aetobatus narinari*; whale shark = *Rhincodon typus*; white shark = *Carcharodon carcharias*.

LITERATURE CITED

1. Motta *et al.*, 2010; 2. Sims, 1999; 3. Motta and Wilga, 2001; 4. Curio, 1976; 5. Heithaus, 2001; 6. Kooyman, 1985; 7. Gunn *et al.*, 1999; 8. Graham *et al.*, 2005; 9. Fouts and Nelson, 1999; 10. Standora and Nelson, 1977; 11. Lowe *et al.*, 1994; 12. Bray and Hixon, 1978; 13. Klimley *et al.*, 1992; 14. Boustany *et al.*, 2002; 15. Klimley *et al.*, 1996a; 16. Klimley, 1994; 17. Klimley *et al.*, 1996b; 18. Motta, 2004; 19. Benton, 2005; 20. Motta and Wilga, 1995; 21. Motta *et al.*, 1997; 22. Wilga and Motta, 1998; 23. Kemp, 1999; 24. Samuel *et al.*, 1989; 25. Kemp and Park, 1974; 26. Ames *et al.*, 1996; 27. Moss, 1977; 28. Bigelow and Schroeder, 1948; 29. Taylor, 1985; 30. Compagno *et al.*, 2005.

RECOMMENDED FURTHER READING

Kemp, N. E. 1999. Integumentary system and teeth. Pp. 43–68 in Hamlett, W. C. (Ed.), Sharks, Skates, and Rays: The Biology of Elasmobranch Fishes. Johns Hopkins University Press, Baltimore.

Klimley, A. P., S. D. Anderson, P. Pyle, and R. P. Henderson. 1992. Spatiotemporal patterns of white shark (*Carcharodon carcharias*) predation at the South Farallon Islands, California. *Copeia*, 1992: 680–690.

Klimley, A. P., P. Pyle, and S. D. Anderson. 1996a. The behavior of white sharks and their pinniped prey during predatory attacks. Pp. 175–191 in Klimley, A. P., and D. G. Ainley (Eds.), Great White Sharks: The Biology of *Carcharodon carcharias*. Academic Press, San Diego.

Lowe, C. G., R. N. Bray, and D. R. Nelson. 1994. Feeding and associated electrical behavior of the Pacific electric ray *Torpedo californica* in the field. *Mar. Biol.*, 120: 161–169.

Motta, P. J., R. E. Hueter, T. C. Tricas, and A. P. Summers. 1997. Feeding mechanism and functional morphology of the jaws of the lemon shark, *Negaprion brevirostris* (Chondrichtheyes, Carcharhinidae). *J. Exp. Biol.*, 200: 2765–2780.

Motta, P. J., M. Maslanka, R. E. Hueter, R. L. Davis, R. de la Parra, S. L. Mulvany, M. L. Habegger, J. A. Strother, K. R. Mara, J. M. Gardiner, J. P. Tyminski, and L. D. Zeigler. 2010. Feeding anatomy, filter-feeding rate, and diet of whale shark, *Rhincodon typus*, during surface ram filter feeding off the Yucatan Peninsula, Mexico. *Zoology*, 113: 199–212.

Motta, P. J., and C. D. Wilga. 1995. Anatomy of the feeding apparatus of the lemon shark, *Negaprion brevirostris*. *J. Morphol.*, 226: 309–329.

Wilga, C. D., and P. J. Motta. 1998. Feeding mechanism of the Atlantic guitarfish, *Rhinobatus lentiginosus*: modulation of kinematic and motor activity. *J. Exp. Biol.*, 2001: 3167–3184.

CHAPTER 13   *Diet and Growth*

A traverse section (left) and longitudinal section (right) of a vertebral centrum of a lemon shark.

Upon my arrival at Scripps Institution of Oceanography in the mid 1970s, Dick Rosenblatt, my doctoral supervisor, asked me to remove a few vertebral centra from one of the largest white sharks caught off the Californian coast then on display at the Sea World Marine Park. He wanted me to estimate the age of this massive female shark over 500 cm from the snout to the tip of the tail—over 17 feet long. She had been preserved in a frozen state in a glass case for public viewing but was now beginning to lose her shape from desiccation, and the display was scheduled to be dismantled. I slowly sawed off half a dozen vertebrae using a hacksaw instead of a scalpel because each vertebral centrum was large, 7 cm across, or 2.75 inches wide—and much larger than the vertebrae of a human being. I looked closely at the largest centrum and noticed many concentric rings like those in the cross-section of a tree. Could each ring indicate one year of growth? However, the vertebrae that I held in my hand were composed of alternating dark and light bands unlike the tree. Dr. Rosenblatt recommended that I soak the largest vertebra in a solution of silver nitrate overnight. This treatment resulted in a clearer distinction between the light and dark bands, particularly when the vertebra was illuminated with ultraviolet light. The light bands really were narrower than the dark rings. The latter bands are thought to be impregnated with more calcium phosphate than the former ones resulting in a denser composition. Might these dark bands be deposited in the late summer and fall? White sharks aggregate at islands along the coast of central California, where they might grow rapidly from feeding on abundant seals and sea lions. Could the lighter rings be deposited when the sharks leave the islands and migrate out into the northwestern Pacific Ocean, where they might feed less often, or visa versa? The mantra at that time was that in temperate and boreal latitudes, thick opaque bands were deposited during the summer, when fish were in warm water, and thin translucent ones during the winter, when the fish were in cold water. What amazed me about this large female was that she had only thirteen pairs of light and dark growth rings. Given one pair of growth band was deposited each year, she reached a large size quite rapidly—about the time taken for a human to mature into a teenager. The idea that they might grow to this size so quickly still bothers me. Maybe, on the other hand, one ring is deposited each year. Supporting this possibility is the following recent published observation. A male white shark with a conspicuous notch on the posterior margin of his dorsal was photographed in the waters near a seal colony in central California in 1987 and sighted again eight times over a period of twenty-two years with its total length estimated to be 4.5 m in 2008.[1] Five white sharks were identified over periods ranging from sixteen to twenty-two years based on photographs of their fins. Some of these sharks were likely four to six years old when first photographed. You must be careful not to simply assume that each pair of these alternating dark and light rings equals one year of growth. Perhaps that large shark was not thirteen years old but twenty-six years old. That would be true if one growth ring were deposited each year in the species. It is really necessary to establish whether two bands are deposited per year with independent indicators of annual growth. Two traditional and one especially innovative technique of age verification will be described later in the chapter.

CARTILAGINOUS FISHES IN THE FOOD CHAIN

While the last chapter was devoted to explaining how cartilaginous fishes capture and swallow their prey, this chapter will reveal what they feed upon, how often they feed and digest their food, their rates of growth, and to what age they live. Perhaps it is best to preface a discussion of diet and growth with a brief introduction to trophic ecology. The plants and animals in the ocean can be classified into different levels, depending on what they consume. On the lowest level are a myriad of single-celled microscopic plants, phytoplankton, which have the pigment chlorophyll in their cells. This converts the light impinging upon them to carbon compounds while absorbing water and carbon dioxide.

Phytoplankton are the largest producers of biomass, living organic material, in the ocean, and all productivity, or growth of populations of animals on higher levels, depends on the size of the bottom level of the trophic pyramid. Planktonic crustaceans, coelenterates, and larval fish, which feed on phytoplankton, comprise the second level, and are considered the primary consumers in a food chain. Mostly small plankton-feeding fish, the secondary consumers, make up the third trophic level, and larger fish and cephalopods, the tertiary consumers, make up the fourth stratum. As you learned in the last chapter, there are three species of sharks—the basking shark (*Cetorhinus maximus*), the megamouth shark (*Megachasma pelagios*), and the whale shark (*Rhincodon typus*)—as well as several mobulid rays of the order Myliobatiformes that are mainly secondary consumers. The tertiary consumers are among the largest predators in the ocean. They are the seals and sea lions, dolphins, largest bony fishes such as the tunas and billfishes, and finally many of the cartilaginous fishes. A few very large sharks, such as the bull (*Carcharhinus leucas*), tiger (*Galeocerdo cuvier*), and white (*Carcharodon carcharias*) sharks, are quaternary consumers, forming the fifth level, along with the killer whale (*Orca orcinus*). These big sharks are the apex predators. At each higher level, there is less biomass. This relationship can be imagined by thinking of each of the successive trophic levels as a step on a sloping pyramid, the first level being the broad base, the next upward layer slightly smaller, the layer above even smaller, and so on until at the top is the pointed apex of the pyramid. Most sharks and rays are tertiary consumers, on the fourth trophic level, feeding on intermediate-sized fishes and squid.

These relationships are not actually as distinct as they might appear. Species feed at different times on prey on a number of trophic levels. Hence, the levels assigned the species in particular families are not integers, but decimal values indicating that they do not feed on prey from one trophic level all of the time. This is apparent from a table of the average trophic level of species in the different families and orders determined from multiple dietary studies on a diversity of species of sharks and rays (table 13.1). [2,3] Note that the bramble shark (*Echinorhinus brucus*) of the family Echinorhinidae and the sandtiger shark (*Carcharias taurus*) of the family Odontaspidae have the highest mean trophic level of 4.4, and this is probably because they have a less diverse diet, feeding mainly on cartilaginous and bony fishes high on the trophic ladder. The basking shark of the family Cetorhinidae and bullhead sharks of the family Heterodontidae have very low mean trophic levels of 3.2. The former feeds primarily on copepods and krill and other planktonic species low on the trophic spectrum, while the latter feed mainly on bivalves, filter feeders of phytoplankton that are also low on the trophic scale. Yet the majority of elasmobranchs have high mean trophic levels, either slightly above or below the forth

TABLE 13.1. Trophic levels of sharks and skates by order and family, where information is available from studies of food contents. Numbers of species and weighted mean trophic levels of orders are given in bold type.

Order	Family	Common Name	Species (N)	Mean Trophic Level
Hexanchiformes			**5**	**4.3**
	Chlamydoselachidae	Frilled sharks	1	4.2
	Hexanchidae	Cow sharks	4	4.3
Squaliformes			**32**	**4.1**
	Echinorhinidae	Bramble sharks	1	4.4
	Squalidae	Dogfish sharks	31	4.1
Squatiniformes			**6**	**4.1**
	Squatinidae	Angelsharks	6	4.1
Pristiophoriformes			**1**	**4.2**
	Pristiophoridae	Sawsharks	1	4.2
Heterodontiformes			**1**	**3.2**
	Heterodontidae	Bullhead sharks	1	3.2
Orectolobiformes			**6**	**3.6**
	Ginglymostomidae	Nurse sharks	2	4.0
	Hemiscyllidae	Longtailed carpet sharks	2	3.6
	Rhincodontidae	Whale shark	1	3.6
	Stegostomatidae	Zebra shark	1	3.1
Lamniformes			**9**	**4.0**
	Alopiidae	Thresher sharks	3	4.2
	Cetorhinidae	Basking shark	1	3.2
	Lamnidae	Mackerel sharks	3	4.3
	Megachasmidae	Megamouth shark	1	3.4
	Odontaspididae	Sandtiger sharks	1	4.4
Carcharhiniformes			**90**	**4.0**
	Carcharhinidae	Reef sharks	39	4.1
	Hemigaleidae	Weasel sharks	2	4.2
	Proscyllidae	Finback catsharks	2	4.1
	Pseudotriakidae	False catsharks	1	4.3

TABLE 13.1. *continued*

Order	Family	Common Name	Species (N)	Mean Trophic Level
	Pseudotriakidae	False catsharks	1	4.3
	Scyliorhinidae	Catsharks	21	3.9
	Sphyrnidae	Hammerhead sharks	6	3.9
	Triakidae	Houndsharks	19	3.8
Rajiformes			**60**	**3.8**
	Anacanthobatidae	Leg skates	1	3.5
	Arhynchobatidae	Longtail skates	19	3.9
	Rajidae	Skates	40	3.8

level, ranging from 3.6 for the carpet sharks of the order Orectolobiformes to 4.3 for the cow and frilled sharks of the order Hexanchiformes. A point to be made is that the skates from the different families in order Rajiformes are on an equivalent mean trophic level, ranging from 3.5 to 3.8, as some of the smaller species of longtailed carpet sharks in the family Hemiscyllidae and houndsharks in the family Triakidae.

DIETARY COMPOSITION

The composition of the diet obviously varies among cartilaginous species, yet the diet of individual species changes as they grow larger and migrate from one geographical region to another. The dietary preferences have been quantified for numerous species of elasmobranchs. There is insufficient space in this volume to devote to a comprehensive description of the dietary habits of every chondrichthyan species, and to explain how their dietary preferences change as they grow from juveniles to adults. You find many references to articles on the dietary habits of different species of sharks in Wetherbee and Cortès (2004) and of skates in Ebert and Bizzarro (2007), Matta and Gunderson (2007), Rinewalt *et al.* (2007), and Robinson *et al.* (2007).

Let us take one species, the white shark, and examine in detail its ontogenetic, or developmental, changes in diet as it grows from a juvenile to an adult. As white sharks grow older, they move from the warm waters off southern California to the cold waters off central California (fig. 13.1a).[4] This is readily apparent from a plot of the latitudes at which sharks of different sizes are cap-

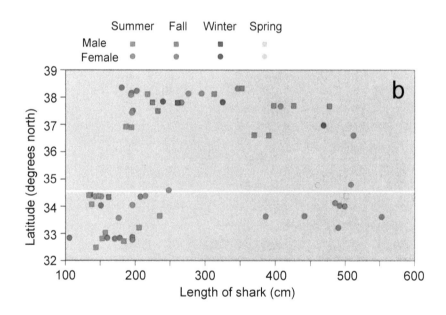

Fig. 13.1 (a) The movements of white sharks from southern to central California as they grow older. (b) Plot of the length of male and female white sharks as a function of latitude. The sex and season of capture are designated by symbols of different shapes and colors. The horizontal line on the plot denotes the latitude of Point Conception.

tured (fig 13.1b). The recently born sharks with a length of roughly 150 cm are captured at latitudes ranging from a little less than 33°00′ to 34° 30′ N south of Point Conception. They are captured during the summer and fall as is apparent from the green and brown symbols. Rather than the points forming a line with a 45° slope, as would indicate that they were slowly moving northward as they grew larger, they form a 90° slope, which indicates a sudden movement northward from Ventura County (34°20′ N) to Monterey Bay (37°00′ N) at a size of 200 cm and age of two years. This movement likely occurs in the late summer and fall, judging from the equal numbers of green and brown points on the graph for individuals caught along the coastline from Monterey to Tomales Bay (37°00′ to 38°30′ N). The large males stay north of 37°00′ N as they grow to a size of 476 cm, as evident from the brown square farthest to the right on the graph. However, the females move southward for at least some time as they reach a length of approximately 380 cm. This is apparent from the downward trend of the capture points over a size range of 380 to 550 cm total length. Note that the circles are colored green and brown due to their being caught during summer and fall when the adult females are at the same latitudes as the newly born sharks. This leads one to believe that the females make a southward migration to give birth to their young in the warm waters of southern California. Direct support for this inference has recently been forthcoming from the record of a satellite tag placed on a large female at Southeast Farallon Island. The record indicated that this female made a southward migration to Catalina Island off the coast of southern California.[5]

Newly born white sharks forage off the western coast of North America along submarine canyons. These are present near La Jolla, Carlsbad, Santa Monica, and Ventura in southern California. These juveniles feed mostly on the Pacific rock crab (*Cancer antennarius*), stingrays, and cabezon (*Scorpaenichthys marmoratus*), all species that live near the bottom (fig. 13.2b).[6] However, they also pursue prey up in the water column such as squid and juvenile dusky smooth-hound sharks (*Mustelus canis*). Small white sharks chase down and seize their prey with pointed teeth resembling those of the shortfin mako shark (*Isurus oxyrinchus*). The young white sharks, from 150–200 cm long, spend two and half years in the waters of southern California before migrating north of Point Conception into the waters of northern California. White sharks of an intermediate size, 200–400 cm long, arrive at central California. They continue to feed on crustaceans and fishes near the bottom as they did as juveniles (fig. 13.2a). However, they can now feed on larger midwater fishes because their mouth is larger. It is possible for them to swallow large bony fishes whole, such Chinook salmon (*Onchorhynchus tshawytscha*), ling cod (*Ophiodon elongatus*), rockfish (*Sebastes* sp.), and green sturgeon (*Acipenser medirostris*), as well as sharks such as adult brown smooth-hound (*Mustelus henlei*), dusky smooth-hound, and tope (*Galeorhinus galeus*). Their teeth become triangular and serrated as they reach a size of 400 cm. At this stage of development, they begin to reside at Año Nuevo, South Farallon Islands, and the Point Reyes Headlands during late summer and fall. At this time, they ambush California sea lions (*Zalophus californianus*), harbor seals (*Phoca vitulina*), and northern elephant seals (*Mirounga angustirostris*) as they return to their coastal colonies.[7]

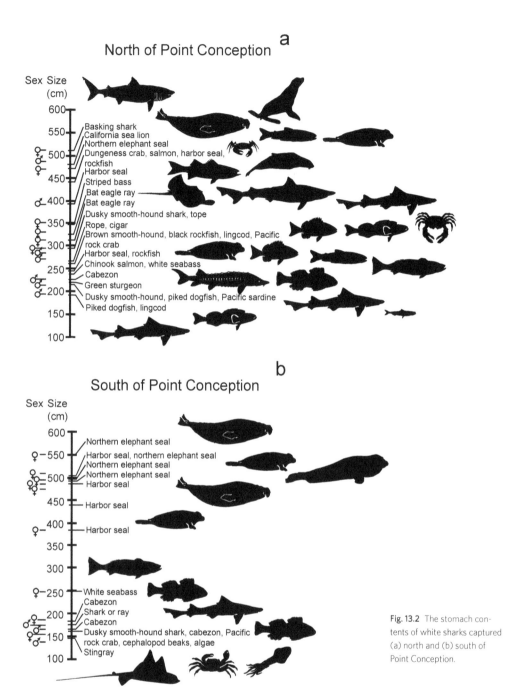

North of Point Conception ᵃ

Sex Size (cm)

- 600
- 550 — Basking shark
  - California sea lion
  - Northern elephant seal
- ♀♀♀ 500 — Dungeness crab, salmon, harbor seal,
  - rockfish
  - Harbor seal
- ♀ 450 — Striped bass
  - Bat eagle ray
- ♂ 400 — Bat eagle ray
  - Dusky smooth-hound shark, tope
- ♀ 350 — Rope, cigar
  - Brown smooth-hound, black rockfish, lingcod, Pacific
- ♀♀♀ 300 — rock crab
  - Harbor seal, rockfish
  - Chinook salmon, white seabass
- 250 — Cabezon
  - Green sturgeon
- ♂♂ 200 — Dusky smooth-hound, piked dogfish, Pacific sardine
  - Piked dogfish, lingcod
- 150
- 100

South of Point Conception ᵇ

Sex Size (cm)

- 600 — Northern elephant seal
- ♀ 550 — Harbor seal, northern elephant seal
  - Northern elephant seal
- ♀♀ 500 — Northern elephant seal
  - Harbor seal
- ♀♀♀
- 450 — Harbor seal
- ♀ 400 — Harbor seal
- 350
- 300
- ♀ 250 — White seabass
  - Cabezon
- ♂♀ 200 — Shark or ray
  - Cabezon
- ♀♀♀ 150 — Dusky smooth-hound shark, cabezon, Pacific
  - ♂ rock crab, cephalopod beaks, algae
  - Stingray
- 100

Fig. 13.2 The stomach contents of white sharks captured (a) north and (b) south of Point Conception.

This dietary shift to marine mammals, whose fatty insulation makes up almost 50% of their body mass and contains twice the energy content of the protein-based muscle of crustaceans and fish, may enable the adult white sharks to more grow rapidly.[4,8] You can appreciate the benefit from this dietary shift if you plot the yearly average growth of individuals of the species as

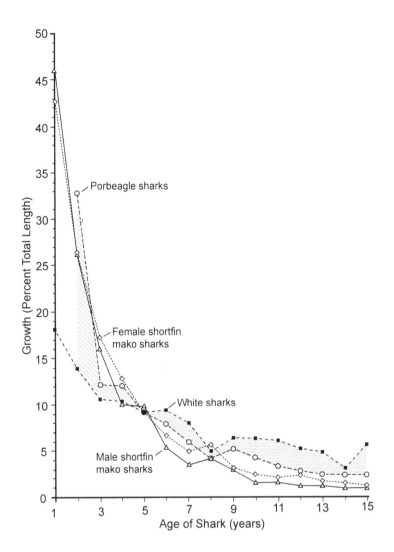

Fig. 13.3 The relative growth rates of white, porbeagle, and male and female shortfin mako sharks. Note that juvenile white sharks grow more slowly than juvenile porbeagle sharks (see coarse cross-hatching indicating difference in growth rates). White sharks grow faster than the porbeagle sharks once the former are feeding on the energy-rich fat of marine mammals and the latter on less nutritious salmonids (see fine cross-hatching).

a percentage of the largest size to which individuals grow relative to their age (fig. 13.3).[9] The rates of growth were determined by counting the number of paired growth rings of sharks with known lengths for three species, the white shark, the porbeagle shark (*Lamna nasus*), and the shortfin mako shark (*Isurus oxyrinchus*). The white shark grows at an average annual rate of 6% of its body length throughout its adulthood.[10] The porbeagle shark, which feeds on less energy-rich salmon in similarly cold temperate and polar waters, grows only at a rate of 2.5% of its body length per year once an adult.[11] Male and female shortfin mako sharks, which are also piscivorous but live in warm water, grow at an annual rate of 2.0% of their body length upon reaching adulthood.[12] This is apparent from the dense stippling between the curves for white sharks (solid squares), and for the porbeagle (clear circles). The curves for the male and female mako sharks (clear triangles and diamonds) are even lower over this range of growth. On the other hand, young porbeagle and mako sharks grow more rapidly than white sharks when the diet of all three species is composed of fishes. This is evident on the graph from the less dense stippling between the growth curves of the porbeagle and white shark over the first three years of their lives. Of course, this comparison is based on the assumption that one pair of growth bands is deposited per year for all three species. Bomb dating has been used to verify that one pair of growth bands is deposited every year in the vertebrae of porbeagle and shortfin mako sharks.[13] This method takes advantage of the incorporation within the vertebrae of cartilaginous fishes of radioactive carbon that rapidly increased in the oceans from the detonation of atomic bombs in the 1950s and 1960 in the earth's atmosphere. This technique will be explained in greater detail later in the chapter.

The diet of sharks also varies with sex. For example, male and female hammerhead sharks live apart from each other much of the time. Hence, they feed on different prey.[14] Females move offshore before males and form large schools at seamounts and islands, and thus feed on more pelagic prey than males. Females less than 160 cm long feed on a higher percentage of pelagic prey than do the males of the same size. Mesopelagic prey form 27.5% and epipelagic prey constitute 5.5% of the female diet, whereas the male diet from these two zones consists of only 18.1% and 3.6% of the total food intake. The females eat mainly squid in the mesopelagic habitat. Furthermore, only 15.1% of the female diet consisted of benthic prey in contrast to 40.9% in males. Most likely, the females occupy offshore seamounts and islands to be closer to the more abundant offshore prey yet experience greater risk of predation, while the males, who need not maximize growth for reproduction, remain in the predator-free inshore waters.

Feeding Frequency and Digestion Rate

The large sharks and rays are thought to forage in short bouts yet digest their prey over long periods of time.[15] They are believed to feed again only after the skeletons of the items are expelled through vomiting or evacuated in their feces. This is supported by the widespread observation of stomachs either empty or containing few prey items in a high state of digestion. The stomachs of juvenile sandbar[16] and lemon[17] as well as adult scalloped hammerhead sharks[14] often contain only a few hard parts such as the skeletons of fish and beaks of cephalopods. Such hard parts take a longer time to digest and evacuate than do soft tissues such as muscle and fat. Hence, their presence in the stomach is evidence of an extended period between meals.[18] If sharks were feeding continuously, their stomachs would contain many different prey items in varying states of digestion and their stomachs would rarely be empty.[15]

The skates forage more frequently than the larger sharks. The sandpaper skate (*Bathyraja kincaidii*), a deepwater ray that inhabits the continental shelf of central California, feeds mainly on euphausiids during periods of upwelling, when nutrients rise to the surface and cause plankton blooms.[19] It redirects its feeding to gammarid amphipods and shrimps when the current direction reverses and the waters are less productive over the shelf. Out of 138 specimens caught in a trawl, only 8 of the individuals (5.8%) had empty stomachs. The longnose skate (*Raja rhina*) is common over soft bottoms in central California in waters ranging from 15 m deep to depths as great as 532 m at the edge of the continental shelf. There is a shift in the diet of the skates as they grow larger.[20] Smaller skates eat small crustaceans; the larger individuals feed upon larger fishes and cephalopods. Of 618 stomachs examined of this species, only 55 (8.9%) were found to be empty. The large percentages of skates with prey in their stomach indicate that they feed more often that the larger sharks.

A foraging bout usually lasts for less than half a day in the larger sharks.[15] This is defined as the period of time necessary for a shark to locate, capture, and ingest its prey. This period has been determined by feeding prey of a known mass to sharks in the laboratory, forcing water into their stomachs to flush out the items at increasingly longer intervals, and recording the mass and appearance of the partially digested prey items as a percentage of their original mass. The time since swallowing an item can then be related to that item's degree of digestion when retrieved. The length of the feeding bout has then been calculated by subtracting the age of the least digested item from the most digested item. The duration of a foraging bout of a juvenile scalloped hammerhead shark ranges from nine to ten hours,[21] a bout of a sandbar shark seven to nine hours,[16] and a bout of a lemon shark eleven hours.[2]

The interval between foraging bouts is on the order of two weeks.[15] Feeding frequency can be estimated by recording the average time between the ingestion and the evacuation of prey in the feces and the proportion of empty stomachs observed in a sample. The frequency of feeding bouts of adult piked dogfish are 10–16 days.[23] The dogfish is a tertiary consumer, as are the majority of sharks, and feeds on a small fish or squid. These middle-sized sharks occupy the same trophic level as the seals, sea lions, and dolphins.[2] However, the large bull, tiger, and white sharks are quaternary consumers and, for that reason, the duration between their feeding bouts exceeds that of the smaller sharks and rays. An adult white shark, 4.6 m long, can satisfy its basic energy needs with an interval of 1.5 months between meals.[24] This interval was established by using calorimetry to ascertain the energetic value of the shark's meal and telemetry to record the metabolic energy expended while swimming. The size of the shark's meal was known because a 30 kg chunk of whale fat was recovered from its stomach. Its basal metabolic rate was calculated based on

The interval between feeding bouts is considerably shorter for juvenile than adult sharks. The juvenile interval is 95 hours for the sandbar shark[16] and 33 to 47 hours for the lemon shark.[17] These two species are ectothermic, with body temperatures the same as those of the waters in which they swim. You would expect them to have a lower metabolic demand than endothermic sharks. Thus, they should be able to feed less frequently than endothermic sharks. This indeed appears to be the case. Chugey Sepulveda, while a graduate student at Scripps Institution of Oceanography, fed chub mackerel with temperature-sensing transmitters concealed within them to three juvenile sharks offshore of La Jolla in the Southern California Bight.[26] He anticipated that the gastric temperature of each of these juveniles would drop as soon as it swallowed one of these fish with the surrounding cold water. One shortfin mako shark was fed a mackerel at the conclusion of the track at 16:40 hours in order to observe the magnitude of the decrease in stomach temperature. You can see

an abrupt dip in the temperature curve toward the end of the seven-hour record of the shark's stomach temperature (fig. 13.4a). The sea surface temperature remained unchanged over this period of time (see dotted line). He next determined the interval between feeding bouts from the depth and temperature records of another shortfin mako shark. This shark fed soon after its release. Its gastric temperature declined abruptly from 24.4° C to 22.5° C after diving to a depth of 30 m at 19:00 hrs. The same individual fed 20 hours later at 13:40 hrs the following day. This is evident from a sawtooth-shaped dip of the thick line, indicating a 5° drop in stomach temperature, on the lower axis on fig. 13.4b. It coincided with the icicle-shaped dip of a thin line underneath on the upper axis on fig. 13.4b, indicating the shortfin mako captured its prey at a depth of 100 m. This little shark was captured soon afterwards, and six Pacific sauries (*Colabis saira*) were found intact within its stomach. The mako had chased down and captured all of these prey in less than half an hour.

the change in its muscle temperature, recorded by a probe lodged in its musculature, as it passed between two water masses with different temperatures. A yearling elephant seal, a favored prey of white sharks off the western coast of North America, would have a body mass of about 140 kg. Given a fat content of 48%,[25] this individual would have 67 kg of fatty tissue in its blubber layer—enough to satisfy the basal metabolic needs of a white shark for three months. The white shark's ability to survive long periods without feeding is consistent with evidence that six white sharks rarely fed only on two seals over a period of a month at a seal colony based on continuous records of their swimming speed, depth, and their proximity to each other.[26]

Finally, how much is eaten during a foraging bout relative to the overall body mass? This can be calculated by formulating an energy budget or balance sheet for each species. The daily energy acquired from ingesting prey must equal the energy used in basal metabolic activity, energy lost either in urine or feces, and the surplus energy devoted to daily growth. The oxygen consumed by a shark swimming within a flow chamber can be measured to determine

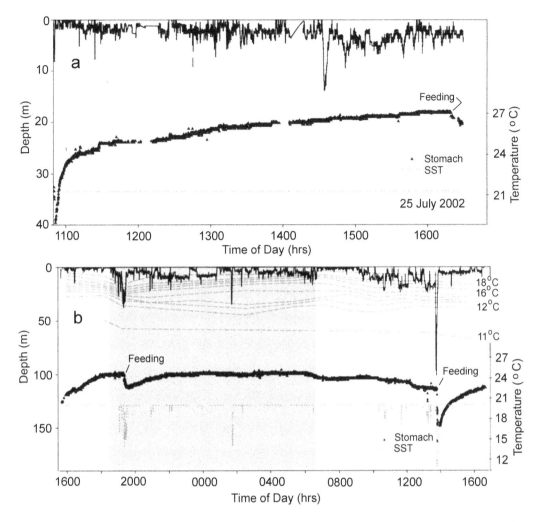

Fig. 13.4 Diving oscillations and stomach temperatures of (a) a juvenile shortfin mako shark of unknown sex that was tagged on 25 July 2002 and (b) a female tagged on 20 September in the Southern California Bight. The bold lines indicate the shark's diving excursions and the dashed lines temperature isocontours in the water column on the upper panels of the two graphs. The bold lines indicate the shark's stomach temperature and the dotted lines the sea surface temperature on the lower panels of the two graphs.

its metabolic energy consumption. The energy lost as feces can be estimated from the efficiency of digestion of the shark's stomach using inert chemical markers. The energy apportioned to growth can then be found by multiplying the energy value of muscle by the rate of growth. The amount of growth is determined from the increment in the number of growth rings counted on the vertebral centra of subjects sacrificed at the beginning and end of the experiment. The adult dogfish, a tertiary consumer, eats a daily ration of 1.3% of its body mass.[22] The juvenile sandbar shark consumes a daily ration of 0.9–1.3% of its body mass;[16] the juvenile lemon shark consumes a daily ration equivalent to 1.5–2.1% of its total body mass.[17] The six sauries recovered from the stomach of the juvenile mako shark weighed 0.4 kg, which would constitute 2.5% of its 16 kg body mass.[26] A quaternary consumer, which is considerably larger than a tertiary consumer, may have a smaller daily ration relative to body mass. For example, a bull shark was estimated to have a daily ration that was 0.5% of its body mass.[27] The southern stingray (*Dasyatis americana*), a tertiary consumer, has a daily ration of 2.5% of its body mass.[28] The plownose chimaera (*Callorhinchus callorynchus*) has a daily ration of 1.4% of its body mass.[29]

ANATOMY OF STOMACH AND INTESTINE

The frequency with which cartilaginous fishes feed is largely influenced by the anatomy of one particular organ, the intestine.[15] Its length is short in comparison to that of the bony fishes, marine turtles, and marine mammals; it rarely exceeds the length of the stomach, against which it lies in the peritoneum or body cavity. Furthermore, it is not coiled as in other marine species. These features are apparent in a diagram and photograph of the stomach and the intestine of the blue skate (*Dipturus batis*) (fig. 13.5).[30] The compact intestine conserves space within the body and leaves extra room for a large liver. The liver, full of less dense lipids, provides cartilaginous fishes with extra buoyancy so they are almost neutrally buoyant in water. Furthermore, the diminutive size of the intestine saves space for a large uterus. Thus, developing embryos can be nourished to a substantial size so they can be released at birth into the environment as fully developed living young.[15] This life history trait, along with the evolution of the cartilaginous skeleton and multiple rows of teeth, are in large part responsible for the evolutionary success of the cartilaginous fishes.

The absorptive surface of the intestine is enhanced along its length by the presence of spiral folds of the mucosa, or mucus-producing tissues that carry out digestion. The anatomy of the stomach differs among the sharks, rays, and chimaeras. All of the squalomorph and many of the galeomorph sharks have an interior surface of the intestine that consists of a flattened spiral mucosum. This tissue, with an enzyme-secreting digestive surface, winds around a central column attached to the outer wall of the stomach (fig. 13.6a).[30] The spiral mucosum of the holocephalans, such the rabbit fish (*Chimaera monstrosa*), has only two or three flattened spiral folds attached to the stomach wall. Members of the hammerhead family Sphyrnidae have a mucosum with folds that form a scroll lengthwise within the intestine (fig. 13.6b).[30] The central margin of the mucosum is free and the outer border attached along the length of the stomach wall. The rays of the skate family Rajidae have a mucosum consisting of a series of interconnecting conical folds (fig. 13.6c).[30] They wind around the central column and are attached to the outer wall of the intestine. The folds of the mucosum provide increased surface area for digestion but at the same time

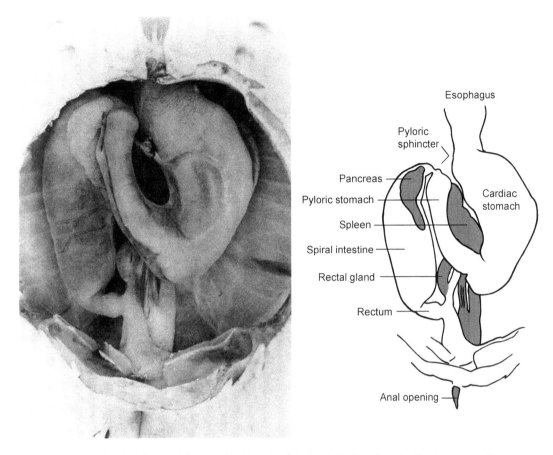

Esophagus

Pyloric
sphincter

Pancreas

Pyloric stomach

Spleen

Spiral intestine

Rectal gland

Rectum

Cardiac
stomach

Anal opening

Fig. 13.5 (a) Photograph and (b) diagram of the stomach and intestine of the blue skate. The peritoneal wall has been removed to expose the organs.

delay the passage of the digestible items through the stomach. They therefore likely increase the time necessary for the breaking down and absorption of prey tissue. The spiral intestine of the lemon shark absorbs 80% of the energy and nutrients that pass through it, and this digestive efficiency is similar to that of carnivorous bony fishes.[31, 32] The time that it takes a meal to pass through the intestine of a lemon shark was determined by feeding a juvenile lemon shark a meal of a blue runner filet containing barium sulfate powder.[33] You can see where the filet is in the stomach and spiral intestine from the white image of the radio-opaque barium on images taken at five successive intervals after it was consumed (fig 13.7). The powder was not apparent after 84 h, when the remains of the blue runner had been fully evacuated from the gut. This is considerably more time than the 24 h periods needed by most carnivorous bony fishes to digest their prey. The folded nature of the intestine likely confers to the cartilaginous fishes both advantages and disadvantages. Its compact nature results in more space available in the peritoneum to accommodate a larger

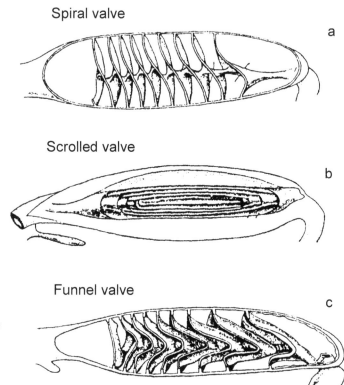

Spiral valve

a

Scrolled valve

b

Funnel valve

c

Fig. 13.6 The anatomy of intestines of the cartilaginous fishes. Shown are (a) the spiral valve of most squaloform and galeoform sharks, (b) the scrolled valve of the hammerhead sharks, and (c) the funnel valve of the rays.

liver and uterus. Yet the economy in the mucosal surface results in a slower digestion rate and prolongs the interval to the next foraging bout. The ultimate impact of this profound tradeoff is slow growth and a lengthy period of time to maturity, and this lowers the reproductive output of cartilaginous fishes relative to other marine species.

GROWTH RATE AND LIFE SPAN

The age of cartilaginous fishes can be determined by counting the number of alternating layers of dense and less dense calcium phosphate in their vertebrae and dorsal spines. These alternating bands are best seen in a section of a vertebral centrum by cutting transversely on a horizontal plane or longitudinally on a vertical plane (fig. 13.8a).[34] A transverse section of a vertebral centrum of a lemon shark (*Negaprion brevirostris*) is shown on the right of the illustration at the beginning of the chapter and a longitudinal section from the same shark on the left. There are on the longitudinal section wide and narrow bands, which extend from the wide arm of the corpus calcareum on one side, across the narrow intermedialia at the center, to the wide opposing arm of

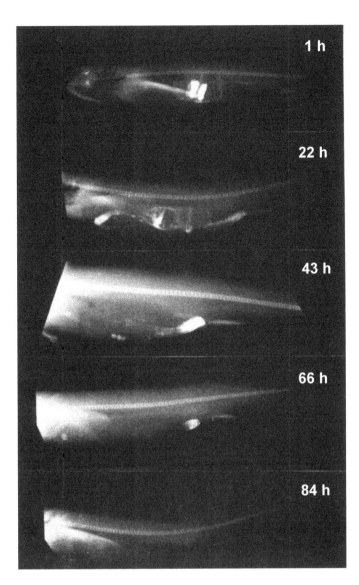

Fig. 13.7 Radiographs of a juvenile lemon shark fed a meal of a blue runner filet containing barium sulfate powder taken at five successive intervals after feeding. The white image is not apparent after 84 h, indicating the gut is empty because the remains of the blue runner have been evacuated.

the corpus calcareum. The bands on a dorsal or caudal fin spine are usually revealed by making a sagittal section (fig. 13.8b). It is generally believed that water temperature regulates the rate of deposition of calcium phosphate in the centrum. Narrower and less dense rings that are translucent are deposited when the water is cold during winter[34] or spring.[35] Wider and denser, opaque rings are depositing when the water is warm during summer.[34] The cyclic seasonal changes in water temperature thus result in the deposition of pairs of concentric growth rings of alternating densities. The age of a cartilaginous fish is usually determined from a count of the narrow bands. One of these translucent bands is termed an annulus.

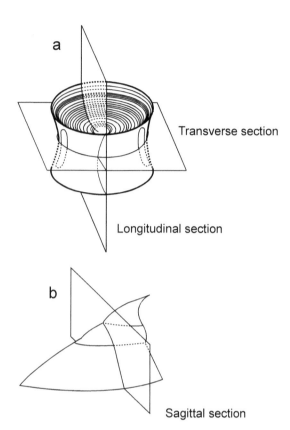

Fig. 13.8 (a) The two sectioning planes, transverse and longitudinal, from which the age of a cartilaginous fish can be determined from its vertebral centrum. (b) The single plane, sagittal, used in sections of a caudal thorn of a skate.

a

Transverse section

Longitudinal section

b

Sagittal section

Verifying Age

To age a cartilaginous fish with certainty, it is necessary to have a time reference within the centrum. This is made possible by injecting oxytetracyline, a general antibiotic, into the muscle of an individual. Here it binds to calcium at the site of active calcification on the centrum, fin spine, or tail thorn. The substance produces a highly visible ring, which can be readily identified when the hard part is viewed under ultraviolet light. An external tag is simultaneously affixed to the individual's body and it is returned to the sea. Upon recapture, the number of concentric growth rings outside the marked ring are counted and related to the time spent by the individual while at large to determine the rate of growth. This technique has been used to age many species of cartilaginous fishes. It has been successfully used to validate that one pair of rings is deposited each year on the centrum of the leopard shark (*Triakis semifasciata*). A marked individual was captured nineteen years after an injection with oxytetracycline.[36] The shark was captured, injected with oxytetracycline, and released in 1979 and was caught again in 1998. Twenty white bands are visible on a photograph of a longitudinal section of a centrum from that shark (fig. 13.9a).

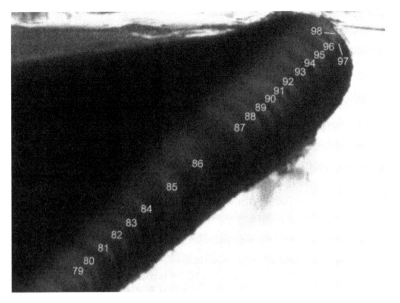

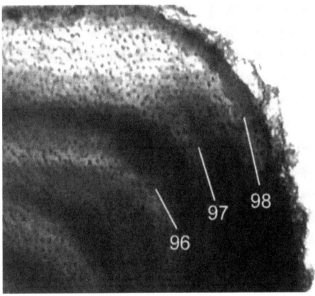

Fig. 13.9 (a) Transverse section of a centrum of a leopard shark with twenty translucent bands identified by year over a twenty-year period from 1979 to 1998. (b) A photographic enlargement of the outermost part of the centrum showing the translucent rings associated with the final three years of growth.

The translucent annuli are seen more clearly in a photographic enlargement of the outermost three rings deposited from 1996 to 1998 (fig. 13.9b).

Marginal increment analysis is another technique used to establish whether the bands are deposited on an annual basis. A measurement is made of the distance from the outer edge of the last fully developed annulus to the centrum's edge of the newly forming pair of growth rings. This is divided by a measurement of the width of the last fully deposited growth ring for sharks captured during all months of the year. You can then calculate what fractional increment

Steven Campana, a fisheries biologist at the Bedford Institute of Oceanography in Dartmouth, Nova Scotia, and
his colleagues used a third, innovative method for verifying
the age of sharks. They utilized an increase in radiocarbon
in a calcified structure, the vertebrae, as a dated marker
for age determination.[13] Radiocarbon (^{14}C) increased in
concentration in the world's oceans as a result of the detonation of atomic bombs during the 1950s and 1960s. This
period of increase can be used as a time reference in calcified structures exhibiting growth bands. The growth bands
are shown for a 264 cm long porbeagle shark collected in
1999 (fig. 13.10). White circles indicate growth rings that
are presumed to be deposited on an annual basis. The
alternating growth bands near the growing edge are very
thin so they are shown in an insert. The estimated age of

this individual, based on the band count, was at minimum
26 years (see white dots in fig. 13.10a). The concentrations of radiocarbon in individual growth bands, expressed
as the difference from a known series of reference values
from fish otoliths (see solid line in fig. 13.10b) can be compared to a curve of the concentrations of radiocarbon
in the northwestern Atlantic (see dashed line). The two
slopes were very similar, indicating that the growth bands
were deposited over the same time span, and that the
porbeagle shark was estimated on the basis of this to be
roughly 23 years old due similarity in the two slopes. The
discrepancy between the two slightly different ages may
be explained by a three-year phase shift of the reference
values obtained from bivalves, corals, and fish otoliths in
the northeastern Atlantic Ocean.

of a pair of rings has been deposited at this particular time of the year. The
fraction will be nearly zero when the pair of rings, starting with the translucent
one, is first deposited. The fraction will become successively larger later in
the year as the ring grows until it reaches a value of 1, when the pair of rings is
fully deposited. A gradual annual increase in the fraction to a value of 1 should
be observed if one pair of growth rings is deposited each single year. The marginal increments for salmon sharks (*Lamna ditropis*), caught off the coast of
Alaska increased from 0.02 in February to 0.92 in January—indicating that
one pair of growth rings is deposited in the centrum of that species per year.[37]

Growth Rates

The rate of growth of a cartilaginous species is usually determined by fit-
ting a curve to a plot of measurements of the lengths of captured individuals
and the corresponding counts of annuli on their vertebral centra. The slope of
this curve indicates the rate of growth at various ages. The curve levels off, or
asymptotes, at a more advanced age because growth decreases with increasing body size. In general, slowing in growth is more pronounced in males than

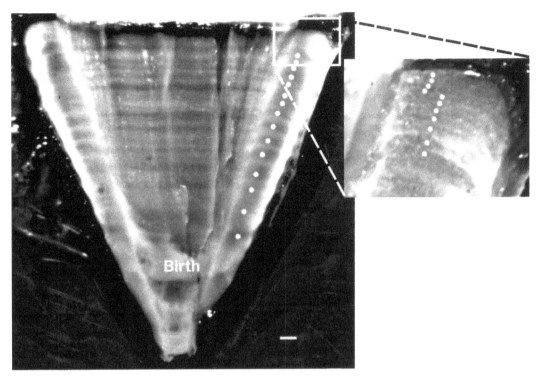

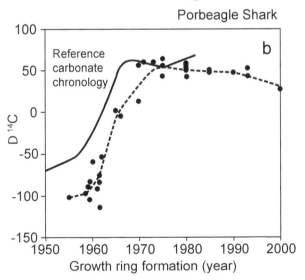

Porbeagle Shark

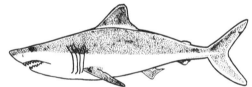

Fig. 13.10 (a) Section of vertebral centrum from a 264 cm fork length porbeagle shark collected in 1999, with insert showing higher magnification of very narrow growth rings near the growing edge. White circles indicate positions of growth bands. (b) Graph comparing reference carbonate chronology (solid line) with $\Delta^{14}C$ concentrations in individual growth bands (solid circles) versus their year of formation based on the count of the bands (dashed line). Note that the slope of a curve fitted to the latter points closely resembles that fitted to the former points.

females. This difference between the sexes is usually attributed to the evolutionary advantage of increased fecundity, or the capability of bearing more young sharks, for females of a greater size.

The rate of growth has been determined for white sharks captured off the western coast of North America.[10] Shown are the centra of a juvenile and adult white shark (fig. 13.11a). Two bands of alternating growth are apparent on the juvenile's centrum, which is only 283 mm across. The 0 indicates the diameter of the centrum at birth, the I the diameter at the end of the first year of growth, and the II the diameter at the end of the second year. The adult centrum is much larger, 70.8 mm across. There are on it thirteen concentric growth rings, each consisting of a light ring, the annulus, separated from the adjacent annulus by a dark ring. These are marked 0 through XIII. The lengths of 20 white sharks are plotted as a function of the number of pairs of growth rings, and a curve is fitted to the distribution of points on the graph (fig. 13.11b). It is apparent from the graph that the size of white sharks at birth is roughly 1400–1800 mm. The counts of annuli on the vertebral centra from the largest sharks indicate that adults roughly 5000 mm long are 14–15 years old.

The fin spines and tail thorns can be used to age cartilaginous fishes as well as their vertebral centra. Many of the dogfishes of the Squaliformes and bull sharks of the Heterodontiformes have spines on the first and second dorsal fins. The skates of the Rajiformes and Myliobatiformes have thorns on their caudal fins. The advantage of using parts other than vertebral centra in aging studies is that the subjects of the analysis do not have to be sacrificed to ascertain their rates of growth and longevity. Furthermore, these other parts can provide a redundant indicator of age and growth rate to accompany simultaneous age determinations using vertebral centra.

The ages of male and female Alaska skates (*Bathyraja parmifera*), the most abundant species of ray on the eastern Bering Sea shelf, were determined from counts of annuli on both the vertebral centrum and caudal thorn.[38] A close correspondence existed between these two counts. The translucent bands are indicated by black dots on a longitudinal section of a centrum (fig. 13.12a) and sagittal section of a thorn (fig. 13.12b) of a seven-year-old 82 cm long skate. The birthmark on the protothorn is apparent from a slight angle change in the boundary between the corpus calcareum on the right relative to the intermedialia at the center of the vertebra. There are seven dots, one for each translucent ring alternating with an opaque ring, on each of the two sections.

Edge and marginal increment analyses were performed to determine whether one pair of growth rings was deposited per year. The nature of the edge of the section, translucent or opaque, was recorded for both thorns and centra throughout the year in the edge analysis. Narrow translucent bands were found at the edge of all of the hard parts during January and February;

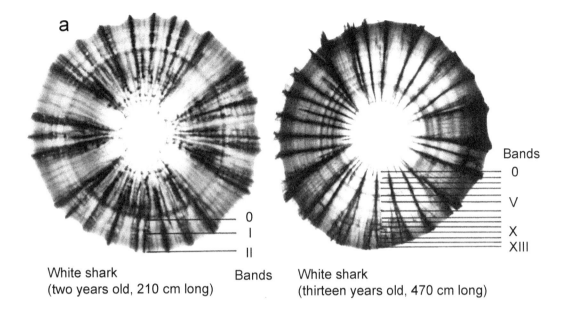

a

Bands
0
V
X
XIII

Bands
0
I
II

White shark
(two years old, 210 cm long)

White shark
(thirteen years old, 470 cm long)

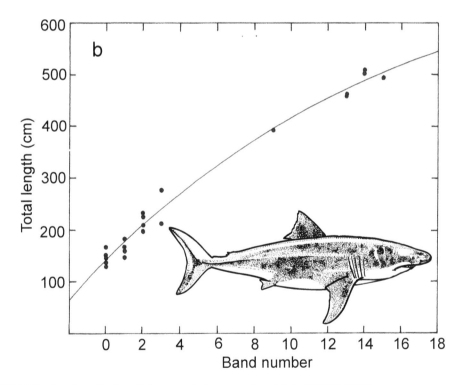

b

Total length (cm)

Band number

Fig. 13.11 (a) The alternating pairs of growth rings on centra from a juvenile and adult white shark (a). The centrum taken from a 2099 mm long juvenile indicates its age was two years; the centrum from the 4609 mm long adult reveals an age of thirteen years. (b) A curve is fitted to the lengths of the sharks plotted as a function of band number.

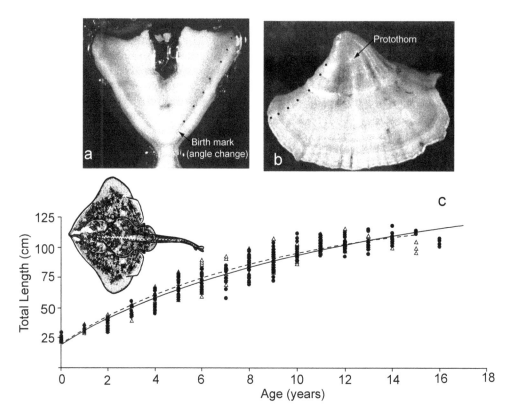

Fig. 13.12 Thin sections of (a) a vertebral centrum and (b) a caudal thorn taken from a seven-year-old Alaska skate. The black circles indicate successive years of growth. (c) Plot with points indicating the size of male (clear triangles) and female skates (solid circles) of different ages. Von Bertalanffy growth curves fitted to the observed distributions of points for males (dashed line) and females (solid line).

narrow opaque bands from June through October, and broad opaque bands from November to December. This indicates a seasonal progression in the formation of the growth bands. A marginal increment analysis was also performed on the vertebral centra removed from rays caught during all months of the year. The ratios of width of the newest outer pair to the previous fully developed pair of growth bands were calculated and plotted on a monthly basis. The smallest mean ratios of 10%–20% were found during January and February, and these ratios increased to 35% in June and 80% and 70% in November and December—additional evidence supporting the annual deposition of the growth rings.

Counts of annuli are plotted against the lengths of the skates with clear triangles for the males and solid circles for the females (fig. 13.12c). The females grow until the age of 17 years whereas males grow until the lesser age of 15 years. The solid line, indicating female growth, continues to rise after 15 years in contrast to the dashed line, indicating male growth. The latter curve reaches an asymptote by the same time. Continued or indeterminate growth is often observed among females in contrast to the determinate growth of males in species of cartilaginous fishes. This growth differential, apparent here only during the 15th year, may enable the female to accommodate more embryos in her uterus.

Growth rates have been estimated for the ghost shark (*Callorhinchus milii*), caught at two widely separated times, 1966–68 and 1983–84, in Pegasus Bay off the coast of the south island of New Zealand.[39] The estimates for growth of these male and female chimaeras are not based on counts of growth bands on their vertebrae but rather on the average sizes of the members of cohorts, groups of individuals born during the same year. They hatch from their egg cases at a size of 10 cm fork length, measured from the tip of the snout to the fork in the tail, during May and July. The females grow to length of roughly 15 cm over a period of four months during the first year, apparent from the peak of ages farthest to the left in the middle histogram (fig. 13.13b) and reach a length of 22 cm by January, evident from the peak of ages farthest to left in the bottom histogram. There are eight to nine peaks to the frequency distribution of fork lengths in the latter two histograms, and these equate to a similar number of cohorts, evident by the inflections, or leveling off, in the growth curves on a plot of fork length as a function of age (fig. 13.13a). The disparity in the slopes for the females (solid lines) and the males (dotted lines) indicates that the females grown quicker than the males after first year. This was observed during both 1966–68 and 1983–84. This is a better example of determinate growth by males and indeterminate growth by females.

The rate of growth and the lifespan varies greatly among cartilaginous fishes. The growth rate, maximum length, and maximum age are shown for representatives of the orders of chimaeras, sharks, and rays from the scientific literature (table 13.2). As you have just learned, female ghost sharks of the order Chimaeriformes grow at a rate of 9.6% of their maximum length of 157 cm in Pegasus Bay in New Zealand.[39] Their maximum longevity is eight years. Males grow much more rapidly, 23.1% per year, but reach a maximum size of only 75 cm. They live as long as the females. The growth rates and maximum sizes vary considerably among geographic locations as evident from the same vital statistics for males and females from Canterbury Bight, New Zealand.

The growth rates and longevities of primitive squaloform sharks vary greatly. Female broadnose sevengill sharks (*Notorynchus cepedianus*) of the order Hexanchiformes grow at an extremely rapid annual growth rate of 29.1% to a maximum length of 192 cm, whereas the males grow at a slower rate of 20.0% to a larger size of 239 cm.[40] The females live from 11 to 21 years while the males live for 5 to 15 years. Female short-spined spurdogs (*Squalus mitsukurii*) of the order Squaliformes grow at a slow rate of only 4.1% of their maximum length

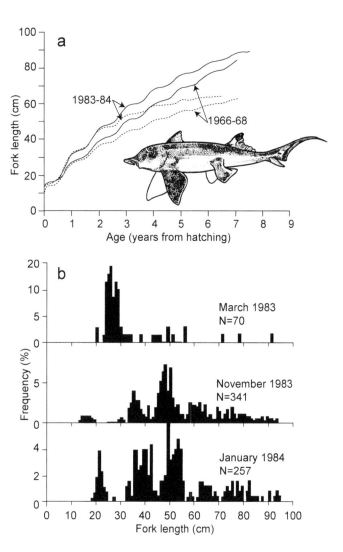

Fig. 13.13 (a) Histogram of the lengths of female ghost sharks captured during January 1984 and (b) growth curves for males (dotted lines) and females (solid lines) during two periods of time, 1966-1969 and 1983-1984. Notice that there are roughly eight peaks to the frequency distribution indicating eight cohorts of different ages.

per year to a size of 107 cm in the North Pacific Ocean.[42] They live to an age of 27 years. Males grow much faster, at a rate of 15.5% per year, but only attain a size of 66 cm, much smaller than that of the female.

The same variability exists among the more advanced galeomorph sharks. Some of the smaller coastal species grow rapidly. For example, female bonnet-head sharks (*Sphyrna tiburo*) grow at an annual rate of 34.0% to a size of 115 cm in the Gulf of Mexico.[55] However, the females are short-lived with a maximum age of 7 years. Male bonnethead sharks grow even more rapidly, at an annual rate 58%, but reach a much smaller size of 89 cm. They are even shorter lived than the females, reaching a maximum age of 6 years. In contrast, female por-beagle sharks grow at an annual rate of only 6.1% and reach a maximum size of 370 cm in the northwestern Atlantic Ocean.[48] They live to a maximum age of

TABLE 13.2. Growth rates of sharks, rays, and chimaeras.

Order	Species	Common Name	Location	Sex	Growth/Yr (% length)	Max. Length (cm)	Max. Age (yrs.)
Chimaeriformes	*Callorhinchus milii*[39]	Ghost shark	Pegasus Bay, New Zealand	Female	9.6	157[b]	8
				Male	23.1	75[b]	8
			Canterbury Bight, New Zealand	Female	6.0	204[b]	8
				Male	8.9	142[b]	6
Hexanchiformes	*Notorynchus cepedianus*[40]	Broadnose sevengill	California	Female	29.1	192[d]	11–21
				Male	20.0	239[d]	5–15
Squaliformes	*Squalus acanthias*[41]	Piked dogfish	Black Sea	Female	17.0	145[d]	13
				Male	20.0	128[d]	14
	Squalus mitsukurii[42]	Shortspine spurdog	North Pacific Ocean	Female	4.1	107[d]	27
				Male	15.5	66[d]	18
Squatiniformes	*Squatina californica*[43]	Pacific angelshark	California	Female	16.2	126[d]	35
				Male	15.2	126[d]	35
Orectolobiformes	*Ginglymostoma cirratum*[44]	Nurse shark		Female			35
				Male			29

continued

TABLE 13.2. *continued*

Order	Species	Common Name	Location	Sex	Growth/Yr (% length)	Max. Length (cm)	Max. Age (yrs.)
	Rhincodon typus[45]	Whale shark	South Africa	Female			19–27
				Male			20–31
Lamniformes	*Alopias superciliosus*[46]	Bigeye thresher	Taiwan	Female	9.2	225[d]	20
				Male	8.8	219[d]	14
	Carcharias taurus[37]	Sandtiger shark	Atlantic Ocean	Female	11.0	296[c]	17
				Male	16.0	250[c]	16
	Lamna nasus[47]	Porbeagle	Northwestern Atlantic Ocean	Female	6.1	370[d]	24
				Male	8.0	258[d]	25
Carcharhiniformes	*Carcharhinus leucas*[48]	Bull shark	South Africa	Female			32
				Male			29
	Carcharhinus plumbeus[49]	Sandbar shark	Northwest Atlantic Ocean	Female	5.9	197[d]	25
				Male	5.9	184[d]	18
	Galeocerdo cuvier[50]	Tiger shark	Northwestern Atlantic Ocean	Both	17.8	337[d]	27
	Mustelus canis[51]	Dusky smooth-hound	Northwest Atlantic Ocean	Female	29.2	124[d]	16
				Male	43.0	105[d]	10
	Prionace glauca[52]	Blue shark	Northwest Atlantic Ocean	Female	13.0	310[d]	15
				Male	18.0	282[d]	16

Order	Species	Common name	Location	Sex			
	Sphyrna lewini[53]	Scalloped hammerhead	Taiwan	Female	24.9	320[d]	14
				Male	22.2	321[d]	11
	Sphyrna tiburo[54]	Bonnethead	Gulf of Mexico	Female	34.0	115[d]	7
				Male	58.0	89[d]	6
Torpediniformes	Torpedo californica[55]	Pacific electric ray	Northeastern Pacific	Female	7.8	137[d]	16
				Male	13.7	93[d]	14
Pristiformes	Pristis pectinata[56]	Sawfish	Northeastern Atlantic	Both	8.0	450[d]	30
Rajiformes	Raja binoculata[57]	Big skate	Monterey Bay, California	Female	37.0	168[d]	12
				Male	43.0	139[d]	11
	Raja rhina[57]	Longnose skate	Monterey Bay, California	Female	16.0	107[d]	12
				Male	25.0	97[d]	13
Myliobatiformes	Dasyatis americana[58]	Southern stingray	Northeastern Atlantic	Female	53.4	150[a]	26
				Male	20.6	113[a]	28
	Urolophus halleri[59]	Haller's round ray	Southwestern Australia	Female	36.9	25[a]	14
				Male	51.4	21[a]	13

[a]Disk width = distance from tip of one pectoral fin to the tip of the other; [b]fork length = distance from tip of the snout to the posterior notch of the caudal fin; [c]precaudal length = distance from tip of snout to bottom of precaudal notch; [d]total length = distance from tip of snout to the posterior tip to the upper lobe of the caudal fin.

24 years. The males grow at a similar rate, 8.0% per year, but reach a smaller maximum size of 258 cm. They have a lifespan of 25 years. Similar slow rates of growth are observed in the sandbar shark. Females and males grow at a rate of 5.9% of their maximum length per year and achieve roughly similar maximum lengths of 197 and 184 cm, respectively.[50] However, the females have a longer life span—25 years, versus 18 for males.

The electric rays appear to grow slowly, but the skates and stingrays grow much more rapidly. Female and male Pacific electric rays (*Torpedo californica*) grow at annual rates of 7.8% and 13.7% in the northeastern Pacific.[56] On the other hand, female and male big skates grow at rate of 37.0% and 43.0% of their maximum length per year in Monterey Bay along the coast of California.[58] Female and male round stingrays grow at rates of 36.9% and 51.4% their maximum length off the coast of southwestern Australia.[60] There is great variability among the growth rates and longevity of the cartilaginous fishes. We sorely need to gain some general understanding of what life history properties are associated with different lifestyles of the species and the habitats and climactic zones in which they live to understand this variation in their growth and longevity.

SUMMARY

We have learned that cartilaginous fishes feed on different prey as they grow from a juvenile to an adult. For example, juvenile white sharks feed mainly on bony fish, intermediate-sized white sharks feed as well on smaller sharks, and adults feed mainly on seals and sea lions off the western coast of North America. This dietary shift follows a migration from the warm waters off the coast of southern California to the colder waters off the coast of central California. Here large colonies of pinnipeds are found on the mainland at the Point Reyes Headlands and at Año Nuevo and the South Farallon Islands. Evidence was presented that in response to this movement northward and switch to a high-energy diet of marine mammals, the growth rate of the white shark accelerates relative to those of other mackerel sharks in the order Lamniformes. The diet can also differ among males and females. This is because in many cartilaginous fishes, the sexes segregate from each other and occupy separate feeding grounds. The cartilaginous fishes are unique in their possession of spiral and scrolled intestines. Their folded nature confers both advantages and disadvantages. Their compactness affords more space in the peritoneum to accommodate a larger liver and uterus. Yet the economy in mucosal surface results in a slower digestion rate and prolongs the interval between foraging bouts. The ultimate impact of this profound tradeoff is slow growth so that maturity is reached only after a lengthy period. This results in a relatively low reproductive output.

The ages of white sharks as well as most other sharks of various sizes are determined by counts of the annuli on their vertebral centra. Similarly, the ages of Alaska skates of various lengths are based on counts of annuli on both their centra and caudal thorns. Techniques exist for verifying that one pair of growth rings is deposited per year. Finally, a brief overview was given of the rates of growth, maximum sizes, and maximum ages reached by members of the different orders of cartilaginous fishes.

* * *
** ** **

KEY TO COMMON AND SCIENTIFIC NAMES

Alaska skate = *Bathyraja parmifera*; basking shark = *Cetorhinus maximus*; bat eagle ray = *Myliobatis californica*; black rockfish = *Sebastes melanops*; blue runner = *Caranx chrysos*; blue skate = *Dipturus batis*; bramble shark = *Echinorhinus brucus*; brown smooth-hound shark = *Mustelus henlei*; bull shark = *Carcharhinus leucas*; cabezon = *Scorpaenichthys marmoratus*; California sea lion = *Zalophus californianus*; Chinook salmon = *Onchorhynchus tshawytscha*; chub mackerel = *Scomber japonicus*; dungeness crab = *Cancer magister*; dusky smooth-hound shark = *Mustelus canis*; ghost shark = *Callorhinchus milii*; green sturgeon = *Acipenser medirostris*; harbor seal = *Phoca vitulina*; lemon shark = *Negaprion brevirostris*; leopard shark = *Triakis semifasciata*; lingcod = *Ophiodon elongatus*; megamouth shark = *Megachasma pelagios*; northern elephant seal = *Mirounga angustirostris*; North Pacific hake = *Merluccius productus*; Pacific rock crab = *Cancer antennarius*; Pacific saury = *Colabis saira*; piked dogfish = *Squalus acanthias*; plownose chimaera = *Callorhinchus callorynchus*; porbeagle shark = *Lamna nasus*; rabbit fish = *Chimaera monstrosa*; rockfish = *Sebastes* sp.; salmon shark = *Isurus paucas*; sandbar shark = *Carcharhinus plumbeus*; sandtiger shark = *Carcharias taurus*; shortfin mako shark = *Isurus oxyrinchus*; southern stingray = *Dasyatis americana*; stingray = *Dasyatis* sp.; striped bass = *Morone saxatilis*; tiger shark = *Galeocerdo cuvier*; tope shark = *Galeorhinus galeus*; whale shark = *Rhincodon typus*; white seabass = *Cynoscion nobilis*; white shark = *Carcharodon carcharias*.

LITERATURE CITED

1. Anderson *et al.*, 2011; 2. Cortés, 1999; 3. Ebert and Bizzarro, 2007; 4. Klimley, 1994; 5. Jorgensen *et al.*, 2009; 6. Klimley, 1985b; 7. Klimley *et al.*, 1992; 8. Klimley *et al.*, 1996a; 9. Klimley, 1990; 10. Cailliet *et al.*, 1985; 11. Aasen, 1963; 12. Pratt and Casey, 1983; 13. Compana *et al.*, 2002; 14. Klimley, 1987; 15. Wetherbee and Cortés, 2004; 16. Medved, 1985; 17. Cortés and Gruber, 1990; 18. San Filippo, 1995; 19. Rinewalt et al., 2007; 20. Robinson *et al.*, 2007; 21. Bush and Holland, 2002; 22. Jones and Geen, 1977; 23. Carey *et al.*, 1982; 24. Webb *et al.*, 1998; 25. Klimley *et al.*, 2001; 26. Sepulveda *et al.*,

Discussion Questions

1. How do you quantify how much a cartilaginous fish has eaten upon examining its stomach? What should you use as an indicator of feeding success: the number of prey items, their combined mass, or their combined volume? A method exists for describing diet based on all of these measures. It is referred to as the index of relative importance (IRI). The percentages of prey given for the male and female scalloped hammerhead sharks caught in different habitats were based on the IRI.[14] Consult Wetherbee and Cortés (2004)[15] for a description of this index.

2. Do sharks feed on a greater variety of prey than other apex predators in the marine environment? There are indices that can be used to quantify prey diversity and make comparisons between species (again see Weatherbee and Cortés, 2004).[15]

3. Do sharks really feed in bouts? Do they really gorge themselves at one time and refrain from hunting while they digest their meal? Humans eat in bouts, termed meals, but also snack between these bouts of feeding. There is a statistical method for identifying bouts from a collection of the time intervals separating feeding events called log survivorship analysis (see Fagan and Young, 1978).[60] You could use this to quantitatively describe bouts of feeding in chondrichthyan fishes.

4. Debate still rages as to whether the alternating translucent and opaque rings are due to a difference in the temperature in the sharks' habitat. A confounding variable is that the cartilaginous fishes may feed more frequently during the summer, and this proclivity might contribute to the denser growth ring deposited at this time. How could you test this hypothesis in the laboratory? How could you evaluate the hypothesis in the field? One approach would be to look for growth rings of species confined to the equator, where the oceanographic changes between seasons are less. This is not always the case because warm and cold currents impinge upon the Galapagos Islands, which are situated at the equator, during different seasons of the year.

5. Can you arrive at some generalizations about which oceanographic conditions and geographical ranges promote certain growth rates, times to maturity, and maximum longevity? You can find these vital statistics for a myriad of species of chimaeras, sharks, and rays in table 14.1 in Cailliet and Goldman (2004).[34]

6. An important objective of the field of physiological energetics is the creation of an energy budget for each species. Necessary for any budget is knowledge of the amount of food consumed, which is converted to the energetic currency of calories. Furthermore, the growth rates, described in this chapter, would be used in any such budget. The budget would express the relationship between the energy acquired from food and the energy consumed by activity and growth as well as that energy lost in feces and urine. Given the following set of variables, can you formulate the standard bioenergetics equation for the cartilaginous fishes? This equation was first published by Ricker (1975).[61]

   C = caloric input from food consumed (hint: left side of equation)
   R = calories released during respiration
   P = calories stored in body growth
   F = calories lost in feces
   U = calories expended in urine

Lowe (2001, 2002) published two exemplary articles on developing an energy budget for juvenile scalloped hammerhead sharks.[62, 63] There is a good review of this subject with regard to elasmobranchs written by Carlson et al. (2004).[64]

2004; 27. Schmid *et al.*, 1990; 28. Bradley, 1996; 29. Di Giácomo *et al.*, 1994; 30. Holgren and Nilsson, 1999; 31. Wetherbee and Gruber, 1990; 32. Wetherbee and Gruber, 1993; 33. Gruber, 1984; 34. Cailliet and Goldman, 2004; 35. Sminkey and Musick, 1995; 36. Smith *et al.*, 2003; 37. Goldman, 2002; 38. Matta and Gunderson, 2007; 39. Francis, 1997; 40. Van Dykhuizen and Mollet, 1992; 41. Avsar, 2001; 42. Wilson and Seki, 1994; 43. Cailliet *et al.*, 1992; 44. Carrier and Luer, 1990; 45. Wintner, 2000; 46. Liu *et al.*, 1998; 47. Natanson *et al.*, 2002; 48. Wintner *et al.*, 2002; 49. Sminkey and Musick, 1995; 50. Natanson *et al.*, 1999; 51. Conrath *et al.*, 2002; 52. Skomal and Natanson, 2003; 53. Chen *et al.*, 1990; 54. Parsons, 1993; 55. Neer and Cailliet, 2001; 56. Simpfendorfer, 2000; 57. Zeiner and Wolf, 1993; 58. Henningsen, 2002; 59. White *et al.*, 2001; 60. Fagan and Young, 1978; 61. Ricker, 1975, 62. Lowe, 2001; 63. Lowe, 2002: 64. Carlson *et al.*, 2004.

RECOMMENDED FURTHER READING

Bush, A. C., and K. N. Holland. 2002. Food limitation in a nursery area: estimates of daily ration in juvenile scalloped hammerhead sharks, *Sphyrna lewini* (Griffith and Smith, 1834) in Kāne'ohe Bay, Ō'ahu, Hawai'i. *Jour. Exp. Mar. Biol. Ecol.*, 278: 157–178.

Cailliet, G. M., and K. J. Goldman. 2004. Age determination and validation in chondrichthyan fishes. Pp. 399–448 in Carrier, J. C., J. A. Musick, and M. R. Heithaus (Eds.), Biology of Sharks and Their Relatives. CRC Press, Boca Raton.

Campana, S. E., L. J. Natanson, and S. Myklevoll. 2002. Bomb dating and age determination of large pelagic sharks. *Can. J. Fish. Aquat. Sci.*, 59: 450–455.

Francis, M. P. 1997. Spatial and temporal variation in the growth rate of elephantfish (*Callorhinchus milii*). *New Zealand Jour. Mar. Freshwater Res.*, 31: 9–23.

Holgren, S., and S. Nilsson. 1999. Digestive system. Pp. 144–173 in Hamlett, W. C. (Ed.), Sharks, Skates, and Rays. Johns Hopkins University Press, Baltimore.

Klimley, A. P. 1985. The areal distribution and autoecology of the white shark, *Carcharodon carcharias*, off the West Coast of North America. *Memoirs, South. Calif. Acad. Sci.*, 9: 15–40.

Klimley, A. P. 1987. The determinants of sexual segregation in the scalloped hammerhead shark, *Sphyrna lewini. Environ. Biol. Fish.*, 18: 27–40.

Matta, M. E., and D. R. Gunderson. 2007. Age, growth, maturity, and mortality of the Alaska skate, *Bathyraja parmifera*, in the eastern Bering Sea. *Environ. Biol. Fishes*, 80: 309–323.

Wetherbee, B. M., and E. Cortès. 2004. Food consumption and feeding habits. Pp. 203–224 *in* Carrier, J. C., J. A. Musick, and M. R. Heithaus (Eds.), Biology of Sharks and Their Relatives. CRC Press, Boca Raton.

CHAPTER 14

Daily Movements, Home Range, and Migration

Upon graduating with a Ph.D. in marine biology, I remained at Scripps Institution of Oceanography initially as a postdoctoral scholar and later as an assistant research scientist. At this time, the Division of Biological Oceanography of the National Science Foundation provided me with a research grant to conduct studies aimed at understanding the means by which scalloped hammerhead sharks migrated in an extraordinarily directional manner in the apparently featureless open ocean between daytime aggregation sites at seamounts and distant feeding grounds where prey were abundant during nighttime. The first objective of the study was to track hammerhead sharks using a small boat at a seamount, El Bajo Espiritu Santo, far from shore in the Gulf of California while obtaining a record of their behavior as a description of the temperatures and light levels in the local underwater environment. In order to attach an ultrasonic transmitter, it was necessary to dive down, pole spear in hand, to the edge of the school of sharks and then release the spear, inserting a dart with an electronic transmitter attached into the thick muscle of the shark. Donald Nelson, a professor at California State University of Long Beach, and I would spend much time searching for the schools of hammerhead sharks and often did not succeed in tagging one until it was almost

Small group of hammerhead sharks shown splitting away from school as dusk approaches, and an individual prior to leaving on it nighttime feeding migration.

dark. The hammerhead sharks swam in compact, polarized schools most of the day when the underwater environment was well illuminated. However, as dusk approached the members of the schools began to stray from their ordered procession. Small groups of three or four individuals then split away and began to swim together at a frenetic pace above the seamount. This behavior reminded me of the heightened activity exhibited by birds prior to leaving their roost and collectively making a long migration as winter approaches. The hammerhead sharks usually left the seamount alone, as you will see from a telemetry study included later in the chapter, and began their nightly feeding migrations. The first time we tracked an individual all night was truly a revelation. We drove our small boat as fast as possible after it, stopping periodically to place our hydrophone in the water and determine its direction away from us. There was no moon out that night, but the shark swam as straight as one might drive a car down a highway between lane markers despite swimming at great depths. It was far from the bottom where it might follow an underwater ridge that led somewhere or near the surface where the beacon-like moon might reveal the distant objective. We glared at the computer screen on the boat that displayed the shark's heading, its swimming depth, the water temperature, and the amount of illumination near the shark. At midnight, the R/V Gordon Sproul, a research vessel positioned between us and the seamount, was barely visible by her navigation nights. Then suddenly to our surprise, the shark reversed its direction and began swimming rapidly back toward the seamount. It passed close to the ship and arrived at the seamount at dawn. We found out later that the shark was nearly 20 km away from the seamount when it decided to return. This degree of mobility is not unusual for sharks and rays. They make daily and seasonal migrations over great distances. In this chapter, we will learn more about their daily movements, seasonal home ranges, yearly migrations, and diving behavior.

The extent of the daily movements of the cartilaginous fishes varies greatly from species to species. Members of some species stay within a confined area and capture their prey by ambushing them. These ambushers are usually cryptic because their coloration matches the color of the bottom, where they remain motionless partially buried in sand or mud with their eyes protruding in order to detect crustaceans or fishes as they pass by. They then accelerate explosively upwards to seize and swallow them using the suction feeding mechanism described earlier in the book (see figs. 12.4b & 12.8). The angelsharks of the order Squatiniformes, some carpet sharks of the order Orectolobiformes, the skates of the order Rajiformes, and some of the stingrays of the order Myliobatiformes have this sedentary lifestyle. However, even the ocean-migrating white sharks (*Carcharodon carcharias*) spend the fall and early winter in central California patrolling a restricted area around seal colonies, where they ambush seals that converge upon these sites to populate the beaches. The movements of Haller's round ray (*Urolophus halleri*) will be described to exemplify the sedentary lifestyle. Then a map will be presented of daily patrolling movements of a white shark at South Farallon Islands, and this is best appreciated if the positions of the shark are compared to the shoreline of the island inhabited by seals and sea lions and the actual locations in the waters around the island where the sharks captured and feed upon them (see fig. 12.3a). Other sharks and rays are more mobile, yet they are confined to the coastal waters or oceanic waters near seamounts or islands. Often they aggregate at a central place, which can be a cave, reef, or seamount. This type of diel space utilization will be described for the blacktail reef shark (*Carcharhinus amblyrhynchos*), but it is common among the bullhead sharks of the order Heterodontiformes and ground sharks of the order Carcharhiniformes. Blacktip reef sharks form social groups near a particular reef within a lagoon during the day, yet the individuals from the schools make excursions at night through a channel to the ocean. Other species do not have a central place aggregation site. The bat eagle rays (*Myliobatis californica*) are common in bays and estuaries along the western coast of North America. They move daily between feeding spots on the mud flats, where they excavate pits in the mud from which they extract mollusks to eat. Nomadic species such as the basking shark (*Cetorhinus maximus*) and whale shark (*Rhincodon typus*) move over great distances in the pelagic environment. The latter have been observed to migrate in the Atlantic Ocean from the northeastern coast of North America to the southeastern coast of South America. The latter have been observed to migrate across the Pacific Ocean from the tip of the Baja Peninsula to the Maldive Islands in the western Pacific.

AMBUSHERS

Haller's round stingrays, which occur from Panama to Eureka, northern California, appear to be particularly common near an electric generating station in southern California. This power station discharges heated seawater into the lower reaches of the San Gabriel River, whose mouth is at Seal Beach, California. The warm water stays close to shore before cooling and provides an ideal habitat for the round stingray, a subtropical species. These small rays, averaging only 56 cm in total length, are present year round in high densities close to shore buried in the sand. The higher frequency of round rays observed here than at other local beaches indicates that they have a preference for the warmer local waters.

Ten of these stingrays were tracked by boat to assess their short-term fine-scale movements at Seal Beach. They were tracked for up to 72 hours. The boat was positioned over a stingray, which was evident when the signal from

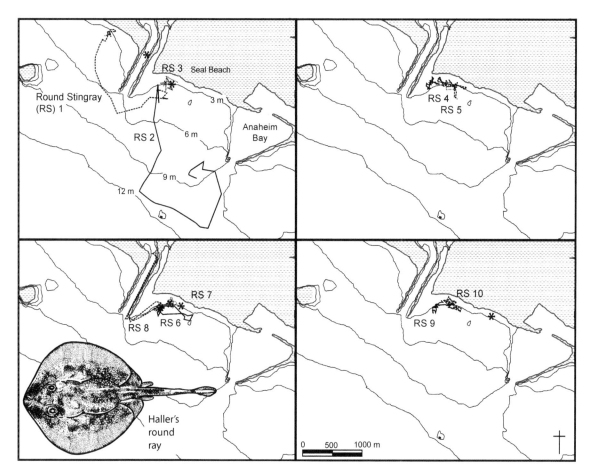

Fig. 14.1 The movements of ten Haller's round rays around Seal Beach in southern California. The asterisks indicate where certain rays were relocated during study.

the transmitter remained the same as the hydrophone was rotated 360°, and an estimate of the individual's position was determined using a global positioning system. It was impossible to see the stingrays in the water because they were often buried within the sand. The round rays stay most of the day in shallow water 2.5 to 3.5 m deep not more than 20 to 30 m offshore of the surf zone. They exhibit little movement most of the day as apparent for seven of the ten rays tracked during the study (RS 3–7, 9, and 10 in fig. 14.1). The rays spent two to four hours buried within the substratum. Only when the tide was ebbing did the rays exhibit any movements. One of the rays moved from one side of the jetty at the river outfall to the other side, a second moved into the river outfall, and a third offshore into waters roughly 9 m deep (RS 1, 2, and 8).

Coded ultrasonic beacons were placed on twenty-five round rays and their seasonal occupation of the beach determined by an array of five tag-detecting monitors on either side of the outfall. Male round stingrays were present at the beach mainly during the summer and autumn. The females spent less time at the beach, only a few weeks during June and July of the two years of the study. More of the rays were detected by the monitors near the outfall than those monitors farther away. Hence, it is likely that the warm effluent from the power plant created a refugium for round rays off Seal Beach.

Another ambusher present on the sandy bottoms off southern California is the Pacific angelshark. It remains motionless and partially buried in one spot most of the day, looking upward waiting for a crustacean to walk or small fish to swim over it. The angelshark then quickly opens its cavernous mouth and expands its buccal cavity to draw water and the prey inside its mouth. Although angelsharks are sedentary during the day, they do move from one feeding location to another at night.[2]

Another ambusher is the white shark. Adults of this species ambush seals and sea lions as they move to and from a colony on the shore of an island. Four white sharks were tracked in a similar manner to the Haller's round rays at the South Farallon Islands, except the locations of the boat positioned over each shark were determined using a theodolite from the island. The track of a 4.6 m long male is shown on a map of the island (fig. 14.2).[3] This male white shark swam back and forth along the western coast of the island. It is likely that at times the shark swam closer to shore because the boat behind the rocky cliffs at the edge of the island would not be visible to the person operating the theodolite from Lighthouse Hill. If you refer back to fig. 12.3a, you will see that there are large colonies of northern elephant seals (*Mirounga angustirostris*) on the southern, western, and northern ends of West End Island, which is adjacent to Southeast Farallon Island (see coastal area with beige color on map). Furthermore, numerous predatory attacks in the waters near the shore were seen by observers at Lighthouse Hill, particularly off the southern and northern ends of West End Island. It was rarely possible to determine the identity of the prey (green circles), but when the prey could be distinguished, it was usually a immature northern elephant seal (beige circles). In one instance, an abalone diver was seized and released off this coastline of the island (pink circle). This white shark was quite site specific in its behavior, feeding exclusively off the western coast of the two islands while ignoring pinnipeds on the beaches and rocks on the rest of the island.

White sharks also constantly patrol the shoreline of Año Nuevo, a small island a kilometer from shore at the northern edge of Monterey Bay, during the fall searching for seals and sea lions. The activity of these sharks was continually recorded by a radio-acoustic positioning (RAP) system consisting of a triangular array of three sonobuoys on the seaward side of the island.[3] The RAP was first operated during October and November 1997. Three white sharks were attracted to the surface by placing a seal-shaped decoy in the water

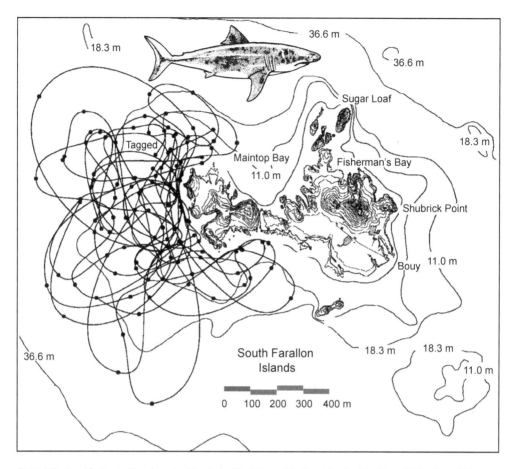

Fig. 14.2 Back-and-forth patrolling of a male white shark off East End and Southeast Farallon Island from 25 October to 1 November 1993. The solid circles indicate when the shark's position was estimated based on the position of the vessel tracking the shark.

where a seal had been seized by a shark near the surface. A beacon was placed on each shark by inserting a barb into its dorsum attached to the beacon on the end of a pole spear. Two more sharks were tagged later the same month. There was little evidence of territoriality by the five tagged sharks. If a particular shark preferred a separate area, it should have been detected within the range of one of the sonobuoys more than the other two. This was not the case. Furthermore, all five sharks passed over the same areas of coast.

Were the sharks social or solitary hunters? The degree to which white sharks are social when hunting was unclear from prior studies. Most observations of white sharks feeding had been of solitary individuals. For example, over a five-year period at the South Farallon Islands, a single shark was observed on 74% of the 195 days when white sharks were seen to feed on pinnipeds or investigating decoys, two sharks on 22% of the days, and three sharks on only 4% of the days.[5] To answer this question, you might ask whether sharks tagged at the same kill were present together more often than sharks tagged on different days. The first three tagged sharks might be members of a hunting squad. The frequencies at which the first shark tagged was accompanied by the second and third were, indeed, higher those for the fourth, which was not present when the other three were observed feeding on a seal. However, there was no consistent evidence of sociality when "true" separation distances were compared to "random" separation distances for each shark. It was obvious from the study that they rarely swam together while near the island. Alternatively, if the sharks were avoiding each other, the true distances would have been greater than the random separation distances. Neither was observed in this comparison. It was apparent from this analysis that the sharks were searching for prey by themselves.

The sharks spent a great amount of time slowly swimming back and forth along the coastline of Año Island as at Southeast Farallon and West End Islands, waiting for the opportunity to ambush a seal or sea lion either leaving or returning to the island. The sharks eventually succeeded in ambushing a seal during its transit through the high-risk zone. It appeared from this continuous monitoring that it is challenging for white sharks to capture pinnipeds at Año Nuevo Island. In this context, a potentially successful feeding strategy would be to remain relatively close to each other so that if one shark catches prey, the others can scavenge on the remainder of its carcass. An energetic analysis indicated that an adult white shark could survive a month and a half between successive meals. The shark's energetic needs were estimated during this period based on the change in muscle temperature as it passed through water masses of different temperature and its energetic supply from the energetic value of a 30 kg portion of whale fat found in the stomach of a captured shark. Each juvenile elephant seal actually contains more energy than required for the short-term sustenance of a single shark. A juvenile elephant seal would provide enough energy to sustain a white shark for three months according to the metabolic model. Thus it is not surprising that the sharks tracked during this study did not remain far apart so they could eavesdrop on each other's activities. Their close proximity would permit them to compete for a share of the carcass after capture by another shark.

CENTRAL-PLACE AGGREGATORS

Other sharks have a central-place social system like some birds and bats that roost at night in great numbers in a clump of trees or a cave and fly great distances during daytime to feed on seeds and fruits in the surrounding fields and forests. Species conserve energy during an inactive period of the day, yet they expend the energy going to and from their feeding grounds and feeding on these grounds during the active period of the day. There is usually a trade-off between the distance traveled and the abundance of food in a particular feeding ground—forage is less common near the roost and more common farther from it. An animal must decide whether to spend less time in travel and more time foraging locally, or more time in travel to a distant location and less time foraging there. The Port Jackson shark, the whitetip reef shark (*Triaenodon obesus*), the blacktail reef shark, and the scalloped hammerhead shark (*Sphyrna lewini*) are all species that have a central-place social system. Divers placed identifying spaghetti-type tags on horn sharks inhabiting small caves at Bondi Reef near the popular bathing beach in Sydney, Australia, and observed that the same sharks returned to the same caves on subsequent days when they visited them.[6] The divers also visited the caves at night. The caves were uninhabited at this time, presumably because the horn sharks were patrolling the surrounding sandy bottom searching for buried prey. Groups of half a dozen whitetip sharks, a reef-dwelling tropical species, remain resting on the bottom in caves during daytime.[7]

The members of those species with a central-place social system do not always rest on the bottom but may mill back and forth within schools at one time of the day while interacting socially. This is a characteristic of the black-tail reef shark, which forms schools of up to fifty members inside the lagoon at Rangiroa Atoll. An individual carrying an ultrasonic beacon was tracked for three successive days (fig. 14.3).[8] During daytime, the reef shark swam in a school that remained within a 500 m² area of the lagoon. The shark was present there from 15:00 to 17:00 hours the first day, 06:00 to 17:00 the second, and 06:00 to 17:00 the third. It moved through the channel leading into the atoll at dusk to forage along the edge of the outer reef. This particular shark passed through the channel at the same time each night, 18:45 hours on the first night, 18:42 the second, and 18:50 the third. It likely fed on fish along the edge of the reef from 19:00 to 22:00 hours before returning to the lagoon in the early morning. The shark spent a few hours in the center of the pass before staying during the early morning hours in very shallow water around a small island on the lagoon side of the pass. However, by sunrise, it moved back into its normal daytime resting area.

Schools of scalloped hammerhead sharks return during daytime to a particular refuging site at the Espiritu Santo Seamount, a ridge 1 km long rising abruptly from surrounding depths exceeding a kilometer. The hammerhead sharks stay on only one side of the seamount despite swimming almost 20 km away to feed at nighttime (see track of shark in fig. 9.8). This was shown by placing coded ultrasonic beacons on eighteen hammerhead sharks.[9] They were detected by two monitors moored at either end of the ridge of the seamount. A small solid circle indicates the location of the monitor at Site 1 on the

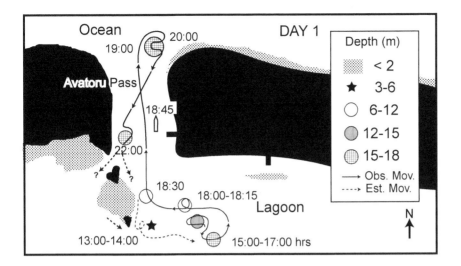

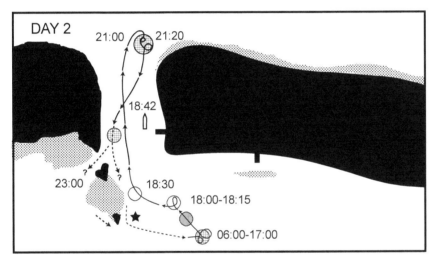

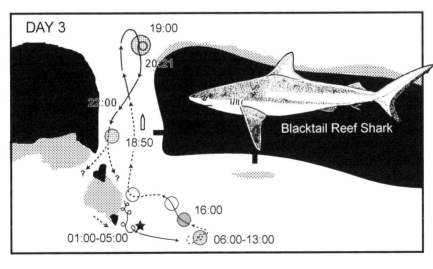

Fig. 14.3 The movements of a blacktail reef shark tracked during three successive days in the lagoon at Rangiroa Atoll.

northwestern end of the ridge. The circle is centered in the large circle with cross-hatching oriented upward and to the right indicating its range of reception (fig. 14.4). The range of this monitor was 150 meters. Site 2, the location of the monitor on the southeastern end of the ridge, is also indicated by a small solid circle with its range indicated by cross-hatching oriented upward to the left. Its range of tag detection was 100 m. The ranges of both monitors slightly overlapped half way along the length of the ridge.

The records of the sharks passing within the range of the two monitors are plotted in clock diagrams with midnight indicated on the top and noon on the bottom of each diagram. There are eighteen concentric rings in each diagram, each containing the detections of a single shark with the shark's presence denoted by solid black. It is apparent from the general observation that the black sections in each concentric ring start and stop together indicating that several sharks entered the range of the monitors together and left within the school that many of the sharks, individuals 3, 6–10, and 11–13, spent most of the day swimming in a school within the range of Monitor 1 at Site 1. Yet the last detection of each shark recorded within each ring at dusk does not line up with the last detections of the other sharks. This suggests that the hammerhead sharks left the seamount alone at night to forage. Likewise, the first detections of each shark recorded during dawn also do not line up. This is consistent with the sharks arriving back at the seamount from their nightly foraging trips alone. What is striking about the clock diagrams with the detection records from the two monitors is the stark contrast between the frequent detections of the sharks by the monitor on the northwestern edge of the ridge (upper circle) and the absence of detections of the same sharks by the monitor on the southeastern edge (lower circle). This difference existed in spite of the monitors being no more than 250 m apart and their detection ranges overlapping for half that distance. This leads you to conclude that the school of hammerheads stayed in a small area on one end of the seamount—unless the monitor at the second site was not functioning properly and failed to detect the coded beacons. This monitor was removed from its mooring and placed at Site 3 on the northwestern edge of the ridge so that its range of detection overlapped that of the Monitor 1 at Site 1 to exclude the latter possibility. The second monitor now recorded the same hammerhead sharks that had been detected by the first monitor. This extreme site fidelity is even more remarkable in the context of the long-distance nocturnal migrations made by individuals described in chapter 9.

Individuals of a species will then move a considerable distance away from one site to establish a new central site from which they make repeated foraging excursions. The blacktip shark (*Carcharhinus limbatus*) displays this pattern of movement. Six individuals carrying individually recognizable beacons were monitored for half a year by an array of tag-detecting monitors distributed

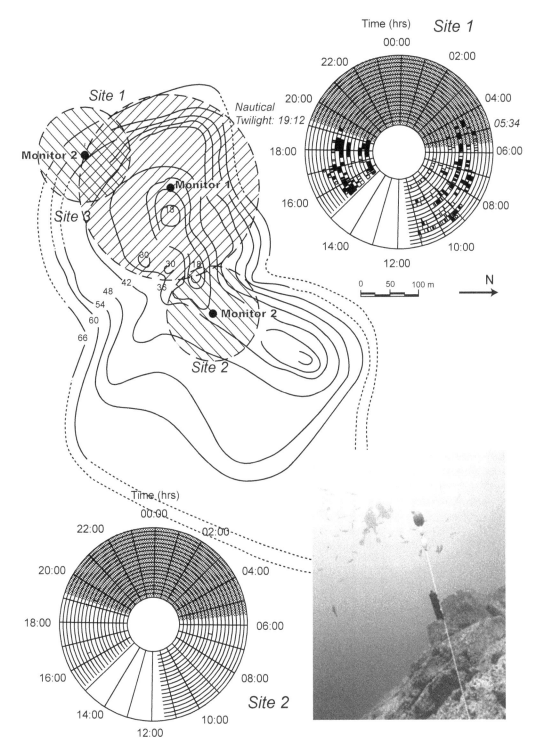

Fig. 14.4 Map of the Espiritu Santo Seamount in the Gulf of California with two circular clock diagrams illustrating the precise site fidelity of scalloped hammerhead sharks at the seamount. Many dark ticks are apparent from 0400 to 1900 hours on the top clock diagram while only two ticks are evident on the bottom diagram. This indicates that the sharks stayed on the northern side of the seamount ridge and not on the southern side. Lower right: A tag detecting monitor moored above a seamount to record the presence of sharks carrying beacons emitting unique individually coded signals.

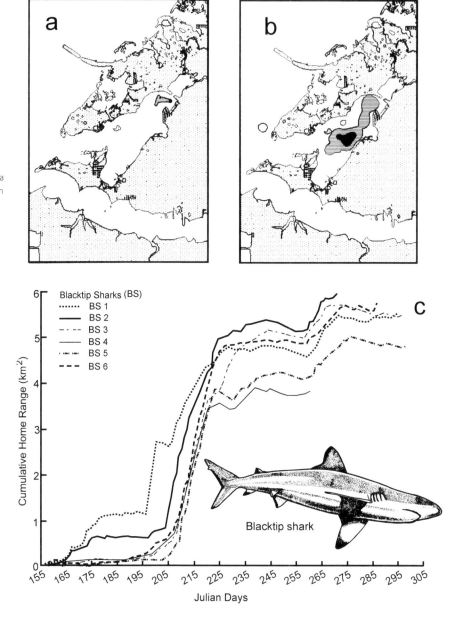

Fig. 14.5 Home range of blacktip shark during (a) June and (b) October in Terra Ceia Bay, Florida. (c) The increase in the cumulative area over which six blacktip sharks traveled over a period of five months in Terra Ceia Bay. Notice the simultaneous increase in their home ranges during late summer from Julian Day 205 to 235. Julian Day is a number incremented from the first day of January.

Blacktip Sharks (BS)
- BS 1
- BS 2
- BS 3
- BS 4
- BS 5
- BS 6

Cumulative Home Range (km^2)

Julian Days

Blacktip shark

throughout Terra Ceia Bay on the western coast of Florida.[10] The home range of a blacktip shark is smaller during June than during October. The area of cross-hatching, indicating the extent of the home range, is smaller in the map of movements of the shark during the former month (fig. 14.5a) than during the latter (fig. 14.5b). Furthermore, the center of activity of the shark changes from a confined space near the isthmus at the northeastern end of the bay to an expansive area in the middle of the bay along the eastern coastline. The rate at which individuals explore new habitats can best be visualized by plotting the cumulative area over which an individual travels on successive days. The curve will rise slowly when an individual stays in one area and will rise steeply when the shark immigrates into a previously unoccupied habitat. The six sharks resided within a small area within the bay from early June through mid-July, expanded their ranges greatly from mid-July to mid-August, and stayed within the expanded range from mid-August to mid-September. This is evident from gradual slopes to the home range curves from Julian Day 155 to 205, their increased steepness from Julian Day 205 to 235, and gradual slopes from Julian Day 235 to 265 (fig. 14.5c). The rapid rise to such a curve and its eventual reaching of an asymptote is characteristic of a central-place forager. The rate of increase in their overall range is limited by the finite distance that they can move during a daily feeding excursion from their resting location.

Some cartilaginous fishes such as the bat eagle ray move continuously over a twenty-four-hour period without returning to a central-place refuging site. These species travel rapidly between foraging areas with highly directional movements and then forage in a confined area with less directional movements. The movements of a bat eagle ray are shown over a day and a half during September 1998 in Tomales Bay, California (fig. 14.6).[11] It exhibited both directed movements as it migrated from the outer to the inner bay and nondirected movements while presumably foraging in the bay. The ray was tagged in the midbay at 11:17 hours and spent from 11:37 to 17:54 hours in a confined circular area of a diameter of only 0.5 km. It then swam in a straight line to a deeper part of the outer bay and remained there in the evening swimming in a less directional manner from 19:00 hours on 26 September to 01:11 hours on 27 September, when it made a highly directional migration over the next four hours to the midbay. It again appeared to forage here within a small area for two hours, before leaving at 07:07 hours. It then swam again in a roughly straight line to a fourth location, where it may have been foraging again. The confined nature to segments of track of the bat eagle ray can be explained by its feeding on a localized food source. Indeed, the concentrated movements observed here were repeated later at an oyster farm, where rays often bite through the protective mesh bags containing oysters and feed upon them.

OCEANIC MIGRATORS

Not only do basking sharks migrate great distances across the North Atlantic Ocean,[12] but they also migrate across the equator into the South Atlantic Ocean.[13] Much of their time is spent near frontal zones when they are feeding at the surface in the temperate latitudes.[14, 15] They also dive occasionally to mesopelagic depths at this time.[12, 14] Pop-up archival tags (PATs) were placed on

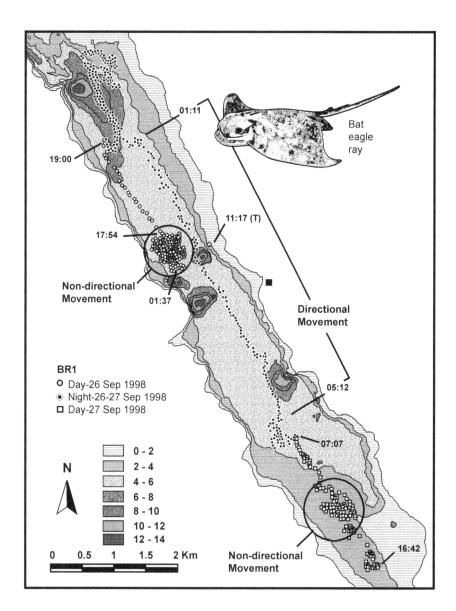

Fig. 14.6 Track of a bat eagle ray in Tomales Bay, just north of San Francisco Bay in central California. The daytime positions of the ray on Day 1 are indicated by clear circles and the nighttime positions by solid circles. The former positions during Day 2 are denoted by clear squares. The clusters of closely spaced positions are likely associated with foraging and the widely spaced positions forming a line indicate the ray is migrating from one prey source to another.

six basking sharks caught off the coast of New England. One of these sharks migrated along the continental shelf to the coast of South Carolina, where it overwintered (A in fig. 14.7a). You will note that it stayed mainly at the surface from 20 July to roughly 20 March, when it began to make dives ranging from 250 to 500 m (A in fig. 14.7c). Three of the sharks crossed the Gulf Stream into

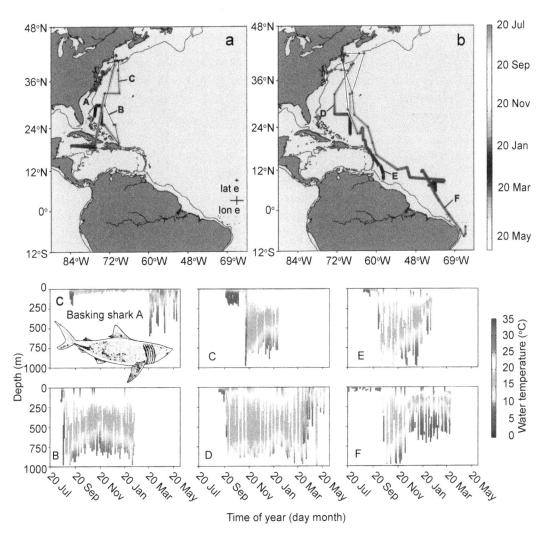

Fig. 14.7 (a, b) The seasonal migration of basking sharks off the eastern coast of North and South America. Bars on left panel indicate longitudinal error (lon e) and latitudinal error (lat e). Notice that two of the basking sharks, E and F, traveled from the coast off Cape Cod southward across the equator to the coast of Brazil. (c) Temperature records for basking sharks A–F.

the deepwaters of the oceanic basin, where they also began making dives from 800 to 100 m (B, C, and D in fig. 14.7c). One of these three migrated as far as the Puerto Rico Trench by October before heading northward to the Bahamas (B in fig. 14.7a). The two others were either north of Puerto Rico or had traversed the Caribbean Sea by early January (C and D in fig. 14.7a & 14.7b). The fifth shark had reached the coast of Venezuela by late January (E in fig. 14.7b). The shark that moved the farthest, F, remained in the shelf waters off New England until September when it moved rapidly across the Gulf Stream, through the Sargasso Sea, past the Windward and Leeward Islands in late November and early December. This shark remained off the mouth of the Amazon River for the month around January before continuing to migrate southward parallel to the coast of Brazil until the PAT popped off the shark in early May (F in fig. 14.7b). The basking sharks tagged off the British Isles and tracked in the eastern North Atlantic may have remained in temperate waters year round because food is abundant due to relatively stable oceanographic conditions. In contrast, basking sharks in the western North Atlantic encounter dramatic seasonal fluctuations in oceanographic conditions during the winter months, and hence migrate during the late fall into the Caribbean Sea or as far as off the coast of Brazil to find sufficient food. These studies have provided an alternative explanation to the absence of observations of basking sharks at the surface during the winter months. Biologists in the past have speculated that they were lying on the bottom hibernating. The tracking studies indicate that they move southward and forage at mesopelagic depths at this time of the year.

Whale sharks are also very nomadic, swimming over large distances in the open ocean. These sharks aggregate seasonally at Los Angeles Bay, midway up the Gulf of California on the eastern shore of the Baja Peninsula, and at the Gorda Seamount, which rises to a depth of 33 m at a distance of 10 km off the southeastern tip of the peninsula. Radio beacons contained within floats were affixed by a long tether to the dorsal fins of whale sharks. The beacons emitted a radio signal detected by the ARGOS satellite if it passed over the shark during its daily orbit around the earth and the shark swam close to the surface so that the float lifted the tag's antenna out of the water. Whale Sharks 1 and 2 moved southward after being tagged at Los Angeles Bay (fig. 14.8).[16] The first shark rounded the tip of the peninsula and stayed offshore in a transition zone between the California Current and the eastern Pacific Ocean. The second shark moved extensively in the equatorial countercurrent south of the Baja Peninsula before making a highly directional movement westward from 125° to 165° W across the Pacific Ocean at a latitude of roughly 15° N to the Federated States of Micronesia. This shark traveled a distance of 13,000 km over the thirty-seven-month period of tracking.

Whale sharks prefer to forage in the vicinity of seamounts and boundary currents, where zooplankton and fish are plentiful and foraging success is high. A whale shark tagged on 19 November 2007 offshore of a river along the coast of Panama exhibited this behavior.[17] The shark remained close to the location where it was tagged initially but then made more long-distance movements to three areas, each near the mouth of a river. The rivers transport nutrients and organic material to the ocean, where phytoplankton grow and attract primary consumers, zooplankton, and secondary consumers, small fishes, both of which are prey of the whale shark. The shark stayed near the mouth of a river northwest of where it was tagged in the presence of high chlorophyll concentrations. The same whale shark moved westward to the area offshore of the

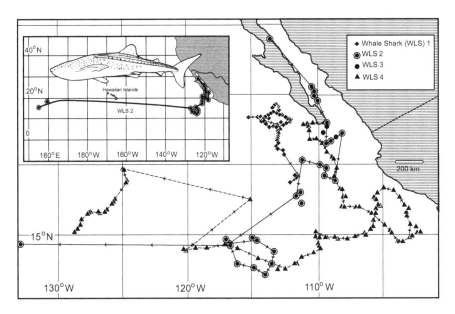

Fig. 14.8 The long-distance movements of four whale sharks tagged in the Gulf of California. Insert shows the long-distance movement of one of the whale sharks after it left the gulf.

mouth another river where the chlorophyll concentration was also high before moving eastward to a third area characterized by similarly high concentrations of chlorophyll.

Similar PATs affixed to white sharks have revealed their seasonal movements off the western coast of North America.[18] White sharks carrying these electronic tags patrolled the large seal colonies at the Point Reyes Headlands and Año Nuevo and Farallon Islands during autumn and winter. They then migrated out into the central Pacific Ocean during spring and summer, moving as far west as the Hawaiian archipelago. For example, white shark WS 1, which was tagged at Southeast Farallon Island on 5 November 2004, remained close to the coast for 103 days before migrating westward on 16 February 2005 (fig. 14.9a). Its migration to a region in the central Pacific Ocean called the Shark Café took only 27 days and covered a distance of 2,234 km—consistent with this shark swimming 90 km/day. It remained at the Shark Café for a period of 127 days before commencing its return migration on 31 July 2005. This took 22 days—consistent with an average transit rate of 75 km/day. The homing migration of this shark can best be understood from the color of the symbols, which denote the month of each position determination. The shark resided from September through December close to shore in central California (green circles), moved offshore during January (purple circles), resided from February through June at the Shark Café (purple and orange circles), and returned during July to central California (olive green circles). It delayed its direct

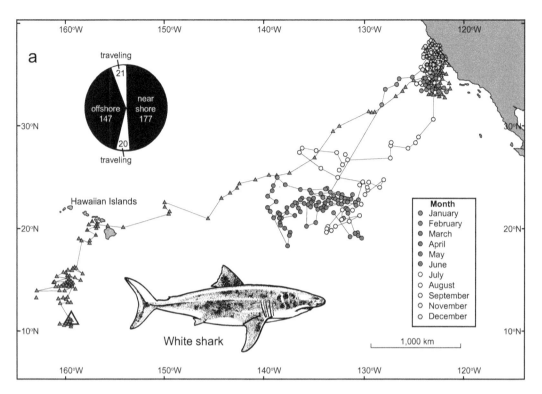

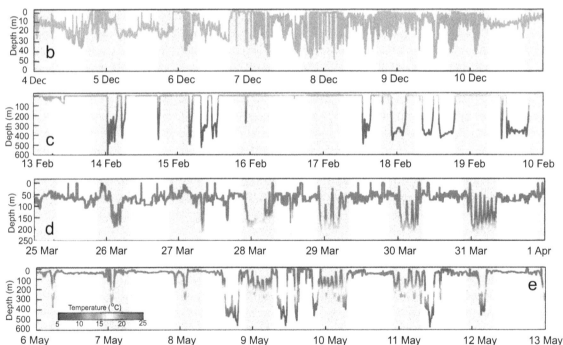

Fig. 14.9 (a) Daily positions of two white sharks, WS 1 (circles) and WS 2 (triangles) indicating their migratory path out into the Pacific Ocean to the Shark Café and Hawaii. The month of the year is indicated by color; the area where the second PAT dislodged from the shark and floated to the surface indicated by a large triangle. One-week time series of depth records for a white shark as it (b) remained within the Gulf of the Farallones, (c) traveled westward through the California Current, (d) continued westward through the subtropical gyre and reached the Hawaiian Islands, proceeded southward and then (e) stayed within a focal area south of the Hawaiian Islands.

return by a short northeasterly movement. The sharks' migratory movements across the vast distances of the ocean are highly directional, in contrast to their movements close the coastline of central California and south of the Hawaiian archipelago.

A second shark WS 2 migrated during the period from January to March in a southwesterly direction from the coast of California farther out into the central Pacific Ocean to the Hawaiian islands (blue and purple triangles) and then veered to the south during April and May before the tag was shed from the shark and stopped emitting a signal (red triangles in fig. 14.9a). Did the shark migrate from one foraging area, a seal colony, to another source of prey? Or did it migrate to a meeting place where males and females court each other as is implied by its appellation, the Shark Café? The evidence available indicates that white sharks engage in courtship along the coast. Mating scars are regularly observed on mature females at the South Farallon and Guadalupe Islands. There are accounts of pairs of white sharks moving in synchrony at Southeast Farallon Island and along the coast in New Zealand.[19] Yet it remains an unanswered question why white sharks make these long-distance migrations out into the center of the Pacific Ocean.

The geographic ranges of the salmon shark (*Lamna ditropis*), shortfin mako shark (*Isurus oxyrinchus*), and white shark overlap in the northeastern Pacific Ocean. The salmon shark has a geographic range of roughly 30° of latitude from 60°N to 30°N in waters ranging in temperature from 3°C to 24°C (blue symbols in fig. 14.10).[20] There are two peaks, one at 5°C and the other at 12°C, to the sea surface temperatures associated with position determinations. The shortfin mako shark is distributed over 30° of latitude from 45°N to 15°N in water ranging in temperature from 10°C to 30°C (red symbols). It has a unimodal distribution to the sea surface temperature across its range. The white shark is distributed from 40°N to 20°N in waters ranging in temperature from 5°C to 28°C (black symbols). There are also two peaks in this thermal distribution, one at 17°C for temperatures encountered in central California and another one at 23°C at the Shark Café. However, there does appear to be a separation in the thermal niche between the salmon and mako sharks, with the former dominating cold boreal and temperate waters with the latter most prevalent in warm subtropical waters. The white shark seems an intermediate between the two, preferring to live at intermediate temperatures at temperate latitudes.

DIVING BEHAVIOR

Two behavioral patterns have often been observed in large sharks during their extended migrations in the open ocean.[21] First, they continuously move up and

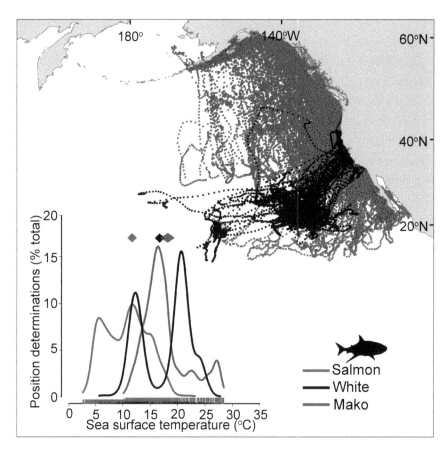

Fig. 14.10 The spatial distributions of salmon shark (blue symbols), white shark (black symbols), and shortfin mako shark (red symbols) in the northeastern Pacific Ocean. With the map is a plot of the distribution of the daily average water temperatures encountered by members of each species over annual periods.

down in the water column. This oscillatory behavior is sometimes called yo-yo swimming. Second, the sharks swim at the surface for long periods of time. Many benefits have been attributed to this oscillatory swimming. An individual that descends into the cold deepwater to capture prey would certainly need to reheat its body upon returning to the warm surface water. Alternatively, an individual might move up and down across two alternating strata of water, one carrying chemical information about its source, and by doing so determine the direction to the origin of that water mass. The fish could then follow it to an area of prey abundance or its spawning site. In addition, a fish could conserve energy during a long migration by quickly propelling itself upward with several strong beats of the tail and slowly glide downward with few beats of the tail. Finally, the shark might descend in the water column to depths at which the magnetic gradients are steeper from the local features in the magnetic field (e.g., the magnetic valleys and ridges associated with lava flows).

Sophisticated electronic tags have provided information about the movements and behavior of sharks and rays in this book. Three types of tags have mainly been used in these studies. The first is the ultrasonic transmitter (fig 14.11a). It emits a signal of a higher frequency than perceptible to the human ear. The transmitters placed on the scalloped hammerhead sharks to study their homing capability were of frequencies between 38 and 42 kHz. To track a shark, you must rotate a hydrophone until the signal from the transmitter is strongest, drive your boat in the direction from which the signal gets stronger, and stop when the signal is equally strong in all directions. You are now located over the hammerhead shark or bat ray carrying the tag. A global positioning system (GPS) within the receiver on the boat continually records the geographical coordinates of the boat (and subject) along with the current time. The receiver also can decode from the signal of the tag measurements made by sensors that record behavioral characteristics such as swimming direction, depth, and speed and properties of the environment such as water temperature and ambient light. An innovation to these tags has been to make them emit a unique burst of pulses, identifying an individual, separated by a long interval of time to increase the life of the tag. The long-lived tags can be detected by a hydrophone-equipped electronic receiver encased within underwater housings moored at a particular location. The study of site fidelity in hammerhead sharks described earlier in this chapter was the first use of this technology in the marine environment.

Another type of electronic tag, attached by a long tether to the dorsal fin of the whale sharks in the Gulf of California, transmits a radio signal with a very stable frequency only when the tag floats on the surface with its antenna rising out of water (fig. 14.11b). This transmitter is located by the ARGOS satellite when it passes overhead by the slight shift in its frequency and the position with a short summary of measurements is downloaded from the satellite to a radio receiver on land. The modern version of this tag is called a SPOT, or smart positioning or temperature, transmitting tag. An email is sent to the investigator daily with information from ARGOS, and the amount of information delivered depends upon how many times the satellite passed overhead of the tagged individual while it was on the surface.

The third type of tag, placed on the basking sharks to track their transoceanic and transequatorial migrations and white sharks to record their migration from the Gulf of the Farallones to the Hawaiian Islands and the Shark Café, is the pop-up archival tag (PAT). It is equipped with a variety sensors, and it stores the measurements recorded by them in memory, and hence is considered an archival tag. This type of electronic tag records daily positions based on continuous measurements of underwater illumination. The idea for how these work originated a meeting in La Jolla of technological nerds arranged by the National Marine Fisheries Service in the mid-1980s. Kim Holland, a researcher at the University of Hawaii, mentioned that ancient mariners determined their longitude by finding the time difference between when the sun was overhead and noon when it was overhead at Greenwich, England. Every hour difference equaled a rotation of 15 degrees westward, or the 360 degree circumference of the earth divided by the complete rotation of the earth in twenty-four hours. It would be impossible for a fish to see the sun at its zenith. I responded by asking why we don't equip the electronic tag with an underwater illumination sensor that would record its level every minute or so. Apparent noon could then be estimated as half the time span between when the light levels increase when the sun rises at dawn and decrease when the sun sets at dusk. Knowing longitude and time of year, the latitudes can be inferred from the length of day. One caveat with these position estimates—they are poor at or near the equinoxes when day length is the same at all latitudes. The accuracy of these geolocations can be improved by taking the temperature measurements stored by the tag when at the surface and comparing them to the surface temperatures of waters near the position determined by the tag. The PAT will release at a pre-set time and float to the surface. A more accurate position of the tag is determined once the ARGOS satellite passes over and this position and a summary of tag measurements is transmitted back to earth as with the SPOT tag. A smalltooth sandtiger (*Odontaspis ferox*) is shown with both a coded ultrasonic beacon and a PAT attached to its dorsum (fig. 14.11c). The PAT is distinguishable from the ultrasonic beacon by its round float made of syntactic foam and antenna for transmitting data to the ARGOS satellite. The three types of electronic transmitters have provided marine scientists with a wealth of knowledge about the movements and behavior of marine fishes over the last thirty years.

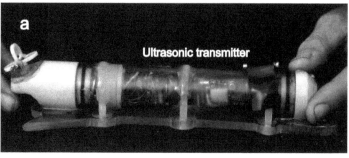

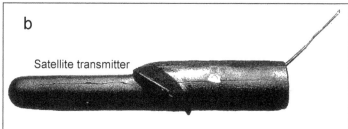

Satellite transmitter

Fig. 14.11 Electronic tags used to record the movements and behavior of cartilaginous fishes. (a) An ultrasonic transmitter used to track hammerhead sharks in the Gulf of California with a paddle-wheel speed sensor (right) and temperature (thumb) and light sensor (forefinger); (b) a satellite-linked radio transmitter used to follow whale sharks in the eastern Pacific Ocean; (c) a coded ultrasonic beacon (right) and pop-up archival tag (PAT) (left) attached to the dorsum of a smalltooth sandtiger shark. The first electronic tag was used to monitor the shark's long-term residency at a seamount near Malpelo Island and the second to track any migratory movement away from the seamount in the eastern tropical Pacific Ocean.

The scalloped hammerhead shark exhibits yo-yo swimming during its highly directional nighttime movements to foraging areas distant in the surrounding pelagic environment. The trajectories of the dives appear steep when depth is plotted over time (fig. 14.12).[22] In reality, the hammerheads descend and rise slowly. There are only eighteen dives made over a period of eleven hours. It is unlikely that members of this species are making the upward excursions to warm themselves. They are ectothermic and adapted to function at ambient water temperatures unlike the white shark, an endotherm that warms its body. Furthermore, the hammerhead sharks do not seem to be moving up and down across the boundary of two water masses, one moving in one direction and the other in another direction. Two water masses are apparent in the graph, in which the yo-yo movements of the sharks are superimposed on temperature contours. The first is evident by the closely spaced isotherms near the surface of the ocean and another by widely separated isotherms deeper in the water column (fig. 14.12c). The shark's excursions do not always cross the boundary between these two contour densities, evidence of the presence of two different water masses. This water mass boundary is further apparent in the change in the directions of currents at the boundary between the masses (see change in direction of the arrows at the same depths in fig. 14.12d). Finally, it would be difficult for a hammerhead shark to return to a seamount or island in the Gulf of California by following the movements of the local currents because the flow is tidally driven. A returning shark would encounter a mass of water flowing in a different direction as the tidal flow changes at six-hour intervals.

It is also unlikely that the hammerhead shark accomplishes these directed migrations to and from a seamount by swimming toward a celestial body or along a long ridge on the seafloor. They make these highly directional movements during nighttime when there is negligible underwater illumination and the shark is moving up and down over a depth range of from 100 m to 400 m, well away from the surface or bottom of the ocean. The level of irradiance near the shark was measured with two distinctly different sensors. The irradiance contours in the uppermost graph (fig. 14.12a) indicate that levels measured with a sensor matched to the low-sensitivity and red-shifted spectral range of the cone receptors in sharks' eyes for daytime vision. The irradiance contours in fig. 14.12b denote the levels measured with a sensor matched to the high-sensitivity and blue-shifted spectral range of the rods that enable hammerhead sharks to see during nighttime when little irradiance is present underwater. Note that during both daytime and nighttime, the sharks swam well below the water that had irradiance levels of 0.0001 μW/cm^2 measured by the photopic (daytime) sensor and 0.001 μW/cm^2 by the scotopic (nighttime) sensor. Irradiance at these intensities is barely perceptible by sharks, and hence it is highly unlikely that they are using visual cues to guide their nighttime migrations.

The magnitude of the geomagnetic gradient was measured at the depth at which a hammerhead shark was swimming highly directionally during a homing movement.[22] This shark was tracked along the boundary between a weak and a strong magnetic lineation. The magnetometer was lowered to points on either side of the shark's track, and measurements were made of the intensity of the total field at 25 m depth increments. The intensity difference was calculated between the intensity of the field at each depth. The shark passed between locations in the evening when its headings were highly directional. It swam at a depth of 175 m on the downward trajectory of an oscillatory swimming path. The magnetic gradient at the depth at which the shark swam was

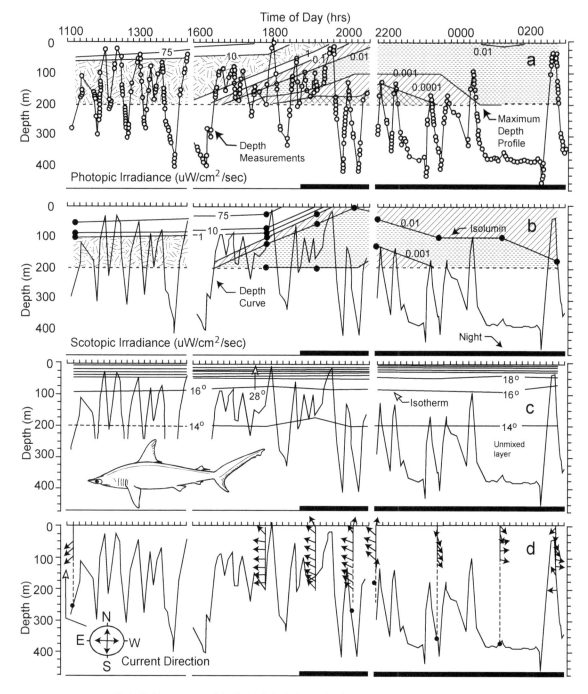

Fig. 14.12 Measurements of depths (circles) of a hammerhead shark superimposed upon contours of irradiance intensity, measured with irradiance sensor matched to (a) the photopic, or high light level red-shifted visual sensitivity and (b) the scotopic, or low light level blue-shifted visual sensitivity. Depths of same hammerhead relative to the distribution of (c) temperature and (d) currents in the water column. Note that the oscillatory swimming movements of the sharks rarely crossed the boundary at 50 m between the two water masses, indicated by the change in spacing between the isotherms.

three times that of the surface. It is possible that sharks make these dives to distinguish and orient to the local pattern of magnetization. The shark was swimming at this time between two magnetic lineations characterized by the maximum change in the local magnetic field gradient.

Finally, the hammerhead shark could be returning to the surface to repay an oxygen debt incurred as it penetrated into the anoxic zone below the surface waters of the Gulf of California.[23] The Humboldt squid, which is a common item in the diet of the hammerhead, makes dives to these depths. The occurrence of prey in this anoxic zone would provide a motivation for the predatory shark to enter this zone.

White sharks also exhibit yo-yo swimming during their long-distance migrations. Their dive patterns differ with the time of the year and geographical location. This is illustrated by a series of week-long dive records of the white shark that traveled from the Gulf of the Farallones to the Hawaiian Islands and back to Southeast Farallon Island.[18] While within the gulf, the shark spent much of its time patrolling the shores of the island searching for seals. At this time it swam at a median depth of roughly 25 m, moving 5 m above and below this median depth but staying within 50 m and the surface (see fig. 14.8b). It spent a significant amount of time (22%) between 1 m and 5 m from the surface. There appeared to be little difference between the daytime and nighttime vertical excursions. The shark began to swim in an oscillatory manner as it traveled westward through the California Current (fig. 14.9c). The oscillations increased in frequency as the shark approached the Hawaiian Islands (fig. 14.9d). The diving record of the shark resembled its movement through the California Current once it moved southward of Hawaii with each of the shark's dives followed by a bout of surface swimming (fig. 14.9e).

It would be difficult for the shark to see the sun or moon while moving away from the surface. This behavior could be the result of the shark orienting to the earth's magnetic field. The white sharks would favor swimming in the surface waters because the earth's dipolar field is most uniform there. The shark could maintain a constant heading simply by keeping the electrical voltage differential induced by its forward movement constant between its ampullary organs. This might enable it to approach the Hawaiian Islands close enough to identify any local magnetic fields associated with the lava flows leading away from the volcanic craters that formed the islands. It then should exhibit oscillatory swimming as it followed these magnetic features to the islands.

The white shark was likely not returning to the surface to warm itself after cooling in the deeper waters because it need only return to the mixed layer to do this and not ascend all of the way to the surface. This was particularly apparent as it traveled past the Hawaiian Islands. The mixed layer in this region is roughly above 200 m. This is apparent from the orange coloration of the upper limbs of the oscillations due to the unchanging temperature of 20°C encountered by the shark as it repeatedly ascended from a depth of 200 m to the surface (fig. 14.9e). One would expect the shark to stop its ascent just above the lower boundary of the mixed layer if the function of the return to the surface were the warming of the body.

Whale sharks make shallow dives at night and deep dives during the day in the Caribbean Sea.[24] These sharks descend to depths ranging from 300 to 1,000 m during daytime. These dives have a periodicity of roughly every 45 minutes. As with the white shark, it is unknown just why they are making these oscillating dives. More information is needed about the vertical distribution of zooplankton to determine whether the sharks are feeding during these dives. The majority of small midwater fishes and crustaceans migrate to the surface at nighttime. Hence, one would expect the whale sharks to swim at the surface at night if they were feeding on them as does the megamouth shark (*Megachasma pelagio*). A megamouth shark was tracked for a period of 50.5 hours in the Southern California Bight between Los Angeles and San Diego. This planktivore stayed above 25 m at nighttime while swimming at depths ranging from 120 to 166 m during the day, altering its depth distribution to match that of its prey.[25]

Juvenile male whale sharks have recently been tracked at Ningaloo Reef off the coast of western Australia carrying multisensor daily diary data loggers with a radio link transmitting the information to the trackers.[25] The tags carry a static accelerometer to record the posture of the shark and a dynamic accelerometer to record the instantaneous change in the shark's swimming velocity. Of particular interest was that the pitches, or the whale shark's downward or upward angles of movement, were derived by calculating the arcsine of the static acceleration measurements and expressed with negative pitch angles indicative of downward-directed movements and with positive pitch angles denoting upward directed swimming. The juvenile whale sharks exhibited a greater diversity of dive profiles than the adult scalloped hammerhead sharks when on foraging excursions away from the seamount. The former often perform the same yo-yo swimming, but also exhibit V-dives, in which the descents and ascents during dives are punctuated by prolonged periods of time swimming at the surface (fig. 14.13a, two upper panels). They distinguished between surface-originating bounce dives and bottom bounce dives that consisted of small oscillatory excursions of less than four meters in depth recorded just above the bottom and V-dives, during which whale sharks remained briefly at the maximum depth, and U-dives, in which the sharks stayed longer at the maximum depth (see lower two panels).

The average pitch angles for descents and ascents are shown for five different behaviors (fig. 14.13b). It is apparent from the similar staircase patterns

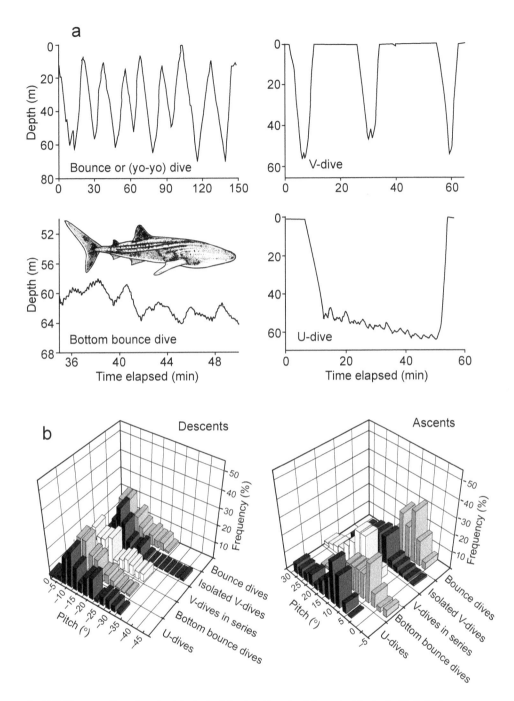

Fig. 14.13 Dive types of juvenile male whale sharks distinguished based on (a) records of the depth of their dives over time and (b) the frequency of pitch measurement recorded by an accelerometer for five different types of dives. The descents are plotted on the left graph with the angles of downward movement indicated by negative values and the ascents shown on the right graph denoted by positive values.

of the histograms that the negative pitch angles to the downward movement of the male whale sharks do not vary much between the five behaviors. Only the isolated V-dives result in angles of descent steeper than 35°, and these measurements are much rarer than the less steep descents in the water column. An estimate of the optimal energy expenditure during bounce dives indicate that the sharks should glide downward at a shallower angle than when swimming upward. Hence, the highest bars on the histogram to the left for the bounce dives correspond to negative pitch angles from 5° to 10°, whereas the highest bars on the histogram to right correspond to positive angles from 10° to 15°. There was little energetic incentive for the sharks to descend at angles steeper than 35°. The average negative angles during descents were roughly the same for all five behaviors, whereas the average angles during the ascents differed significantly.

The shallow angles during descents indicate that the whale sharks are exceptional gliders, and this ability is possible only with the amount of lift produced by dorsoventrally flattened and outward extending pectoral fins. In contrast to the small variation in angles of descent, the ascent angles for the five different behaviors varied substantially (see left histogram in fig. 14.13b). Horizontal travel is most efficient at low ascent pitch angles, which are evident in the bounce dives and bottom bounce dives. The sharks may utilize this energy saving to reduce the cost of filter feeding at greater depths. Evidence exists that some of the prey species at Ningaloo Reef aggregate near the bottom.[26] Bounce dives over a greater depth range may serve dual functions of both scanning the water column for prey and traveling in an energy-efficient manner. The V-dives, characterized by steeper ascent angles, would increase the ability of the shark to detect prey throughout the water column. The depth distribution of prey does vary on the reef, even over a restricted geographic range. The prolonged periods at the surface might serve to increase travel distance and locate widely separated food patches in the environment. In this study, the estimates of the energetic costs of different dive types and the distribution of prey are consistent with the behavior of the whale sharks on the continental shelf of western Australia being closely linked with prey availability and distribution.

SUMMARY

There are diverse movement patterns exhibited by sharks and rays. Some elasmobranchs are ambushers, and their movements are restricted to a confined area. Many skates, some stingrays, and the angelsharks remain in wait on the bottom for long periods of time to ambush crabs and shrimps or fishes that walk or swim close to the bottom. The white shark also assumes the role of ambusher at colonies of seals or sea lions, moving back and forth in the waters adjacent to the coast. Members of other species of sharks actively search for their prey over intermediate distances yet return to a single location to rest and interact with other members of the species. Such is the case of the blacktail reef shark and scalloped hammerhead shark. Those sharks that are planktivorous such as the basking and whale sharks are more nomadic, moving rapidly over large distances in the oceans while staying for prolonged periods of time where plankton is abundant such as at the mouths of rivers, seamounts, and the boundaries between currents.

Many cartilaginous fishes exhibit two types of travel modes, performing highly directional and rapid movements as they migrate between resting or foraging locations and then exhibiting randomly directed slow movements when interacting socially or foraging at their feeding grounds. During their migrations, these species display two types of swimming behaviors, oscillatory diving and surface swimming. The function of these behaviors is not yet understood with certainty. Recent studies of whale sharks on the continental shelf off western Australia indicate that the behavior of juvenile male whale sharks is closely linked with prey availability and distribution.

* * *

KEY TO COMMON AND SCIENTIFIC NAMES

Basking shark = *Cetorhinus maximus*; bat eagle ray = *Myliobatis californica*; blacktail reef shark = *Carcharhinus amblyrhynchos*; blacktip shark = *Carcharhinus limbatus*; California sea lion = *Zalophus californianus*; Haller's round ray = *Urolophus halleri*; Humboldt squid = *Dosidicus gigas*; megamouth shark = *Megachasma pelagios*; northern elephant seal = *Mirounga angustirostris*; Pacific angelshark = *Squatina californica*; Port Jackson shark = *Heterodontus portusjacksoni*; salmon shark = *Lamna ditropis*; scalloped hammerhead shark = *Sphyrna lewini*; smalltooth sandtiger shark = *Odontaspis ferox*; whale shark = *Rhincodon typus*; white shark = *Carcharodon carcharias*; whitetip reef shark = *Triaenodon obesus*.

LITERATURE CITED

1. Vaudo and Lowe, 2006; 2. Standora and Nelson, 1977; 3. Goldman and Anderson, 1999; 4. Klimley *et al.*, 2001; 5. Klimley *et al.*, 1992; 6. McLoughlin and O'Gower, 1971; 7. Randall, 1977; 8. Johnson, 1978; 9. Klimley *et al.*, 1988; 10. Simpfendorfer and Heupel, 2004; 11. Klimley *et al.*, 2005; 12. Gore *et al.*, 2008; 13. Skomal *et al.*, 2009; 14. Sims *et al.*, 2003; 15. Skomal *et al.*, 2004; 16. Eckert and Stewart, 2001; 17. Guzman, 2008; 18. Weng *et al.*, 2007; 19. Francis, 1996; 20. Block *et al.*, 2011; 21. Klimley *et al.*, 2002; 22. Klimley, 1993; 23. Jorgensen *et al.*, 2009; 24. Graham *et al.*, 2005; 25. Nelson *et al.*, 1997; 25. Gleiss *et al.*, 2011; 26. Wilson *et al.*, 2003.

Discussion Questions

1. Little is known about the movements of chimaeras. They are present in deep water at the temperate latitudes yet in shallow waters in the high temperate and polar waters. Hence, they are rarely viewed while diving on SCUBA. Can you design a study to utilize electronic tags to study their movements and behavior? What type of electronic transmitter would you place on them in the first place?

2. Can you design an experiment to test whether a pelagic shark returns to the surface to warm itself after foraging in the cold waters of the deep sea? One idea might be to place on the head of a shark (where the cranial rete is located) a package containing a heat-emitting chemical that would be activated upon reaching a considerable depth.

3. What other experiments can you think of to yield insight into why pelagic sharks either swim at the surface or make repeated dives? Why do they not stay at depths where prey may be more abundant?

4. How different is the diving behavior of endothermic sharks from that of ectothermic sharks? A comparison such as this might yield insight into the function of yo-yo diving.

RECOMMENDED FURTHER READING

Block, B. A., J. D. Jonsen, S. J. Jorgensen, A. J. Winship, S. A. Shaffer, S. J. Bograd, E. L. Hazen, D. G. Foley, G. A. Breed, A. L. Harrison, J. E. Ganong, A. Swithenbank, M. Castleton, H. Dewar, B. R. Mate, G. L. Shillinger, K. M. Schaefer, S. R. Benson, M. J. Weise, R. W. Henry, and D. P. Costa. 2011. Tracking apex marine predator movements in a dynamic ocean. *Nature*, 475: 86–90.

Domeier, M. L. (Ed.). 2012. Global Perspectives on the Biology and Life History of the White Shark. CRC Press, Boca Raton.

Gleiss, A. C., B. Norman, and R. P. Wilson. 2011. Moved by that sinking feeling: variable diving geometry underlies movement strategies in whale sharks. *Functional Ecol.*, 25: 595–607.

Graham, R. T., C. M. Roberts, and J. C. R. Smart. 2005. Diving behaviour of whale sharks in relation to a predictable food source. *Jour. Royal Soc. Interface*, 2003: 1–8.

Jorgensen, S. J., A. P. Klimley, and A. Muhlia-Melo. 2009. Diving of scalloped hammerhead in anoxic zone in Gulf of California. *Jour. Fish Biol.*, 74:1682–1687

Klimley, A. P., R. L. Kihslinger, and J. T. Kelly. 2005. Directional and non-directional movements of bat rays, *Myliobatis californica*, in Tomales Bay, California. *Environ. Biol. Fishes*, 74: 79–88.

Skomal, G. B., S. I. Zeeman, J. H. Chisholm, E. L. Summers, H. J. Walsh, K. W. McMahon, and S. R. Thorrold. 2009. Transequatorial migrations by basking sharks in the western North Atlantic Ocean. *Current Biol.*: 19: 1–4.

Weng, K. C., A. M. Boustany, P. Pyle, S. D. Anderson, A. Brown, and B. A. Block. 2007. Migration and habitat of white sharks (*Carcharodon carcharias*) in the eastern Pacific Ocean. *Mar. Biol.*, 152: 877–894.

Vaudo, J. J., and C. G. Lowe. 2006. Movement patterns of the round stingray *Urobatis halleri* (Cooper) near a thermal outfall. *J. Fish Biol.*, 68: 1756–1766.

Cartilaginous Fishes and Humans

I have yet to be bitten or even witness a shark bite someone despite spending an enormous amount of time underwater viewing sharks in their natural habitat. However, I once did witness a young girl impaled by the poisonous barb of a stingray while playing in the surf off the western coast of Baja California Sur. Having driven many miles in dry, cactus-covered desert on the Baja Peninsula, I saw on the distant horizon the inviting blue color of the Pacific Ocean not far from Guerrero Negro in Mexico. I pulled off the main road with my motorcycle and drove for miles and miles over a bumpy dirt road to reach a lovely white, sandy beach. Here I put on my bathing suit and took a refreshing swim. Suddenly, there was an unexpected loud shriek, followed by several screams, and then I saw not far from me a young girl struggling to swim in the water. I shuffled over to her, grabbed her arm, and lifted her up only to hear her scream, "Mi pie! El dolor en mi pie es terrible!" (My foot hurts! The pain is terrible!). Then her mother suddenly appeared on the beach running toward us screaming, "Mi hija, mi hija! Se ha hecho daño Por favor, ayúdenos!" (my daughter! My daughter is hurt! Please help us!). I carried the girl to shore, sat her down, and looked at the foot that was giving her so much pain. Upon examination, I could see sticking out of her leg just above the ankle a small notched spine of a stingray. I carefully pulled the spine from her body, but she still was hurting terribly and crying. I then told her mother to do what had been taught me many years earlier in a first aid class required for a merit badge in the boy scouts. She should heat water on her stove as warm as she felt her daughter could tolerate and pour the water into a bucket, into which her daughter should put her aching foot. The young girl was still crying and in great pain when her mother put the water in the bucket. However, once her foot had been in the hot water for about five minutes, she stopped crying and said, "Ya no me duele tanto" (The pain is less). We were stunned by this quick reversal. The teacher in the first aid class was correct—the hot water had denatured the venom, a high-molecular weight enzyme, just as one uses a chemical tenderizer to breaks the bonds of the proteins in meat before cooking it. You really have to shuffle your feet when moving over a sandy bottom so you do not step on stingrays hiding in the sand to ambush their prey. If you do step on one, it will quickly elevate its tail and aim its poisonous barb upward to impale a presumed predator.

This chapter is devoted to the interactions between the cartilaginous fishes and humans. The first topic to be discussed is the risk of being bitten by a shark or being impaled by the barb of a stingray; the second topic that will be discussed is the growing interest of ecologically aware tourists in viewing sharks underwater. A few infamous shark attacks will be recounted to illustrate the severe economic repercussions that ensue from such a highly publicized mishap, an explanation will follow of the motivations behind this behavior, and finally the modest risk of shark attack will be placed in the context of many other greater risks in our everyday life. Toward the end of the chapter there will be a discussion of the economic value of shark and ray ecotourism, and information will be presented as to what species can now be viewed underwater and where to go if you want to see them. These days, when someone shouts "There are sharks in the water!," bathers do not frantically run out of the water as in the movie *Jaws*, but those who enjoy the sea put on their skin diving or SCUBA gear to observe these beautiful creatures underwater.

SHARK ATTACKS ON HUMANS

A shark attack on a human usually attracts immediate and extensive attention by the media. The newspaper headlines and articles reporting the attack are usually sensational and frightening. They may have a negative impact on tourism in coastal recreational areas. The magnitude of the impact is amplified greatly when more than one attack occurs at roughly the same place and time. For example, a shark swimming in the waters near Beach Haven, New Jersey, killed Charles Vansant on 1 July 1916. Five days later, on 6 July, a shark killed

Charles Bruder at Spring Lake, New Jersey, only 70 km from Beach Haven. Just six days later, on 12 July, Lester Stilwell and Stanley Fisher were fatally attacked, and Joseph Dunn was seriously mauled by a shark 75 km away up Matawan Creek in New Jersey.[1, 2] The identity of the species responsible for these attacks has never been determined for certain. However, some experts think that the first attacker was a white shark (*Carcharodon carcharias*) and the second a bull shark (*Carcharhinus leucas*), as individuals of the latter species are well known to enter estuaries and migrate upstream in the fresh waters within rivers. These five attacks happened near the Fourth of July holiday, when public beach recreation is characteristically at its peak. The attacks were widely covered by the media. The bad press resulted in bathers avoiding their customary recreational destinations for the rest of the summer in the New Jersey and New York metropolitan areas. This notorious serious of events, which adversely affected the economy of many coastal communities in New Jersey, was the real-life event that served as the basis for Peter Benchley's *Jaws*.

The sensation caused by shark attacks is not confined to the United States. Michael Hely, a sixteen-year-old boy, was bitten in the hip by a sandtiger shark (*Carcharias taurus*) at the town of Amazintoti in the province of Natal, South Africa, in April 1960.[3] Although this incident was not fatal, the shark attack was reported widely in the South African press. This was followed by a fatal attack on Petrus Sithold, a twenty-five-year-old male, 138 km away at Margate, a major resort town also in Natal, in December of the same year. These attacks and two additional ones on young boys, thirteen and fifteen years old, on 6 and 22 January 1961[4, 5] created widespread fear in the local community, and the news of these attacks spread from South Africa across Europe. The negative publicity resulted in a dramatic reduction in the number of tourists visiting Durban, a popular tourist destination, over the next five years. The need to learn more about a shark's motivation to attack humans and to develop ways of avoiding shark attack led to the creation of the South African Association for Marine Biological Research, whose first members were shark researchers, and the building of the Durban Aquarium, in which sharks were held captive for their research studies. Furthermore, the bad publicity also led to the South Africans developing a practical solution to alleviate the shark attack problem. They set protective nets along the coast of South Africa to capture sharks and reduce the risk of shark attack in the local waters.[5]

Even today a particularly gruesome shark attack or several less serious attacks occurring over a short period of time will invariably attract great media attention and cause widespread fear among the public. Recently five people were attacked by sharks, one fatally, in an 8 km section of the coast near the resort town of Sharm El-Sheikh, in Egypt between 30 November and 5 December 2010. One of these attacks involved Olga Martsinko, who was snorkeling with her daughter, with the shark grabbing her outstretched arm and severing it below the elbow. Only a few minutes later, Lyudmila Stolyarova, an elderly woman, had her foot severed by an oceanic whitetip shark (*Carcharhinus longimanus*). These and the other attacks at this time were highly publicized by the British Broadcasting Corporation. Shark experts from the United States were also invited to visit the resort and provide an explanation for the attacks. Their general conclusion was that these attacks were the result of dive guides and tourists feeding sharks in the area.

Motivation

One can minimize the chances of being attacked by reducing the motivation of the shark to attack. Hunger is one obvious motivation for a shark to attack a human. Some attack wounds do resemble the bites made by a shark when feeding. On the other hand, the magnitude and shape of attack wounds do not always bear this out. If attacks were entirely hunger motivated, you would expect the prey to be consumed either partially or wholly. Indeed, this is the case in some instances. During October 1939 two divers were witnessed being attacked off a beach in New South Wales, Australia. On the following day their remains were recovered from the stomach of a 3.5 m long tiger shark.[8] More often the evidence for a shark attack is less conclusive, such as when the remains of a victim are found in a shark's stomach but the attack is not witnessed. The victims may well have drowned before being consumed. These inconclusive shark attacks are usually classified as scavenging events rather than attacks.

In most instances, victims of shark attacks have experienced the loss of only small amounts of flesh. One such example occurred off West Palm Beach in 1968.[7] A young boy was attacked close to shore while snorkeling. Teeth marks in the shape of a half-moon were impressed upon the bottom of both of his swimming fins. These marks could only result from a slashing movement made by outstretched jaws rather than a vertical bite. Severe wounds were inflicted between the boy's knee and ankle. These also lacked the characteristic puncture wounds of the pointed gripping teeth of the shark's lower jaw. Although close to 1,000 sutures were needed to close the child's wounds, very little flesh was lost. This observation led David Baldridge, a physician with the US Navy, to publish an important scientific article, entitled "Shark attack: feeding or fighting?"[8] He questioned whether the traditional hunger-motivated explanation for shark attacks was correct and suggested that attack behavior might rather be defensive in nature—the attack might be triggered by the victim's intrusion into an area protected by the shark.

Defense

When animals feel threatened, they experience conflicting instincts—one is to escape and the other is to fight. If they are unable to flee because their opportunity to escape is blocked, they do not always fight but often perform an agonistic, or aggressive, display. Agonistic displays are conspicuous and exaggerated postures and movements that convey the threatened animal's ill ease due to the presence of another and advertise the capacity to inflict harm should the intruder come any closer.[9] The hunching posture, erection of hair, baring of teeth, and hissing of a cat when confronted by a barking dog is a display that conveys to the dog the cat's readiness to inflict bodily harm should the dog move any closer. The display enables both the cat and the dog to avoid the injuries that they would sustain in a fight. The holding upward of clenching fists in front of our chest and revealing our teeth is a human display that is directed at those that we perceive as dangerous. The purpose of an agonistic display is to discourage an intruder from approaching any closer without resort to the actual physical contact that occurs during a fistfight.

An agonistic display was first described for the blacktail reef shark (*Carcharhinus amblyrhynchos*).[10] There are five postural components to its aggressive display. They can be identified by comparing the body posture of a normally swimming shark with that of a displaying shark from three different perspectives, from its side, front, and above (fig. 15.1). The first postural component is the upward pointing of the snout, the second is a bend between the chondrocranium and vertebral column caused by the elevation of the snout, the third is a lowering of the pectoral fins to bring them close to each other, the fourth is the arching of the shark's back, and the fifth is lateral bending of the body with the tail pushed to either side in an exaggerated arc. In a sense, the shark is attempting to increase the size of its profile when viewed from the side. This is similar to a Siamese fighting fish unfurling its dorsal and ventral fins to intimidate a competitor or a cat erecting the hair on its body and tail to appear larger than normal. The difference is apparent in photographs of a displaying (fig. 15.2a) and a nondisplaying shark (fig. 15.2b).

The best way to distinguish this aberrant form of locomotion is to study a time series of the postures of a displaying shark. A normally swimming shark moves its tail back and forth over an arc of roughly 60° while keeping the forward part of its torso pointed directly ahead. This is best appreciated when the shark is viewed from above (fig. 15.3a). In exaggerated swimming, the shark moves its tail from one side to the other over a larger arc, 90°, while its forward torso swings slightly to one side and then to the other in a compensatory manner (fig. 15.3b). The trajectory of the displaying shark is best viewed from

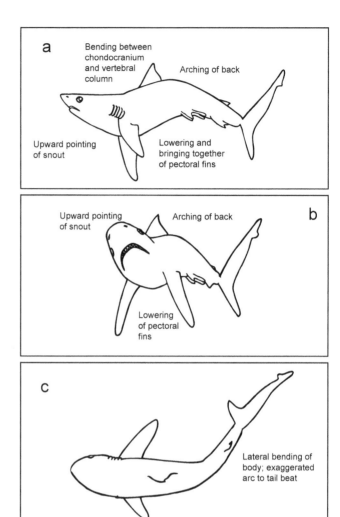

Fig. 15.1 The postural components exhibited during a display of the black-tail reef shark from three perspectives, viewed from (a) the side, (b) the front, and (c) the top.

the side (fig. 15.3c). The shark moves slightly upward and rolls onto its side while moving its tail to one side (postures 1–2) and then accelerates upward and rolls to the other side while moving its tail in an exaggerated manner to the other side (postures 3–4). The shark may alternatively continue upward and then downward in an even more pronounced mode of swimming, termed spiral looping (postures 3–6).

The intensity of the display can vary from mild to high. Its intensity is evident by the degree of horizontal compression to the looping trajectory of the shark while swimming. A diver who swims slowly toward a blacktail reef shark but remains at a distance will find that the shark slowly turns and swims directly away. A person who continues to move closer to the shark may find that the shark swims in a sinusoidal path while rolling from one side to the other and moving up and down in the swimming column. The zealous underwater

Fig. 15.2 Photographs of a shark (a) performing agonistic display and (b) normally swimming.

photographer that charges rapidly toward the shark in his eagerness to get a better picture will find that the shark now swims in a trajectory comprised of wide spiral loops. This shark may inflict bodily harm if the photographer continues forward and arrives in close proximity to the shark. It is now likely to swim in highly compressed spiral loops, turn around, and then charge directly toward the diver. The final dash toward the surprised witness usually culminates in the shark's rapidly inflicting a slashing bite. Let me repeat this important point—the shark will attack almost invariably if you do not retreat

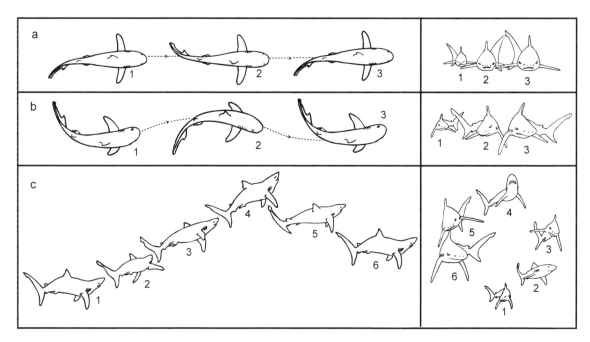

Fig. 15.3 A shark (a) swimming normally, (b) exhibiting exaggerated swimming, and (c) performing spiral looping shown from the side and from the front.

but continue to approach the shark. Similar displays have been observed by the blacknose (*Carcharhinus acronotus*), lemon (*Negaprion brevirostris*), silky (*Carcharhinus falciformis*), and bonnethead (*Sphyrna tiburo*) sharks.[11]

The sequential relationship between the display and an attack has been demonstrated experimentally.[12] A submersible the size of a large shark was steered toward blacktail reef sharks in attempts to evoke the agonistic display. This intrusive behavior often stimulated the shark to strike the submersible. The sharks were closely followed with the submersible, keeping it immediately behind them as they swam slowly in a looping trajectory with exaggerated tail beats. The chance that a display would be elicited was greatest when a shark became cornered against the reef. Of the 38 times that the submersible approached reef sharks closely, 27 approaches resulted in agonistic displays of varying intensity; 5 of these displays were mild displays, 12 approaches resulted in moderate to strong displays, and on 10 occasions the shark struck the submersible. The display behavior, once begun, often continued for 30 seconds or more.

The displaying shark performed one of three behaviors in response to the submersible. The shark slowly accelerated to gradually outdistance the submersible, swam in a circle around it, or slowed down while intensifying its display as the submersible drew closer. This last behavior led most often to an attack. This was initiated when the submersible came within a critical distance of the shark. Immediately prior to the attack, the shark appeared tense. Its body axis was rolled over 45° from the horizontal plane, its snout was lifted high, the pectorals depressed, and its back was arched so that the rear torso and tail were oriented downward. Also, the shark ceased to move forward and began to sink in the water. The ensuing attack was begun from a position just below the approaching submersible (see third frame down in first column in fig. 15.4). The attack occurred after 5–20 seconds of pursuit. It consisted of a sudden acceleration directly toward the submersible (see fourth frame down). The strike was initiated when the shark was 1.5 to 5.0 meters from the submersible. The final approach was rapid. In one instance, it took the shark only 0.33 seconds to reach the submersible and bite the propeller on the fin of the submersible (see bottom three frames in second column). The shark rapidly swam around the side or behind the submersible and accelerated into a high-speed strike from there outside the view of the pilot in three instances. In almost all of the attacks, the shark rapidly accelerated and struck the submersible with great force. The shark delivered bites to the forward end of the submersible. Four strikes were directed at the acrylic dome, four on the upper fiberglass hull, one on the motor on the right side fin, and another on the left forward fin. In two cases, the shark twice bit the submersible—the second bite delivered in less than a second from the first.

Feeding

There are three species of sharks that have been observed to occasionally prey on humans, although humans are not a large component of their normal diets. These species are the bull, tiger, and white sharks. Adult tiger sharks feed on a diverse diet of large vertebrates such as sea turtles, albatross fledglings, and juvenile dugongs. Humans have also been found in their stomachs. Adult bull sharks occasionally prey on marine mammals including sea lions[7] and dolphins,[13] and are fully capable of consuming humans. Adults of this species enter estuaries and rivers. One was possibly responsible for the attacks on the three bathers in Matawan Creek in July 1916. Additionally, hunger is likely the motivation behind white sharks attacks on humans off the central Pacific coast of the United States, southwestern coast of Australia, and southern coast of Africa.[14]

A great deal of attention has been devoted in the popular media to what happens when a white shark attacks a human because of the occasional attacks that are directed toward divers and surfers on the Pacific coast of North America. Insight into attacks on humans has been provided by observations of natural

Fig. 15.4 The agonistic display and frontal attack directed at pursuing submersible. The time sequence of frames increments downward to bottom on right column and follows the same order on the left column. Note that the shark was in a very tense posture, rolling over to a 45° angle relative to the horizontal axis (see frames 2-3 in left column) before accelerating forward and abruptly turning to bite the right forward motor twice, and breaking the plastic propeller.

feeding of white sharks on northern elephant seals (*Mirounga angustirostris*) made from the vantage of Lighthouse Hill in the waters surrounding Southeast Farallon Island.[15] The initial strike of the shark was rarely witnessed during an attack. This is likely because the prey is seized while swimming underwater. Rather, the observers most often were alerted of the attack by a large blood-stained area of water and seagulls circling above. The bloodstain elongated in one direction, and then the shark reappeared at the surface with the seal beside it. The shark was often seen swimming with wide tail beats. Such beats would be necessary for the shark to propel itself forward when carrying a heavy seal in its jaws. After a prolonged interval the seal then rose to the surface and floated in an immobile state with a bite of flesh missing. Significantly, the wound, in most cases, was no longer bleeding. The shark then surfaced and swam to the carcass and seized it. This scenario is consistent with the suggestion that the white shark kills its prey by exsanguination, or blood deprivation. The shark likely holds the seal tightly in its jaws until it no longer bleeds.

A commonly held belief is the sharks bite seals and then spit them out in a wounded but intact state to die before again attacking them.[16] It is true that they spit them out, but they do not wait for them to die at the Farallon Islands.[15] The sharks chased and captured 64% of the seals that survived the original attack. Furthermore, the sharks did not return immediately to feed on the carcass once the seal died. Seal carcasses were observed floating in immobile states for as long as 140 minutes, before sharks returned to feed on them.

Two puzzling questions—and two that are important to people who might wish to survive a shark attack—is how a shark decides, first, what it is going to attack, and second, what it is going to eat. There are some similarities and some important differences when a shark bites a human being as opposed to when it bites a seal. At 2 p.m. on 9 September 1989, a white shark attacked Mark Tisserand, a commercial abalone diver, at South Farallon Islands. The attack began like many attacks on seals. A white shark seized Tisserand's leg while he was 5–8 m below the surface, nearly 200 m from shore. At this time, he was pausing in a prone position to clear his ears. He said that "the shark swam up from underneath, seized me, carried me down for five to seven seconds, and suddenly let me go and swam off." He lost much blood at this time. The shark carried him underwater for some distance as it would do with a seal or sea lion. However, Tisserand was able to strike the shark with the butt of a bang stick three times and it released him. What is inconsistent with the idea that the shark exsanguinates a human is that it released the abalone diver in an intact state and swam off. In a seal attack, the shark would have carried its prey longer underwater and removed a bite from its body. The course of events reported by Tisserand is typical of white shark encounters with people off California. A white shark was also observed to seize and release a brown pelican, even though it was quickly disabled and unable to resist further attack. The bird was left bleeding profusely and struggling at the surface until it died two minutes later. The shark never returned to feed on the pelican. Furthermore, many sea otters are found whole but dead along the California coast with fragments of white shark teeth embedded in their open wounds.[17] A sea otter has yet to be found in the stomach of a white shark.[18]

There may be a connection between all of these seemingly unrelated observations. Humans, birds, and sea otters are composed mainly of muscle, whereas the preferred prey of sharks—sea lions, dolphins, and whales—are composed mainly of fat. Sharks may prefer energy-rich marine mammals to other comparatively energy-deficient species. Supporting this is the propensity of white sharks to selectively remove the blubber from whale carcasses floating on the surface of the water and not to eat the muscle underneath.[19] This may explain why white sharks rarely completely consume humans after they attack them. Finally, white sharks are found to readily consume chunks of fat removed from seals in which electronic tags are hidden, but refuse to eat when offered the muscle from the interior body of a seal.[15]

Further evidence of the reluctance of white sharks to consume humans wholly comes from the International Shark Attack File. In 56.8% of the 125 attacks on humans recorded worldwide in 1910–1995, the white shark bit only once and then departed.[14] No flesh was bitten off in 10.4% of the attacks. Two such bites were observed in only 16.8% and one in 6.2% of the attacks. In striking contrast, only 1 sea lion escaped from a predatory white during the 129 attacks on pinnipeds recorded by video at the Farallon Islands, resulting in a predatory success rate of 99.2%.[16] The intact remains of a whole human have yet to be recovered from the stomach of the attacking shark after its capture.[14] Yet the possibility exists that a highly hunger-motivated white shark might consume a human in the absence of more nutritious prey.

International Shark Attack File

Information on all shark attacks on humans worldwide is collected in the International Shark Attack File,[20] maintained by George Burgess and his associates at the Florida Museum of Natural History Museum of the University of Florida. More than 4,000 records of shark attacks on humans have been entered into this database since its establishment in 1958. They have been found in the scientific literature as early as during the mid-1500s. It includes statistics on the frequency of occurrence of unprovoked attacks, defined as incidents where the shark attacks a live human being in its natural habitat without provocation.

The number of unprovoked attacks by all species of sharks worldwide on humans per five-year period has risen from 8 during 1900-1904, of which 2 were fatal, to 330 during 2000-2004, of which 29 were fatal (fig. 15.5). The number has risen steadily. This is likely due to the continuous growth of the human population and the growing amount of human recreation and commerce on the ocean. The increase in attacks is also, in part, an artifact of the increased reporting efficiency of the International Shark Attack File. Over the last 20 years, it has expanded greatly through Internet communication with cooperating scientific observers around the world. Yet the average yearly number of attacks by all sharks during the five-year period from 2000-2004 was only 66 per year, of which 5.8 were fatal. This was slightly greater than the 266 total attacks, 53.2 attacks per year, of which 6.4 were fatal, during the prior five years from 1995 to 1999. The percentage of attacks that are fatal has steadily decreased from the 60% fatality rate of 1905-1909. Only 12% and 9% of the attacks were fatal during the two most recent five-year reporting periods. The low percentage of fatal attacks in recent years is the result of vastly improved medical response and treatment for traumatic injuries like those received from shark bites.

The risk of being attacked by a shark is truly negligible compared to other risks encountered by humans on a daily basis. The National Safety Council tallies and makes public statistics on the average number of deaths caused by various other injuries over three-year periods. The yearly average of 66 attacks on humans, of which 5.8 were fatal, can be compared to the average yearly rates of fatalities caused by other everyday risks over the same period. The number of fatalities in motor vehicle accidents within the United States averaged 42,593 per year over a three-year period from 1993-1995 (fig. 15.6a). The risk of dying from an automobile in the United States is six thousand times greater than that of dying from a shark attack in any place in the world. The deaths due to a fall, a common cause of the loss of life among the elderly, averaged 13,524 per year—a risk exceeding that of dying from an attack by a shark by a factor of greater than 2,000 during that time period. The risk of dying from a shark attack anywhere in the world was far less than that of drowning during recreation, which averages 791 deaths per year, drowning in a bathtub, 296 deaths occurring per year, or dying by being hit by a bolt of lightning, which averages 72 deaths per year, in the United States. For example, the number of attacks worldwide was less than the 220 deaths per year in the United States of people falling off a horse (fig. 15.6b). Stings from a hornet, wasp, or bee result in 49 deaths per year. Finally, the risk of being killed by sharks anywhere in the world was less than that of dying after being attacked by a pack of dogs, which results in 16 deaths per year. In conclusion, given that you avoid swimming near a seal or sea lion colony, the risk of being attacked is minuscule in comparison with other hazards you encounter everyday.

Avoiding Harm from Sharks

To avoid being attacked by a shark, it is best to take some basic precautions. These precautions include avoiding areas where sharks may be feeding and their prey is abundant such as at seal colonies and near schools of fish, and not acting in a manner that may mimic their natural prey. Do not enter the water if bleeding, or if there are other chemical attractants in the water caused by fishing activities or sewage effluents. Do not enter the water alone, as sharks tend to target solitary individuals. Avoid murky waters or swimming in low light conditions when a shark's vision is limited. If you are in the presence of

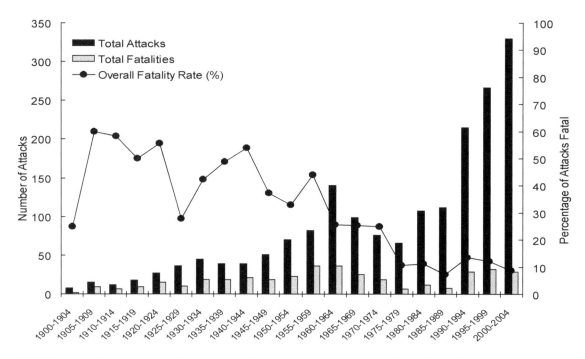

Fig. 15.5 The total number of attacks from all species of sharks, the number of those that are fatal, and the overall rate of fatality plotted during five-year periods from 1900 to 2004.

a shark, it is obviously prudent not to harass or otherwise threaten it, and you should leave the water as quickly and calmly as possible.[21]

When confronted by a threatening shark, you want to recognize the motivation of the attacker—that is whether it is attacking in order to defend itself or to feed upon you. For that reason, a good practice would be to wear underwater goggles when in waters where sharks are common. Then you can recognize the agonistic display of a shark and act to reduce the likelihood of being attacked. It is imperative not to continue swimming toward the shark, particularly at a rapid rate, or to force it into confinement such as against a coral or rocky reef. When confined, the shark no longer has the option of flight and is likely to fight. Its likelihood of doing so is apparent from the speed at which it swims and the compressed nature of its trajectory. The shark likely will not accelerate directly toward your torso to deliver a bite. Rather it will dash by your side and quickly turn around to seize another part of your body such as the arm or leg. Its bite will likely result in a slashing wound, not a bite wound.

When confronted by a large predatory shark such as a bull, tiger, or white shark, it is important to face the shark to indicate your awareness of its presence. Predators often eschew attacking animals aware of their presence because they risk injury such as being bitten by their teeth or scratched by their claws. If with a buddy, you should assume a position back to back so that you

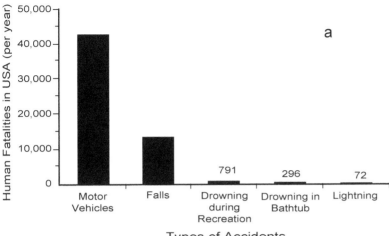

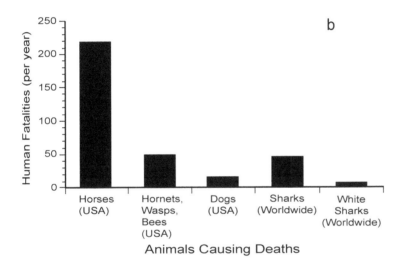

Fig. 15.6 Risks to humans from hazards other than sharks. (a) The average rates of fatalities per year due to accidents over a three-year period from 1993 to 1996. (b) The average rates of fatalities per year from interactions with other species compared with the annual rate of attacks by all species of sharks worldwide or the white shark worldwide.

can collectively see in all directions. It is essential not to appear frightened. Do not swim rapidly away from the predatory shark as this withdrawal-type behavior most often elicits a strike from predators.

STINGRAY INJURIES TO HUMANS

The stingrays of the order Myliobatiformes are present in the tropical and temperate oceans worldwide as well as in estuaries and freshwater rivers. They are named for the poisonous barb on their tails. There are nine families of stingrays. The barbs of the whip stingrays of the family Dasyatidae have the longest barbs, reaching up to 37 cm in length; the butterfly rays of the Gymnuridae the next longest with a maximum length of 12 cm, while the eagle rays of the

Myliobatidae and round rays of the Urolophidae have the smallest barbs with maximum lengths of 4 and 2.5 centimeters, respectively.[22] Fig. 15.7a shows the tail of whip stingray with its long barb covered by skin and fig. 15.7b shows a barb with its many serrations held in a human hand. Stingrays are common in shallow waters, where they remain buried just under the sand or mud with their eyes exposed waiting to seize crustaceans and fishes as they pass by them near the bottom. If a predator approaches too closely, either alerting the ray by water motion or contact, the stingray will raise its tail in a whip-like manner and its spine will be thrust into the predator, releasing the venom stored within the spine. Stingrays also can move quickly and agilely in a circle while thrusting their barb sideways or backwards over their bodies to hit the predator.[23] Bathers step on stingrays while in the water at the beach because they are hard to see on the bottom through the sea surface, and when one steps on the pectoral disk of the stingray, it reflexively lifts its tail and stabs the barb into the ankle or lower leg of the bather. The venom along with fragments of the barb are often left in the body, and they cause the victim great pain. Fishermen are also frequent victims of stingray wounds because they are stung in the hands as they remove the rays caught in their fishing nets.[24]

The stingray spine is made of vasodentin, a bone-like cartilaginous substance, and has serrated edges along both sides with a sharp, pointed distal end that easily penetrates the skin. Notice that the serrations on the barb held in the hand are angled away from the pointed end and hence increase the size of the laceration when pulled out of tissue in the same way that a series of metal leafs at the opening to a one way entrance to parking lot will puncture a tire if driven across in the opposite direction. The saw-edged spine is covered by a thin layer, the integumentary sheath (see fig. 15.7a). It has secretory cells within the epithelium that contain cylindrical or elliptical vesicles that produce a proteinaceous venom that covers the serrated barb.[25] When the spine is thrust into the victim, the epithelial sheath comes apart, and the venom is released into the wound.[26] The mucus causes great pain to the victim following the sting.

The location of the barb on the tail, the number of serrations on either side of the barb, and the number of barbs varies greatly with the species of stingray, its sex, and its habitat.[27] The location of the venom secretory cells varies among species. In some species, they are either concentrated around or inside the serrations, and this difference in location results in more or less severe envenomation. Their location in the epithelial layer also differs among species. Multiple rows of secretory vesicles are either interspersed with epithelial cells or occur in separate layers below the integumentary sheath.[25] The number of serrations is dependent upon gender, with males having up to twice as many serrations as females. Stingrays that inhabit open water have more serrations

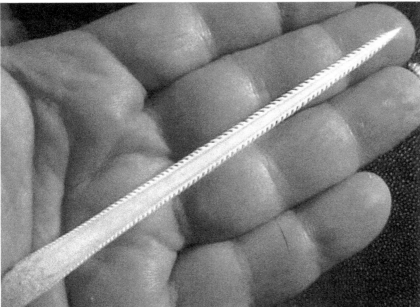

Fig. 15.7 (a) The tail of a whip stingray of the family Dasyatidae with the barb covered by the integumentary sheaf. (b) The barb of a whip stingray held in a human hand.

on their barbs than those that reside on the bottom. The freshwater stingrays of the family Pomatotrygonidae have secretory cells located along the entire length of the barb in contrast to the marine species that have them located only within and around the ventral, or lower, grooves on the side of the barb. The larger number of secretory cells, distributed along the length of the barbs of freshwater stingrays, explains why these species are capable of inflicting more severe envenomations than the marine stingrays.[28] Freshwater stingrays

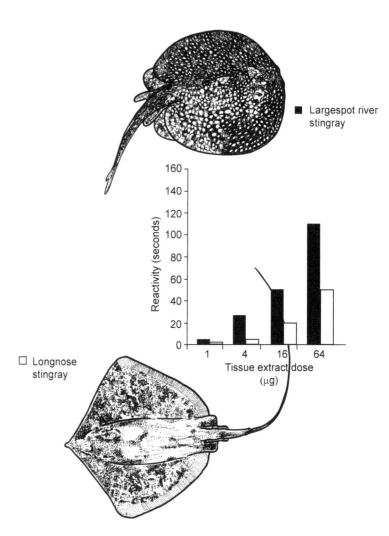

Largespot river stingray

Longnose stingray

Fig. 15.8 The responses of mice to envenomations of four increasing doses of tissue extract from the largespot freshwater ray and the longnose ray.

also have specialized secretory vesicles that have higher protein concentrations with the result that their venom is more toxic than that of marine stingrays.[29] This is apparent in the difference in the duration of reactivity of mice to venom in four concentrations of the largespot river stingray (*Pomatotrygon falkneri*) and the longnose stingray (*Dasyatis guttata*) (fig. 15.8).[29] Mice were injected with the four increasing doses of venom, and the reactivity was measured by the amount of time spent licking and biting the injected area during a thirty-minute period. The dark left bars, indicating the reactive periods for the largespot, are higher than the white right bars, denoting the reactive periods for the longnose for all four tissue extracts of 2 μg, 4 μg, 16 μg, and 64 μg, indicating a greater reactivity to the freshwater stingray envenomations. At a dosage of 4 μg, the freshwater stingray's venom was four times as potent as the marine stingray's venom.

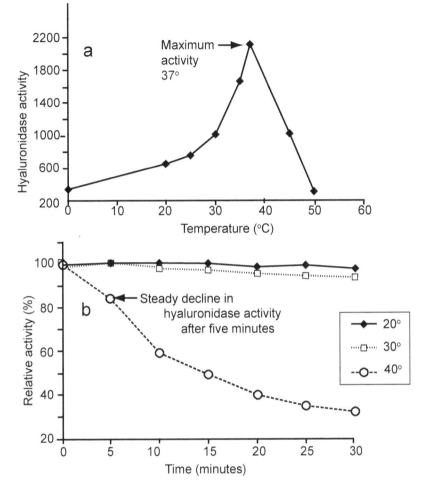

Fig. 15.9 (a) The degree of hyaluronidase activity at different temperatures. (b) The relative hyaluronidase activity occurring at three temperatures. The activity of this active constituent of stingray venom begins to decrease in five minutes when exposed to a temperature of 40°C with a concomitant decrease in pain.

The venom of marine stingrays is composed mainly of three components: large-molecular mass polypeptides, serotonin, and hyaluronidase. Serotonin is a neurotransmitter that causes smooth muscle to severely relax in the body. Hyaluronic acid comprises a large portion of the extracellular matrix in the tissues in the human body. The hyaluronidase in stingray venom is an enzyme that breaks up hyaluronic acid in order to maximize absorption and diffusion rates of the venom through the tissues of the body. This enzymatic protease has a broad substrate specificity, and this permits it to attack multiple types of proteins in the body and cause extensive tissue damage. The hyaluronidase activity is greatest at a temperature of 37°C (see peak to curve in fig. 15.9a).[30] The venom is very heat sensitive, becoming inactive at high temperatures. At temperatures of 20 and 30°C, the venom is stable for 30 minutes (see upper two curves in fig. 15.9b). However, at a temperature of 40°C, the hayaluronidase activity begins to decline after 5 minutes to 30% of its initial activity after

Fatalities from the venom of stingrays are extremely rare. Although the stingray's venom causes great pain, rarely does it cause a death. As mentioned earlier, the stings usually occur to the ankles or lower legs when a bather accidentally steps on a stingray buried in the sand and the frightened ray suddenly lifts its tail upward and to the side impaling the victim's lower leg with the barb on its tail. A fatality would only occur if the barb penetrated the abdomen or heart of a person. Steve Irwin was snorkeling in less than two meters of water while filming a documentary, titled *The Ocean's Deadliest*, off the coast of Australia. According to witnesses, he swam above the front of an Australian bull ray (*Myliobatis australis*) while his cameraman was positioned behind the ray to block its escape. Irwin and the cameraman exposed the ray to the same fight-or-flight situation as the photographer swimming toward a shark with escape prevented by a coral reef surrounding it. This situation would elicit an intense agonistic display from a shark and perhaps trigger an attack on the diver. The same situation elicited a defensive response from the large ray, over a meter wide and two meters long—it suddenly lifted its tail with the barb at the base of the tail lifted upward. The ray drove the poisonous barb through Irwin's chest and into his heart, and this likely caused complete and irreversible cardiac standstill, resulting in his immediate death.[25, 30]

30 minutes (see lower curve in fig. 15.9b). This explains why the little girl in Baja California began to feel better as little as five minutes after her envenomated foot was placed in a bucket of hot water.

ECOTOURISM

Ethological studies in the early 1980s dispelled the perception that all sharks attack humans.[31, 32, 33] Nature documentaries featured scientists swimming among schools of hammerhead sharks at an offshore seamount in the Gulf of California without being attacked. People did not distinguish between the scalloped hammerhead (*Sphyrna lewini*) and the more dangerous great hammerhead (*Sphyrna mokarran*) at that time, and the generic hammerhead was considered to be the third most dangerous shark species.[7] Scalloped hammerhead sharks feed primarily on fishes and squid in the Gulf of California[34] and thus should not be expected to feed on humans, in contrast to the white,[35, 36] tiger,[37] and bull sharks,[38] which feed on mammalian prey and might be expected to attack humans. It is safer to swim among schools of these piscivorous sharks, which are not as aggressive and are solely concerned with maintaining their uniform spacing and common heading with other school members,[39] than confronting a solitary shark that is likely to exhibit an agonistic display if frightened by you.[12]

Shark ecotourism is immensely popular now. There are many opportunities for viewing sharks and rays in their natural environment in the world's oceans (table 15.1).[40] The number of ecotourism operators varies at each destination as well as the shark or ray species that are main attractions and their ecological lifestyles. The ecotourism sites are presented for seven regions. Many SCUBA divers know about the best sites for viewing sharks such as Wolf and Darwin Islands in the Galapagos Archipelago, where six boats take divers to view immense schools of scalloped hammerhead sharks, or the reefs of Utila in Hondurus, where over twenty boats take tourists to see whale sharks (*Rhincodon typus*), or Dangerous Reef off Port Lincoln, Australia, where four boats take tourists to see white sharks. They were first filmed here for the nature documentary *Blue Water, White Death*. Divers are also familiar with Stingray City off the Cayman Islands, where twenty boats bring tourists to feed southern stingrays (*Dasyatis americana*). However, few of us know that there are many other lesser-known ecotourist destinations such as the Isle of Man between the islands of Great Britain and Ireland, where you can view basking sharks (*Cetorhinus maximus*) from a boat while they swim at the surface, or Jardines de la Reina off the coast of Cuba, where you view bull and silky sharks underwater, or Ponta D'Ouro in Mozambique, where scalloped hammerheads and whale sharks can also be observed while SCUBA diving, The most ecotourist destinations, 22% of the total number, are found in Oceania (see fig. 15.10a). This region includes South Australia and Chatham Island off New Zealand, where white sharks are the main attraction, Queensland, Australia, where the tropical blacktail reef and silvertip sharks can be viewed while SCUBA diving on the Great Barrier Reef, and French Polynesia, the Solomon Islands, and Fiji, where the same species can be viewed on coral atolls. Fig. 15.11 shows the number of operators at each of these ecotourism destinations in each of the world's oceans. North America and the Greater Caribbean are tied at 16% for the second most tourist destinations. Central and South America and Southern and Eastern Africa are the next most frequent destinations for viewing sharks. The most common main attractions are reef shark species and the whale shark, followed closely by the scalloped hammerhead shark (fig. 15.10b).

Economic Value

The economic value of shark and ray ecotourism has been estimated only at a few locations. For example, the economic value of ecotourism for white sharks in Gansbaai, South Africa, is estimated to be roughly $4.4 million every year (table 15.2). The earnings of whale shark ecotourist operators at Ningaloo Reef in Western Australia and in the Seychelles Islands are $5.9 and $4.9 million,

TABLE 15.1. Summary of established global shark ecotourism operators from examination of internet websites.

Location	Region	Operators	Main attraction	Life-style
Rhode Island, USA	NA	2	Blue, shortfin mako sharks	P
Long Island, USA	NA	1	Blue, shortfin mako sharks	P
Stellwagon Bank, MA, USA	NA	2	Basking sharks	H
Nantucket, MA, USA	NA	1	Blue, shortfin mako sharks	P
Outer Banks, NC, USA	NA	8	Sandtiger	CR
Venice, Louisiana	NA	1	Dusky, silky sharks	CR
Palm Beach, FL, USA	NA	2	Caribbean reef	CR
Farallon Islands, CA, USA	NA	5	White sharks	H
Catalina Island, CA, USA	NA	3	Blue, shortfin mako sharks	P
San Diego, CA, USA	NA	1	Blue, shortfin mako sharks	P
Hornsby Island, BC, CA	NA	1	Bluntnose sixgill sharks	CR
Haleiwa, HI, USA	NA	2	Galapagos sharks	CR
Honolulu, HI, USA	NA	4	Galapagos, whitetip reef sharks	CR
Kona, HI, USA	NA	8	Giant manta rays	CR
Isla Guadalupe, Mexico	CSA	5	White sharks	H
Sea of Cortex, Mexico	CSA	4	Scalloped hammerhead, whale sharks, giant mantas	CR,H
Revilligigedos Islands, Mexico	CSA	4	Scalloped hammerhead, whale sharks, giant mantas	CR,H
Playa del Carmen, Mexico	CSA	1	Bull sharks	CR
Isla Holbox, Mexico	CSA	>20	Whale sharks	H
Ambergris Caye, Belize	CSA	4	Whale sharks, stingrays	H
Gladden Spit, Belize	CSA	4	Whale sharks, stingrays	H
Bat Islands, Costa Rica	CSA	5	Bull, whitetip reef sharks	CR
Cocos Island, Costa Rica	CSA	4	Scalloped hammerhead sharks, stingrays, gt. mantas	CR
Utila, Hondurus	CSA	>20	Whale sharks	H
Galapagos Islands, Ecuador	CSA	6	Scalloped hammerhead, whale sharks	CR,H
Malpelo Island, Ecuador	CSA	3	Scalloped hammerhead, whale sharks	CR,H
Canara Islands, Spain	E	3	Angel sharks	CR
Cornwall, UK	E	3	Basking sharks	H
Isle of Man, UK	E	>10	Basking sharks	H

TABLE 15.1. *continued*

Location	Region	Operators	Main attraction	Life-style
The Azores, Portugal	E	2	Blue, shorfin mako sharks	P
St. Maarten, Dutch Antilles	GC	3	Caribbean reef	CR
Tiger Beach, the Bahamas	GC	3	Tiger, hammerhead sharks	CR
Nassau, the Bahamas	GC	3	Caribbean reef sharks	CR
Long Island, the Bahamas	GC	1	Caribbean reef sharks	CR
Bimini, the Bahamas	GC	1	Caribbean reef sharks	CR
Grand Bahama, the Bahamas	GC	2	Caribbean reef sharks	CR
Exuma Cays, the Bahamas	GC	3	Caribbean reef, whale sharks	CR,H
Eleuthera, the Bahamas	GC	1	Caribbean reef, blacktip sharks	CR
San Salvador Island, the Bahamas	GC	1	Caribbean reef, scalloped hammerhead sharks	CR
Walker's Cay, the Bahamas	GC	2	Caribbean reef sharks	CR
Turks and Caicos, West Indies	GC	1	Caribbean reef, nurse sharks, stingrays	CR
Playa Santa Lucia, Cuba	GC	1	Bull sharks	CR
Jardines de la Reina, Cuba	GC	2	Bull, silky sharks	CR,H
Grand Cayman Island	GC	>10	Southern stingrays	CR
Fish Rock, NSW Australia	O	3	Sandtiger sharks	CR
Coffs Harbor, NSW, Australia	O	2	Sandtiger sharks	CR
Forster, NSW, Australia	O	3	Sandtiger sharks	CR
Ningaloo Reef, WA, Australia	O	>10	Whale sharks, giant mantas	H
Port Lincoln, SA, Australia	O	4	White sharks	H
West Rock, QLD, Australia	O	4	Sandtiger sharks	CR
Barrier Reef, QLD, Australia	O	>20	Blacktail reef, silvertip sharks, stingrays, giant mantas	CR
Coral Sea Islands, QLD, Australia	O	3	Blacktail reef, silvertip sharks	CR
Moorea, French Polynesia	O	4	Blacktip reef, lemon sharks	CR
Fakarava, French Polynesia	O	2	Blacktail reef, blacktip reef sharks	CR
Rangiroa, French Polynesia	O	6	Blacktail reef, blacktip reef sharks	CR
Bora Bora, French Polynesia	O	2	Blacktail reef, Blacktip reef sharks, stingrays, mantas	CR
New Georgia, Solomon Islands	O	3	Blacktip reef, whitetip reef sharks	CR
Vanua Levu, Fiji	O	5	Blacktail reef, blacktip reef sharks	CR

TABLE 15.1. *continued*

Location	Region	Operators	Main attraction	Life-style
Mana Island, Fiji	O	4	Blacktail reef, blacktip reef sharks	CR
Bequa Lagoon, Fiji	O	1	Tiger, bull sharks	CA
Chatham Islands, New Zealand	O	1	White sharks	H
North Island, New Zealand	O	2	White, shortfin mako sharks	H,P
Beirut, Lebanon	AME	2	Sandtiger sharks	CR
Hurghada, Egypt	AME	4	Blacktail reef sharks, whitetip reef sharks	CR
Daedalus Reef, Egypt	AME	10	Oceanic whitetip sharks	P
Coastal Reefs, Sudan	AME	3	Silvertip, scalloped hammerhead sharks	H, CR
Tubbatha Reef, Philippines	AI	3	Whitetip reef, blacktip reef sharks, stingrays	CR
Donsol, Philippines	AI	>10	Whitetip reef, blacktip reef sharks, stingrays	CR
Pescador Island, Philippines	AI	7	Pelagic thresher sharks	P
Phuket, Thailand	AI	5	Whale sharks	H
Similan, Thailand	AI	>5	Whale sharks	H
Palau Archipelago, Palau	AI	>10	Blacktail reef, whitetip reef sharks	CR
The Maldives	AI	>20	Whale, blacktail reef sharks, stingrays	H, CR
False Bay, South Africa	SEA	2	White sharks	H
Mossel Bay, South Africa	SEA	1	White sharks	H
Gansbaai, South Africa	SEA	7	White sharks	H
Protea Banks, KZN, South Africa	SEA	7	Sandtiger, bronze whaler sharks	CR
Aliwal Shoal, KZN, South Africa	SEA	>10	Tiger, sandtiger sharks	CR
Sardine Run, KZN, South Africa	SEA	>10	Blacktip, bronze whaler sharks	CR
Sodwana Bay, KZN, South Africa	SEA	3	Sandtiger, whale sharks	CR, H
Dar es Salaam, Tanzania	SEA	3	Whale sharks	H
Inhambane, Mozambique	SEA	5	White, scalloped hammerhead sharks	H, CR
Bassas da India, French Territory	SEA	2	White, scalloped hammerhead sharks	H, CR
Ponta D'Ouro, Mozambique	SEA	4	Scalloped hammerhead, whale sharks	CR, H
The Seychelles	SEA	4	Whale sharks	H

Notes: Regions consist of NA, North America; CSA, Central and South America; E, Europe; GC, Greater Caribbean; O, Oceania; AME, North Africa and Middle East; AI, Asia and Indonesia; SEA, Southern and Eastern Africa. Categories: H, highly migratory; CR, coastal and reef associated; and P, pelagic.

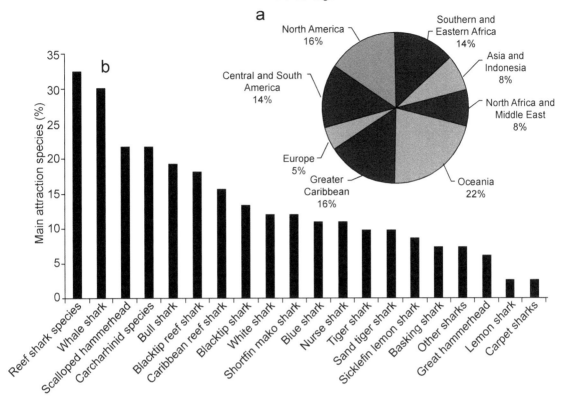

Percentage of Shark Ecotourism Locations

a

North America
16%

Southern and
Eastern Africa
14%

Asia and
Indonesia
8%

Central and South
America
14%

North Africa and
Middle East
8%

Europe
5%

Greater
Caribbean
16%

Oceania
22%

b

Fig. 15.10 (a) Pie chart showing the percentages of the total number of ecotourism destinations occurring in each of seven regions of the world. (b) Histogram indicating the percentages of ecotourism sites at which particular species or broad taxonomic categories are the main attractions. Note that two categories, the reef shark species and Caribbean reef sharks, are inclusive and include many species from family Carcharhinidae.

respectively. The earnings in the Bahamas, an archipelago of many islands, are even greater, $78 million in 2008. This industry appears to be growing, as is evident from the near tripling of earnings in the Maldive Islands from $2.3 million in 1993 to $6.3 million in 1998. Carwardine and Watterson have written an informative book, *The Shark Watchers Handbook: A Guide to Sharks and Where to See Them,* providing directions on how to visit diverse locations where sharks are abundant and advising on the diving conditions.[41]

The whale shark, scalloped hammerhead shark, and white shark are species that are often main attractions at ecotourism destinations. Whale sharks often swim close to the surface and are easily viewed while snorkeling (see picture at the beginning of chapter). Hence, tourists do not need to be skilled in making long breath-hold dives or be certified in SCUBA to view them. Whale sharks can be viewed in the Atlantic Ocean at Isla Holbox off the coast

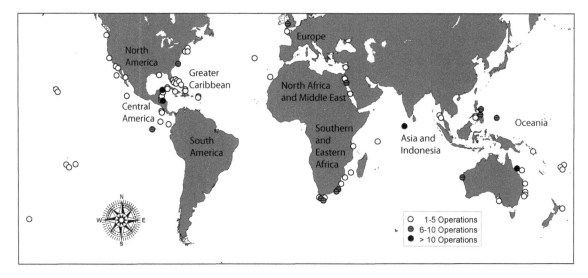

Fig. 15.11 The numbers of operators at shark ecotourism destinations indicated on map of the world's oceans.

of Quintana Roo, Mexico, and along the coasts of Belize and Hondurus and in the eastern Pacific Ocean at Cocos Island off Costa Rica, Malpelo Island off Colombia, and Darwin and Wolf Islands off Ecuador. There are numerous resorts where whale sharks can be viewed in the western Pacific Ocean such as in Western Australia, the Philippines, and Thailand.

Divers from all over the world now take expeditions to seamounts and islands inhabited by schools of hammerhead sharks to see these sharks in their own habitat. The schools of sharks tend to stay at a greater distance from SCUBA than breath-hold divers because they are frightened by the hissing sounds emitted as air is drawn through the regulator and the pulsed, low-frequency sounds and reflection of light associated with the bubbles of air oscillating back and forth as they rise to the surface. Most sharks can be approached more closely by making breath-holding dives or using rebreathers that recirculate air with the addition of oxygen and avoid the production of bubbles. There are dive operations with boats that take ecotourists to view schools of hammerhead sharks at Espiritu Santo Seamount off the southeastern coast of the Baja Peninsula in the Gulf of California, the Revilligigedos Islands off Mexico, Cocos Island, Malpelo Island, and Wolf and Darwin Islands.

Public viewing of white sharks from the safety of a shark cage has also become very popular. They must be attracted to its vicinity with a mixture of fish body fluids, oils, and macerated tissues—called chum. There are expeditions to view white sharks at the South Farallon Islands off central California, Guadalupe Island off the Pacific coast of Baja California, Dyer Island off the southern coast of South Africa, and Dangerous Reef and the Neptune

TABLE 15.2. Economic impact of established ecotourism markets.

Location	Value ($ million/per year)	Species
Gansbaai, South Africa	4.4	White sharks
Maldive Islands	2.3 (1993)	Various reef species
Maldive Islands	6.6 (1998)	Various reef species
Ningaloo Reef, Australia	5.9	Whale sharks
Seychelles	4.9 (2004)	Whale sharks
The Bahamas	78.0 (2004)	Various reef species

Islands in Spencer Gulf off the southern coast of Australia. There is concern that food given to these sharks that are attracted for viewing is a positive reinforcement and conditions them to approach any boat, regardless of whether shark diving is carried aboard it. Hence, ecotourism operators are prohibited from feeding sharks once they arrive at tourist boats in South Africa. Feeding of sharks does occur at Guadalupe Island and in Spencer Gulf. The likelihood that sharks learn to associate boats with the opportunity to feed is dependent on the frequency with which the same sharks are attracted and fed near a boat. In Spencer Gulf, less than 5% of the sharks viewed from protective cages have been observed to return five or more times. This is the believed threshold for making the association between food and a boat.[42]

The face-to-face interactions between divers and sharks at these sites has generally promoted a greater understanding and appreciation of sharks and helped to improve their much-maligned reputation. It has also helped to promote conservation measures to protect the many threatened shark species around the world. Tourists should always, however, keep in mind that sharks are potentially dangerous and should always be treated with care. Shark ecotourism is providing the public with an observational experience that can be as pleasurable as whale watching. It is also a cost-effective alternative source of employment for fishermen. This could lead to reduced shark fishing in certain regions of the world, which could enable shark populations to recover to their former levels of abundance.

SUMMARY

Hunger is one obvious motivation for a shark to attack a human. Some attack wounds do resemble the bites made by a shark when feeding. On the other hand, the magnitude and shape of attack wounds do not always bear this out. In most instances, victims of shark attacks have experienced the loss of only small amounts of flesh. When animals feel threatened, they experience conflicting instincts—one is to escape and the other is to fight. If they are unable to flee because their opportunity to escape is blocked, they do not always fight but often perform an agonistic, or aggressive, display. This was described first for the blacktail reef shark. There are five postural components to its aggressive display. The first is the upward pointing of the snout, the second a bend between the head and back caused by the elevation of the snout, the third the lowering of the pectoral fins to bring them close to each other, the fourth the arching of the back, and the fifth the lateral bending of the body with the tail pushed to either side in an exaggerated arc. The shark moves its tail from one side to the other over a larger arc, 90°, while its forward torso swings slightly to one side and then to the other in a compensatory manner. There are three species of sharks—the bull, tiger, and white sharks—that have been observed to occasionally prey on humans, although humans are not a large component of their normal diets. Information on all shark attacks on humans worldwide is collected in the International Shark Attack File. The risk of being attacked by a shark is truly negligible compared to other risks encountered by humans on a daily basis.

Stingrays are present in the tropical and temperate oceans worldwide as well as in estuaries and freshwater rivers. They are named for the poisonous barb on their tails. If a predator approaches too closely, either alerting the ray by water motion or contact, the stingray will raise its tail in a whip-like manner with its spine to be thrust into the predator, releasing venom stored within the notches of the spine. Bathers step on stingrays while at the beach because they are hard to see on the bottom through the surface of the water, and upon stepping on the pectoral disk of the stingray cause it to reflexively lift its tail and stab the barb into the foot or lower leg of the bather. The venom along with fragments of the barb are often left at the penetration site, and they cause the victim great pain. The venom can be denatured by placing the ankle or hand in hot water.

Shark ecotourism has become immensely popular during the 1990s and 2000s and has become worldwide in its scale. A diversity of opportunities exists for viewing sharks and rays in their natural environment. The face-to-face interactions between divers and sharks at these sites have generally promoted a greater understanding and appreciation of sharks, and helped to improve their much-maligned reputation. It has also helped to promote conservation measures to protect the many threatened shark species around the world.

Discussion Questions

1. How would you minimize the risk of being attacked when near a large gray shark that appeared to be swimming erratically during your next vacation in Tahiti.

2. Where is it safe to go diving? Where is it unsafe to dive? You should be able to dive in most places in the ocean given the knowledge of shark behavior gained from reading this chapter.

3. How should you walk on the bottom if you hear that stingrays are common at a location? How would you treat a stingray wound to yourself?

KEY TO COMMON AND SCIENTIFIC NAMES

Basking shark = *Cetorhinus maximus*; blacknose shark = *Carcharhinus acronotus*; blacktail reef shark = *Carcharhinus amblyrhynchos*; blacktip reef shark = *Carcharhinus melanopterus*; blacktip shark = *Carcharhinus limbatus*; blue shark = *Prionace glauca*; bluntnose sixgill shark = *Hexanchus griseus*; bonnethead shark = *Sphyrna tiburo*; bronze whaler shark = *Carcharhinus brachyurus*; bull shark = *Carcharhinus leucas*; Caribbean reef shark = *Carcharhinus perezi*; dusky shark = *Carcharhinus obscurus*; Galapagos shark = *Carcharhinus galapagensis*; giant manta ray = *Manta birostris*; great hammerhead shark = *Sphyrna mokarran*; largespot river stingray = *Pomatotrygon falkneri*; lemon shark = *Negaprion brevirostris*; longnose stingray = *Dasyatis guttata*; northern elephant seal = *Mirounga angustirostris*; oceanic whitetip shark = *Carcharhinus longimanus*; pelagic thresher shark = *Alopias pelagicus*; reef shark species = *Carcharhinus* spp.; sandtiger shark = *Carcharias taurus*; scalloped hammerhead shark = *Sphyrna lewini*; shortfin mako shark = *Isurus oxyrinchus*; sicklefin lemon shark = *Negaprion acutidens*; silky shark = *Carcharhinus falciformis*; silvertip shark = *Carcharhinus albimarginatus*; southern stingray = *Dasyatis americana*; tiger shark = *Galeocerdo cuvier*; whale shark = *Rhincodon typus*; white shark = *Carcharodon carcharias*; whitetip reef shark = *Triaenodon obesus*.

LITERATURE CITED

1. Schultz and Mallin, 1975; 2. Fernicola, 2001; 3. Davis and D'Aubrey, 1961a; 4. Davis and D'Aubrey, 1961b; 5. Davis and D'Aubrey, 1961c; 6. Cliff and Dudley, 1991; 7. Klimley, 1974; 8. Baldridge and Williams, 1969; 9. Burghart, 1970; 10. Johnson and Nelson, 1973; 11. Myrberg and Gruber, 1974; 12. Nelson *et al.*, 1986; 13. Heithaus, 2001; 14. Burgess and Callaghan, 1996; 15. Klimley *et al.*, 1996a; 16. McCosker, 1985:17. Ames *et al.*, 1996; 18. Klimley, 1985a; 19. Curtis *et al.*, 2006; 20. International Shark Attack File Statistics, 2007; 21. Caldicott *et al.*, 2001; 22. Germain *et al.*, 2000; 23. Diaz, 2008; 24. Haddad *et al.*, 2003; 25. Dehghani *et al.*, 2010; 26. Evans and Davies, 1996; 27. Schwartz, 2007; 28. Pedroso *et al.*, 2007; 29. Barbaro *et al.*, 2007; 30. Magalhaes *et al.*, 2008; 31. Klimley, 1981; 32. Klimley and Nelson, 1981; 33. Klimley, 1982; 34. Klimley, 1987; 35. Tricas and McCosker, 1984; 36 Klimley, 1985a; 37. Lowe *et al.*, 1996; 38. Heithaus, 2001; 39. Klimley, 1985b; 40. Gallagher and Hammerschlag, 2011; 41. Carwardine and Watterson, 2002; 42. Robbins, 2004.

Baldridge, J. R., and J. Williams. 1969. Shark attack: feeding or fighting? *Military Medicine*, 134: 130–133.

Carwardine, M., and K. Watterson. 2002. The Shark Watchers Handbook: A Guide to Sharks and Where to See Them. Princeton University Press, Princeton, N.J.

Gallagher, A. J., and N. Hammerschlag. 2011. Global shark currency: the distribution, frequency, and economic value of shark ecotourism. *Current Issues in Tourism*, 2011: 1–16.

Johnson, R. H., and D. R. Nelson. 1973. Agonistic display in the gray reef shark, *Carcharhinus menisorrah*, and its relationship to attacks on man. *Copeia*, 1973: 76–84.

Nelson, D. R., R. R. Johnson, J. N. McKibben, and G. G. Pittenger. 1986. Agonistic attacks on divers and submersibles by gray reef sharks, *Carcharhinus amblyrhynchos*: antipredatory or competitive? *Bull. Mar. Sci.*, 38: 68–88.

Pedroso, C. M., C. Jared, P. Charvet-Almeida, M. P. Almeida, D. G. Neto, M. S. Lira, V. Haddad, Jr., K. C. Barbaro, and M. M. Antoniazzi. 2007. Morphological characterization of the venom secretory epidermal cells in the stinger of marine and freshwater stingrays. *Toxicon*, 50: 688–697.

Schwartz, F. J. 2007. A survey of tail spine characteristics of stingrays frequenting African, Arabian to Chagos-Maldive archipelago waters. *Smithiana Bulletin*, 8: 41–52.

CHAPTER 16

Fisheries and Conservation

In the last chapter, you learned how sharks and rays are now becoming tourist attractions. However, just as this is happening, they are vanishing from the oceans. Adult hammerheads have become rare at the Espiritu Santo Seamount in the Gulf of California. Only one small group of eight hammerheads were seen during a two-week cruise to study pelagic fishes at this seamount during the summer of 1998, when we spent most of our time searching for hammerheads by making free and scuba dives. During a similar cruise eighteen years earlier, I observed enormous schools swimming around this underwater ridge less than a kilometer long. The size of the population in the vicinity of the seamount was estimated based on the ratio of tagged to nontagged sharks to be 525 sharks. We visited Isla Pardito, a small island that was home to the local fishermen, during the summer of 2003 in another abortive search for hammerheads and found that the family fishing operation of forty-two people run by three brothers and their offspring had dwindled to a vestigial community of three. The last of the brothers to stay on the island complained to his granddaughter about the gradual disappearance of the hammerhead sharks, which he attributed not to the impact of his family-run business but to that of the large commercial fishing operations from the mainland. By co-

The fishery for scalloped hammerhead sharks in the Gulf of California was initially small-scale, but has grown to the point that only a few small sharks are caught by the local fishermen in the Gulf of California.

incidence, her Ph.D. research project was to be supervised by Felipe Galvan-Magaña, who had been on one of our first trips to the seamount in the summer 1980 with me. The study entailed catching and releasing what few sharks were left while placing spaghetti-type tags on them to estimate the size of the remnant population of sharks.

WHO'S EATING WHOM?

If certain sharks are threatened, it is not out of fear or fascination by ecotourists but rather by the fact that humans like to eat seafood. Many of the traditional food fishes have become scarce from overfishing, and in recent years fishermen have switched to lower-grade market fish, which include the sharks and rays. The life-history characteristics of the elasmobranchs make populations particularly vulnerable to overfishing, and the wide travels of many species make fisheries management difficult. There has been a concomitant growth of fisheries for sharks worldwide as more traditional fisheries have declined. Directed fisheries for sharks have often shown a boom-and-bust pattern. Shark populations have decreased sharply after periods of intense fishing pressure. The precipitous decrease in the numbers of hammerhead sharks observed at diving sites at seamounts and islands throughout the Gulf of California occurred simultaneously with the explosive growth of the Mexican shark fisheries during the 1980s and 1990s.

HISTORICAL DECLINE OF SHARK FISHERIES

Fisheries biologists consider a stock, or population of a species, unsustainable when the rate of exploitation by fishermen reduces its standing crop, or its abundance, to a level at which the current rate of capture can no longer be sustained on an annual basis. For productive species, the level of exploitation can be increased substantially to produce annually sustainable catches before substantially reducing the species' standing crop. However, for longer-lived and less productive species, the initial rate of exploitation may exceed the rate at which the stock can replenish itself through reproduction, and right from the beginning the catches are unsustainable. Even for the stocks of vulnerable species, the catch rate may hold steady or even rise for a few years as individuals, having grown to adulthood over decades, are depleted. It is only when they have been harvested and the standing crop, consisting of each year's production or contribution of new individuals, begins to decline that the fishery collapses.

Many shark fisheries have collapsed in the past. Examples are the fates of the fishery for the tope shark (*Galeorhinus galeus*) off the western coast of

North America, the basking shark (*Cetorhinus maximus*) off Europe and Canada, and the porbeagle shark (*Lamna nasus*) off the eastern coast of North America. The tope shark, a ground shark in the family Carcharhinidae, is a prime example. The landings of sharks off the coast of California in 1930–1936, most of them tope sharks, remained low and constant at about 270 tons per year.[1] The fishery expanded greatly after the appearance of a new market for the oil in the shark's liver. This high-grade oil was in demand primarily as a rich source of high-potency vitamin A and secondarily as a lubricant for machinery used by the military before and during World War II. The liver, impregnated with oil, can reach a third of the body mass of an adult tope shark. The price of this high-grade oil rose from $50 per ton in 1937 to $2,000 per ton in 1941. The catch of sharks rose from 270 tons per year in the early 1930s to a peak of 4,185 tons per year in 1939. Tope landings were distinguished from those of other sharks from 1941 onwards, and the catch of 2,172 tons in 1941 dropped to 287 tons by 1944. The time span between boom and bust was only eight years for this fishery, and although demand dropped after vitamin A was synthesized in 1947, catches and effort in the fishery had already declined dramatically by that time. At the present, over sixty years later, there are no fisheries specifically targeting the species, and thus it is uncertain whether the stock has recovered.

The target of one of the oldest fisheries for sharks was the large planktivorous basking shark. Records were first kept of basking sharks caught over Sunfish Bank, a shallow bank west of Ireland, during the late eighteenth century.[2] This fishery lasted for many decades while fishing effort was minimal. However, the rising demand for liver oil resulted in increased fishing pressure, and the catch had begun to decline by 1830 and had collapsed by the end of the nineteenth century. There was no commercial harvesting of basking sharks during the early twentieth century, but fishing for these sharks commenced again during 1947 near Achill Island. The average number of sharks caught per year declined from an average of 1,067 per year in 1949–1958 to 119 in 1959–1968, and further dropped to 40 per year during the last seven years of the fishery. Today basking sharks are occasionally observed in the region but not in as large schools as in the past.

The best studied historical decline is that of the fisheries for the porbeagle shark in the northwest Atlantic[3]. This is a large oceanic species that occupies the cold waters at temperate and polar latitudes. In the northwest Atlantic, this shark is present over the continental shelf from Newfoundland to New Jersey but is most abundant in the waters off the eastern coast of Canada between the Gulf of Maine and Newfoundland. The fishery for porbeagle sharks commenced in 1961; at that time, Norwegian vessels began long-line fishing for them off the eastern coast of North America. The fishermen had been fish-

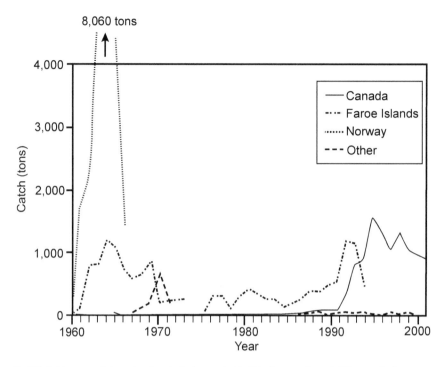

Fig. 16.1 Catch per year in tons of porbeagle sharks reported for fisheries operating out of Norway, the Faroe Islands, Canada, and other locations.

ing previously for porbeagle sharks in the northeastern Atlantic off Europe. The Norwegian vessels caught 1,900 tons of sharks during the first year of the fishery; the catch rose to over 8,000 tons three years later during 1964 and dropped to less than 1,300 tons in 1966 (see dotted line on graph in fig. 16.1). Fishermen from the Faroe Islands began to fish for porbeagle sharks during 1962 and captured roughly 800 tons of them by the end of the first fishing season (see dotted and dashed line). Their catch peaked at 1,200 tons during 1964 but declined to roughly 200 tons by 1970 and remained at that low level of less than 500 tons until 1989. At that time, the Faroe vessels increased their effort, resulting in a catches during 1991–1992 exceeding 1,000 tons, yet they phased out their fishery over the next two years. The Canadians began setting long lines to catch porbeagle sharks in the mid 1980s. By 1994 the catch of three offshore long-line vessels and a few inshore vessels reached 1,600 tons per year (see solid line). The catch, which was uncontrolled, declined to 1,066 tons in 1998. Landings over the next three years were restricted to a quota below 1,000 tons. They were further restricted to 250 tons after 2002.

This fishery was well studied by fishery biologists. They detected the most obvious symptom of overfishing—the yield from the fishery could not be sustained from year to year at the current fishing effort. This observation led biologists to reduce the fishing effort by specifying quotas, or maximum limits to the catch. One impact of intense fishing pressure is to rapidly limit the size (or age) distribution of the individuals captured in the fishery. As the older sharks are removed from the population over time, only the younger individuals remain who have yet to be captured and removed from the population of sharks. Fishermen in the offshore fleet recorded the median fork length of each shark caught in their gear. The fork length is the distance from the snout to central notch between the upper and lower lobes of the caudal fin. The median fork length of porbeagle sharks captured by the Canadian fishermen declined from 200 cm in 1980 to 185–162 cm from 1985 to 1995 and finally to 142 cm in 2000 (fig. 16.2a).

A second indicator of a fishery's impact on a population of sharks is a steady decrease in the catch per unit effort (CPUE), which is expressed as the fraction of sharks captured per thousand hooks on the line. Hence, the first point in a plot of the number of mature sharks, greater than 200 cm in fork length, or 0.004 during 1989 indicates that between three and four sharks were caught for every thousand hooks deployed during the year (fig. 16.2b). The bars with horizontal lines on either end, corresponding to values of 0.002 and 0.006, denote that 95% of the periods over which the gear was deployed resulted in catching between two and six sharks per thousand hooks. The CPUE for mature sharks caught increased during the first three years of the intensive fisheries from 1989–1992 but declined sharply afterward as effort increased. Following was a decrease in the mean CPUE during the next year to roughly 0.004 where it stayed during the next two years. The CPUE further declined to less than 0.001 in 1997 and remained low during the following three years until 2000—a catch rate only 15% that of the peak CPUE during 1992. The CPUE for the immature sharks peaked during 1991, but the rates prior to 1995 seemed to hover around 0.080 while the rates afterward stayed around 0.050 (fig. 16.2c).

A fishery management plan was implemented in Canada during 1994, requiring that fishermen record the number of sharks caught, the identity of each shark captured as well as its length and weight, and finally the fishing effort. This information was required to be kept in logbooks provided to the fishery managers. The analysis of these records led to yearly stock assessments and a fishing quota 1,000 tons of porbeagle shark per year from 1998 to 2001 and 250 tons of sharks captured per year after 2003. In addition to reducing the catch, an outcome of the stock assessment in 2002 was that fishing was prohibited on the mating grounds until the stock of the species rebounds with the reduced fishing pressure.

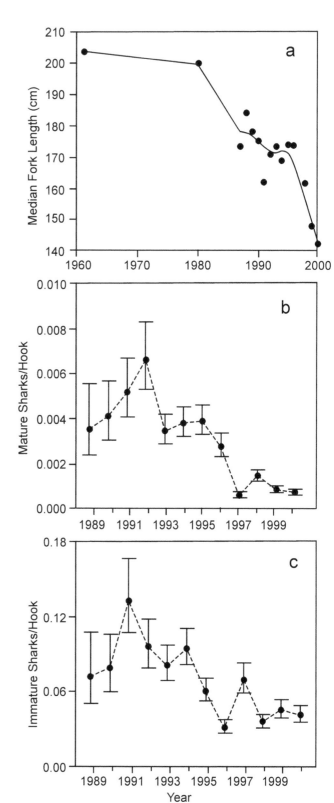

Fig. 16.2 (a) The median fork lengths of mature porbeagle sharks, which are larger than 200 cm fork length, and immature sharks, which are smaller than 200 cm fork length, that were caught by the off-shore fleet during September and October between 1960 and 2000. The rate of the capture of sharks, expressed as the fraction of sharks per single hook, captured from 1989–2000 for (b) mature and (c) imma-ture sharks. Note the decline in the median length of sharks, which corresponds to a decrease in the CPUE of adult sharks.

CURRENT STATUS OF SHARK POPULATIONS
IN THE NORTH ATLANTIC OCEAN

Catch records indicate that the populations of all of the large species of sharks have declined in the North Atlantic Ocean off the eastern coast of North America since 1970, when commercial fisheries began to target the large sharks.[4] In contrast, the sizes of the populations of the smaller sharks and rays have increased over the same period. These species occupy two different trophic levels, the larger sharks (apex predators) being at the top of the food chain and the smaller sharks and rays (meso-predators) at the next level down. Population trends were estimated for these species from scientific research surveys and records of catch over time. These were obtained from fishery observers and fisher's logbooks. The curves (fig. 16.3) indicating the rate of capture for the large sharks slope downward dramatically. A major decline has been observed in the population sizes of the sandbar shark (*Carcharhinus plumbeus*), blacktip shark (*Carcharhinus limbatus*), bull shark (*Carcharhinus leucas*), dusky shark (*Carcharhinus leucas*), scalloped hammerhead shark (*Sphyrna lewini*), and tiger shark (*Galeocerdo cuvier*). This is apparent from the steep downward slopes to the curves in each of the plots in the upper row of the graphs in fig. 16.3. The longest continuous survey of sharks has been conducted annually since 1977 off the coast of North Carolina. It indicates that these apex predators have been virtually eliminated from the ecosystem. Declines in the catch of the six species of large sharks were 87% for the sandbar, 93% for the blacktip, 99% for the bull, 99% for the dusky, 98% for the scalloped hammerhead, and 97% for the tiger shark. This survey is conducted at a junction between the migratory route of these species, which migrate into northern waters during the summer and reside in southern waters during winter. Hence, this survey is likely indicative of the change in abundance of all of the species along the entire eastern coast of North America.

However, concomitant with the reduction of the large sharks has been the dramatic rise in the population size of the small sharks and rays—those species the large sharks prey upon and that occupy the next trophic level down the food chain. The curves fitted to the catch rates of five of these species—the little skate (*Leucoraja erinacea*), Atlantic sharpnose shark (*Rhizoprionodon terraenovae*), chain catshark (*Scyliorhinus rotifer*), smooth butterfly ray (*Gymnura micrura*), and cownose ray (*Rhinoptera bonasus*)—all exhibit steep rising slopes, which indicate a sudden rise in their populations (see lower row of plots). The commercial shark fisheries, largely targeting the large sharks, have removed the apex predators from the ecosystem. A likely conclusion from these trends is that a reduction of predation by large sharks, due to their capture, on smaller sharks and rays has resulted in an increase in their abundance over time.

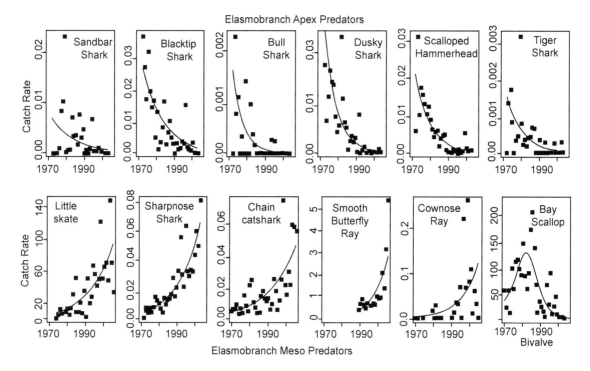

Fig. 16.3 Change in the abundance of species over time, estimated from scientific surveys and the catch of commercial fisheries, over a period of twenty-eight years for the larger sharks in the higher trophic level (upper row), the smaller sharks and rays in the trophic level below them (lower row), and the bay scallop (last plot on lower row).

One of the most dramatic population explosions among the smaller meso-predators has been that of the cownose ray, which inhabits the waters from southeast Florida to Raritan Bay in New Jersey—although it has recently been observed off the coast of Long Island in New York. There has been a tenfold increase in the number of cownose rays off the eastern coast of North America since the mid-1970s. This can be appreciated by looking at the next to last plot in lower row of graphs. Aerial surveys indicate that there may now be over 40 million individuals in the population. In autumn, these rays leave the northern estuaries, which they inhabit during the summer, and migrate southward along the continental shelf to the water off Florida. Here they remain during the winter. During this migration, they enter the bays and sounds all along the eastern coast of North America. They feed on bay scallops (*Argopecten irradians*), soft-shell clams (*Mya arenaria*), hard-shell clams (*Mercenaria mercenaria*), and oysters (*Crassostrea virinica*) buried in the sandy bottoms of these shallow waterways. The rays swim back and forth close the bottom and detect their prey either by chemoreception or electroreception. The harvest of bay scallops has decreased greatly since 1980. This is indicated by the steep righthand slope to the bell-shaped curve indicating the yearly catch of scallops in the last plot on the lower row of graphs. This population decline coincided with the increase in the catch of the rays. The inverse relationship between the abundance of this predator and its prey suggests that increased predation by the former species has reduced the abundance of the latter. This relationship has been supported experimentally by building stockades around the scallop beds to exclude the rays. The densities of scallops in the unprotected control areas decreased during the fall migration of the rays while the densities of scallops in the stockades did not decrease substantially. This is an example of what is termed a trophic cascade, in which the removal of a species higher on the food chain has a cascading impact on the abundances of species at multiple trophic levels lower in the food chain. In this case, the removal of the large predatory sharks from the ecosystem is likely to have reduced the level of predation on cownose rays—a species occupying the next lower trophic level. The greater numbers of cownose rays now deplete the numbers of their prey, the bay scallop, which is on the next lower trophic level in the ecosystem. Another well-known example of a trophic cascade was the impact that the protection of sea otters had on the prevalence of abalone, sea urchins, and kelp off the western coast of North America. Once protected from harvest, sea otters became more abundant along the coast. The impact of the growing population of otters was to limit the local abundance of abalone and sea urchins, two algal grazers. Their reduction enabled the seaweed kelp to increase its local distribution and expand its range over most of the western coast of North America.

WHAT MAKES CARTILAGINOUS FISHES VULNERABLE TO OVERFISHING?

The abundance of species remains at a relatively constant level over a long period of time within a stable ecosystem. The population has already expanded to the maximum size that can be supported in that particular environment at that time. This population level is commonly called the ecosystem's carrying capacity and designated K in the equation that describes the exponential growth of a population until it reaches an equilibrium state. If a catastrophic climatic event occurs at some point in time, the population may suddenly decline to a small fraction of its prior equilibrium level but then grow exponentially. This positive rapid growth is usually designated by the exponential term r, which describes the instantaneous rate of population growth, which slows to zero upon reaching the environment's carrying capacity in a given ecosystem. A fire in a forest is one example of a catastrophic event, and it may burn all the grasses, shrubs, and trees, leaving just charred dirt where vegetation had been before. Some species of fast-growing grasses and shrubs colonize the floor of the forest after less than a year and remain in high abundance until other species such as the trees appear only after four of five years. The other species grow more slowly but eventually become the dominant species in the forest after one or two decades because their lofty canopy provides their leaves access to sunlight while the same canopy results in shade underneath that eliminates many of the light-requiring grasses and shrubs that had originally colonized the area. There are two broad classes into which species can be classified with life history properties that are very different—*r-selected* species such as the grasses and shrubs that are adapted best for a changing environment and *K-selected* species such as trees that are adapted best for a constant environment. The former species grow fast, reach maturity quickly, and have many seeds or young; the latter species grow slower, reach maturity later, and have fewer young. Of course, this distinction is based on extremes, and the majority of species have life history properties that fall somewhere in between. Most of the cartilaginous fishes are *K* species. They are adapted to the constant conditions within the oceans and succeed in these conditions because they are efficient predators. They exhibit slow growth, late maturity, and low fecundity. In contrast, many bony fishes are *r* species, adapted to a changing environment, and as such they grow faster, mature earlier, and are very fecund, producing many eggs and young.

A prime example of a *K* species is the white shark (*Carcharodon carcharias*) (fig. 16.4). Individuals of these species grow more slowly, reach maturity at a later age, and have fewer young than bony fishes. The rate of growth for both the cartilaginous and bony fishes can be determined by counting the growth rings on a vertebral centrum. Males of the white shark reach maturity at a total length of 390 to 410cm at an age of 8–10 years, females at a total length of 450–500 cm at an age of 12 years. They are thought to live to a maximum age of 36 years. Females are believed to give birth in alternating years. The embryos take roughly a year to develop within the female before 7 to 14 are released into the ocean (fig. 16.5).[6] Given these life history properties, it is not surprising that there are few white sharks. Members of species occupying the pinnacle of the food chain are never very numerous. The white shark feeds on seals and sea lions, which in turn feed on smaller prey such as fish and squid; these feed on even smaller planktonic animals, which in turn feed on planktonic plants. At each link in the chain some energy is lost, resulting in less biomass, or collective body mass of species, at each successive level up the food chain. When four large white sharks were caught on 5 October 1982, close to shore at the South Farallon Islands near San Francisco, the number of attacks on seals and sea lions in the local waters dropped by half during the next two years—indicating that there might only have been twice that number of sharks present at the island at that time.[7] Nine to fourteen white sharks were observed over a five-year period during October and November at the same place.[8] The sizes

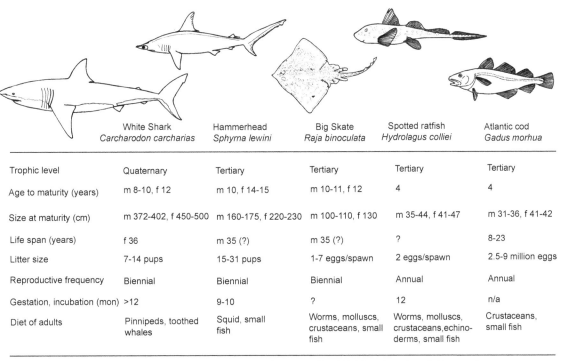

	White Shark *Carcharodon carcharias*	Hammerhead *Sphyrna lewini*	Big Skate *Raja binoculata*	Spotted ratfish *Hydrolagus colliei*	Atlantic cod *Gadus morhua*
Trophic level	Quaternary	Tertiary	Tertiary	Tertiary	Tertiary
Age to maturity (years)	m 8-10, f 12	m 10, f 14-15	m 10-11, f 12	4	4
Size at maturity (cm)	m 372-402, f 450-500	m 160-175, f 220-230	m 100-110, f 130	m 35-44, f 41-47	m 31-36, f 41-42
Life span (years)	f 36	m 35 (?)	m 35 (?)	?	8-23
Litter size	7-14 pups	15-31 pups	1-7 eggs/spawn	2 eggs/spawn	2.5-9 million eggs
Reproductive frequency	Biennial	Biennial	Biennial	Annual	Annual
Gestation, incubation (mon)	>12	9-10	?	12	n/a
Diet of adults	Pinnipeds, toothed whales	Squid, small fish	Worms, molluscs, crustaceans, small fish	Worms, molluscs, crustaceans, echino-derms, small fish	Crustaceans, small fish

Note that references are available for life history characteristics in Ebert (2003)[5] and FishBase (http://www.fishbase.org).

Fig. 16.4 Life history traits such as trophic level and diet of adults, age to and size of maturity, reproductive frequency, gestation, litter size, and life span of four cartilaginous fishes, the white shark, scalloped hammerhead shark, big skate, and spotted ratfish, and a bony fish, the Atlantic cod. Note that the chondrichthyans have a *K* reproductive strategy, giving birth less often, at a more advanced age, and to fewer young than the cod, a bony fish that has an *r* reproductive strategy of rapidly reaching maturity, reproducing often, and producing many eggs.

of local white shark populations have been estimated in three geographic areas: the southern coast of South Africa,[9] Spencer Gulf in South Australia[10], and the Gulf of the Farallones in California.[11] The estimated population sizes in these three areas were 1,279, 192, and 219 adults, respectively. The white shark is a quaternary consumer, one trophic level above the majority of sharks, which are tertiary consumers. It shares this level with the bull and tiger sharks, both of which also feed on smaller sharks. Their life history properties are similar to those of the white shark.

The tertiary consumers are more common because they feed on small fish and squid, which are the secondary consumers that feed on zooplankton, the primary consumers in the food chain. The majority of cartilaginous fishes are tertiary consumers. Three examples are the scalloped hammerhead, the big skate (*Raja binoculata*), and the spotted ratfish (*Hydrolagus colliei*) (fig. 16.4). They are still on the *K* side of the continuum of physiological life history characteristics existing between these two very different life history strategies, being adapted for a stable environment as are the white, bull, and tiger sharks.

Fig. 16.5 The white shark has a *K* reproductive strategy with few young. Here is a female white shark shown at the Tokyo Fish Market in Japan with her nine near-term embryos.

The scalloped hammerhead feeds on squid and small fishes and is distributed in semitropical and tropical waters in all of the oceans. It is one of the most abundant of the sharks worldwide. Males of the scalloped hammerhead mature in 10 years and females 14–15 years. This species is thought to live as long as 35 years and gives birth to 15–31 pups on alternating years. The big skate is common off the western coast of North America, where it is present in shallow bays and on the continental shelf to a depth of 800 m. Individuals feed on polychaete worms and mollusks buried in the mud or crustaceans and small fishes that live near the bottom. Males and females become mature at roughly the same time as the scalloped hammerhead, 10–11 years for males and 12 years for females, and they live to a similar age. They do not appear to have a distinct mating season but produce very large egg cases all year round. The spotted ratfish is found over rocky bottoms in the intertidal zone and at shallow depths off British Columbia in the northernmost waters of its range, but it lives in progressively deeper waters exceeding 30 m on reef slopes in the southernmost extent of its range off southern California. These chimaeras feed on polychaete worms and mollusks buried in the mud and sand as well crustaceans, echinoderms, and small fishes that may be finding refuge in rocky crevices within reefs. Males and females reach maturity at a length of 35–44 cm and 41–47 cm after a period of four years. It is unknown just how long members of this species live. The female spotted ratfish extrudes two bullet-shaped egg cases from her uterus that remain attached to her body and hang freely in the water below her for four to six days by an elastic capsular filament that is a long and slender extension of the egg case. The slender egg stalks are later broken and eggs are deposited with an upward orientation in the mud or gravel bottom. The embryos remain within the egg case for up to twelve months before breaking out of the egg case. The young of these tertiary consumers are also advanced in development upon parturition, and the young are thus less likely to be eaten by a predator and more likely to locate, capture, and ingest their prey. Hence, there is comparatively little loss of life after birth for these species.

Many of the species of bony fishes that are tertiary consumers differ vastly from the cartilaginous fishes in their life history characteristics. They are *r* species. They grow to maturity quickly, and are more fecund and reproduce more frequently than the elasmobranchs. For example, male and female Atlantic cod (*Gadus morhua*) reaches maturity in 4 years at lengths of only 31–36 and 41–42 cm, at which time the female releases a multitude of tiny eggs into the water to be fertilized by males within her proximity. Spawning may continue throughout much of the day and is often repeated on successive days and years—a mode of reproduction referred to as iteroparity. A female cod is believed to release 2.5–9 million eggs into the environment during the first year after she reaches maturity and may continue to annually produce this large number of eggs. This reproductive strategy is not as successful as it might seem. Only a few of the eggs released by the females are fertilized by sperm from the males, who must actively seek out the females during the spawning season and remain in close proximity while timing the release of their sperm to coincide with that of the females. The vast majority of eggs are eaten by other bony fishes, which may aggregate at the mating grounds of the cod to gorge themselves on the abundance of passively floating and defenseless prey. Once the larva absorbs the nutrients within its yolk sac, it must immediately find and ingest phytoplankton and minute zooplankton or starve to death. The vast number of propagules produced by a female bony fish is dramatically reduced during this short critical period through starvation. The larva then needs to grow to a size at which it no longer is helplessly transported as plankton in the currents, but can swim independent of the currents and seek a refuge from predators at the bottom. It still must grow for a considerable period of time until it reaches the size of the ray or chimaera after leaving the egg case and even more time to reach the size of the free-swimming shark once being expelled from the uterus.

The *K* reproductive strategy of the cartilaginous fishes was a very successful strategy in the past, resulting in their being the dominant predators over a period close to 400 million years. However, the advantage to this mode of reproduction was based on the strategy of producing a few large and fully developed young. In the past, there was little risk of loss of life to these young due to starvation or predation. Their advanced swimming ability upon birth together with their fully developed jaws ensured that they would have a high probability of survival during the first few years of their lifespan. Yet the young produced by cartilaginous fishes are now often captured by fishermen during their first few years of their life using longlines or gill nets that can extend over many kilometers of the ocean surface.

MANAGEMENT AND PROTECTION OF SHARK POPULATIONS

The recently observed decline in the stocks of fishes off the eastern coast of North America led to the enactment of legislation controlling unrestricted fishing for sharks. Each state manages both its recreational and commercial fishery for sharks in the waters from its coastline to a distance of 3 miles from shore. Exceptions to this rule are Texas and the other states on the Gulf Coast, which have jurisdiction over the fisheries for sharks to a distance of ten miles from shore. The federal government manages fisheries over a zone encompassing the rest of the continental shelf from 3 miles (or 10 miles on the gulf coast) to an outer limit of 200 miles from shore. The current federal plans for managing sharks are aimed at rebuilding and keeping certain populations of sharks at constant and sustainable levels (see http://www.nmfs.noaa.gov /sharks/FS_management.htm). For example, in the northwestern Atlantic Ocean and Gulf of Mexico, fishing effort is regulated by control of the number of annual permits given to commercial vessels and the limits to their annual catch. Commercial fishing for sharks is contingent upon obtaining a permit from the National Marine Fisheries Service (NMFS). The commercial fishermen who are issued these permits are required to report the numbers of sharks caught identified by species as well as the amount of fishing gear used such as the number of hooks or length of gill net to catch the sharks. The fishermen are required to carry observers aboard their vessels to witness fishing operations and ensure that catch information is accurately entered into a logbook by the captain. The information in the logbooks must be provided to NMFS twice each year, over two six-month periods. Fisheries biologists at NMFS periodically determine the annual CPUE, make an annual assessment of the stock size, and may recommend a decrease the number of permits and annual quota of sharks to be captured next year if the CPUE of the current year has declined from the prior year. Furthermore, NMFS has prohibited the practice of removing just the fins of the sharks, discarding the rest of the body, and selling the fins at a high price to make shark soup—a delicacy in China. Finally, certain sharks that are designated protected must be released upon capture in a manner that maximizes their chances of survival. These species

Shark fins are a highly valued commodity in Asia. A single fin can fetch a price of over $1,200 in Hong Kong, while a bowl of shark fin soup can cost close to $100 at a local restaurant within the city. The value of shark meat is much less than that of fins, roughly $12 per kilogram. This is only 2.4% of the $500 cost of a kilogram of shark fins.[12] Due to this stark difference in price, fishermen cut off the fins of sharks that they capture and often throw the sharks overboard alive to die as they sink to the bottom. The fins are then stored aboard the fishing vessel and subsequently exported to Hong Kong, where they are sold in the largest international fin trade market. This inhumane and wasteful practice is commonplace worldwide. Close to 10 million kilograms of shark fins were imported into Hong Kong from eighty-seven countries in 2008.[12] Spain, Singapore, Taiwan, Indonesia, the United Arab Emirates, and Costa Rica have been the top exporters of frozen and dried shark fins to Hong Kong. The fins of the scalloped hammerhead constitute one of the most popularly traded items in the Asian fin market, fetching a high commercial price.[13] This is because their large fins contain many supporting carti-laginous ceratotrichia that have a high fin "needle" content, which is the gelatinous product used to make shark fin soup.

The global exports of shark fins, being roughly 20 million kilograms per year, have remained constant since 2000 with a slight decline during 2008 and 2009. This reduction in the trade may be in reaction to the international consternation aroused upon learning of the mass finning of sharks at Malpelo Island off the coast of Colombia, where many scalloped hammerhead carcasses were observed on the bottom without their fins by scuba divers visiting this International Heritage Site on an ecotourism vessel to view the large schools of hammerhead sharks. The Shark Conservation Act, which went into effect at the beginning of 2010, requires that all sharks captured in United States waters, with an exemption for the dusky smooth-hound (*Mustelus canis*), be landed with their fins naturally attached to their bodies. This act ends the practice of removing the fins of sharks while in the territorial waters of United States.

include the megamouth (*Megachasma pelagios*), white shark (*Rhincodon typus*), basking shark, tiger shark, and bigeye sandtiger shark (*Odontaspis noronhai*).

The Highly Migratory Species Division of NMFS has authority over seventy-two species of coastal and pelagic sharks off the eastern coast of North America and in the Gulf of Mexico. The large coastal sharks such as the sandbar, blacktip, bull, tiger, and hammerheads are considered overfished; the small coastal sharks such as the Atlantic sharpnose, finetooth (*Carcharhinus isodon*), and bonnethead (*Sphyrna tiburo*) are considered fully fished; and the population status is unknown for pelagic sharks such as the blue, shortfin mako, porbeagle, and thresher (*Alopias vulpinus*). It is imperative that international assessments of the population status of the pelagic species in the Atlantic Ocean continue to be made.

The status of most shark species in the Pacific Ocean is unknown. Recent assessments found that the blue shark population is healthy, while the populations of thresher and Pacific angelsharks (*Squatina californica*) are in recovery.

There is a Pacific Highly Migratory Species Fishery Management Plan currently under development for North Pacific waters offshore of California, Oregon, and Washington. The draft of the plan proposes harvest guidelines for the shortfin mako and thresher sharks and coastwide protection for the basking, megamouth, and white sharks. The Western Pacific Pelagic Fisheries Fishery Management Plan covers the waters off Hawaii and Guam and the oceanic populations of blue, shortfin mako, and thresher sharks. In the North Pacific off Alaska, the salmon (*Lamna ditropis*), Greenland (*Somniosus microcephalus*), and dogfish sharks of the order Squaliformes are covered under the Groundfish Fishery Management Plan.

There are fisheries in many countries, such as Mexico and the countries of Central and South America, that target the coastal species of sharks. However, the pelagic sharks such as the blue, silky (*Carcharhinus falciformis*), shortfin mako, and oceanic whitetip (*Carcharhinus longimanus*) are most often caught as bycatch or in mixed species fisheries. They are frequently captured in longline, purse-seine, and drift-net fisheries targeting tuna and billfish within the countries' exclusive economic zones that extend out to 200 nautical miles as well as in the international waters beyond the zone. However, even these sharks are now being targeted themselves because of the high monetary value of their fins.

Beyond the outer limit of 200 miles to federal jurisdiction, there are some insular countries with regional fishery management councils that include shark species in their management plans. However, the development of these plans in countries other than the United States has been slow. Now there are international efforts underway to ensure the proper management of local shark fisheries and protect threatened stocks of sharks. The World Conservation Union publishes a Red List, which categorizes the present status of the populations of each species of oceanic shark, listing them as "Least Concern," "Vulnerable," "Near Endangered," and "Critically Endangered." The Convention on International Trade in Endangered Species (CITES) has established a framework for protecting shark species at risk and preventing trade in those species that are judged endangered.[14] This international organization has recommended a review of the status of shark populations and the effects of international trade. It is concerned about the expanding fisheries for sharks, the absence of oversight and management, the vulnerability of shark species to sustained fishing pressure, and evidence of a decline in these stocks worldwide. In response to these recommendations, the United Nations Food and Agriculture Organization (FAO) has developed a plan for the conservation and management of shark populations. FAO has recommended that the 113 countries that report their landings of sharks adopt this management plan. Of these countries, only

8 have adopted the plan, yet 16 more are in the process of drafting plans. Finally, the International Commission for Conservation of Atlantic Tunas (ICCAT) has prepared stock assessments for three species of sharks—the blue (*Prionace glauca*), porbeagle (*Lamna nasus*), and shortfin mako (*Isurus oxyrinchus*)—outside the 200 mile outer limit to the jurisdiction of the federal government in the North Atlantic Ocean (see http://www.iccat.es/en/). This international agency compiles fishery statistics from its members and from all entities fishing for tuna and tuna-like species in the Atlantic Ocean. The organization produces stock assessments and provides science-based advice for the management of these species for the members of the organization.

Management without Catch and Effort from Fishers

The intrinsic rate of increase, r, of each shark and ray species varies with its unique life history characteristics. A method has been developed to calculate r for a variety of sharks as a measure of rebound potential $(r_{Z(msy)})$—the rate of population growth achieved by a species hypothetically exposed to a maximum sustainable level of fishing. This technique estimates how different shark species are likely to respond to the same level of fishing pressure. This turns out to be a useful criterion for comparing species according to their capacity to withstand exploitation by fisheries.[15] You can estimate for each species the rate of population expansion at a particular level of mortality, or loss due to fishing, that achieves the maximum sustainable yield—the largest catch that can be harvested from year to year while still allowing replacement of the stock. The populations of different species of sharks and rays grow at different rates based on when the females reach maturity, how often they give birth, how many young they produce during pregnancy, and how long the female is reproductively active after reaching maturity and before death. These rebound potentials are an index of each species' ability to sustain a particular level of exploitation. They can be used to develop precautionary management plans for each species, especially when used in conjunction with catch records in those few shark and ray fisheries that have been monitored closely over time.

The intrinsic rate of growth, r, is calculated using a fundamental equation describing the growth of populations[16]:

$$\sum_{x=a}^{w} l_x e^{-rx} m_x = 1$$

A population will be in an equilibrium state when the above equation is balanced and equals a value of 1. This occurs when the survival rate to each age (l_x) and the addition of newborn females per adult female (m_x) at each age sum between the ages of first maturity ($x = \alpha$) and the maximum age of reproduction ($x = w$) to produce a unique balancing rate r of exponential population growth.

The rebound potential is determined in three steps. They consist of calculating under different conditions the net reproductive rate, which is a female's expected reproductive output during her lifetime and is expressed mathematically as $\Sigma l_x m_x$. In the first step, this reproductive rate is determined for an unfished population with adult mortality equal to the natural mortality M. The population is now in a stationary equilibrium. This relationship is illustrated in a three-dimensional graph of the net reproductive rate (l_x) and fecundity (m_x) as a function of age (x) (fig. 16.6a). The net reproductive rate for each female is depicted on the graph as a shaded volume delimited on the x-axis by the age at maturity α and maximum age of reproduction w, on the z-axis by the average number of female pups per adult female b, and on the y-axis by the adult survival from mortality M, beginning at $l_{a,M}$. In the second step, adult females are made to suffer a total mortality, Z, equal to the fishing mortality, F, plus the natural mortality, M. The resulting decrease in the population's reproductive output is compensated for by a higher rate of survival to the age of maturity that brings the population again into stable equilibrium. This is given that this added level of mortality is equal to $2M$. This higher, compensatory preadult survival is $l_{a,2M}$ on the y-axis and results in a steeper slope to the surface indicating the net reproductive rate (fig. 16.6b). In the third step, the fishing mortality is removed so that the population rebounds under natural mortality M with the new compensated preadult survival rate in place. The rebound rate r_Z is found from the above equation describing the net reproductive rate of that increasing population (fig. 16.6c).

The logistic equation can alternatively be expressed in a form that enables r to be estimated using five life history characters. The five parameters are age at maturity, α, maximum reproductive age, w, adult instantaneous mortality, M, the average number of female pups per adult female, b, and survival to age at maturity, l_a. The equation used to determine the rebound potential, r, is given below.

$$e^{-(M+r)} + l_a b e^{-ra}[1 - e^{-(M+r)(w-a+1)}] = 1$$

The values for each of the parameters were obtained from the scientific literature on the biology of each particular shark species. The age at maturity was based on the average of estimates published in growth and reproductive studies. The maximum reproductive age was also determined from these studies. The natural mortality was obtained from an equation relating mortality to maximum age of reproduction for females of fish and mammalian species. The average number of pups per adult female was estimated by dividing the

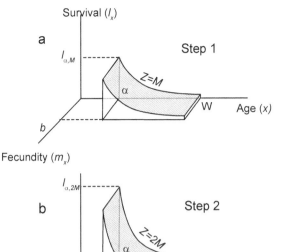

Survival (l_x)

a

$l_{\alpha,M}$

Step 1

$Z=M$

α

W

Age (x)

b

Fecundity (m_x)

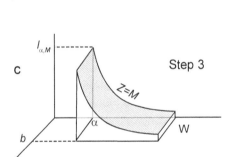

$l_{\alpha,2M}$

b

Step 2

$Z=2M$

α

W

b

$l_{\alpha,M}$

c

Step 3

$Z=M$

α

W

b

Fig. 16.6 Three steps for determining the intrinsic rebound rate, r_{2M}, based on the effect of increased fishing mortality on the net reproductive rate (NRR) of the species. (a) Step one involves expressing the NRR in the unfished state; (b) step 2 imposes an increased preadult survival that enables the population to sustain the effect of added fishing mortality; and (c) step 3 involves reducing the mortality back to the natural level while keeping the pre-adult survival the same, permitting the NRR to rebound.

fecundity by two, assuming half of the individuals in each species were female. Two levels of b were used for each species—the second increased by a factor of 25% to allow for females growing larger and having more young in the absence of fishing pressure.

The intrinsic rebound potential, r, is an index of a population's ability to rebound when fishing mortality is removed. It varies among species because each has unique life history characteristics. The value r is the growth rate of the population necessary to sustain the maximum harvest for each species at a population level approximately half that of the virginal population size, or half the carrying capacity of the environment for the species (fig. 16.7).[17] The net production of sharks in each population is thought to increase slowly at first, grow more rapidly to a maximum rate at half the carrying capacity of the environment, and then grow slower as the environment becomes saturated with individuals. This form of logistic population growth can be expressed with a helmet-shaped production curve consisting of an upward slope, rounded

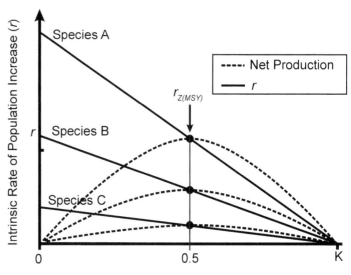

Fig. 16.7 Schematic illustrating the intrinsic rate of population increase and the net production as a function of the population size relative to the environment's carrying capacity.

peak, and downward slope (see dashed lines). The intrinsic rate of population increase, r, a function of the life history characteristics, decreases linearly producing a downward sloping straight line for each species (see solid lines). The point where the r-line intersects the peak of the helmet-shaped curve is the maximum sustainable yield, or $r_{Z(MSY)}$. Species B and C, which are less productive than Species A, have lower maximum sustainable yields. Thus, their populations will decline less before they become depleted than the more productive species.

The intrinsic rebound potential varies considerably among different species of sharks and rays (fig. 16.8).[17] Two levels of fishing mortality relative to natural mortality have been used, values of 1.5 and 2.0—the former is now thought to be the most appropriate estimate of the maximum sustainable level of mortality for sharks.[18] The lowest rebound rates were characteristic of a group of predatory sharks as well as a planktivore—the piked dogfish (*Squalus acanthias*), dusky shark, basking shark, sevengill shark (*Notorhynchus cepedianus*), bull shark, scalloped hammerhead, sandbar shark, and bigeye thresher (*Alopias superciliosus*). The basking and the bigeye thresher shark are among the least fecund species of sharks, producing only 1.0 and 1.5 female pups per litter per year, and these species take longer to mature than most other sharks. Many intermediate- and large-sized coastal sharks share the late-maturing and long-lived reproductive strategy of these two species. The highest rebound rates were calculated for the pelagic stingray (*Pteroplatytrygon violacea*), brown (*Mustelus henlei*), and gray smooth-hound (*Mustelus californicus*). The pelagic stingray was among the most fecund of the elasmobranchs. It produces 3.0

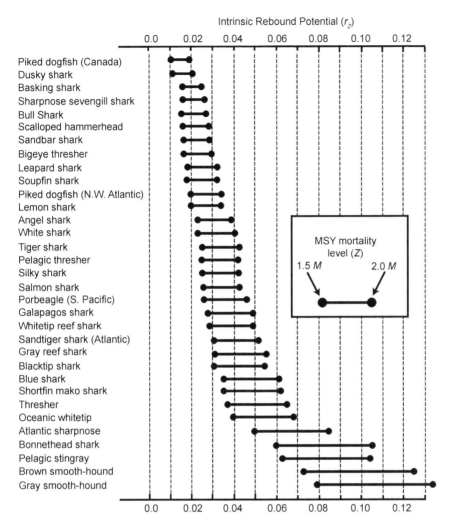

Fig. 16.8 The intrinsic rebound potentials resulting in the maximum sustainable yield given for two rates of mortality for thirty-three species of sharks.

female pups per litter per year. It also reaches maturity quickly over a period of three years. The pelagic stingray resembles in its life history the many small inshore sharks, which are fast growing and mature early. The species that were more productive had a greater range in rebound rates under the two rates of mortality, and this indicates that they have a higher sensitivity to changes in rates of harvest.

The intrinsic rates of rebound potential for the cartilaginous fishes are low compared to those of many bony fishes, yet they are comparable to those of many marine mammals. Species with low productivity are particularly vulnerable to fisheries as evident from the flattest yield curve (see point on curve for Species C in fig. 16.7). Even a slight reduction from their production peak can

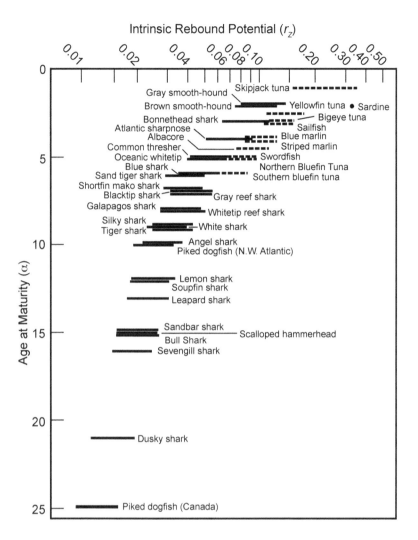

Fig. 16.9 The intrinsic rebound potentials of twenty-seven species of sharks eleven species of bony fishes.

lead to the incipient collapse of the fisheries for that species. The more productive species are less sensitive to small increases in fishing effort, but their faster generation time makes them more sensitive to changes in total mortality and vital rates such as age to maturity that can bring about relatively large changes in productivity.

The age of maturity affects the rebound potentials greatly, and hence it is instructive to view them as a function of age of maturity when comparing the potentials of cartilaginous and bony fishes (fig. 16.9).[19] The length of each bar indicates the range of the rebound potentials between two rates of fecundity, one when the population is fully exploited and the second when the population has rebounded so that there are more females of reproductive age. Only the small short-lived coastal sharks with high rates of production, due to their rapidly reaching maturity and giving birth to many young, have rebound potentials that overlap those of bony fishes. For example, the gray smooth-hound shark has a relatively high potential of 0.08–0.14. It reaches maturity in 2 years, reproduces to an age of 12 years, and has 1.6 female pups per year. The bonnet-head shark has a similarly high rebound potential. It reaches maturity slightly later than the gray smooth-hound shark in 3 years, reproduces at the same age, but has more pups, 4.5 females per year, than the gray smooth-hound shark. Finally, the Atlantic sharpnose shark has a potential of 0.05–0.09. It reaches maturity even later, at 4 years, yet reproduces to an advanced age of 10 years and produces 2.5 female pups per year. The rebound potentials of the gray smooth-hound, bonnethead, and sharpnose sharks overlap the potentials of some tunas and billfishes. For example, the southern bluefin tuna (*Thunnus maccoyii*) has a rebound potential of 0.06–0.09. It reaches maturity at an age of 6 years and reproduces to an age of 20 years. The northern bluefin tuna (*Thunnus orientalis*) has potential of 0.07–0.10. It reaches maturity at an age of 5 years and reproduces to an age of 20 years, as does its congener *T. maccoyii*. The swordfish (*Xiphias gladius*) has a rebound potential of 0.07–0.96. It reaches reproductive maturity and spawns to the same age as the northern bluefin tuna. However, the reproductive potentials of the bulk of the sharks are lower than the majority of bony fishes. The fast-growing tropical tunas are the most resilient to overfishing. For example, the skipjack tuna (*Katsuwonus pelamis*) has a rebound potential of 0.16–0.34. It reaches maturity at the end of its first year and reproduces over a period of 5 years. The yellowfin tuna (*Thunnus albacares*) has a rebound potential of 0.10–0.18, matures in 2.5 years, and reproduces annually until an age of 8 years. Due to their early age of maturity, these species of teleosts have over twice the rebound potential of the small, fast-growing, and short-lived inshore sharks and three times the potential of the slower-growing but longer-lived medium to large-size coastal sharks.

ESTABLISHMENT OF MARINE RESERVES

It is very difficult to manage highly migratory sharks effectively over the wide expanse of oceans bordered by so many countries, each of which must coordinate its fishery regulation with other countries to ensure that the shark populations are protected effectively. This is necessary because many sharks migrate over long distances from one country's jurisdictional waters to another's — each zone extending 200 miles out from its coasts. It will be a challenging but not impossible task to induce many countries to work together to produce global management based on annual assessments of sharks.

An alternative to the traditional method of fisheries management for sharks confined to specific areas during crucial stages in their life histories is the establishment of marine reserves. They can protect species that concentrate in cer-

One can ensure protection against a population collapse of most sharks by permitting females a few reproductive seasons once they become mature before the fisheries begin to target them. The small coastal species such as the gray smoothhound, bonnethead, and sharpnose sharks are quick to grow to maturity. Maximum exploitation rates, or collapse thresholds, can be calculated from their maximal rebound and preadult survival rates. Affording three years of pupping protection to females of the first two species would result in a sustainable exploitation rate of 1.00 (see third column in table 16.1). This would mean that females 4 years or more past the age of maturity could be taken without causing a collapse of the population. Females of the medium-sized to large species, such as the thresher and blue sharks, reach maturity later. They would require 4 additional years of protection. Other species such as the mako, leopard (*Triakis semifasciata*), and bull sharks would require 5 more years of protection. Note that the spiny dogfish, which takes 25 years to reach maturity, would need to be protected for 10 additional years to maintain a sustainable fishery. Hence, recreational fisheries for sharks should specify a size limit for keeping females after capture. The minimum size of females should exceed that enabling them to have reproduced adequately to sustain a stable population.

tain areas for mating, pupping, or feeding where they become highly vulnerable to fishing. As shown earlier, sharks of some species such as the scalloped hammerhead remain resident in small enclaves along the coast of offshore islands and seamounts and then may make relatively rapid migratory movements over a fixed and traditional path to another aggregation site many hundreds of kilometers away. There is an ongoing effort to create reserves at these biological hotspots and to prohibit commercial fishing within them. It may be possible to prohibit fishing in certain areas so that individuals can grow and reproduce sufficiently to ensure against collapse of the fishery for the species.

The Galapagos Archipelago off the coast of Ecuador, Cocos Island off Costa Rica, Malpelo Island off Colombia, and Coiba Island off Panama are some of the last known outposts where large numbers of sharks and other marine predators still remain in the eastern tropical Pacific Ocean. The Galapagos Marine Reserve and other marine protected areas of the region offer some protection, but their effectiveness is limited because they were created without knowledge of the spatial dynamics of the predators within them. Information about the spatial distribution of two key species is currently being used to improve the design of the reserves around these islands. Satellite and shipboard tracking have led to the description of the movements of the scalloped hammerhead shark and the Galapagos shark (*Carcharhinus galapagensis*) on insular (less than 50 km), interisland (50 to 400 km), and oceanic (greater than

TABLE 16.1. The maximum exploitation rates, or collapse thresholds for ten species of sharks according to the number of times adult females are permitted to reproduce before being removed from the population by the fishery.[19] Extending the protection beyond t_c max causes the remaining age classes to be completely expendable, thus ensuring against fisheries collapse at any exploitation rate.

Species	Common Name	α-age	1	2	3	4	5	6	7	8	9	10	t_c max
Mustelus californicus	Gray smooth-hound	2	0.31	0.74	1.00								3.3
Sphyrna tiburo	Bonnethead	3	0.32	0.76	1.00								4.1
Rhizoprionodon terraenovae	Atlantic sharpnose	4	0.36	1.00									4.9
Alopias vulpinus	Thresher	5	0.22	0.38	0.85	1.00							7.1
Prionace glauca	Blue	6	0.21	0.35	0.74	1.00							8.2
Isurus oxyrinchus	Shortfin mako	7	0.16	0.23	0.36	0.70	1.00						10.3
Carcharodon carcharias	White	9	0.13	0.17	0.24	0.36	0.67	1.00					13.4
Triakis semifasciatus	Leopard	13	0.15	0.21	0.32	0.56	1.00						16.6
Carcharhinus leucas	Bull	15	0.16	0.24	0.38	0,76	1.00						18.2
Squalus acanthias	Piked Dogfish	25	0.07	0.08	0.14	0.11	0.14	0.17	0.22	0.31	0.48	1.00	34.0

500 km) spatial scales. Furthermore, a significant degree of migration between the islands has been shown for members of both species carrying individually coded beacons. The species stay around the islands for substantial periods of time but move outside the boundaries of the current marine protected areas. A network of reserves has been proposed for the eastern tropical Pacific Ocean (see map of reserves in fig. 16.10).[20] The size and boundaries are based on knowledge of the extent of the home ranges of the two species, their socio-economic value, ecological functionality, vulnerability to fisheries, and per-sistence over time in an abundant state. The dashed lines on the map indicate the economic exclusive zones (EEZ) of the countries of Central and South America. The tan areas are the home ranges, defined by a kernel algorithm, of members of the two species tracked by boat and satellite at the Galapagos Islands. Red arrows indicate the movements of both species. Marine protected areas (MPAs) already exist at the Galapagos, Cocos, and Malpelo Islands; their boundaries are indicated by solid white lines. There is a proposal to alter the boundaries of the current reserves by creating two new types of MPAs. The first MPA would consist of a no-take zone at each of the islands. Three such zones are proposed with the first centered at Darwin and Wolf Islands within the Galapagos Archipelago, the second around Cocos Island, and the third around Malpelo Island (see red circles with cross-hatching). The second MPA would be a special management zone. Fishing would be regulated in this area, which extends from the coast of Central America and the northwestern coast of South America outward into the eastern Pacific Ocean and around all three islands (see white line surrounding vertical hatched lines). Note that two additional kernels of the home ranges of the two species would be contained within the special management zone (see two tan areas).

Species such as the scalloped hammerhead and Galapagos sharks are par-ticularly vulnerable to fishing near seamounts and islands, not in the open ocean. Hence, it is at these sites that they are best protected. It is more feasible to protect these small areas where sharks are abundant than throughout the vast ocean basins. Yet the protection given to these vulnerable populations must be real, and this will require the presence of ocean reserve wardens. An effort was attempted during 2009–2010 at the Galapagos Islands to keep a vessel with researchers and reserve wardens aboard at Darwin Island through-out the entire year. Many tourists view the large schools of scalloped hammer-head sharks and smaller aggregations Galapagos sharks at the Rockfall at Wolf Island and Bus Stop at Darwin Island. They are charged a substantial fee to be brought to these sites by pleasure boats on weeklong expeditions. The viability of the populations of many species of cartilaginous fishes may in the future depend upon the creation of reserves at biological hotspots where they can be protected and the exerting of the will to protect them at these sites. The recent

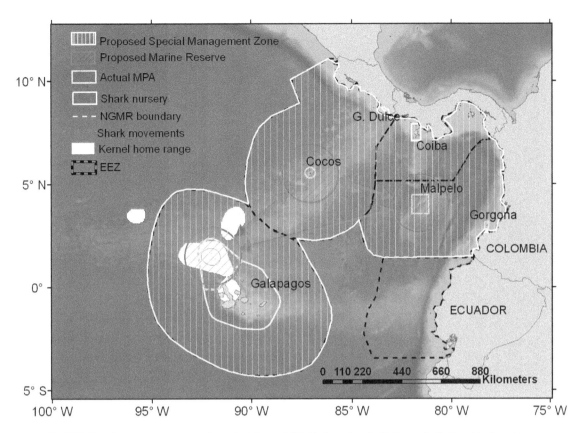

Fig. 16.10 Map showing two proposed marine protected areas (MPAs) in the eastern Pacific Ocean. The first consists of a series of three localized no-take zones with circular boundaries (see cross-hatching in red circles); the second is much larger, a special management zone extending from the coast of Central America and the northwestern coast of South America outward into the eastern Pacific Island around all three islands (see white vertical hatching with white boundaries).

expansion of this highly profitable ecotourism industry may provide a financial incentive for the protection to sharks worldwide in the future.

SUMMARY

The life history characteristics of the elasmobranchs make populations particularly vulnerable to overfishing, and the wide travels of many species make fisheries management difficult. There has been a concomitant growth of fisheries for sharks worldwide as more traditional fisheries have declined. Directed fisheries for sharks have often shown a boom-and-bust pattern. Shark populations have decreased sharply after periods of intense fishing pressure. Examples are the fates of the fishery for the tope shark off the western coast of North America, the basking shark off Europe and Canada, and the porbeagle shark off the eastern coast of North America. Catch records indicate that the populations of all of the large species of sharks have declined in the

North Atlantic Ocean off the eastern coast of North America since 1970, when commercial fisheries began to target the large sharks. In contrast, the sizes of the populations of the smaller sharks and rays have increased over the same period. The most dramatic population explosion of smaller meso-predators has been that of the cownose ray, and large schools have depleted commercial stocks of bivalves along the eastern coast of North America. This is an example of a trophic cascade, in which the removal of a species higher on the food chain has a cascading impact on the abundances of species at multiple trophic levels lower in the food chain.

The cartilaginous fishes are *K-selected* species. They are adapted to the constant conditions within the oceans and succeed in these conditions because they are efficient predators. They exhibit slow growth, late maturity, and low fecundity. The *K* reproductive strategy was very successful for the cartilaginous fishes in the past, resulting in their being the dominant predators over a period close to 400 million years. Their advanced swimming ability upon birth together with their fully developed jaws ensured that they would have a high probability of survival during the first few years of their lifespan. Yet the young produced by cartilaginous fishes are now often captured by fishermen during the first few years of their life using longlines or gill nets that can extend over many kilometers of the ocean surface.

Each state manages both its recreational and commercial fisheries for sharks in the waters from its coastline to a distance of 3 miles from shore. Exceptions to this rule are Texas and the other states on the Gulf Coast, which have jurisdiction over the fisheries for sharks to a distance of 10 miles from shore. The federal government manages fisheries over a zone encompassing the rest of the continental shelf from 3 miles (or 10 miles on the Gulf Coast) to an outer limit of 200 miles from shore. The current federal plans for managing sharks are aimed at rebuilding and keeping certain populations of sharks at constant and sustainable levels.

The intrinsic rate of increase, r, of each shark and ray species varies with its unique life history characteristics. A method has been developed to calculate r for a variety of sharks as a measure of rebound potential ($r_{Z(msy)}$) — the rate of population growth achieved by a species hypothetically exposed to a maximum sustainable level of fishing. This technique permits one to estimate how different shark species are likely to respond to the same level of fishing pressure. This turns out to be a useful criterion for comparing species according to their capacity to withstand exploitation by fisheries. The rate of population expansion for each species can be estimated at a particular level of mortality, or loss due to fishing, that achieves the maximum sustainable yield. An alternative to the traditional method of fisheries management for sharks confined to specific areas during crucial stages in their life histories is the establishment

1. Do you believe that a fishery for sharks can succeed over the long run? Species with what types of life histories would you permit fishermen to catch? How would you regulate such a fishery to sustain the yield of sharks over time?

2. Are marine reserves a viable method of protecting shark populations? How would you go about designing a marine reserve? What information would you need so that it successfully protects resources within it? What research studies would you conduct to obtain information critical to its design?

3. How would you estimate the size of a population of sharks, for which there is no fishery? You should read Chapple *et al.* (2011),[11] "A first estimate of white shark, *Carcharodon carcharias*, abundance off Central California," for an idea of how to do this.

of marine reserves, which can protect species that concentrate in certain areas where they become highly vulnerable to fishing, allowing mating, pupping, or feeding.

* * *

KEY TO COMMON AND SCIENTIFIC NAMES

Atlantic sharpnose shark = *Rhizoprionodon terraenovae*; basking shark = *Cetorhinus maximus*; bay scallop = *Argopecten irradians*; bigeye sandtiger shark = *Odontaspis noronhai*; bigeye thresher shark = *Alopias superciliosus*; bigeye tuna = *Thunnus obesus*; big skate = *Raja binoculata*; blackfin reef shark = *Carcharhinus amblyrhynchos*; blacktip shark = *Carcharhinus limbatus*; blue shark = *Prionace glauca*; bonnethead shark = *Sphyrna tiburo*; brown smooth-hound shark = *Mustelus henlei*; bull shark = *Carcharhinus leucas*; chain catshark = *Scyliorhinus rotifer*; cownose ray = *Rhinoptera bonasus*; dusky shark = *Carcharhinus obscurus*; dusky smooth-hound shark = *Mustelus canis*; Galapagos shark = *Carcharhinus galapagensis*; gray smooth-hound shark = *Mustelus californicus*; Greenland shark = *Somniosus microcephalus*; hard-shell clam = *Mercenaria mercenaria*; lemon shark = *Negaprion brevirostris*; leopard shark = *Triakis semifasciata*; little skate = *Leucoraja erinacea*; megamouth shark = *Megachasma pelagios*; northern bluefin tuna = *Thunnus orientalis*; oceanic whitetip shark = *Carcharhinus longimanus*; oyster = *Crassostrea virinica*; Pacific angelshark = *Squatina californica*; pelagic stingray = *Pteroplatytrygon violacea*; pelagic thresher shark = *Alopias pelagicus*; piked dogfish = *Squalus acanthias*; porbeagle shark = *Lamna nasus*; salmon shark = *Lamna ditropis*; sandbar shark = *Carcharhinus plumbeus*; sandtiger shark = *Carcharias taurus*; sardine = *Anchoa sardinops*; scalloped hammerhead shark = *Sphyrna lewini*; sevengill shark = *Notorhynchus cepedianus*; shortfin mako shark = *Isurus oxyrinchus*; silky shark = *Carcharhinus falciformis*; skipjack tuna = *Katsuwonus pelamis*; smooth butterfly ray = *Gymnura micrura*; softshell clam = *Mya arenaria*; southern bluefin tuna = *Thunnus maccoyii*; spotted ratfish = *Hydrolagus colliei*; swordfish = *Xiphias gladius*; thresher shark = *Alopias vulpinus*; tiger shark = *Galeocerdo cuvier*; tope shark = *Galeorhinus galeus*; whale shark = *Rhincodon typus*; white shark = *Carcharodon carcharias*; whitetip reef shark = *Triaenodon obesus*; yellowfin tuna = *Thunnus albacares*.

LITERATURE CITED

1. Camhi *et al.*, 1998; 2. Berrow and Heardman,1994; 3. Campana *et al.*, 2008; 4. Myers *et al.*, 2007; 5. Ebert, 2003; 6. Uchida, *et al.*,1996; 7. Ainley *et al.*, 1985; 8. Klimley and Anderson, 1996; 9. Cliff *et al.*, 1996; 10. Strong *et al.*, 1996; 11. Chapple *et al.*, 2011; 12. Oceana, 2010; 13. Abercrombie *et al.*, 2005; 14. Cavanagh *et al.*, 2008; 15. Smith *et al.*, 1998; 16. Stearns, 1992; 17. Smith *et al.*, 2008; 18. Cortés, 2008; 19; Au *et al.*, 2008; 20. Ketchum, 2011.

RECOMMENDED FURTHER READING

Au, D. W., S. E. Smith, and C. Show. 2008. Shark productivity and reproductive protection, and a comparison with teleosts. Pp. 298–308 *in* Camhi, M. D., E. K. Pikitch, and E. A. Babcock (Eds.), Sharks of the Open Ocean: Biology, Fisheries & Conservation. Blackwell Publishing Ltd., Oxford.

Camhi, M., S. Fowler, J. Musick, A. Bräutigam, and S. Fordham. 1998. Sharks and their relatives: ecology and conservation. *Occas. Paper, IUCN Species Surv. Com.*, 20: 1–39.

Camhi, M. D., E. K. Pikitch, and E. A. Babcock. 2008. Sharks of the Open Ocean: Biology, Fisheries, and Conservation. Blackwell Science, Oxford.

Chapple, T. K., S. J. Jorgensen, S. D. Anderson, P. E. Kanive, A. P. Klimley, L. W. Botsford, and B. A. Block. 2011. A first estimate of white shark, *Carcharodon carcharias*, abundance off Central California. *Biological Letters*, doi: 10.1098/rsbl.2011.0124.

Compana, S. E., W. Joyce, L. Marks, P. Hurley, L. J. Natanson, N. E. Kohler, C. F. Jensen, J. J. Mello, H. L. Pratt, Jr., S. Myklevoll, and S. Harley. 2008. The rise and fall (again) of the porbeagle shark population in the northwest Atlantic. Pp. 445–446 *in* Camhi, M. D., E. K. Pikitch, and E. A. Babcock (Eds.), Sharks of the Open Ocean: Biology, Fisheries and Conservation. Blackwell Science, Oxford.

Oceana. 2010. The international trade of shark fins: endangered shark populations worldwide. Available at: http://oceana.org/sites/default/files/reports/OCEANA_inter national_trade_shark_fins_english.pdf.

Smith, S. E., D. W. Au, and C. Show. 1998. Intrinsic rebound potential of 26 species of Pacific sharks. *Mar. Freshwater Res.*, 49: 663–678.

Smith, S. E., D. W. Au, and C. Show. 2008. Intrinsic rates of increase in pelagic elasmobranchs. Pp. 288–297 *in* Camhi, M. D., E. K. Pikitch, and E. A. Babcock (Eds.), Sharks of the Open Ocean: Biology, Fisheries & Conservation. Blackwell Publishing Ltd., Oxford.

References

Aasen, O. 1963. Length and growth of the porbeagle (*Lamna nasus* Bonnaterre) in the northwest Atlantic. *Fiskeridir. Skr. Ser. Havunders.*, 13: 20–37.

Abercrombie, D. L., S. C. Clarke, and M. S. Shivji. 2005. Global-scale genetic identification of hammerhead sharks: applications to assessment of the international fin trade and law enforcement. *Cons. Genetics*, 6: 775–799.

Ahlberg, P., K. Trinajstic, Z. Johanson, and J. Long. 2009. Pelvic claspers confirm chondrichthyan-like internal fertilization in arthrodires. *Nature*, 460: 888–889.

Ainley, D. G., R. P. Henderson, H. R. Huber, R. J. Boekelheide, S. G. Allen, and T. L. McElroy. 1985. Dynamics of white shark/pinniped interactions in the Gulf of the Farallones. *South. Calif. Acad. Sci, Mem.*, 9: 109–122.

Alexander, R. L. 1995. Evidence of counter-current heat exchanger in the ray, *Mobula tarapacana* (Chondrichthyes: Elasmobranchii: Batoidea: Myliobatiformes). *J. Zool., London*, 237: 377–384.

Alexander, R. L. 1996. Evidence of brain-warming in the mobulid rays, *Mobula tarapacana* and *Manta birostris* (Chondrichthyes: Elasmobranchii: Batoidea; Myliobatiformes). *Zool. J. Linnean Soc.*, 118: 151–164.

Alexander, R. L. 1998. Blood supply to the eyes and brain of lamniform sharks (Lamniformes). *J. Zool., London*, 245: 363–369.

Alexander, R. M. 1965. The lift produced by the heterocercal tails of Selachii. *J. Exp. Biol.*, 40: 23–56.

Ali, M. A., and M. Anctil. 1976. Retinas of Fishes: An Atlas. Springer-Verlag, Berlin.

Altringham, J. D., P. H. Yancey, and I. A. Johnston. 1982. The effects of osmoregulatory solutes on tension generation by dogfish skinned muscle fibres. *J. Exp. Biol.*, 96: 443–446.

Ames, J. A., J. J. Geibel, F. E. Wendell, and C. A. Pattison. 1996. White shark–inflicted wounds of sea otters in California, 1996-1992. Pp. 309–316 *in* Klimley, A. P., and D. G. Ainley (Eds.), Great White Sharks: The Biology of *Carcharodon carcharias*. Academic Press, San Diego.

Anctil, M., and M. A. Ali. 1974. Giant ganglion cells in the retina of the hammerhead shark (*Sphyrna lewini*). *Vision Res.*, 14: 903–904.

Anderson, S. D., T. K. Chapple, S. J. Jorgensen, A. P. Klimley, and B. A. Block. 2011. Long-term individual identification and site fidelity of white sharks, *Carcharodon carcharias*, off California using dorsal fins. *Mar. Biol.*, 158: 1233–1237.

Anderson, S. D., R. P. Henderson, P. Pyle, and D. G. Ainley. 1996. White shark reactions to unbaited decoys. Pp. 223–228 *in* Klimley, A. P., and D. G. Ainley (Eds.), Great White Sharks: The Biology of *Carcharodon carcharias*. Academic Press, San Diego.

Anderson, S. D., A. P. Klimley, P. Pyle, and R. P. Henderson. 1996. Tidal height and white shark predation at the Farallon Islands, California. Pp. 275–279 in Klimley, A. P., and D. G. Ainley (Eds.), Great White Sharks: The Biology of *Carcharodon carcharias*. Academic Press, San Diego.

Antkowiak, D., and J. W. Boylan. 1974. Glomerular populations in kidney of *Raja erinacea* and *Squalus acanthias*. *Bull. Mt. Desert Isl. Biol. Lab.*, 14: 1–3.

Applegate, S. P. 1967. A survey of shark hard parts. Pp. 37–67 *in* Gilbert, P. W., R. F. Matthewson, and D. P. Rall (Eds.), Sharks, Skates, and Rays. Johns Hopkins University Press, Baltimore.

Aronson, L. R., F. R. Aronson, and E. Clark. 1967. Instrumental conditioning and light-dark discrimination in young nurse sharks. *Bull. Mar. Sci.*, 17: 249–256.

Au, D. W., S. E. Smith, and C. Show. 2008. Shark productivity and reproductive protection, and a comparison with teleosts. Pp. 298–308 *in* Camhi, M. D., E. K. Pikitch, and E. A. Babcock (Eds.), Sharks of the Open Ocean: Biology, Fisheries & Conservation. Blackwell Publishing Ltd., Oxford.

Avsar, D. 2001. Age, growth, reproduction and feeding of the spurdog (*Squalus acanthias* Linnaeus, 1758) in the south-eastern Black Sea. *Estuarine Coastal Shelf Sci.*, 52: 269–278.

Baldridge, J. R., and J. Williams. 1969. Shark attack: feeding or fighting? *Military Medicine*, 134: 130–133.

Banner, A. 1967. Evidence of sensitivity to acoustic displacements in the lemon shark, *Negaprion brevirostris* (Poey). Pp. 265–273 *in* Cahn, P. H. (Ed.), Lateral Line Detectors. Indiana University Press, Bloomington.

Banner, A. 1972. Use of sound in predation by young lemon sharks, *Negaprion brevirostris* (Poey). *Bull. Mar. Sci.*, 22: 251–283.

Barbaro, K. C., M. S. Lira, M. B. Malta, S. L. Soares, D. G. Neto, J. L. C. Cardoso, M. L. Santoro, and V. Haddad, Jr. 2007. Comparative study on extracts from the tissue covering the stingers of freshwater (*Potamotrygon falkneri*) and marine (*Dasyatis guttata*) stingrays. *Toxicon*, 50: 676–687.

Beatty, D. D. 1969. Visual pigments of three species of cartilaginous fishes. *Nature*, 222: 285.

Bechert, D. W., M. Bartenwerfer, G. Hoppe, and W. E. Reif. 1986. Drag reduction mechanisms derived from shark skin. Pp. 1044–1068 *in* Proceedings of the 15th ICAS Congress. *Am. Inst. Aeron. Astron.*

Bechert, D. W., G. Hoppe, and W. E. Reif. 1985. On the drag reduction of the shark skin. *Am. Inst. Aeron. Astron., Shear Flow Control Conference*, 1985: 1–18.

Beit, B. E. 1977. Secretion of rectal gland fluid in the Atlantic stingray, *Dasyatis sabina*. *Copeia*, 1977: 585–587.

Benton, M. J. 2005. Vertebrate Paleontology. Blackwell Publishing, Oxford.

Berrow, S. D., and C. Heardman. 1994. The basking shark *Cetorhinus maximus* (Gunnerus) in Irish waters: patterns of distribution and abundance. *Proc. Roy. Irish Acad.*, 94B: 101–107.

Bertin, L. 1985. Denticules cutanés et dents. Pp. 505–531 in Grassé, P. P. (Ed.), Traité de Zoologie. Anatomie-Systematique Biologie, Vol. 13, Pt. 1., Masson, Paris.

Bigelow, H. B., and W. C. Schroeder. 1948. Fishes of the Western North Atlantic: Lancelets, Cyclostomes, Sharks. Sears Foundation for Marine Research, Yale University.

Block, B. A., and F. G. Carey. 1983. Warm brain and eye temperatures in sharks. *J. Comp. Physiol. B*, 156: 229–236.

Block, B. A., and J. R. Finnerty. 1994. Endothermy in fishes: a phylogenetic analysis of constraints, predispositions, and selection pressures. *Environ. Biol. Fish.*, 40: 283–302.

Block, B. A., J. D. Jonsen, S. J. Jorgensen, A. J. Winship, S. A. Shaffer, S. J. Bograd, E. L. Hazen, D. G. Foley, G. A. Breed, A. L. Harrison, J. E. Ganong, A. Swithenbank, M. Castleton, H. Dewar, B. R. Mate, G. L. Shillinger, K. M. Schaefer, S. R. Benson, M. J. Weise, R. W. Henry, and D. P. Costa. 2011. Tracking apex marine predator movements in a dynamic ocean. *Nature*, 475: 86–90.

Bone, Q. 1999. Muscular system: microscopical anatomy, physiology, and biochemistry of elasmobranch muscle fibers. Pp. 115–143 *in* Hamlett, W. C. (Ed.), Sharks, Skates, and Rays. Johns Hopkins University Press, Baltimore.

Bonting, S. L. 1966. Studies on the sodium-potassium-activated adenosinetriphosphatase: XV. The rectal gland of the elasmobranches. *Comp. Biochem. Physiol.*, 17: 953–966.

Boustany, A. M., S. F. Davis, P. Pyle, B. J. Le Boeuf, and B. A. Block. 2002. Expanded niche for white sharks. *Nature*, 415: 35–36.

Bozzano, A., and S. P. Collin. 2000. Retinal ganglion cel topography in elasmobranches. *Brain Behav. Evol.*, 55: 191–208.

Bozzano, A., R. Murgia, S. Vallerga, J. Hirano, and S. N. Archer. 2001. The photoreceptor system in the retinae of two dogfishes, *Scyliorhinus canicula* and *Galeus melastomus*: possible relationship with depth distribution and predatory life style. *J. Fish Biol.*, 59: 1258–1278.

Bradley, J. L. 1996. Prey energy content and selection, habitat use and daily ration of Atlantic stingray, *Dasyatis sabina*. Thesis, Florida Institute of Technology, Melbourne, Florida.

Bray, R. N., and M. A. Hixon. 1978. Night-shocker: predatory behavior of the Pacific electric ray (*Torpedo californica*). *Science*, 200: 333–334.

Bridges, C. D. 1965. The grouping of fish visual pigments about preferred positions in the spectrum. *Vision Res.*, 5: 223–238.

Brown, C. 1980. Survivors of the USS Indianapolis still relive their ordeal. Los Angeles Times, 27 August 1980.

Budker, P. 1938. Les cruptes sensorieles et les denticles cutanés de Plagiostomes. *Ann. Inst. Oceanogr.*, 18: 207–288.

Burger, J. W. 1967. Problems in the electrolyte economy of the spiny dogfish, *Squalus acanthias*. Pp. 177–185 *in* Gilbert, P. W., R. F. Mathewson, and D. P. Rall (Eds.), Sharks, Skates, and Rays. Johns Hopkins University Press, Baltimore.

Burger, J. W., 1972. Rectal gland secretion in the stingray, *Dasyatis sabina*. *Comp. Biochem. Physiol.*, 42: 31–32.

Burger, J. W., and W. N. Hess. 1960. Function of the rectal gland of the spiny dogfish. *Science*, 131: 670–67.

Burgess, G. H., and M. Callaghan. 1996. Worldwide patterns of white shark attacks on humans. Pp. 457–469 *in* Klimley, A. P., and D. G. Ainley (Eds.), Great White Sharks: the Biology of *Carcharodon carcharias*. Academic Press, San Diego.

Burghart, G. M. 1970. Defining "communication." Pp. 5–18 *in* Johnston, W. H., Jr. D. G. Moulton, and A. Turk (Eds.), Communication by Chemical Signals. New Appleton-Century-Crofts, New York.

Bush, A. C., and K. N. Holland. 2002. Food limitation in a nursery area: estimates of daily ration in juvenile scalloped hammerhead sharks, *Sphyrna lewini* (Griffith and Smith, 1834) in Kāne'ohe Bay, Ō'ahu, Hawai'i. *J. Exp. Mar. Biol. Ecol.*, 278: 157–178.

Cailliet, G. M., and K. J. Goldman. 2004. Age determination and validation in chondrichthyan fishes. Pp. 399–448 *in* Carrier, J. C., J. A. Musick, and M. R. Heithaus (Eds.), Biology of Sharks and Their Relatives. CRC Press, Boca Raton.

Cailliet, G. M., H. F. Mollet, G. G. Pittenger, D. Bedford, and L. J. Natanson. 1992. Growth and demography of the Pacific angel shark (*Squatina californica*), based on tag returns off California. *Aust. J. Mar. Freshwater Res.*, 43: 1313–1330.

Cailliet, G. M., L. J. Natanson, B. A. Welden, and D. A. Ebert. 1985. Preliminary studies on the age and growth of the white shark, *Carcharodon carcharias*, using vertebral bands. *Memoirs, South. Calif. Acad. Sci.*, 9: 49–60.

Cairns, D. K., A. J. Gaston, and F. Huettmann. 2008. Endothermy, ectothermy, and the global structure of marine vertebrate communities. *Mar. Ecol. Progr. Ser.*, 356: 239–250.

Caldicott, D. G. E., R. Mahanani, and M. Kuhn. 2001. The anatomy of a shark attack: a case report and review of the literature. *Injury (Int. J. Of the Care of the Injured)*, 32: 445–453.

Callard, I. P., O. Putz, M. Paolucci, and T. J. Koob. 1995. Elasmobranch reproductive life-histories: endocrine correlates and evolution. Pp. 2004–2008 *in* Goetz, F., and P. Thomas (Eds.), Proceedings of the Fifth International Symposium on the Reproductive Physiology of Fish, University of Texas, Austin.

Camhi, M., S. Fowler, J. Musick, A. Bräutigam, and S. Fordham. 1998. Sharks and their relatives: ecology and conservation. *Occas. Paper, IUCN Species Surv. Com.*, 20: 1–39.

Campana, S. E., W. Joyce, L. Marks, P. Hurley, L. J. Natanson, N. E. Kohler, C. F. Jensen, J. J. Mello, H. L. Pratt, Jr., S. Myklevoll, and S. Harley. 2008. The rise and fall (again) of the porbeagle shark population in the northwest Atlantic. Pp. 445–446 *in* Camhi, M. D., E. K. Pikitch, and E. A. Babcock (Eds.), Sharks of the Open Ocean: Biology, Fisheries and Conservation. Blackwell Science, Oxford.

Campana, S. E., L. J. Natanson, and S. Myklevoll. 2002. Bomb dating and age determination of large pelagic sharks. *Can. J. Fish. Aquat. Sci.*, 59: 450–455.

Capetta, H. 1987. Chondrichthyes II, Mesozoic and Cenozoic Elasmobranchii, Vol. 3B. 1–193 *in* Schultze, H. P. (Ed.), Handbook of Paleoichthyology. Verlag Friedrich Pfeil, Munich.

Capetta, H., C. Duffin, and J. Zidela. 1993. Chondrichthyes. Pp. 593–609 *in* Benton, M. J. (Ed.), The Fossil Record. Chapman and Hall, London.

Carey, F. G. 1973. Fishes with warm bodies. *Sci. Am.*, 228: 36–44.

Carey, F. G. 1982. Warm fish. Pp. 216–233 *in* Taylor, C. R., K. Johansen, and L. Bolis (Eds.), A Companion to Animal Physiology. Cambridge University Press, Cambridge.

Carey, F. G., J. G. Casey, H. L. Pratt, D. Urquhart, and J. E. McCosker. 1985. Temperature, heat production and heat exchange in lamnid sharks. *Memoirs, South. Calif. Acad. Sci.*, 9: 92–108.

Carey, F. G., J. W. Kanwisher, O. Brazier, G. Gabrielson, J. G. Casey, and H. L. Pratt. 1982. Temperature and activities of a white shark, *Carcharodon carcharias*. *Copeia*, 1982: 254–260.

Carey, F. G., and J. V. Scharold. 1990. Movements of blue sharks (*Prionace glauca*) in depth and course. *Mar. Biol.*, 106: 329–342.

Carey, F. G., J. M. Teal, and J. W. Kanwisher. 1981. The visceral temperatures of mackerel sharks (Lamnidae). *Physiol. Zool.*, 54: 334–344.

Carlson, J. K., K. J. Goldman, and C. G. Lowe. 2004. Metabolism, energetic demand, and endothermy. Pp. 203–224 *in* Carrier, J. C., J. A. Musick, and M. R. Heithaus (Eds.), Biology of Sharks and Their Relatives, CRC Press, Boca Raton.

Carr, W. E. S. 1988. The molecular nature of chemical stimuli in the aquatic environment. Pp. 3–27 *in* Atema, J., R. R. Fay, A. N. Popper, and W. N. Tavolga (Eds.), Sensory Biology of Aquatic Animals. Springer-Verlag, Berlin.

Carrier, J. C., M. R. Heithaus, and John A. Musick. 2009. Sharks and Their Relatives: Physiological Adaptations, Behavior, Ecology, Conservation, and Management. CRC Press, Boca Raton.

Carrier, J. C., and C. A. Luer. 1990. Growth rates in the nurse shark, *Ginglymostoma cirratum*. *Copeia*, 1990: 686–692.

Carrier, J. C., H. L. Pratt, Jr., J. I. Castro. 2004. Reproductive biology of elasmobranchs. Pp. 269–285 *in* Carrier, J. C., J. A. Musick, and M. R. Heithaus (Eds.), Biology of Sharks and Their Relatives. CRC Press, Boca Raton.

Carrier, J. C., H. L. Pratt, Jr., and L. K. Martin. 1994. Group reproductive behaviors in free-living nurse sharks, *Ginglymostoma cirratum*. *Copeia*, 1994: 646–656.

Carroll, R. L. 1988. Vertebrate Paleontology and Evolution. W. H. Freeman, New York.

Carwardine, M., and K. Watterson. 2002. The Shark Watchers Handbook: A Guide to Sharks and Where to See Them. Princeton University Press, Princeton, New Jersey.

Castro, J. E., P. M. Bubucis, and N. A. Overstrom. 1988. The reproductive biology of the chain dogfish, *Scyliorhinus retifer*. *Copeia*, 1988: 749–746.

Cavanagh, R. D., S. L. Fowler, and M. D. Camhi. 2008. Pelagic sharks and the FAO international plan of action for the conservation and management of sharks. Pp. 478–492 *in* Camhi, M. D., E. K. Pikitch, and E. A. Babcock (Eds.), Sharks of the Open Ocean: Biology, Fisheries and Conservation. Blackwell Publishing Ltd., Oxford.

Chapple, T. K., S. J. Jorgensen, S. D. Anderson, P. E. Kanive, A. P. Klimley, L. W. Botsford, and B. A. Block. 2011. A first estimate of white shark, *Carcharodon carcharias*, abundance off central California. *Biological Letters*, doi: 10.1098/rsbl.2011.0124.

Chen, C. T., T. C. Leu, S. J. Joung, and N. C. H. Lo. 1990. Age and growth of the scalloped hammerhead, *Sphyrna lewini*, in northeastern Taiwan waters. *Pac. Sci.*, 44; 156–170.

Cioni, A. L. 1996. The extinct genus of *Notidanodon* (Neoselachii, Hexanchiformes). Pp. 63–73 *in* Arratia, G., and G. Viohl (Eds.), Mesozoic Fishes: Systematics and Paleoecology. Verlag, Dr. Friedrich Pfiel, Munich.

Clark, E. 1963. The maintenance of sharks in captivity with a report on their instrumental conditioning. Pp. 115–149 *in* Gilbert, P. W. (Ed.), Sharks and Survival. D. C. Heath, Boston.

Clark, E. 1974. The Red Sea's sharkproof fish. *Nat. Geogr. Mag.*, 145: 718–727.

Clark, E. 1983. Shark repellent effect of the Red Sea Moses sole. Pp. 135–150 in Zahuranec, B. J. (Ed.), Shark Repellents from the Sea. Westview Press, Boulder.

Clark, E., and K. Von Schmidt. 1965. Sharks of the central Gulf Coast of Florida. *Bull. Mar. Sci.*, 15: 13–83.

Cliff, G., and S. F. J. Dudley. 1991. Sharks caught in the protective gill nets off Natal, South Africa. 4. The bull shark *Carcharhinus leucas* Valenciennes. *South African J. Mar. Sci.*, 10: 253–270.

Cliff, G., R. P. Van Der Elst, A. Govender, T. Witthuhn, and E. M. Bullen. 1996. First estimates of mortality and population size of white sharks on the South African coast. Pp. 393–400 *in* Klimley, A. P., and D. G. Ainley (Eds.), Great White Sharks: The Biology of *Carcharodon carcharias*. Academic Press, San Diego.

Cohen, J. L. 1980. Functional organization of the retina of the lemon shark (*Negaprion brevirostris*, Poey); an anatomical and electrophysiological approach. Dissertation, University of Miami, Coral Gables.

Cohen, J. L., R. E. Hueter, and D. T. Organisciak. 1990. The presence of porphyropsin-based visual pigments in the juvenile lemon shark (*Negaprion brevirostris*). *Vision Res.*, 30: 1949–1953.

Collin, S. P. 1988. The retina of the shovel-nosed ray, *Rhinobatus batillum* (Rhinobatidae); morphology and quantitative analysis of the ganglion, amacrine and bipolar cell populations. *Exp. Biol.*, 47: 195–207.

Combes, S. A., and T. L. Daniel. 2001. Shape, flapping and flexion; wing and fin design for forward flight. *J. Exp. Biol.*, 204: 2073–2085.

Combes, S., and J. C. Montgomery. 1999. The enigmatic lateral line system. Pp. 319–362 *in* Fay, R. R., and A. N. Popper (Eds.), Comparative Hearing: Fish and Amphibians. Springer-Verlag, New York.

Compagno, L. J. V. 1999. Endoskeleton. Pp. 69–92 *in* Hamlett, W. C. (Ed.). Sharks, Skates, and Rays. Johns Hopkins University Press, Baltimore.

Compagno, L., M. Dando, and S. Fowler. 2005. Sharks of the World. Princeton University Press, Princeton.

Conrath, C. L., J. Gelsleichter, and J. A. Musick. 2002. Age and growth of the smooth dogfish (*Mustelus canis*) in the northwest Atlantic Ocean. *Fish. Bull.*, 100: 674–682.

Conrath, C. L., and J. A. Musick. 2002. Reproductive biology of the smooth dogfish, *Mustelus canis*, in the northwest Atlantic Ocean. *Environ. Biol. Fish.*, 64: 367–377.

Cortés, E. 1999. Standardized diet compositions and trophic levels of sharks. *ICES J. Mar. Sci.*, 56: 707–717.

Cortés, E. 2008. Comparative life history and demography of pelagic sharks. Pp. 307–322 *in* Camhi, M. D., E. K. Pkitch, and E. A. Babcock (Eds.), Sharks of the Open Ocean: Biology, Fisheries & Conservation. Blackwell Publishing Ltd., Oxford.

Cortés, E., C. Manire, and R. E. Hueter. 1996. Diet, feeding habits, and diel feeding chronology of the bonnethead shark, *Sphyrna tiburo*, in southwest Florida. *Bull. Mar. Sci.*, 58: 353–367.

Corwin, J. T. 1977. Morphology of the macula neglecta in the shark genus *Carcharhinus*. *J. Morph.*, 152: 341–361.

Corwin, J. T. 1978. The relation of inner ear structure to the feeding behavior in sharks and rays. *Scanning Electron Microscopy*, 11: 1105–1112.

Corwin, J. T. 1981. Peripheral auditory physiology in the lemon shark: evidence of parallel otolithic and non-otolithic sound detection. *J. Comp. Physiol.*, 142A: 379–390.

Crescitelli, F., M. McFall-Ngai, and J. Horwitz. 1985. The visual pigment sensitivity hypothesis: further evidence from fishes of varying habitats. *J. Comp. Physiol.*, 157A: 323–333.

Cummings, W. C., and P. O. Thompson. 1971. Gray whales, *Eschrichtius robustus*, avoid the underwater sounds of killer whales, *Orcinus orca*. *Fish. Bull.*, 69: 525–530.

Curio, E. 1976. The Ethology of Predation. Springer-Verlag, Berlin.

Curtis, T. H., J. T. Kelly, K. L. Menard, R. K. Laroche, R. E. Jones, and A. P. Klimley. 2006. Observations on the behavior of white sharks scavenging from a whale carcass at Point Reyes, California. *Calif. Fish Game*, 93: 112–124.

Davis, D. H., and J. D. D'Aubrey. 1961a. Shark attack off the east coast of South Africa on 24 December, 1960 with notes on the species of shark responsible for the attack. Investigational Report #2, Oceanographic Research Institute, Durban, South Africa.

Davis, D. H., and J. D. D'Aubrey. 1961b. Shark attack off the east coast of South Africa, 6 January, 1961. Investigational Report #3, Oceanographic Research Institute, Durban, South Africa.

Davis, D. H., and J. D. D'Aubrey. 1961c. Shark attack off the east coast of South Africa, 22nd January, 1961. Investigational Report #4, Oceanographic Research Institute, Durban, South Africa.

Dean, B. 1906. Chimaeroid Fishes and Their Development. Carnegie Institution of Washington, Pub. 32. Wilkens-Sheiry Printing, Washington, D.C.

Dean, M. N., W. Chiou, and A. P. Summers. 2005. Morphology and ultrastructure of prismatic calcified cartilage. *Microsc. Microanl.*, 11 (Suppl. 2): 1196–1197.

Dean, M. N., and A. P. Summers. 2006. Mineralized cartilage in the skeleton of chondrichthyan fishes. *Zoology*, 109: 164–168.

DeBose, J. L., S. C. Lema, and G. A. Nevitt. 2008. Dimethylsulfoniopropionate as a foraging cue for reef fishes. *Science*, 319: 1356.

DeBose, J. L., and G. A. Nevitt. 2007. Investigating the association between pelagic fish and dimethylsuloniopropionate in a natural coral reef system. *Mar. Freshw. Res.*, 58: 720–724.

Dehghani, H., M. M. Sajjadi, P. Parto, H. Rajaian, and A. Mokhlesi. 2010. Histological characterization of the special venom secretory cells in the stinger of rays in the northern waters of Persian Gulf and Oman Sea. *Toxicon*, 55: 1188–1194.

DeLoach, N. 1999. Reef Fish Behavior: Florida, Caribbean, Bahamas. New World Publications, Jacksonville.

Dempster, R. P., and E. S. Herald. 1961. Notes on the hornshark, *Heterodontus francisci*, with observations on mating activities. *Occ. Pap. Calif. Acad. Sci.*, 33: 1–7.

Demski, L. S., and R. G. Northcutt. 1996. The brain and cranial nerves of the white shark: an evolutionary perspective. Pp. 121–130 *in* Klimley, A. P., and D. G. Ainley (Eds.), Great White Sharks: The Biology of *Carcharodon carcharias*. Academic Pres, San Diego.

Denton, E. J., and J. A. C. Nicol. 1964. Visual pigments of deep-sea elasmobranchs. *J. Mar. Biol. Assoc., U.K.*, 44: 65–70.

Denton, E. J., and T. I. Shaw. 1963. The visual pigments of some deep-sea elasmobranchs. *J. Mar. Biol. Assoc., U.K.*, 43: 65–70.

Diaz, J. H. 2008. The evaluation, management, and prevention of stingray injuries in travelers. *J. Travel Medicine*, 15: 102–109.

Dick, J. R. F. 1981. *Diplodoselachi woodi* gen. et. sp. Nov., an early Carboniferous shark from the Midland Valley of Scotland. *Trans. Royal Soc. Edinburgh, Earth Sci.*, 72: 99–113.

Didier, D. A. 2004. Phylogeny and classification of extant Holocephali. Pp. 115–135 *in* Carrier, J. C., J. A. Musick, and M. R. Heithaus (Eds.), Biology of Sharks and Their Relatives. CRC Press, Boca Raton.

Di Giácomo, E., A. M. Parma, and J. M. Orensanz. 1994. Food consumption by the cock fish, *Callorhynchus callorhynchus* (Holocepahali: Callorhynchidae) from Patagonia (Argentina). *Environ. Biol. Fish.*, 40: 199–211.

Dingerkus, G., S. Séret, and E. Guibert. 1991. Multiple prismatic calcium phosphate layers in the jaws of present-day sharks (Chondrichthyes: Selachii). *Experimentia*, 47: 38–40.

Domeier, M. L. (Ed.). 2012. Global Perspectives on the Biology and Life History of the White Shark. CRC Press, Boca Raton.

Douady, C. J., M. Dosay, M. S. Shivji, and M. J. Stanhope. 2003. Molecular phylogenetic evidence refuting the hypothesis of Batoidea (rays and skates) as derived sharks. *Molecular Phylogen. Evol.*, 26: 215–221.

Drew, E. A. 1983. Light. Pp. 10–57 *in* Earll, R., and D. G. Erwin (Eds.), Sublittoral Ecology. Clarendon Press, Oxford.

Duffin, C. J. 1988. The upper Jurassic Selachian *Paleocarcharias* de Beaumont (1960). *Zool. Journ. Linnean Soc.*, 94: 271–286.

Ebert, D. A. 2003. Sharks, Rays, and Chimaeras of California. University of California Press, Berkeley.

Ebert, D. A., and J. J. Bizzarro. 2007. Standardized diet compositions and trophic levels of skates (Chondrichthyes: Rajiformes: Rajoidei). *Environ. Biol. Fish.*, 80: 221–237.

Eckert, S. A., and B. S. Stewart. 2001. Telemetry and satellite tracking of whale sharks, *Rhincodon typus*, in the Sea of Cortez, Mexico, and the North Pacific Ocean. *Environ. Biol. Fish.*, 60: 299–308.

Elasser, W. M. 1946. Induction effects in terrestrial magnetism. *Theor. Phys. Rev.*, 69: 106–116.

Evangelista, C., M. Mills, U. E. Siebeck, and S. P. Collin. 2010. A comparison of the external morphology of the membranous inner ear in elasmobranchs. *J. Morph.*, 271: 483–495.

Evans, D. H., P. M. Piermarini, and K. P. Choe. 2004. Homeostasis, osmoregulation, pH regulation, and nitrogen excretion. Pp. 247–268 *in* Carrier, J. C., J. A. Musick, and M. R. Heithaus (Eds.), Biology of Sharks and Their Relatives. CRC Press, Boca Raton.

Evans, R. J., and R. S. Davies. 1996. Stingray injury. *J. Accident and Emergency Medicine*, 13: 224–225.

Fagan, R. M., and D. Y. Young. 1978. Temporal patterns of behaviors: durations, intervals, latencies, and sequences. Pp. 79–114 *in* Colgan, P. W. (Ed.), Quantitative Ethology. John Wiley and Sons, New York.

Fay, R. R. 1988. Hearing in Vertebrates: A Psychophysics Databook. Heffernan Press, Worchester.

Fay, R. R., J. I. Kendall, A. N. Popper and A. L. Tester. 1974. Vibration detection by the macula neglecta of sharks. *Comp. Biochem. Physiol.*, 47A: 1235–1240.

Feder, H. M., C. H. Turner, and C. Limbaugh. 1974. Observations on fishes associated with kelp beds in southern California. *Cal. Fish Game Bull.*, 160: 1–144.

Fernicola, R. G. 2001. Twelve Days of Terror: A Definitive Investigation of the 1916 New Jersey Shark Attacks. Lyons Press, Guilford, Connecticut.

Ferry, L. A., and G. V. Lauder. 1996. Heterocercal function in leopard sharks: a three-dimensional kinematic analysis of two models. *J. Exp. Biol.*, 199: 2253–2268.

Flock, Å. 1967. Transducing mechanism in the lateral line canal organ receptors. *Cold Spring Harbor Symp. Quant. Biol.*, 30: 133–144.

Fouts, W. R., and D. R. Nelson. 1999. Prey capture by the Pacific angel shark, *Squatina californica*: visually mediated strikes and ambush-site characteristics. *Copeia*, 1999: 304–311.

Francis, M. 1996. Observations on a pregnant white shark with a review of reproductive biology. Pp. 157–172 *in* Klimley, A. P., and D. G. Ainley (eds.), Great White Sharks: The Biology of *Carcharodon carcharias*. Academic Press, San Diego.

Francis, M. P. 1997. Spatial and temporal variation in the growth rate of elephantfish (*Callorhinchus milii*). *New Zealand J. Mar. Freshwater Res.*, 31: 9–23.

Friedman, P. A., and S. C. Hebert. 1990. Diluting segment in kidney of dogfish shark. I. Localization and characterization of chloride absorption. *Amer. J. Physiol.*, 258: 398–408.

Gallagher, A. J., and N. Hammerschlag. 2011. Global shark currency: the distribution, frequency, and economic value of shark ecotourism. *Current Issues in Tourism*, 2011: 1–16.

Gardiner, J. M., and J. Atema. 2010. The function of bilateral odor arrival time differences in olfactory orientation of sharks. *Current Biol.*, 20: 1187–1191.

Gelsleichter, J., L. E. L. Rasmussen, C. A. Manire, J. Tyminski, B. Chang, and L. Lombardi-Carlson. 2002. Serum steroid concentrations and development of reproductive organs during puberty in male bonnethead sharks, *Sphyrna tiburo. Fish Physiol. Biochem.*, 26: 389–401.

Germain, M., K. J. Smith, and H. Skelton.2000. The cutaneous cellular infiltrate to stingray envenomization contains increased TIA+ cells. *Brit. J. Dermatology*, 143: 1074–1077.

Ghouse, H. M., B. Parsa, J. W. Boylan, and J. C. Brennan. 1968. The anatomy, microanatomy and ultrastructure of the kidney of the dogfish *Squalus acanthias. Bull. Mt. Desert Island Biol. Lab.*, 8: 22–39.

Gilbert, P. W., and G. W. Heath. 1972. The clasper-siphon sac mechanism in *Squalus acanthias* and *mustelus canis. Comp. Biochem. Physiol.*, 42A: 97–119.

Gilbert, P. W., and S. Springer. 1963. Testing shark repellents. Pp. 477–494 *in* Gilbert, P. W. (Ed.), Sharks and Survival. D. C. Heath, Boston.

Gilmore, R. G., Jr., O. Putz, and J. W. Dodrill. 2005. Oophagy, intrauterine cannibalism and reproductive strategy in lamnoid sharks. Pp. 435–462 *in* Hamlett, W. C. (ed.), Reproductive Biology Biology and Phylogeny of Chondrichthyes: Sharks, Batoids and Chimaeras. Science Publishers, Enfield, New Hampshire.

Gleiss, A. C., B. Norman, and R. P. Wilson. 2011. Moved by that sinking feeling: variable diving geometry underlies movement strategies in whale sharks. *Functional Ecol.*, 25: 595–607.

Goldman, K. J. 1997. Regulation of body temperature in the white shark, *Carcharodon carcharias. J. Comp. Physiol., B*, 167: 423–429.

Goldman, K. J. 2002. Aspects of age, growth, demographics and thermal biology of two Lamniform shark species. Dissertation, College of William and Mary.

Goldman, K. J., and S. D. Anderson. 1999. Space utilization and swimming depth of white sharks, *Carcharodon carcharias*, at the South Farallon Islands, central California. *Environ. Biol. Fishes*, 56: 351–364.

Goodwin, A., J. Wyles, and A. Moreley. 2001. Paleofiles: The Permo-Triassic extinction. Http://palaeo.gly.bris.ac.uk/Palaeofiles/Permian/intro.html. University of Bristol.

Gordon, I. 1993. Pre-copulatory behavior of captive sand tiger sharks, *Carcharias taurus. Environ. Biol. Fishes*, 38: 159–165.

Gore, M. A., D. Rowat, D. Hall, F. R. Gell, and R. F. Ormond. 2008. Transatlantic migration and deep mid-ocean diving by basking shark. *Biol. Lett.*, 4: 395–398.

Gottfried, M. D., L. J. V. Compagno, and S. C. Bowman. 1996. Size and skeletal anatomy of the giant "megatooth" shark *Carcharodon megalodon*. Pp 55–66 *in* Klimley, A. P., and D. G. Ainley (Eds.), Great White Sharks: The Biology of *Carcharodon carcharias*. Academic Press, San Diego.

Govardovski, V. I., and L. V. Lychakov. 1977. Visual cells and visual pigments in the Black Sea elasmobranchs. *Zh. Evol. Biokhim. Fiziol.*, 13: 162–166.

Gove, P. B. 1966. Webster's Third New International Dictionary of the English Language, G&C Merriam Co., Springfield, MA.

Graham, R. T., C. M. Roberts, and J. C. R. Smart. 2005. Diving behaviour of whale sharks in relation to a predictable food source. *J. Royal Soc. Interface*, 2003: 1–8.

Green, C. 1986. Single nephron structure and function, and renal effects of catechol-amines in the dogfish *Scyliorhinus canicula*. Dissertation, University of Hull, England.

Griffith, P. W., P. K. T. Pang, A. K. Srivàstava, and G. E. Pickford. 1973. Serum composi-tion of freshwater stingrays (Potamotrygonidae) adapted to fresh and dilute seawater. *Biol. Bull.*, 144: 304–320.

Grogan, E. D., and R. Lund. 2004a. Diversity and evolution of Paleozoic chondrichthyes. Presentation, American Elasmobranch Society.

Grogan, E. D., and R. Lund. 2004b. The origin and relationships of early Chondrichthyes. Pp. 3–31 *in* Carrier, J. C., J. A. Musick, and M. R. Heithaus (Eds.), Biology of Sharks and Their Relatives. CRC Press, Boca Raton.

Grove, A. J., and G. E. Newell. 1936. A mechanical investigation into the effectual action of the caudal fin of some aquatic chordates. *Annals Mag. Nat. Hist.*, 17: 280–290.

Gruber, S. H. 1984. Bioenergetics of the captive and free-ranging lemon shark (*Negaprion brevirostris*). *Proc. Am. Assoc. Parks Aquar.*, 1984: 341–373.

Gruber, S. H., and J. L. Cohen. 1976. Visual system of the elasmobranches: state of the art 1960–1975. Pp. 11–105 *in* Hodgson, E. S., and R. F. Mathewson (Eds.), Sensory Biology of Sharks, Skates, and Rays. U.S. Government Printing Office, Washington, D.C.

Gruber, S. H., and J. L. Cohen. 1985. Visual system of the white shark, *Carcharodon car-charias*, with emphasis on retinal structure. *S. Calif. Acad. Sci., Mem.*, 9: 61–72.

Gruber, S. H., R. L. Gulley, and J. Brandon. 1975. Duplex retina in seven elasmobranch species. *Bull. Mar. Sci.*, 25: 25: 353–358.

Gruber, S. H., D. I. Hamasaki, and C. D. B. Bridges. 1963. Cones in the retina of the lemon shark (*Negaprion brevirostris*). *Vision Res.*, 3: 397–399.

Gunn, J. S., J. D. Stevens, T. L. O. Davis, and B. M. Norman. 1999. Observations on the short-term movements and behaviour of whale sharks (*Rhincodon typus*) at Nigaloo Reef, western Australia. *Mar. Biol.*, 1335: 553–559.

Guzman, H. 2008. Personal communication. Smithsonian Research Laboratory, Panama City.

Haddad, V., Jr., D. G. Neto, J. Batista de Paula Neto, J., F. Portella de Luna Marques, and K. C. Barbaro. 2003. Freshwater stingrays: study of epidemiologic, clinical and thera-peutic aspects based on 84 envenomings in humans and some enzymatic activities of the venom. *Toxicon*, 43: 287–294.

Hamasaki, D. I., and S. H. Gruber. 1965. The photoreceptors of the nurse shark, *Gingly-mostoma cirratum*, and the sting ray, *Dasyatis sayi*. *Bull. Mar. Sci.*, 15: 1051–1059.

Hamlett, W. C. 1993. Ontogeny of the umbilical cord and placenta in the Atlantic sharp-nose shark, *Rhizoprionodon terraenovae*. *Environ. Biol. Fish.*, 38: 253–267.

Hamlett, W. C., and M. K. Hysell. 1998. Uterine specializations in elasmobranchs. *J. Exp. Zool.*, 282: 438–459.

Hamlett, W. C., and T. J. Koob. 1999. Female reproductive system. Pp. 398–443 *in* Ham-lett, W. C. (Ed.), Sharks, Skates, and Rays. Johns Hopkins University Press, Baltimore.

Hara, T. J. 1973. Olfactory responses to amino acids in rainbow trout, *Salmo gairdneri*. *Comp. Biochem. Physiol.*, 44A: 31–36.

Hara, T. J. 1975. Molecular structure and stimulatory effectiveness of amino acids in fish olfaction. Pp. 223–225 *in* Denton, D. A., and J. P. Coghlin (Eds.), Olfaction and Taste. Academic Press, New York.

Hart, N. S., T. J. Lisney, and S. P. Collin. 2006. Visual communication in elasmobranchs. Pp. 337–392 *in* Ladich, F., S. P. Collin, P. Moller, and B. G. Kapoor (Eds.), Communi-cation in Fishes. Science Publishers, Enfield, New Hampshire.

Hart, N. S., T. J. Lisney, N. J. Marshall, and S. P. Collin. 2004. Multiple cone visual pigments and the potential for trichromatic colour vision in two species of elasmobranch. *J. Exp. Biol.*, 207: 4587–4594.

Hart, N. S., S. M. Theiss, B. K. Harahush, and S. P. Collin. 2011. Microspectrophotometric evidence for cone monochromacy in sharks. *Naturwissenschaften*, 98: 193–201.

Haywood, G. P. 1975. Indication of sodium chloride and water exchange across the gills of the striped dogfish *Poroderma africanus*. *Mar. Biol.*, 29: 265–276.

Heiligenberg, W. H. 1993. Electrosensation. Pp. 137–160 in Evans, D. H. (Ed.), The Physiology of Fishes. CRC Press, Boca Raton.

Heithaus, M. R. 2001. Predator-prey and competitive interactions between sharks (order Selachii) and dolphins (order Odontoceti): a review. *Proc. Zool. Soc. London*, 253: 53–68.

Heithaus, M. R. 2001. The biology of tiger sharks, *Galeocerdo cuvieri*, in Shark Bay, western Australia: sex ratio, size distribution, diet, and seasonal changes in catch rates. *Environ. Biol. Fish.*, 61: 25–36.

Henderson, I. W., L. B. O'Toole, and N. Hazon. 1988. Kidney function. Pp. 201–214 *in* Shuttleworth, T. J. (Ed.), Physiology of Elasmobranch Fishes. Springer-Verlag, Berlin.

Henningsen, A. D. 2002. Age and growth in captive southern stingrays, *Dasyatis americana*. Abstract, Meeting of American Elasmobranch Society, Kansas City.

Hentschel, H. 1988. Renal blood vascular system in the elasmobranch, *Raja erinacea* Mitchill, in relation to kidney zones. *Am. J. Anat.*, 183: 130–147.

Hodgson, E. S. 1974. Chemoreception. Pp. 127–167 *in* Rockstein, M. (Ed.), The Physiology of the Insecta. Academic Press, New York.

Hodgson, E. S., and R. F. Mathewson. 1978. Electrophysiological studies of chemoreception in elasmobranches. Pp. 227–267 *in* Hodgson, E. S., and R. F. Mathewson (Eds.), Sensory Biology of Sharks, Skates, and Rays. U.S. Government Printing Office, Washington, D.C.

Holgren, S., and S. Nilsson. 1999. Digestive system. Pp. 144–173 in Hamlett, W. C. (Ed.), Sharks, Skates, and Rays. Johns Hopkins University Press, Baltimore.

Holt, W. F., and D. R. Idler. 1975. Influence of the interrenal gland on the rectal gland of a skate. *Comp. Biochem. Physiol.*, 50C: 111–119.

Hueter, R. E. 1991. Adaptations for spatial vision in sharks. *J. Exp. Zool. Suppl.*, 5: 130–141.

Hueter, R. E., and C. A. Manire. 1994. Bycatch and catch-release mortality of small sharks in the Gulf coast nursery ground of Tampa Bay and Charlotte Harbor. Final Report, NOAA/NMFS, MARFIN Project, NA17FF0378.

Hueter, R. E., D. A. Mann, K. P. Maruska, J. A. Sisneros, and L. S. Demski. 2004. Sensory biology of elasmobranches. Pp. 325–368 *in* Carrier, J. C., J. A. Musick, and M. R. Heithaus (Eds.), Biology of Sharks and Their Relatives. CRC Press, Boca Raton.

International Shark Attack File Statistics. 2007. http://www.flmnh.ufl.edu/fish/Sharks/Statistics/statistics.htm

Janech, M. G., and P. M. Piermarini. 2002. Renal water and solute excretion in the Atlantic stingray in fresh water. *J. Fish Biol.*, 50: 1–5.

Johnsen, P. B., and J. H. Teeter. 1985. Behavioral responses of bonnethead sharks (*Sphyrna tiburo*) to controlled olfactory stimulation. *Mar. Behav. Physiol.*, 11: 283–291.21.

Johnson, R. H. 1978. Sharks of Polynesia. Les Éditions de Pacifique, Papeete, Tahiti.

Johnson, R. H., and D. R. Nelson. 1973. Agonistic display in the gray reef shark, *Carcharhinus menisorrah*, and its relationship to attacks on man. *Copeia*, 1973: 76–84.

Johnson, R. H., and D. R. Nelson. 1978. Copulation and possible olfaction-mediated pair formation in two species of carcharhinid sharks. *Copeia*, 1978: 539–542.

Jones, G. C., and G. H. Geen. 1977. Food and feeding of spiny dogfish (*Squalus acanthias*) in British Columbia waters. *J. Fish. Res. Board Can.*, 34: 2067–2078.

Jorgensen, S. J., A. P. Klimley, and A. Muhlia-Melo. 2009. Diving of scalloped hammerhead in anoxic zone in Gulf of California. *J. Fish Biol.*, 74: 1682–1687

Kajiura, S. M. 2001. Head morphology and electrosensory pore distribution of carcharhinid and sphyrnid sharks. *Environ. Biol. Fishes*, 61: 124–133.

Kajiura, S. M., J. B. Fornia, and A. P. Summers. 2005. Olfactory morphology of carcharhinid and sphyrnid sharks: does the cephalofoil confer a sensory advantage? *J. Morphology*, 264: 253–263.

Kajiura, S. M., and K. N. Holland. 2002. Electroreception in juvenile and sandbar sharks. *J. Exp. Biol.*, 205: 3609–3621.

Kajiura, S. M., and T. C. Tricas. 1996. Seasonal dynamics of dental sexual dimorphism in the Atlantic stingray, *Dasyatis sabina*. *J. Exp. Biol.*, 199: 2297–2306.

Kalmijn, A. J. 1971. The electric sense of sharks and rays. *J. Exp. Biol.*, 55: 371–383.

Kalmijn, A. J. 1974. The detection of electric fields from inanimate and animate sources other than electric organs. Pp. 147–200 *in* Fessard, A. (Ed.), Handbook of Sensory Physiology. Springer-Verlag, New York.

Kalmijn, A. J. 1982. Electric and magnetic field detection in elasmobranch fishes. *Science*, 218: 916–918.

Kelly, J. C., and D. R. Nelson. 1975. Hearing threshholds of the horn shark, *Heterodontus francisci*. *J. Acoust. Soc. Amer.*, 58: 905–909.

Kemp, N. E. 1999. Integumentary system and teeth. Pp. 43–68 in Hamlett, W. C. (Ed.), Sharks, Skates, and Rays: The Biology of Elasmobranch Fishes. Johns Hopkins University Press, Baltimore.

Kemp, N. E., and J. H. Park. 1974. Ultrastructure of the enamel layer in developing teeth of the shark *Carcharhinus menisorrah*. *Arch. Oral Biol.*, 19: 633–644.

Ketchum, J. T. 2011. Movement patterns and habitat use of scalloped hammerhead sharks (*Sphyrna lewini*) in the Galapagos Islands: implications for design of marine reserves. Dissertation, University of California, Davis.

Kleerekoper, H. 1978. Chemoreception and its interaction with flow and light perception in the locomotion and orientation of some elasmobranches. Pp. 269–329 *in* Hodgson, E. S., and R. F. Mathewson (Eds.), Sensory Biology of Sharks, Skates, and Rays. U.S. Government Printing Office, Washington, D.C.

Kleerekoper, H., and D. Gruber. 1975. Accuracy of localization of a chemical stimulus in flowing and stagnant water by the nurse shark, *Ginglymostoma cirratum*. *J. Comp. Physiol.*, 98: 257–275.

Klimley, A. P. 1974. An inquiry into the causes of shark attack. *Sea Frontiers*, 20: 66–75.

Klimley, A. P. 1978. Nurses at home and school. *Marine Aquarist*, 8: 5–13.

Klimley, A. P. 1980. Observations of courtship and copulation in the nurse shark, *Ginglymostoma cirratum*. *Copeia*, 1980: 878–882.

Klimley, A. P. 1981. Grouping behavior in the scalloped hammerhead. *Oceanus*, 24: 65–71.

Klimley, A. P. 1982. Social organization of schools of scalloped hammerhead shark, *Sphyrna lewini* (Griffith and Smith), in the Gulf of California. Dissertation, University of California, San Diego.

Klimley, A. P. 1985a. Schooling in the large predator, *Sphyrna lewini*, a species with low risk of predation: a non-egalitarian state. *Ethology*, 70: 297–319.

Klimley, A. P. 1985b. The areal distribution and autoecology of the white shark, *Carcharodon carcharias*, off the west coast of North America. *South. Calif. Acad. Sci., Memoirs*, 9: 15–40.

Klimley, A. P. 1987. The determinants of sexual segregation in the scalloped hammerhead shark, *Sphyrna lewini*. *Environ. Biol. Fish.*, 18: 27–40.

Klimley, A. P. 1990. Do white sharks select their prey on the basis of fat content? Talk, meeting of American Association of Ichthyologists and Herpetologists, San Francisco (text, tables, and figures available from author).

Klimley, A. P. 1993. Highly directional swimming by scalloped hammerhead sharks, *Sphyrna lewini*, and subsurface irradiance, temperature, bathymetry, and geomagnetic field. *Mar. Biol.*, 117: 1–22.

Klimley, A. P. 1994. The predatory behavior of the white shark. *Am. Sci.*, 82: 122–133.

Klimley, A. P., and S. D. Anderson. 1996. Residency patterns of white sharks at the South Farallon Islands, California. Pp. 365–379 *in* Klimley, A. P., and D. G. Ainley (Eds.), Great White Sharks: The Biology of *Carcharodon carcharias*. Academic Press, San Diego.

Klimley, A. P., S. D. Anderson, P. Pyle, and R. P. Henderson. 1992. Spatiotemporal patterns of white shark (*Carcharodon carcharias*) predation at the South Farallon Islands, California. *Copeia*, 1992: 680–690.

Klimley, A. P., S. C. Beavers, T. H. Curtis, and S. J. Jorgensen. 2002. Movements and swimming behavior of three species of sharks in La Jolla Canyon, California. *Environ. Biol. Fish.*, 63: 117–135.

Klimley, A. P., S. B. Butler, D. R. Nelson, and A. T. Stull. 1988. Diel movements of scalloped hammerhead sharks (*Sphyrna lewini* Griffith and Smith) to and from a seamount in the Gulf of California. *J. Fish Biol.*, 33: 751–761.

Klimley, A. P., R. L. Kihslinger, and J. T. Kelly. 2005. Directional and non-directional movements of bat rays, *Myliobatis californica*, in Tomales Bay, California. *Environ. Biol. Fish.*, 74: 79–88.

Klimley, A. P., B. J. Le Boeuf, K. M. Cantara, J. E. Richert, S. F. Davis, S. Van Sommeran, and J. T. Kelly. 2001. The hunting strategy of white sharks (*Carcharodon carcharias*) near a seal colony. *Mar. Biol.*, 138: 617–636.

Klimley, A. P., and A. A. Myrberg. 1979. Acoustic stimuli underlying withdrawal from a sound source by adult lemon sharks, *Negaprion brevirostris*. *Bull. Mar. Sci.*, 29: 447–458.

Klimley, A. P., and D. R. Nelson. 1981. Schooling of scalloped hammerhead, *Sphyrna lewini*, in the Gulf of California. *Fish. Bull.*, 79: 356–360.

Klimley, A. P., and D. R. Nelson. 1984. Diel movement patterns of the scalloped hammerhead shark (*Sphyrna lewini*) in relation to El Bajo Espiritu Santo: a refuging central-position social system. *Behav. Ecol. Sociobiol.*, 15: 45–54.

Klimley, A. P., P. Pyle, and S. D. Anderson. 1996a. The behavior of white sharks and their pinniped prey during predatory attacks. Pp. 175–191 in Klimley, A. P., and D. G. Ainley (Eds.), Great White Sharks: The Biology of *Carcharodon carcharias*. Academic Press, San Diego.

Klimley, A. P., P. Pyle, and S. D. Anderson. 1996b. Tail slap and breach: agonistic displays among white sharks. Pp. 241–255 *in* Klimley, A. P., and D. G. Ainley (Eds.), Great White Sharks: The Biology of *Carcharodon carcharias*. Academic Press, San Diego.

Kohbara, J., H. Niwa, and M. Oguri. 1987. Comparative light microscopic studies on the retina of some elasmobranch fishes. *Bull. Jpn. Soc. Sci. Fish.*, 53: 2117–2125.

Kooyman, G. 1985. Personal communication. Scripps Institution of Oceanography.

Kritzler, H., and L. Wood. 1961. Provisional audiogram for the shark, *Carcharhinus leucas*. *Science*, 133: 1480–1482.

Lacy, E. R., and E. Reale. 1999. Urinary system. Pp. 353–397 *in* Hamlett, W. C. (Ed.), Sharks, Skates, and Rays: The Biology of Elasmobranch Fishes. Johns Hopkins University Press, Baltimore.

Lacy, E. R., E. Reale, D. S. Schlusselburg, W. K. Smith, and D. J. Woodward. 1985. A renal countercurrent system in marine elasmobranch fish: a computer assisted reconstruction. *Science*, 227: 1351–1354.

Liem, K. F., and A. P. Summers. 1999. Muscular system: gross anatomy and functional morphology of muscles. Pp. 93–114 *in* Hamlett, W. C. (Ed.), Sharks, Skates, and Rays. Johns Hopkins University Press, Baltimore.

Lisney, T. J., K. E. Yopak, J. C. Montgomery, and S. P. Collin. 2008. Variation in brain organization and cerebellar foliation in Chondrichthyans, *Brain, Behav., Evol.*, 72: 262–282.

Litherland, L. 2001. Retinal topography in elasmobranchs: interspecific and ontogenetic variation. Hons. Thesis, University of Queensland, Brisbane.

Liu, K. M., P. J. Chiang, and C. T. Chen. 1998. Age and growth estimates of the bigeye thresher shark, *Alopias superciliosus*, in northeastern Taiwan waters. *Fish. Bull.*, 96: 482–491.

Long, D. J., and R. E. Jones. 1996. White shark predation and scavenging on cetaceans in the eastern North Pacific Ocean. Pp. 293–307 *in* Klimley, A. P., and D. G. Ainley (Eds.), Great White Sharks: The Biology of *Carcharodon carcharias*. Academic Press, San Diego.

Lowe, C. G. 2001. Metabolic rates of juvenile scalloped hammerhead sharks (*Sphyrna lewini*). *Mar. Biol.*, 139: 447–453.

Lowe, C. G. 2002. Bioenergetics of free-ranging scalloped hammerhead sharks (*Sphyrna lewini*) in Kaneohe Bay, Oahu, HI. *J. Exp. Mar. Biol. Ecol.*, 278: 141–156.

Lowe, C. G., R. N. Bray, and D. R. Nelson. 1994. Feeding and associated electrical behavior of the Pacific electric ray *Torpedo californica* in the field. *Mar. Biol.*, 120: 161–169.

Lowe, C. G., B. M. Wetherbee, G. L. Crow, and A. L. Tester. 1996. Ontogenetic shifts and feeding behavior of the tiger shark, *Galeocerdo cuvier*, in Hawaiian waters. *Environ. Biol. Fish.*, 47: 203–212.

Lowenstein, O., and T. D. M. Roberts. 1951. The localization and analysis of the responses to vibrations from the isolated elasmobranch labyrinth. *J. Physiol.*, 114: 471–489.

Lucas, Z., and W. T. Stobo. 2000. Shark-inflicted mortality on a population of harbour seals (*Phoca vitulina*) at Sable Island, Nova Scotia. *J. Zool., London*, 252: 405–414.

Luer, C. A., and P. W. Gilbert. 1985. Mating behavior, egg deposition, incubation period and hatching in the clearnosed skate, *Raja eglanteria. Environ. Bio. Fish.*, 13: 161–171.

Lutton, B. V., J. St. George, C. R. Murrin, L. A. Fileti, and I. P. Callard. 2005. The elasmobranch ovary. Pp. 237–281 *in* Hamlett, W. C. (Ed.), Reproductive Biology and Phylogeny of Chondrichthyes, Science Publishers, Enfield, New Hampshire.

Magalhaes, K. W., C. Lima, A. A. Piran-Soares, E. E. Marques, C.A Hiruma-Lima, and M. Lopes-Ferreira.2006). Biological and biochemical properties of the Brazilian Potamotrygon stingrays: *Potamotrygon* cf. *scobina* and *Potamotrygon* gr. *orbignyi. Toxicon*, 47: 575–580.

Maisey, J. G. 1996. Discovering Fossil Fishes. Westview Press, Oxford.

Maisey, J. G., G. J. P. Naylor, and D. J. Ward. 2004. Mesozoic elasmobranchs, neoselachian phylogeny and rise of modern elasmobranch diversity. Pp. 17–56. *In* Arratia, G., and A. Tintori (Eds.), Mesozoic Fishes, Vol. 3, Systematics and Fossil Record, Proceedings of the 3rd International Meeting, Serpiano. Verlag, Dr. Friedrich Pfiel, Munich.

Manire, C. A., L. E. L. Rasmussen, D. L. Hess, and R. E. Hueter. 1995. Serum steroid hormones and the reproductive cycle of the female bonnethead shark, *Sphyrna lewini. Gen. Comp. Endochrinol.*, 97: 366–376.

Maruska, K., E. Cowie, and T. Tricas. 1996. Periodic gonadal activity and protracted mating in elasmobranch fishes. *J. Exp. Zool.*, 276: 219–232.

Maruska, K. P. 2001. Morphology of the mechanosensory lateral line system in elasmobranch fishes: ecological and behavioral considerations. *Environ. Biol. Fish.*, 60: 57–75.

Matta, M. E., and D. R. Gunderson. 2007. Age, growth, maturity, and mortality of the Alaska skate, *Bathyraja parmifera*, in the eastern Bering Sea. *Environ. Biol. Fishes*, 80: 309–323.

McConnaghey, R. 2000. Personal communication. Birch Aquarium, Scripps Institution of Oceanography.

McCosker, J. E. 1985. White shark attack behavior: observations of and speculations about predator and prey strategies. *South. Calif. Acad. Sci., Mem.*, 9: 123–135.

McEachran, J. D., and H. Konstantinou. 1996. Survey of variation of alar and malar thorns in skates (Chondrichtheyes: Rajoidea). *J. Morphol.*, 228: 165–178.

McFarland, W. N. 1986. Light in the sea: correlations with behaviors of fishes and invertebrates. *Amer. Zool.*, 26: 389–401.

McLoughlin, R. H., and A. K. O'Gower. 1971. Life history and underwater studies of a heterodont shark. *Ecol. Monogr.*, 41: 271–289.

Medved, R. J. 1985. Gastric evacuation in the sandbar shark, *Carcharhinus plumbeus. J. Fish Biol.*, 26: 239–235.

Metten, H. 1941. Studies on the reproduction of the dogfish. *Phil. Trans. Royal Soc. of Lond., Ser. B*, 230: 217–238.

Meyer, C. G., K. N. Holland, and Y. P. Papastamatiou. 2005. Sharks can detect changes in the geomagnetic field. *J. Royal Soc. Interface*, 2: 129–130.

Michael, S. 1993. Reef Sharks and Rays of the World. Sea Challengers, Monterey.

Miller, R. F., R. Cloutier, and S. Turner. 2003. The oldest articulated chondrichthyan from the early Devonian period. *Nature*, 425: 501–504.

Montgomery, J. C. 1988. Sensory physiology. Pp. 79–98 *in* Shuttleworth, T. J. (Ed.), Physiology of Elasmobranch Fishes. Springer Verlag, Berlin.

Moss, S. A. 1977. Feeding mechanisms in sharks. *Am. Zool.*, 17: 355–364.

Motta, P. J. 2004. Prey capture behavior and feeding mechanics of elasmobranch. Pp. 165–202 *in* Carrier, J. C., J. A. Musick, and M. R. Heithaus (Eds.), Biology of Sharks and Their Relatives, CRC Press, Boca Raton.

Motta, P. J., and C. D. Wilga. 1995. Anatomy of the feeding apparatus of the lemon shark, *Negaprion brevirostris. J. Morphol.*, 226: 309–329.

Motta, P. J., and C. D. Wilga. 2001. Advances in the study of feeding behaviors, mechanisms, and mechanics of sharks. *Environ. Biol. Fish.*, 60: 131–156.

Motta, P. J., M. Maslanka, R. E. Hueter, R. L. Davis, R. de la Parra, S. L. Mulvany, M. L. Habegger, J. A. Strother, K. R. Mara, J. M. Gardiner, J. P. Tyminski, and L. D. Zeigler. 2010. Feeding anatomy, filter-feeding rate, and diet of whale shark, *Rhincodon typus* during surface ram filter feeding off the Yucatan Peninsula, Mexico. *Zoology*, 113: 199–212.

Motta, P. J., R. E. Hueter, T. C. Tricas, and A. P. Summers. 1997. Feeding mechanism and functional morphology of the jaws of the lemon shark, *Negaprion brevirostris* (Chondrichtheyes, Carcharhinidae). *J. Exp. Biol.*, 200: 2765–2780.

Muñoz-Chápuli, R. 1999. Circulatory system: anatomy of the peripheral circulatory system. Pp. 198–217 *in* Hamlett, W. C. (Ed.), Sharks, Skates, and Rays: The Biology of Elasmobranch Fishes. Johns Hopkins University Press, Baltimore.

Munz, F. W. 1965. Adaptation of visual pigments to the photic environment. In DeReuck, A. V. S., and J. Knight (Eds.), Colour Vision: Physiology and Experimental Psychology. Little, Brown, Boston.

Munz, F. W., and W. N. McFarland. 1973. The significance of spectral position in the rhodopsins of tropical marine fishes. *Vision Res.*, 13: 1829-1874.

Musick, J. A., and J. K. Ellis. 2005. Reproductive evolution of chondrichthyans. Pp.45-79 *in* Hamlett, W. C. (ed.), Reproductive Biology and Phylogeny of Chondrichthyes: Sharks, Batoids, and Chimaeras. Science Publishers, Enfield, New Hampshire.

Musick, J. A., M. M. Harbin, and L. J. V. Compagno. 2004. Historical zoogeography of the Selachii. Pp. 33-78 *in* Carrier, J. C., J. A. Musick, and M. R. Heithaus (Eds.), Biology of Sharks and Their Relatives. CRC Press, Boca Raton.

Myers, R. A., J. K. Baum, T. D. Shepherd, S. P. Powers, and C. H. Peterson. 2007. Cascading effects of the loss of apex predatory sharks from a coastal ocean. *Science*, 315: 1846-1850.

Myrberg, A. A. 1978. Ocean noise and the behavior of marine animals: relationships and implications. Pp. 169-208 in Fletcher, J. L., and R. G. Busnel (Eds.), Effects of Noise on Wildlife. Academic Press, San Diego.

Myrberg, A. A. 2001. The acoustical biology of elasmobranchs. *Environ. Biol. Fish.*, 60: 31-45.

Myrberg, A. A., and S. H. Gruber. 1974. The behavior of the bonnethead shark, *Sphyrna tiburo. Copeia*, 1974: 358-374.

Nash, J. 1931. The number and size of kidneys with observations on the morphology. *Am. J. Anatomy*, 47: 425-445.

Natanson, L. J., J. G. Casey, N. E. Kohler, and T. Colket. 1999. Growth of the tiger shark, *Galeocerdo cuvieri*, in the western North Atlantic based on tag returns and length frequencies: and a note on the effects of tagging. *Fish. Bull.*, 97: 944-953.

Natanson, L. J., J. J. Mello, and S. E. Campana. 2002. Validated age and growth of the porbeagle sharks (*Lamna nasus*) in the western North Atlantic Ocean. *Fish. Bull.*, 100: 266-278.

Neer, J. A., and G. M. Cailliet. 2001. Aspects of the life history of the Pacific electric ray, *Torpedo californica* (Ayers). *Copeia*, 2001: 842-847.

Nelson, D. R. 1969. The silent savages. *Oceans Magazine*, 1: 8-22.

Nelson, D. R. 1983. Shark attack and repellency research: an overview. Pp. 11-74 *in* Zahuranec, B. J. (Ed.), Shark Repellents from the Sea. Westview Press, Boulder.

Nelson, D. R., and S. H. Gruber. 1963. Sharks: attraction by low-frequency sounds. *Science*, 142: 975-977.

Nelson, D. R., J. N. McKibben, W. R. Strong, C. G. Lowe, J. A. Sisneros, D. M. Schroeder, and R. J. Lavenberg. 1997. An acoustic tracking of a megamouth shark, *Megachasma pelagios*: a crepuscular vertical migratory. *Environ. Biol. Fish.*, 49: 389-399.

Nelson, D. R., R. R. Johnson, J. N. McKibben, and G. G. Pittenger. 1986. Agonistic attacks on divers and submersibles by gray reef sharks, *Carcharhinus amblyrhynchos*: antipredatory or competitive? *Bull. Mar. Sci.*, 38: 68-88.

Nelson, J. S. 2006. Fishes of the World. John Wiley & Sons, Hoboken.

Northcutt, R. G. 1978. Brain organization in the cartilaginous fishes. Pp. 117-193 *in* Hodgson, E. S., and R. F. Mathewson (Eds.), Sensory Biology of Sharks, Skates, and Rays. U.S. Government Printing Office, Washington.

Oceana. 2010. The international trade of shark fins: endangered shark populations worldwide. Available at: http://oceana.org/sites/default/files/reports/OCEANA_international_trade_shark_fins_english.pdf.

Olson, K. R. 1999. Rectal gland and volume homeostasis. Pp. 329-352 *in* Hamlett, W. C. (Ed.), Sharks, Skates, and Rays: The Biology of Elasmobranch Fishes. Johns Hopkins University Press, Baltimore.

Pang, P. K.T, R. W. Griffith, and J. W. Atz. 1977. Osmoregulation in elasmobranchs. *Am. Zool.*, 17: 365–377.

Parsons, G. R. 1993. Age determination and growth of the bonnethead shark *Sphyrna tiburo*: a comparison of two populations. *Mar. Biol.*, 117: 23–31.

Payan, P., and J. Maetz. 1970. Balance hydrique et minerale chez les elasmobranches: arguments en faveur d'uncontrôle endocrinien. *Bull. Inf. Sci. Tech. CEA*, 146: 77–96.

Peach, M. B., and N. J. Marshall. 2009. The comparative morphology of pit organs in elasmobranchs. *J. Morph.*, 270: 688–701.

Pedroso, C. M., C. Jared, P. Charvet-Almeida, M. P. Almeida, D. G. Neto, M. S. Lira, V. Haddad, Jr., K. C. Barbaro, and M. M. Antoniazzi. 2007. Morphological characterization of the venom secretory epidermal cells in the stinger of marine and freshwater stingrays. *Toxicon*, 50: 688–697.

Piermarini, P. M., and D. H. Evans. 1998. Osmoregulation of the Atlantic stingray (*Dasyatis sabina*) from the freshwater Lake Jesup of the St. Johns River, Florida. *Physiol. Zool.*, 71: 553–560.

Pough, F. H., C. M. Janis, and J. B. Heiser. 2002. Vertebrate Life. Prentice Hall, Upper Saddle River, New Jersey.

Pratt, H. L. 1979. Reproduction in the blue shark, *Prionace glauca*. *Fish. Bull.*, 77: 445–470.

Pratt, H. L. 1988. Elasmobranch gonad structure: a description and survey. *Copeia*, 1988: 719–729.

Pratt, H. L., and J. C. Carrier. 2001. A review of elasmobranch reproductive behavior with a case study on the nurse shark, *Ginglymostoma cirratum*. *Environ. Biol. Fish.,* 60: 157–188.

Pratt, H. L., and J. C. Carrier. 2005. Elasmobranch courtship and mating behavior. Pp. 129-169 *in* Hamlett, W. C. (Ed.), Reproductive Biology and Phylogeny of Chondrichthyes: Sharks, Batoids, and Chimaeras. Science Publishers, Plymouth, Massachusetts.

Pratt, H. L., and J. G. Casey. 1983. Age and growth of the shortfin mako, *Isurus oxyrinchus*, using four methods. *Can. J. Fish. Aquat. Sci.*, 40: 1944–1957.

Pridmore, P. A. 1995. Submerged wading in the epaulette shark *Hemiscyllium oscellatum* (Hemiscyllidae) and its implications for locomotion in rhipidistian fishes and early tetrapods. *Zoology*, 98: 278–297.

Primor, N., J. Parness, and E. Zlotkin. 1978. Pardaxin: the toxic factor from the skin secretion of the flat-fish *Pardachirus marmoratus* (Soleidae). Pp. 539–547 *in* Rosenberg, P. (Ed.), Toxins: Animal, Plant, and Microbial. Pergamon Press, Oxford.

Purdy, R. W. 1996. Paleoecology of fossil white sharks. Pp. 67–78 *in* Klimley, A. P., and D. G. Ainley (Eds.), Great White Sharks: The Biology of *Carcharodon carcharias*. Academic Press, San Diego.

Randall, D., W. Burggren, and K. French. 2002. Eckert Animal Physiology: Mechanisms and Adaptations. W. H. Freeman, New York.

Randall, J. E. 1977. Contribution to the biology of the whitetip reef shark (*Triaenodon obesus*). *Pac. Sci.*, 31: 143–164.

Raschi, W., and C. Tabit. 1992. Functional aspects of placoid scales: a review and update. *Aust. J. Mar. Freshwater Res.*, 43: 123–147.

Raschi, W,, and J. Elsom. 1986. Comments on the structure and development of the drag reduction–type placoid scale. Pp. 409–436 *in* Uyeno, T., R. Arai, T. Taniuchi, and K. Matsuura (Eds.), Indo-Pacific Fish Biology: Proceedings the Second International Conference on Indo-Pacific Fishes. The Ichthyological Society of Japan, Tokyo.

Rasmussen, L. E. L., D. L. Hess, and C. A. Luer. 1999. Alterations in serum steroid concentrations in the clearnose skate, *Raja eglanteria*: correlations with season and reproductive status. *J. Exp. Zool.*, 284: 575–585.

Read, L. J. 1971. Chemical constituents of body fluids and urine of the holocephalan *Hydrolagus colliei*. *Comp. Biochem. Physiol.*, 39A: 185–192.

Reese, A. M. 1910. The lateral line system of *Chimaera colliei*. *J. Exp. Zool.*, 9: 349–370.

Reif, W. E. 1985. Morphology and hydrodynamic effects of the scale of fast swimming sharks. *Fortschr. Zool.*, 30: 483–485.

Rhodes, D., and R. L. Smith. 1983. Body temperature of the salmon shark, *Lamna ditropis*. *J. Mar. Biol. Assoc. United Kingdom*, 63: 343–244.

Rinewalt, C. S., D. A. Ebert, and G. M. Cailliet. 2007. Food habits of the sandpaper skate, *Bathyraja kincaidii* (Garman, 1908) of central California: seasonal variation in diet linked to oceanographic conditions. *Environ. Biol. Fish.*, 80: 147–163.

Rivera-Vicente, A. C., J. Sewell, and T. C. Tricas. 2011. Electrosensitive spatial vectors in elasmobranch fishes: implications for source localization. *PLoS* ONE, 6 (1): e1608; doi: 10.1371, *pone* 0016008.

Robbins, R. 2004. Environmental factors affecting sexual and size segregation, and the effect of baiting on the natural behaviours of great white sharks, *Carchardon carcharias*, at the Neptune Islands, south Australia. Dissertation, University of Technology, Sydney.

Robertson, J. D. 1975. Osmotic constituents of the blood plasma and parietal muscle of *Squalus acanthias*. *L. Biol. Bull.*, 148: 303–39.

Robertson, J. D. 1976. Chemical composition of the body fluids and muscle of the hagfish *Myxine glutinosa* and the rabbit-fish *Chimaera monstrosa*. *Jour Zool.*, 178: 261–277.

Robinson, H. J., G. M. Cailliet, and D. A. Ebert. 2007. Food habits of the longnose skate, *Raja rhina* (Jordan and Gilbert, 1880), in central California waters. *Environ. Biol. Fish.*, 80: 165–179.

Rosenberger, L. J. 2001. Pectoral fin locomotion in batoid fishes: undulation versus oscillation. *J. Exp. Biol.*, 204: 379–394.

Samuel, N., P. Bringas, Jr., V. Santos, A. Nanci, and H. C. Slavkin. 1989. Selachian tooth development: I. Histogenesis, morphogenesis, and anatomical features of *Squalus acanthias*. *J. Craniofac. Genet. Dev. Biol.*, 3: 29–41.

San Filippo, R. A. 1995. Diet, gastric evacuation and estimates of daily ration of the gray smoothhound, *Mustelus californicus*. Thesis, San Jose State University.

Satchell, G. H. 1999. Circulatory system: distinctive attributes of the circulation of elasmobranch fish. Pp. 218–237 *in* Hamlett, W. C. (Eds.), Sharks, Skates, and Rays: The Biology of Elasmobranch Fish. Johns Hopkins University Press, Baltimore.

Saville, K. J., A. M. Lindley, E. G. Maries, J. C. Carrier, and H. L. Pratt. 2002. Multiple paternity in the nurse shark, *Ginglymostoma cirratum*. *Environ. Biol. Fish.*, 63: 347–351.

Schluessel, V., M. B. Bennett, H. Bleckmann, S. Blomberg, and S. P. Collin. 2008. Morphometric and ultrastructural comparison of olfactory system in elasmobranchs: the significance of structure-functional relationships based on phylogeny and ecology. *J. Morph.*, 269: 1365–1386.

Schmid, T. H., F. L. Murru, and F. McDonald. 1990. Feeding habits and growth rates of a bull (*Carcharhinus leucas* [Valenciennes]), sandbar (*Carcharhinus plumbeus* [Nardon]), sandtiger (*Eugomphodus taurus* [Rafinesque]) and nurse (*Ginglymostoma cirratum* [Bonnaterre]) sharks maintained in captivity. *J. Aquar. Aquat. Sci.*, 5: 100–105.

Schuijf, A. 1975. Directional hearing of cod (*Gadus morhua*) under approximate free field conditions. *J. Comp. Physiol.*, 98: 307–332.

Schultz, L. P., and M. H. Mallin. 1975. A list of shark attacks for the world. Pp. 529–551 *in* Gilbert, P. W. (Ed.), Sharks and Survival. D. C. Heath, Lexington.

Schwartz, F. J. 2007. A survey of tail spine characteristics of stingrays frequenting African, Arabian to Chagos-Maldive archipelago waters. *Smithiana Bulletin*, 8: 41–52.

Scotese, C. R. 2006. Paleomap project. http://www.scotese.com/climate.htm.

Sepulveda, C. A., S. Kohin, C. Chan, R. Vetter, and J. B. Graham. 2004. Movement patterns, depth preferences, and stomach temperatures of free-swimming juvenile mako sharks, *Isurus oxyrinchus*, in the Southern Bight. *Mar. Biol.*, 145: 191–199.

Shadwick, R. E. 2005. How tunas and lamnid sharks swim: an evolutionary convergence. *Am. Sci.*, 93: 524–531.

Shannon, J. A. 1940. On the mechanism of the renal tubular excretion of creatinine by the dogfish *Squalus acanthias*. *J. Cell. Comp. Physiol.*, 16: 285–291.

Shephard, K. L. 1994. Functions of fish mucus. *Rev. Fish Biol. Fisheries*, 4: 401–429.

Shirai, S. 1996. Phylogenetic interrelationships of neoselachians (Chondrichthyes: Euselachii). Pp. 9–34 *in* Stiassny, M. L. J., L. R. Parenti, and G. D. Johnson (Eds.), Interrelationships of Fishes. Academic Press, San Diego.

Shuttleworth, T. J. 1988. Salt and water balance – extrarenal mechanisms. Pp. 171–199 *in* Shuttleworth, T. J. (Ed.), Physiology of Elasmobranch Fishes. Springer-Verlag, Berlin.

Sillman, A. J., G. A. Letsinger, S. Patel, E. R. Loew, and A. P. Klimley. 1996. Visual pigments and photoreceptors in two species of shark, *Triakis semifasciata* and *Mustelus henlei*. *J. Exp. Zool.*, 276: 1–10.

Simons, J. R. 1970. The direction of the thrust produced by the heterocercal tails of two dissimilar elasmobranches; the Port Jackson shark, *Heterodontus portusjacksoni* (Meyer) and the piked dogfish, *Squalus megalops* (Macleay). *J. Exp. Biol.*, 52: 95–107.

Simpfendorfer, C. A. 2000. Predicting population recovery rates for endangered western Atlantic sawfishes using demographic analysis. *Environ. Biol. Fish.*, 58: 371–377.

Simpfendorfer, C. A., and M. R. Heupel. 2004. Assessing habitat use and movements. Pp. 553–572 *in* Carrier, J. C., J. A. Musick, and M. R. Heithaus (Eds.), Biology of Sharks and Their Relatives. CRC Press, Boca Raton.

Sims, D. W. 1999. Threshold foraging behavior of basking sharks on zooplankton: life on an energetic knife-edge? *Proc. R. Soc. Lond.*, B, 266: 1437–1443.

Sims, D. W., E. J. Southall, A. J. Richardson, P. C. Reid, and J. D. Metcalfe. 2003. Seasonal movements and behaviour of basking sharks from archival tagging: no evidence of winter hibernation. *Mar. Ecol. Progr. Ser.*, 248: 187–196.

Sims, D. W., V. J. Wearmouth, E. J. Southall, J. M. Hill, P. Moore, K. Rawlingson, N. Hutchinson, G. D. Budd, D. Righton, J. D. Metcalfe, J. P. Nash, and D. Morritt. 2006. Hunt warm, rest cool; bioenergetic strategy underlying diel vertical migration of a benthic shark. *J. Anim. Ecol.*, 75: 176–190.

Sisneros, J. A., and D. R. Nelson. 2001. Surfactants as chemical shark repellents: past, present, and future. *Environ. Biol. Fish.*, 60: 117–129.

Skiles, D. D. 1985. The geomagnetic field; its nature, history, and biological relevance. Pp. 43–102 *in* Kirschvink, J. L., D. S. Jones, and B. J. MacFadden (Eds.), Magnetite Biomineralization and Magnetoreception in Organisms. Plenum Press, New York.

Skomal, G. B., and L. J. Natanson. 2003. Age and growth of the blue shark, *Prionace glauca*, in the North Atlantic Ocean. *Fish. Bull.*, 101: 627–639.

Skomal, G. B., G. Wood, and N. Caloyianis. 2004. Archival tagging of a basking shark, *Cetorhinus maximus*, in the western North Atlantic. *J. Mar. Biol. Assoc., U.K.*, 84: 795–799.

Skomal, G. B., S. I. Zeeman, J. H. Chisholm, E. L. Summers, H. J. Walsh, K. W. McMahon, and S. R. Thorrold. 2009. Transequatorial migrations by basking sharks in the western North Atlantic Ocean. *Current Biol.*: 19: 1–4.

Skulberg, H. 2009. Sharks with arms – exploring the transformation from natural phenomenon to innovation object, via applied bionics. *Engaging Artifact*s, http://ocs.sfu.ca/nordes/index.php/nordes/2009/paper/view/228/171

Sminkey, T. R., and J. A. Musick. 1995. Age and growth of the sandbar shark, *Carcharhinus plumbeus*, before and after population depletion. *Copeia*, 1995: 871–883.

Smith, S. E., D. W. Au, and C. Show. 1998. Intrinsic rebound potential of 26 species of Pacific sharks. *Mar. Freshwater Res.*, 49: 663–678.

Smith, S. E., D. W. Au, and C. Show. 2008. Intrinsic rates of increase in pelagic elasmobranches. Pp. 288–297 *in* Camhi, M. D., E. K. Pikitch, and E. A. Babcock (Eds.), Sharks of the Open Ocean: Biology, Fisheries & Conservation. Blackwell Publishing Ltd., Oxford.

Smith, S. E., R. A. Mitchell, and D. Fuller. 2003. Age validation of a leopard shark (*Triakis semifasciata*) recaptured after 20 years. *Fish. Bull.*, 1001: 194–198.

Sobatka, H. 1965. Comparative biochemistry of marine animals. *Bioscience*, 1965: 583–584.

Springer, S. 1960. Natural history of the sandbar shark, *Eulamia milberti*. *U.S. Fish. Bull.*, 61: 1–37.

Stahl, B. J. 1989. Non-autostylic Pennsylvanian iniopterygian fishes. *Palaeontology*, 23: 315–324.

Standora, E. A., and D. R. Nelson. 1977. A telemetric study of the behavior of free-swimming Pacific angel sharks, *Squatina californica. Bull. South. Calif. Acad. Sci.*, 76: 193–201.

Stearns, S. C. 1992. The Evolution of Life Histories. Oxford University Press, New York.

Stell, W. K. 1972. The structure and morphologic relations of rods and cones in the retina of the spiny dogfish, *Squalus. Comp. Biochem. Physiol.*, 42A: 141–151.

Stevens, E. D., and F. G. Carey. 1981. One why of the warmth of warm-bodied fish. *Am. J. Physiol.*, 240: 151–155.

Stevens, J. D. 1974. The occurrence and significance of tooth cuts on the blue shark (*Prionace glauca*) from British waters. *J. Mar. Biol. Assoc., U.K.*, 54: 373–378.

Stolte, H. R., G. Galaske, G. M. Eisenbach, C. Lechene, B. Schmidt-Nielson, and J. W. Boylan. 1977. Renal tubule ion transport and collecting duct function in the elasmobranch little skate *Raja erinacea. J. Exp. Zool.*, 199: 403–410.

Strong, Jr., W. R. 1996. Shape discrimination and visual predatory tactics in white sharks. Pp. 229–240 *in* Klimley, A. P., and D. G. Ainley (Eds.), Great White Sharks: The Biology of *Carcharodon carcharias*. Academic Press, San Diego.

Strong, Jr., W. R., B. D. Bruce, D. R. Nelson, R. D. Murphy. 1996. Population dynamics of white sharks in Spencer Gulf, south Australia. Pp. 401–414 *in* Klimley, A. P., and D. G. Ainley (Eds.), Great White Sharks: The Biology of *Carcharodon carcharias*. Academic Press, San Diego.

Summers, A. P., and J. H. Long. 2006. Skin and bones, sinew and gristle: the mechanical behavior of fish skeletal tissues. Pp. 141–177 *in* Shadwick, R. E., and G. W. Lauder (Eds.), Fish Biomechanics. Elsevier, Amsterdam.

Taylor, L. 1985. White sharks in Hawaii: historical and contemporary records. *South. Calif. Acad. Sci., Mem.*, 9: 41–48.

Thies, D. 1983. Jahrzeitliche neoselachian aus Deutschland und S. England. *Cour. Forsch. Inst. Senckenberg*, 58: 1–116.

Thomerson, J. E. 1977. The bull shark, *Carcharhinus leucas*, from the upper Mississippi River near Alton, Illinois. *Copeia*, 1971: 166–168.

Thompson, K. S. 1976. On the heterocercal tail in sharks. *Palaeobiol.*, 2: 19–38.

Thompson, K. S., and D. E. Simanek. 1977. Body form and locomotion in sharks. *Am. Zool.*, 17: 343–354.

Thorson, T. B. 1972. The status of the bull shark, *Carcharhinus leucas*, in the Amazon River. *Copeia*, 1972; 601–605.

Thorson, T. B., C. M. Cowan, and D. E. Watson. 1967. Body fluid solutes of juveniles and adults of the euryhaline bull shark *Carcharhinus leucas* from freshwater and saline environments. *Physiol. Zool.*, 46: 29–42.

Thorson, T. B., R. M. Wotton, and T. A. Gorgi. 1978. Rectal gland of freshwater stingrays, *Potamotrygon* spp. (Chondrichthyes: Potamotrygonidae). *Bio. Bull.*, 154: 508–516.

Tricas, T. C. 1980. Courtship and mating-related behaviors in myliobatid rays. *Copeia*, 1980: 553–556.

Tricas, T. C., and E. M. Lefeuvre. 1985. Mating in the reef white-tip shark *Triaenodon obesus. Mar. Biol.*, 84: 233–237.

Tricas, T. C., and J. A. Sisneros. 2004. Ecological functions and adaptations of the elasmobranch electrosense. Pp. 308–329 *in* Von der Emde, G., J. Mogdans, and B. G. Kapoor (Eds.), The Senses of Fish: Adaptations for the Reception of Natural Stimuli. Kluwer Academic Publishers, Boston.

Tricas, T. C., and J. E. McCosker. 1984. Predatory behavior of the white shark (*Carcharodon carcharias*), with notes on its biology. *Proc. Calif. Acad. Sci.*, 43: 221–238.

Tricas, T. C., S. W. Michael, and J. A. Sisneros. 1995. Electrosensory optimization to conspecific phasic signals for mating. *Neurosci. Letters*, 2002: 129–132.

Tuve, R. L. 1963. Development of the U.S. Navy "Shark Chaser" chemical shark repellent. Pp. 455–464 in Gilbert, P. W. (Ed.), Sharks and Survival. D. C. Heath, Boston.

Uchida, S, M. Toda, K. Teshima, and K. Yano. 1996. Pregnant white sharks and full-term embryos from Japan. Pp. 139–155 *in* Klimley, A. P., and D. G. Ainley (Eds.), Great White Sharks: The Biology of *Carcharodon carcharias*. Academic Press, San Diego.

Uchida, S., M. Toda, and Y. Kamei. 1990. Reproduction of elasmobranches in captivity. Pp. 211–237 *in* Pratt, H. L., Jr., S. H. Gruber, and T. Taniuchi (Eds.), Elasmobranchs as Living Resources: Advances in the Biology, Ecology, Systematics, and Status of the Fisheries. NOAA Technical Report, 90.

Urist, M. R. 1962. Calcium and other ions in blood and skeleton of Nicaraguan freshwater shark. *Science*, 137: 985–986.

Van Dykhuizen, G. 2011. Description of *Hydrolagus* mating for text/reference book. Email to author, 5 October 2011, with three photographs. Aquatic Creations Group.

Van Dykhuizen, G., and H. F. Mollet. 1992. Growth, age estimation and feeding of captive sevengill sharks, *Notorynchus cepedianus*, at the Monterey Bay Aquarium. *Aust. J. Mar. Freshwater Res.*, 43: 297–318.

Vaudo, J. J., and C. G. Lowe. 2006. Movement patterns of the round stingray *Urobatis halleri* (Cooper) near a thermal outfall. *J. Fish Biol.*, 68: 1756–1766.

Wainwright, S. A., F. Vosburgh, and J. H. Hebrank. 1978. Shark skin: function in locomotion. *Science*, 202: 747–749.

Walker, T. I. 2005. Reproduction in Fisheries Science. Pp. 82–127 *in* Hamlett, W. C. (ed.), Reproductive Biology Biology and Phylogeny of Chondrichthyes: Sharks, Batoids and Chimaeras. Science Publishers, Enfield, New Hampshire

Walsh, M. J. 1980. Drag characteristics of V-grove and transverse curvature riblets. Pp. 168–184 in Hough, G. R. (Ed.), *Progr. Astronaut Aeronaut*, 72: 168–184.

Walsh, M. J. 1982. Turbulent boundary layer drag reduction using riblets. *American Institute of Aernautics and Astronautics*, Paper No., 82-0169.

Walsh, M. J., and L. M. Weinstein. 1978. Drag and heat transfer on surfaces with small longitudinal fins. *Am. Inst. Aeron. Astron.*, Paper No. 78-1161, AIAA 11th Fluid and Plasma Dynamics Conference, Seattle.

Waltman, B. 1966. Electrical properties and fine structure of the ampullary canals of Lorenzini. *Acta. Physiol. Scand, Suppl.*, 66: 1–60.

Webb, P. 1984. Form and function in fish swimming. *Sci. Am.*, July: 72–82.

Webb, P. M., D. E. Crocker, S. B. Blackwell, D. P. Costa, and P. J. Le Boeuf. 1998. Effects of buoyancy on the diving behaviour of northern elephant seals. *J. Exp. Biol.*, 201: 2349–2358.

Weng, K. C., A. M. Boustany, P. Pyle, S. D. Anderson, A. Brown, and B. A. Block. 2007. Migration and habitat of white sharks (*Carcharodon carcharias*) in the eastern Pacific Ocean. *Mar. Biol.*, 152: 877–894.

Weng, K. C., P. C. Castilho, J. M. Morrissette, A. M. Landeira-Fernandez, D. G. Holts, R. J. Schallert, K. J. Goldman, and B. A. Block. 2005. Satellite tagging and cardiac physiology reveal niche expansion in salmon sharks. *Science*, 310: 104–106.

Wetherbee, B. M., and E. Cortès. 2004. Food consumption and feeding habits. Pp. 203–224 *in* Carrier, J. C., J. A. Musick, and M. R. Heithaus (Eds.), Biology of Sharks and Their Relatives. CRC Press, Boca Raton.

Wetherbee, B. M., and S. H. Gruber. 1990. The effect of ration level on food retention time in juvenile lemon sharks, *Negaprion brevirostris*. *Environ. Biol. Fish.*, 29: 59–65.

Wetherbee, B. M., and S. H. Gruber. 1993. Absorption efficiency of the lemon shark *Negaprion brevirostris* at varying rates of energy intake. *Copeia*, 1993: 416–425.

White, W. T., M. E. Platell, and I. C. Potter. 2001. Relationship between reproductive biology and age composition and growth in *Urolophus lobatus* (Batoidea: Urolophidae). *Mar. Biol.*, 138: 135–147.

Wilga, C. A. D., and G. V. Lauder. 2004. Biomechanics of locomotion in sharks, rays, and chimeras. Pp. 139–164 in Carrier, J. C., J. A. Musick, and M. R. Heithaus (Eds.), Biology of Sharks and Their Relatives. CRC Press, Boca Raton.

Wilga, C. D., and G. V. Lauder. 2000. Three-dimensional kinematics and wake structure of the pectoral fins during locomotion in leopard sharks *Triakis semifasciata*. *J. Exp. Biol.*, 2003: 2262–2278.

Wilga, C. D., and G. V. Lauder. 2002. Function of the heterocercal tail in sharks: quantitative wake dynamics during steady horizontal swimming and vertical maneuvering. *J. Exp. Biol.*, 2005: 2365–2374.

Wilga, C. D., and P. J. Motta. 1998. Feeding mechanism of the Atlantic guitarfish, *Rhinobatus lentiginosus*: modulation of kinematic and motor activity. *J. Exp. Biol.*, 2001: 3167–3184.

Wilson, C. D., and M. P. Seki. 1994. Biology and population characteristics of *Squalus mitzukuii* from a seamount in the central North Pacific Ocean. *Fish. Bull.*, 92: 851–864.

Wilson, J. M., and P. Laurent. 2002. Fish gill morphology: inside out. *J. Exp. Zool.*, 293: 192–213.

Wilson, W. G., Carleton, J. H., and M. G. Meekan. 2003. Spatial and temporal patterns in the distribution of macrozooplankton on the southern North West Shelf, western Australia. *Estuarine., Coastal, Shelf Sci.*, 56: 897–908.

Winchell, C. J., A. P. Martin, and J. Mallatt. 2004. Phylogeny of elasmobranches based on LSU and SSU ribosomal RNA genes. *Mol. Phylog. Evol.,* 31: 214–224.

Wintner, S. P. 2000. Preliminary study of vertebral growth rings in the whale shark, *Rhincodon typus*, from the east coast of South Africa. *Environ. Biol. Fish.*, 59: 153–441–451.

Wintner, S. P., S. F. J. Dudley, N. Kistnasamy, and B. Everett. 2002. Age and growth estimates for the Zambezi shark, *Carcharhinus leucas*, from the east coast of South Africa. *Mar. Freshwater Res.*, 53: 557–566.

Wolf, N. G., P. R. Swift, and F. G. Carey. 1988. Swimming muscle helps warm the brain of lamnid sharks. *J. Comp. Physiol. B*, 157: 709–715.

Wong, T. M., and D. K. O. Chan. 1977. Physiological adjustments to dilution of the external medium in the lip-shark *Hemiscyllium plagiosum* (Bennett). II. Branchial, renal and rectal gland function. *J. Exp. Zool.*, 200: 85–96.

Wood, C. M., A. Y. Matsuo, R. J. Gonzalez, R. W. Wilson, M. L. Patrick, and A. L. Val. 2002. Mechanisms of ion transport in *Potamotrygon*, a stenohaline freshwater elasmobranch native to the ion-poor black-waters of the Rio Negro. *J. Exp. Zool.*, 205: 3039–3059.

Wourms, J. P. 1977. Reproduction and development in chondrichythyan fishes. *Am. Zool.*, 17: 379–410.

Yancey, P. H., and G. N. Somero. 1980. Methylamine osmoregulatory solutes of elasmobranch fishes counteract urea inhibition of enzymes. *J. Exp. Zool.*, 212: 205–213.

Yano, K., F. Sata, and T. Takahashi. 1999. Observations of the mating behavior of the manta ray, *Manta birostris*, at the Ogasawara Islands. *Ichthyol. Res.*, 46: 289–296.

Yopak, K. E., T. J. Lisney, S. P. Collin, and J. C. Montgomery. 2007. Variation in brain organization and cerebellar foliation in Chondrichthyans: sharks and holocephalans. *Brain, Behav., Evol.*, 69: 280–300.

Zahuranec, B. J. 1978. The Office of Naval Research and shark research in retrospect. Pp. 647–655 in Hodgson, E. S., and R. F. Mathewson (Eds.), Sensory Biology of Sharks, Skates, and Rays. U.S. Government Printing Office, Washington.

Zeiner, S. J., and P. Wolf. 1993. Growth characteristics and estimates of age at maturity of two species of skate (*Raja binoculata* and *Raja rhina*) from Monterey, California. *NOAA Tech. Rep.*, 115: 87–99.

Zeiske, E., J. Caprio, and S. H. Gruber. 1986. Morphological and electrophysiological studies on the olfactory organ of the lemon shark, *Negaprion brevirostris* (Poey). Pp. 381–391 *in* Uyeno, T., R. Arai, T. Taniuchi, and K. Matsuura (Eds.), Indo-Pacific Fish Biology: Proceedings of the Second International Conference on Indo-Pacific Fishes. Ichthyological Society of Japan, Tokyo.

Zieske, E., B. Theisen, and S. H. Gruber. 1987. Functional morphology of the olfactory organ of two carcharhinid shark species. *Can. J. Zool.*, 65: 2406–2412.

Illustration Credits

1.0: Photograph of white shark courtesy of Chris and Monique Fallows, photograph of basking shark courtesy of Sean Van Sommeran/Pelagic Shark Research Foundation; 2.0: Illustration courtesy of John Maisey and Westview Press; 2.1: Histogram courtesy of Wikipedia; 2.2: Maps courtesy of C. R. Scotese; 2.3a, 2.3b, 2.3f, 2.3g, 2.3h, 2.3i, 2.6a, 2.6b, 2.6c, 2.6d, 2.7a, 2.7b, 2.7c, 2.7d, 2.7e, 2.7f, 2.10a, 2.10b, 2.10c, 2.10d, 2.11a, 2.11b, 2.11c: Drawings courtesy of W. H. Freeman Company; 2.3c, 2.3d, 2.3e: Drawings courtesy of Blackwell Publishing; 2.4a, 2.4b, 2.12a, 2.12b, 1.12c: Photographs courtesy of the American Natural History Museum; 2.4c: Photograph courtesy of Blackwell Publishing; 2.5a, 2.5b: Drawings courtesy of CRC Press; 2.13: Diagram courtesy of CRC Press; 2.14: Maps courtesy of Academic Press; 2.15: Drawings courtesy of John Wiley and Sons; 3.0: Photographs courtesy of Donald Nelson; 3.1a: Modified drawing courtesy of Leonard Compagno; 3.1b, 3.3a, 3.7b, 4.3b: Modified drawing courtesy of Johns Hopkins University Press; 3.1c: Drawing based on photograph courtesy of Academic Press; 3.2a, 3.2b, 3.2c, 3.2d, 3.2e, 3.2f: Photographs courtesy of the Microscopic Society of America; 3.3c: Drawing courtesy of Wilkens-Sheiry Printing; 3.4: Drawing courtesy of Johns Hopkins University Press; 3.5: Modified drawings courtesy of Sigma Xi Society; 3.6a, 12.9b: Photographs courtesy of Elsevier Science; 3.6b, 4.4c: Modified drawing courtesy of Science Publishers; 3.7a, 3.7c, 6.2c: Modified drawings courtesy of the Ichthyological Society of Japan; 3.8: Photograph courtesy of *Australian Journal of Marine and Freshwater Research*; 3.9a, 3.9b, 3.9c, 3.9d, 3.9e, 3.9f, 12.12a, 12.12b, 12.12c, 12.12d, 12.12e: Drawings courtesy of Sears Foundation for Marine Research; 3.10b: Photograph converted to drawing courtesy of *Journal of Experimental Biology;* 3.11a: Modified drawing courtesy of CRC Publishing; 3.11b, 3.11c: Modified drawings and photographs courtesy of CRC Publishing; 4.3c: Photograph courtesy of Springer-Verlag; 4.4a: Photograph courtesy of *American Journal of Anatomy;* 4.4b: Modified drawing courtesy of CRC Press; 4.5: Modified diagram courtesy of CRC Press; 5.0: Photograph courtesy of Jeremiah Sullivan; 5.1a, 5.1b, 5.2, 5.4a, 5.4b: Modified drawings courtesy of Donald Garber; 5.3a, 5.3b, 5.3c: Photographs courtesy of Southern California Academy of Sciences; 5.5a, 5.5b: Modified depth profiles courtesy of American Society of Ichthyologists and Herpetologists; 5.5c: Modified depth profile courtesy of *Physiological Zoology;* 5.6: Modified diagram courtesy of *Physiological Zoology;* 5.7a: Modified diagram courtesy of *Journal of Zoology;* 5.7b, 5.7c: Photographs courtesy of *Journal of Zoology;* 5.8a: Photograph courtesy of *Zoological Journal of Linnaean Society;* 5.8b: Modified diagram courtesy of *Zoological Journal of Linnaean Society;* 5.9: Modified graph courtesy of Science Publishers; 6.0, 6.8a, 11.17a, 11.17b, 11.17c, 14.11b: Photographs courtesy of Kluwer Academic Publishers and Westview Press; 6.1a, 6.1b: Modified diagram and map courtesy of Academic Press; 6.2a, 6.2b: Modified diagrams and photographs courtesy of *Canadian Journal of Zoology;* 6.2d, 9.4, 9.8b: Modified diagrams courtesy of Springer-Verlag; 6.3a, 6.3b, 6.3c, 6.3d, 6.4a, 6.4b, 6.4c, 6.4d, 6.4e, 6.4f, 6.4g, 6.4h, 6.4i, 6.4j, 6.4k, 6.4l, 6.4m, 6.4m, 6.4n, 6.4o, 6.4p, 6.4q, 6.4r, 6.4s, 6.4t, 6.4u: Photographs courtesy of Wiley; 6.5: Modified histogram courtesy of Wiley; 6.6: Modified graph courtesy of Wiley; 6.7: Modified diagram and waveform records courtesy of Government Printing Press; 6.8b: Photograph courtesy of Westview Press; 6.9: Drawing courtesy of *Marine Behavior and Physiology;* 6.10: Diagram courtesy of *Marine Behavior and Physiology;* 6.11a, 6.11b, 6.11c: Diagrams and graph courtesy of Wiley; 6.12a, 6,12b, 6,12c: Graphs courtesy of *Current Biology;* 7.0: Photograph courtesy of Arthur Myrberg; 7.1b, 7.1c: Sound spectrograms courtesy of *Science* magazine; 7.2a: Sound spectrogram courtesy of Bulletin of Marine Science; 7.3a, 7.3b: Modified drawings courtesy of Springer-Verlag; 7.4a, 7.4d, 7.4e, 7.4f, 7.4g, 7.5a, 7.5b, 7.5c, 7.5d, 7.7a, 7.7b: Drawings courtesy of Kluwer Academic Publishers; 7.6a, 7.6b: Micrograph and histogram courtesy of Kluwer Academic Publishers; 7.8a, 7.8b, 7.8c: Modified drawings courtesy of Kluwer Academic Publishers; 7.9a, 7.9c, 7.9d: Modified drawings courtesy of AMF, O'Hare, Ohio; 7.9b: Drawings courtesy of *Journal of Comparative Biochemistry and Physiology;* 7.10: Drawings courtesy of AMF, O'Hare, Ohio; 7.11: Diagrams and micrographs courtesy of Wiley; 7.12: Dendogram courtesy of Wiley; 7.13a, 7.13b: Graphs courtesy of Acoustical Society of America; 8.0: Photograph courtesy of Ted Rulison; 8.1a: Modified graph courtesy of Clarendon Press; 8.1b: Modified diagram courtesy of American Society of Zoologists; 8.2, 8.4d, 8.5b, 8.6c, 8.12, 11.16: Photographs courtesy of Science Publishers; 8.3a: Modified diagram courtesy of Les Éditions du Pacifique; 8.3b, 8.3c, 9.1b, 9.1c, 12.5a, 12.5c: Modified diagrams courtesy of CRC Press; 8.4a, 8.4b, 8.4c: Photographs courtesy of U.S. Government Printing Office; 8.5a: Photograph courtesy of U.S. Government Printing Office; 8.6a: Photograph and drawing courtesy of U.S. Government Printing Office; 8.7a, 8.7b: Modified graphs courtesy of Science Publishers; 8.8: Modified graph courtesy of *Journal of Vision Research;* 8.9a, 8.9b: Modified graphs courtesy of U.S. Government Printing Office; 8.10: Photograph courtesy of U.S. Government Printing Office; 8.11: Modified graph courtesy of Springer; 9.2a, 9.2b, 9.2c: Diagrams courtesy of *PLoS One;* 9.3a, 9.3b, 9.3c: Modified diagrams courtesy of *PLoS One;* 9.5a, 9.5b, 9.5c, 9.5d, 9.5e, 9.5f: Diagrams courtesy of *Journal of Experimental Biology;* 9.6a, 9.6b, 9.6c, 9.6d: Diagrams courtesy of Company of Biologists Limited; 9.6e, 9.6f: Histograms courtesy of Company of Biologists Limited; 9.7a, 9.7b, 9.7c: Diagrams courtesy of Elsevier Science; 9.8a: Diagram courtesy of Sigma Xi Society; 9.9a, 9.9b, 11.12a, 11.12b: Modified diagrams courtesy of Kluwer Academic Publishers; 10.0: Photograph by A. Peter Klimley; 10.1: Modified diagram courtesy of Prentice Hall; 10.2a, 10.2b, 10.2c, 10.3a, 10.3c: Modified diagrams courtesy of U.S. Printing Office; 10.3b: Modified diagram courtesy of Academic Press; 10.4: Modified diagram courtesy of U.S. Printing Office; 10.5a, 10.b: Photographs courtesy of *Brain, Behavior, and Evolution;* 10.6a, 10.6b: Modified histograms courtesy of *Brain, Behavior, and Evolution;* 10.7, 10.8: Modified dendrograms courtesy of *Brain, Behavior, and Evolution;* 10.9a, 10.9f: Modified drawings and diagrams courtesy of Science Publishers; 10.10a, 10.10b: Graphs courtesy of U.S. Printing Office; 10.11a, 10.11b: Modified diagram and graph courtesy of *Science* magazine; 10.12a, 10.12b: Graphs courtesy of *Bulletin of Marine Science;* 11.0: Photographs courtesy of Alan Baldridge; 11.1a, 11.1b, 11.8: Modified diagrams courtesy of *Fisheries Bulletin;* 11.2a, 11.2b, 11.2c, 11.2d, 11.2e: Modified diagrams courtesy of Pergamon Press; 11.3, 11.9, 12.6a, 12.6b: Modified diagrams courtesy of Wiley-Liss, Inc.; 11.4a, 11.4c, 11.4e, 15.2a, 15.2b: Photographs courtesy of American Society of Ichthyologists and Herpetologists; 11.4b, 11.4d, 11.4f: Photographs courtesy of *Fisheries Bulletin;* 11.5a, 11.5b, 11.5c, 11.5d, 11.7a, 11.7b, 11.7d: Modified graphs courtesy of Kluwer Academic Press; 11.6: Micrographs courtesy of Kluwer Academic Press; 11.10: Modified dendrogram courtesy of Science Publishers; 11.11a, 11.11c: Photographs courtesy of the Johns Hopkins University Press; 11.11b, 12.4a, 12.4b, 12.10, 12.11a, 12.11b, 12.11c, 12.11d, 12.11e: Photographs courtesy of CRC Press; 11.13a, 11.13b, 11.13c, 11.13d, 11.13e: Modified diagram and graphs courtesy of Academic Press; 11.14a, 11.14b, 11.14c, 11.14d, 11.14e: Modified diagram and graphs courtesy of Wiley-Liss, Inc.; 11.15, 15.3a, 15.3b, 15.3c: Diagrams courtesy of American Society of Ichthyologists and Herpetologists; 11.18a, 11.18b, 11.18c, 11.18d: Modified diagrams courtesy of NOAA Technical Reports; 11.19a, 11.19b, 11.19c, 11.19d: Modified diagram and photographs courtesy of Wiley-Liss, Inc.; 12.0: White shark photograph courtesy of Peter Pyle, whale shark photograph courtesy of Phil Motta; 12.1a, 12.1b, 12.1c, 12.1d, 12.1e: Drawings and photographs courtesy of Elsevier; 12.2a, 12.2b, 12.2c, 12.2d, 12.2e, 12.2f: Photographs courtesy of *Science* magazine; 12.3a, 12.3b, 12.3c: Map and diagrams courtesy of Sigma Xi; 12.5b, 12.5d: Modified diagrams courtesy of Blackwell Publishing; 12.7a, 12.7b, 12.7c, 12.8a, 12.8b, 12.8c, 12.8d: Modified diagrams courtesy of the Company of Biologists Limited; 12.9a: Photograph courtesy of Pablo Brindas; 12.13a, 12.13b, 12.13c, 12.13e: Modified diagrams and photograph courtesy of CRC Press; 12.13d: Photograph courtesy of Southern California Academy of Science; 13.0: Photographs courtesy of J. M. Hoenig; 13.1a, 13.1b: Diagram and map courtesy of Sigma Xi Society; 13.2a, 13.2b: Modified diagrams courtesy of Sigma Xi Society; 13.4a, 13.4b: Modified diving profiles courtesy of Springer-Verlag; 13.5a, 13.5b: Photograph and modified drawing courtesy of Johns Hopkins University Press; 13.6a, 13.6b, 13.6c: Diagrams courtesy of Johns Hopkins University Press; 13.7: Photograph courtesy of AAZPA; 13.8a, 13.8b: Diagrams courtesy of CRC Press; 13.9a, 13.9b: Photographs courtesy of *Fisheries Bulletin;* 13.10a, 13.10b: Photograph and modified diagram courtesy of N.R.C. Press; 13.11a, 13.11b, 13.11c: Photographs and graph courtesy of Southern California Academy of Science; 13.12a, 13.12b, 13.12c: Photographs and graph courtesy of Springer-Verlag; 13.13a, 13.13b: Graph and histograms courtesy of *New Zealand Journal of Marine and Freshwater Research;* 14.0: Photographs courtesy of Howard Hall; 14.1a, 14.1b, 14.1c: Modified track diagrams courtesy of Fisheries Society of Great Britain; 14.2a: Modified track diagram courtesy of Kluwer Academic Publishers; 14.3a, 14.3b, 14.3c: Track diagrams courtesy of Les Éditions de Pacifique; 14.4: Map and circular diagrams courtesy of Fisheries Society of Great Britain, photograph courtesy of German Soler; 14.5a, 14.5b, 14.5c: Home range diagrams and graph courtesy of Kluwer Academic Publishers; 14.6a: Modified track diagram courtesy of Springer-Verlag; 14.7a, 14.7b, 14.7c, 14.7d, 14.7e, 14.7f, 14.7g, 14.7h: Modified track diagram and dive profiles courtesy of *Current Biology*; 14.8: Track diagram courtesy of Kluwer Academic Publishers; 14.9a,

14.9b, 14.9c, 14.9d, 14.9e: Modified track diagram and depth profiles courtesy of Springer-Verlag; 14.10: Modified diagram courtesy of *Nature*; 14.11c: Photograph courtesy of Fred Buyle; 14.12a, 14.12b, 14.12c: Dive profiles courtesy of Springer-Verlag; 14.13a, 14.13b, 14.13c, 14.13d, 14.13e, 14.13f: Dive profiles and graphs courtesy of British Ecological Society; 15.1a, 15.1b, 15.1c: Modified diagrams courtesy of American Society of Ichthyologists and Herpetologists; 15.4: Photographs courtesy of *Bulletin of Marine Science;* 15.8: Modified histogram courtesy of *Toxicon;* 15.9a, 15.9b: Modified graphs courtesy of *Toxicon;* 15.10a, 15.10b: Modified pie diagram and histogram courtesy Austin Gallagher and Neil Hammerschlag; 15.11. Modified map courtesy of Austin Gallagher and Neil Hammerschlag; 16.0: Photographs by A. Peter Klimley; 16.1, 16.2a, 16.2b, 16.2c: Modified graphs courtesy of Blackwell Publishing; 16.3: Modified graphs courtesy of *Science* magazine; 16.4: Modified diagram courtesy of Sigma Xi Society; 16.5: Photograph courtesy of Academic Press; 16.6a, 16.6b, 16.6c: Modified diagrams courtesy of CSIRO; 16.7, 16.8, 16.9: Modified diagrams courtesy of Blackwell Publishing.

Index

cartilaginous fishes (*continued*)

5, 213, 216, 234; *areae* of, 205; *area centralis* of, 206; *area horizontalis* of, 206; body temperature of, 101–2; brain-to-body ratios, 254–55, 258, 260; brains of, 5, 244; brain sizes of, 258, 262; central-place social system of, 385; chemical orientation of, 141, 143–47, 149–50; chondrocranium, and pineal organ, 242; cilia, and movements of water, 159–60; cladistic analysis of, 33; cold-bodied variety of, 103–4; continental shelf, expansion of, response to, 14; continental shelf, movement of, 28–29; courtship behaviors of, 5–6; diets of, 7, 340, 343–48, 368; digestion of, 7; diversity of, and evolutionary tree, *29*; and DMSP, 126; early examples of, 12, 26, 29; egg cases, 286–88; Elasmobranchii of, 2; electric sense of, 5, 215–16, 227, 234–35; as euryhaline, 88; evolution of, 46; external receptors of, 159–60; eyes of, 5, 194–95, 198, 204; feeding frenzy, and digestion rates of, 348–51; in food chain, 340–41, 343; in freshwater, 88; geomagnetic topotaxis of, 230; growth rates of, 340, 354–55, 358, 360, 363, 368, 446; as Holocephali subclass, 2; as ionized, 227; intestines, anatomy of, 352–54, 368–69; as isosmotic with, 3; jaws and teeth, anatomy of, 6–7, 321, 325; as *K-selected* species, 446, 449–50, 464; labyrinth (inner ear) of, 170–72, 174–76; learning capabilities of, 5, 232; life spans of, 354–55; mate, detection of, by electric sense, 224–27; mechanoreception of, 159, 180; metabolic processes of, as aerobic, 78–79; metabolic processes of, as anaerobic, 78–79; migration patterns of, 374, 401; movement patterns of, 7; muscles of, 59, 78–79; myomeres of, 59–60; nares of, 3–4; nasal organs of, 127–28; and odors, 126–27; optic tectum of, 258; orders of, 40, 42–46; oscillatory swimming of, 401; and osmoregulation, 85; oviparity of, 281; as parasite, 334; photoreceptors of, 205; phylogenetic relationships among, 33, 35; pit organs of 161; as predators, 313; prey, detection of, by electric sense, 214, 224, 234; prey, seizing of, 208; as quaternary consumers, 313; and radioactive carbon, 348; rebound potential of, 457–58; reproductive biology of, 6, 304, 307; reproductive cycles of, 290, 307; reproductive modes of, 5, 266; senses,

diversity of, 3–4; skeletal elements of, 53, 56; skeletons and organs of, *6*, 33; skin of, 64–65; smell and hearing of, 125; sound detection, and macula neglecta, importance of, 171; sound pressure organs of, 4–5; as stenohaline, 85, 91; stomachs, anatomy of, 352–54; subclasses of, 2; surface swimming of, 401; swimming modes of, 78–79, 401; tapetum of, 5, 198, 208; temperature of, 103; as tertiary consumers, 312–13, 333, 341, 447; TMAO in, 89; toothlike scales of, 65–66; tubules of, 216; and underwater sounds, 153–54, 156; urea in, 86, 89; vertebral column of, 55–56; visual capabilities of, 5; warm-bodied variety of, 104, 106–10, 112–15; water, advantages and disadvantages of on, 66–67. *See also* chimaeras; chondrichthyans fishes; rays; sharks

Catalina Island, 345

catch per unit effort (CPUE), 441, 450

catfish, 170

catsharks (*Scyliorhinidae*), 37, 77, 117–18, 252, 280, 286; foliation index of, 247

Cayman Islands, 425

Cenozoic Era, 13, 40

Central America, 8, 103, 425, 452; economic exclusive zones (EEZ) of, 462

central-place refuging system, 117

cephalofoils, 146

Cetorhinidae family, 341

chain catsharks (*Scyliorhinus rotifer*), 251–53, 281; brain of, 246; population increase of, 443

Chatham Island, 425

chemical shark repellent, 140; and pardaxin, 141

chemoreception, 125, 232, 252

Chiloscyllium genus, 57

China, 9

chimaeras, 3, 5, 8–9, 12, 29, 30, 33, 46, 52, 88, 153, 156, 448; ancestors of, 24, 40; angular canal, 168; brains, anatomy of, 242–45, brain sizes of, 246; canals of, 168; and cartilage, 52–53; cephalic tentaculum of, 58–59; chondrocranium of, 58; claspers of, 59, 271; courtship of, 303, 307; and denticles, 65; dorsal fins of, 59; eyes of, 191; features of, 45; fins of, 59; gills of, 45; growth rates of, 363; hearing sensitivity of, 176, 180; hyoid arch of, 58; internal hydromineral balance of, 84–85; jaws of, 7, 323; jaw types of, *322*; jugular canal, 168; lateral line canal, 168; locomotion of, 61;

Woods Hole Oceanographic Institution, 140
World Conservation Union: Red List of, 452
World War II, 9, 126, 140

Xenacanthiformes, 20; fins of, 21–22; skulls
    of, 21; teeth of, 22
Xenacanthus genus, 21, 26; teeth of, 330

yearling elephant seals, 350
yellowfin tuna (*Thunnus albacares*): rebound
    potential of, 459

zooplankton, 334, 447, 449